KV-025-409

Wildlife Habitat Management of Wetlands

Wildlife Habitat Management of Wetlands

Neil F. Payne
College of Natural Resources
University of Wisconsin—Stevens Point
Stevens Point, Wisconsin

KRIEGER PUBLISHING COMPANY
MALABAR, FLORIDA
1998

Original Edition 1992
Original Title: Techniques for Wildlife Habitat Management of Wetlands
Reprint Edition 1998 with new preface and corrections

Printed and Published by
KRIEGER PUBLISHING COMPANY
KRIEGER DRIVE
MALABAR, FLORIDA 32950

Copyright © 1992 by McGraw-Hill, Inc.
Transferred to author 1997
Reprinted by arrangement

All rights reserved. No part of this book may be reproduced in any form or by any means, electronic or mechanical, including information storage and retrieval systems without permission in writing from the publisher.
No liability is assumed with respect to the use of the information contained herein.
Printed in the United States of America.

FROM A DECLARATION OF PRINCIPLES JOINTLY ADOPTED BY A COMMITTEE OF THE AMERICAN BAR ASSOCIATION AND A COMMITTEE OF PUBLISHERS:
This publication is designed to provide accurate and authoritative information in regard to the subject matter covered. It is sold with the understanding that the publisher is not engaged in rendering legal, accounting, or other professional service. If legal advice or other expert assistance is required, the services of a competent professional person should be sought.

Library of Congress Cataloging-In-Publication Data

Payne, Neil F.
[Techniques for wildlife habitat management of wetlands]
Wildlife habitat management of wetlands / Neil F. Payne.
p. cm.
Originally published: Techniques for wildlife habitat management of wetlands. New York: McGraw-Hill, c1992. (McGraw-Hill biological resource management series). With new pref. and corrections.
Includes bibliographical references and index.
ISBN 1-57524-089-0 (alk. paper)
1. Wildlife habitat improvement. 2. Wetlands—Management.
3. Wetland conservation. I. Title.
SK355.P39 1998b
639.9'2—dc21 98-13107
CIP

10 9 8 7 6 5 4 3 2

Dedicated to
Burd S. McGinnes, my mentor,
and the memory of
Henry S. Mosby, my teacher

Contents

Foreword

With continental duck populations at historically low numbers, largely due to habitat alteration and nesting ground failure from extended drought conditions, this book points the way to improve the habitat, not only for waterfowl but also for other species of wildlife living within wetlands. Indeed, quality habitat is the answer to our waning waterfowl populations, and without it nothing else matters.

By definition, habitat can be described as the place where an organism lives and its surroundings, both living and nonliving. To survive, every species of wildlife must have its habitat which supplies the basic requirements of food, cover, and water. The lack of any one component means the failure of the species to thrive or survive. Wetlands are no exception to the needs of each species of wildlife we are interested in preserving, perpetuating, or increasing in number.

Acre for acre, wetlands exceed all other land types in wildlife productivity. With the loss approaching close to a half million acres of wetlands each year (over 60 acres every hour) on the continent to drainage for industrial, urbanization, and agricultural pursuits, it is small wonder that our waterfowl populations are at such a low ebb. Of the original 221 million acres of wetlands in the contiguous United States, 117 million acres have been drained.

But there is hope for brighter days with recent developments and programs that recognize the seriousness of habitat losses. From the halls of Congress to the many federal and state waterfowl development programs, to the many interested conservation organizations, and down to the private individual landowner, we hear more attention being paid to habitat. To save our precious wetlands, it will take the efforts of all interested citizens.

Foremost among these recent programs is the North American Waterfowl Management Plan (NAWMP) by the United States and Canadian governments (1986), whereby the emphasis is placed on habitat development for all stages of life for waterfowl: on nesting, migration, and wintering areas. We can achieve the goals of the NAWMP (62 million breeding ducks, 100 million migrating ducks, and 6 million wintering geese) only with adequate habitat acquisitions and development. The implementation of the goals of President George

Bush's "No net loss" federal wetlands policy will aid greatly in restoring and rejuvenating our wetlands. Implemented by the Wetlands Action Plan of the U.S. Fish and Wildlife Service, this policy seeks to stimulate a long-term increase in the quality and quantity of wetlands across the nation.

Additionally, in the private sector, the Conservation Reserve, Swamp Buster, and Sod Buster programs of the U.S. Department of Agriculture, and Section 404 of the Clean Water Act with new features, will go a long way toward enhancing and preserving our beautiful wetlands so vital to our waterfowl. The picture becomes even brighter with at least 95 universities and other colleges and training institutions teaching the science of wildlife and fisheries management to more students than ever before, including in much detail the necessity and desirability of managing wetlands for particular species of wildlife.

This book will be used by ever-increasing numbers of wildlife managers, teachers, biologists, administrators, and private citizens for the amplification and benefit of wildlife. There is no quick, easy, or cheap solution to the basic habitat requirements of our waterfowl. As the author so plainly points out, ultimately wildlife habitat management will be a compilation of the best judgment of the best trained minds to produce the desired results.

And, finally, the latest technology and techniques for habitat development, management, and perpetuation of wetlands are given by the author of *Techniques for Wildlife Habitat Management of Wetlands* in a concise and clear manner. The author has skillfully described ways in which quality habitat can be improved and preserved or developed to obtain the maximum results.

So—you want to build a wetland ecosystem, a waterfowl habitat, a dike, a marsh, a pothole, a nesting island—read on.

Jessop B. Low
Logan, Utah

Preface to the Reprint Edition

The book *Wildlife Habitat Management of Wetlands* and the book *Wildlife Habitat Management of Forestlands, Rangelands, and Farmlands* were designed initially as one volume, but length demanded the division into two companion volumes. The first printing used the titles *Techniques for Wildlife Habitat Management of Wetlands* and *Techniques for Wildlife Habitat Management of Uplands*. These titles were a bit cumbersome, so they were modified. And although the word *uplands* means forestlands, rangelands, and farmlands, many professionals did not identify the *uplands* book as such, nor did computer searches of key words. So the title was clarified, making it longer than desirable but, we hope, more recognizable, hence more useful. Through lack of exposure, field biologists, that is, the professional on-the-ground natural resource land managers with federal and state/provincial natural resource agencies, and also those with private nongovernment organizations, are essentially unaware that these two how-to field guides and reference books exist, as are some university instructors.

The demand for the two books results in this reprint edition. With 1500 references in the forestland, rangelands, and farmlands book, and 700 in the wetlands book, the material in them remains unchanged, for little change has occurred to warrant a second edition. Errors in the first printing have been corrected in this reprint edition.

These two books are comprehensive how-to reference books for land managers, as well textbooks for teachers. As such, they are designed to be an important part of each land manager's personal library for handy reference. As the first printing states in the Preface, "in the wildlife profession, habitat means everything; without it nothing else matters."

Neil F. Payne
February 1998

Preface

In the wildlife profession, habitat means everything; without it, nothing else matters. Wildlife habitat is the very basis for wildlife management. Intensive land use from a growing human population is reducing wildlife habitat at an alarming rate. The highest priority in the wildlife profession is habitat protection and the improvement of reduced amounts of habitat, coupled with improving the environmental literacy of society. A healthy ecosystem is the foundation of society. Wildlife managers are ecologists and recreationists with increasingly important roles in society. The demand for a healthy ecosystem and the recreational uses of wildlife and its habitat are influencing legislation and helping to boost tourism as one of the top industries of many states and provinces of the United States and Canada. Habitat improvement will help maintain a healthy ecosystem in our stressed environment and will produce maximum populations of wildlife to help meet society's increasing demands for consumptive and nonconsumptive wildlife recreation.

In 1961 I began my training in wildlife at Virginia Tech in Blacksburg as the greenest of all M.S. candidates, with Wildlife Unit Leader Burd S. McGinnes as my advisor and Henry S. Mosby as my teacher and committee member. Burd put me on a cottontail project in the Coastal Plains of Tidewater Virginia on Hog Island Wildlife Refuge on the shore of the James River—an excellent training opportunity for disciplined field research and for exposure to wetland habitat management and a variety of wildlife. My relationship with Burd continues as imprint of his personal and professional influence on me. Henry Mosby was a pillar of technique and habitat improvement. He was the editor of the first two editions of *The Techniques Manual*, as it is called. I observed his struggle to eliminate the errors in the first edition (1960) which resulted in so much criticism, as he produced the second edition (1963), for which I helped review the Literature Cited section. (Editions of *The Techniques Manual* now are spaced 10 years apart.) The habitat chapter in that book has properly evolved into the largest chapter by far, in testimony to its relative importance.

Many universities with wildlife programs teach a course in habitat management. The habitat chapter in *The Techniques Manual* is excellent, but limited by space restrictions. Most of us wildlife teachers have suffered the frustration of shotgunning our students throughout the library to obtain literature sources on wildlife habitat improvement and trying to make some of these sources available for reserve reading in the library. The sources occur mainly in (1) professional journals; (2) federal, state, and provincial documents of the United States and Canada; (3) proceedings of conferences, symposia, and workshops; (4) books; (5) reports from private organizations; and (6) wildlife magazines. No single volume on applied habitat management of wetlands exits as a sourcebook for professional wildlife managers or as a textbook for college students.

Can a book on wildlife habitat management that incorporates information from across North America be useful locally? I think so, especially if wildlife managers are not too provincial in outlook and avoid the temptation to dismiss summarily the techniques successfully used elsewhere with the notion that they don't apply locally and then proceed to reinvent the wheel. If not directly applicable, techniques usually can be modified for local use or at least to stimulate ideas. Although many published wildlife maps seem to depict North America as ending ecologically along the 49th parallel between the United States and Canada, this is hardly the case, as migratory and even resident wildlife readily attest. Nonetheless, local conditions must be understood in order to practice the art of accurate and efficient management. Thus we have the compromise of implementing the broad approach modified by local conditions, because wetlands vary in response to changes in vegetation management, especially from water-level manipulation, even within the same state or province.

Initially I considered organizing the book by region or wetland class in the United States and Canada. A review of the literature convinced me that techniques for habitat management of wetlands did not vary substantially by region or wetland class. Techniques for dikes, emergency spillways, mechanical spillways, blasted and dug potholes, prescribed burns, etc., were basically similar, as were those for water manipulation, although to a lesser extent. I tried to indicate important differences.

I tried a table of contents similar to the present one, then changed it to include a chapter on controlled marshes and another on greentree reservoirs, dredge fill marshes, and beaver ponds. But that organization resulted in more duplication and cross-referencing, so I returned to my original organization. I struggled with organization.

I also struggled with consistency of detail and what to call the wetland types. Although the U.S. Fish and Wildlife Service's *Classification of Wetlands and Deepwater Habitats of the United States* (1979) and the Canadian Wildlife Service's *The Canadian Wetland Classification System* (1987) are the most current systems, terminology of the U.S. Fish and Wildlife Service's *Wetlands of the United States* (1956) usually was used, if needed, because of its descriptive brevity (e.g., Type 4), use in legislation, and continued use in the literature and by wetland managers. But comparisons are made in the Appendix.

This book is a techniques manual on wetland habitat improvement. It is a how-to book, emphasizing technique and minimizing principle. It is meant to serve as (1) a practical guide and sourcebook of ideas and techniques for resource managers to adapt and modify regionally in the United States and Canada, and elsewhere where applicable, as they plan and develop habitat improvement projects and (2) a training guide for aspiring wildlife managers. Details impractical to include are provided in the references.

Some time ago someone asked me how the book was coming. I said that there seemed to be no end to it. She said that she had heard that authors don't end; they just quit. She is right. I quit. Maybe I should have quit sooner.

In writing this book, I used about 700 references. The names of some people appear repeatedly in the narrative because of their outstanding contribution to the techniques for wildlife habitat management of wetlands. I acknowledge them: I. J. Ball, F. C. Bellrose, E. G. Bolen, R. H. Chabreck, H. F. Duebbert, L. H. Fredrickson, K. F. Higgins, J. A. Kadlec, H. A. Kantrud, P. L. Knutson, A. F. Linde, W. A. Mitchell, W. H. Neely, L. M. Smith, E. S. Verry, M. W. Weller, W. W. Woodhouse. Other outstanding contributions were made by anonymous authors from the Atlantic Waterfowl Council, U.S. Army Corps of Engineers, U. S. Fish and Wildlife Service, U.S. Forest Service, and U.S. Soil Conservation Service.

I thank Mickey E. Heitmeyer of Ducks Unlimited, Sacramento, CA, and both Dick Hunt and Tom Meier of the Wisconsin Department of Natural Resources for reviewing the manuscript. I thank UW-SP faculty typists Virginia Crandell and Lorraine Swanson who typed the manuscript and tolerated my many changes and hieroglyphics, Dorothy Snyder of the UW-SP College of Natural Resources who constantly requested materials for me, Marg Whalan and Victoria Billings who found documents in the UW-SP library and tolerated my absconding with them beyond reasonable time, Kathy Halsey and Christine Neidlein who obtained innumerable UW-SP interlibrary loans promptly for me, UW-SP College of Natural Resources graduate students Gus Smith and Matthew Lovallo who checked scientific names and proofread the literature cited, and Dean Alan Haney and Associate Dean Rick Wilke of the UW-SP College of Natural Resources who provided encouragement and administrative support. And I thank my wife, Jan, whose moral support and tolerance of a constantly messy dining room table made this work possible.

Neil F. Payne

Conversion Table

Relationship of SI Units (International System of Modernized Metric Units) to U.S. Customary System Units.

SI (metric) unit	Equivalent U.S. Customary System unit
Centimeter (cm)	0.394 inch (in)
Decimeter (dm)	0.328 foot (ft)
Meter (m)	1.094 yards (yd) or 3.281 feet (ft)
Kilometer (km)	0.622 mile (mi)
Square centimeter (cm^2)	0.155 square inch (in^2)
Square meter (m^2)	1.197 square yards (yd^2)
Hectare (ha)	2.473 acres*
Square kilometer (km^2)	0.386 square mile (mi^2)
Cubic centimeter (cm^3)	0.061 cubic inch (in^3)
Cubic meter (m^3)	1.309 cubic yards (yd^3) or 35.320 cubic feet (ft^3)
Liter (L)	0.908 quart (qt) or 0.227 gallon (gal) or 0.028 bushel (bu)
Milliliter (mL)	0.068 tablespoon (tblsp)
Gram (g)	0.035 ounce (oz)
Kilogram (kg)	2.205 pounds (lb)
Metric ton (mt)	1.103 tons or 2205 pounds (lb)
Degrees Celsius (°C)	$\frac{9}{5}$ + 32 = degrees Fahrenheit (°F)

*1 acre-foot (acre · ft) of water = 43,560 ft^3 = 1233 m^3.

Chapter

1

Introduction

Wildlife can be managed through manipulation of its population or habitat. Because of the limited size of areas owned by management agencies and the expense involved in extensive habitat management, most wildlife management is accomplished through population management, i.e., alterations of seasons and quotas.

Habitat management for game and nongame species can be active (direct) or passive (indirect). Most is passive, involving incidental alteration of habitat through activities of agencies or other landowners pursuing goals such as logging or cattle grazing. Indirect habitat management can be beneficial to wildlife if accomplished in a proper multiple-use concept so often ignored. Direct habitat management focuses mainly on wildlife. It is accomplished usually on land owned by a public wildlife agency and occasionally by other agencies and private landowners. Because of its cost, direct habitat improvement generally is undertaken only when (1) it provides the nucleus for improving a larger area of habitat, (2) it is the only way to provide a missing essential habitat factor, or (3) it restores habitat damaged or altered by human activity or catastrophic weather which cannot be restored naturally within a reasonable time.

Two broad categories of habitat are wetlands and uplands. Most habitat management procedures imposed on wetlands are direct if they are owned by public wildlife agencies (U.S. Fish and Wildlife Service, Canadian Wildlife Service, state and provincial wildlife agencies) or wildlife interest groups (e.g., Ducks Unlimited). Most wetlands are not managed for other products. Moreover, direct habitat management intensifies management efforts on remaining wetlands and restores or creates others, because little competition exists to manage for other products. (Competition can result from cranberry or rice production, e.g., but most competition results from drainage or filling for other purposes, after which the area is no longer a wetland unless it is restored.) Habitat management on uplands has involved mainly indirect habitat management to produce wildlife as a byproduct in multiple-use emphasis of other products such as logs, cattle, or crops.

Whether habitat management is direct or indirect, some species will benefit and others will not. The goal can be featured species management to maximize the population density of a key species or community (systems or species richness) management to emphasize biodiversity. Often little difference exists between the two goals because the key species of wildlife to be featured for habitat management is usually one with broad habitat requirements, so that the management measures benefit many other wildlife species too, unless the featured species has special status, such as endangered or threatened, and a narrow niche. Although some key species of wildlife are featured at times in this book, the wide diversity of wildlife species and requirements generally requires habitat management by guild, i.e., groups of wildlife species with similar or overlapping general habitat needs, which is similar to management for species richness.

Any development that increases the supply of shelter (hiding and thermal cover), food, water, and space or improves their distribution will tend to increase an area's carrying capacity. Habitat evaluation procedures developed by the U.S. Fish and Wildlife Service (1980) can be used to determine missing ingredients or used in environmental impact statements and assessments, mitigation measures, and litigation. Methods to improve habitat can be structural or nonstructural. Whatever method is chosen, ecological implications must be addressed. Untampered nature is priceless. Improving the dynamic processes of natural systems might be impossible. Protection might be a better management goal than manipulation.

Habitat management involves the manipulation of landforms and successional stages of plant communities to benefit associated animal communities. Such manipulation produces diversity and interspersion from the structure and species composition of the plant community and its juxtaposition to other plant communities. Most wildlife species and densities are associated with edges of habitat types—but not all. Too much edge effect results in extensive fragmentation of habitat types, which causes interior wildlife populations to decline. Most wildlife species and highest densities also are found in midsuccessional stages, which have some structural similarity to edge. Management schemes that produce and maintain mainly midsuccessional fragmented stages of habitat at the exclusion of other stages must be avoided.

Wetlands are tremendously valuable to society and the environment for a variety of reasons. Perhaps the most comprehensive overview of major wetland functions and values is that by Sather and Smith (1984), who reviewed the literature for five categories of wetland values: hydrology, water quality, food chain support and nutrient cycling, habitat, and socioeconomics. Foster (1986) recognized intrinsic wetland values and four categories of socioeconomic wetland values: economic, scientific and educational, experiential (ethical, spiritual, physical, aesthetic, recreational, inspirational), and ecological (groundwater recharge or discharge, flood control, erosion control, water purification, primary production, secondary production, threatened and endangered species, wildlife refuge). Adamus et al. (1987) outlined a wetland evaluation technique to assess wetland functions and values.

To compensate for the extensive losses of natural wetlands, mitigation is needed (Table 1.1), and most remaining wetlands need intensive management. Such management usually is expensive and generally involves constructing impoundments for altering ecological succession through water control. Managers must be skilled to interpret this complex and rapid succession of several hundred species of plants and must manipulate habitats to encourage the several dozen species most desired (Baldwin 1968). The dominant forces causing dy-

TABLE 1.1 Options for Mitigation

Mitigation types	Recommended area
Restoration—former wetland, no or few functions	1.5:1; 1:1 up front
Creation—made from different community	2:1; 1:1 up front
Enhancement—increase certain functions	3:1; 1:1 up front
Exchange—enhancement to the extreme	Case by case
Preservation—purchase and donation	Case by case
Timing of Mitigation	
Before—most prudent; require if unknowns	
Concurrent—encouraged for typical projects	
After—discouraged	
Location of Mitigation	
On site—same locale in watershed or ecosystem	
Off site—different locale or different ecosystem	
Community Type	
In-kind—same species	
Out-of-kind—different species	

SOURCE: Kruczynski (1989).

namic change in marshes are water-level fluctuation, herbivore use of plants, ice action, and possibly fire and nutrient turnover (Weller 1987). But little is known about many basic aspects of wetlands.

Although wetland types vary (App. A), management of brackish marshes is similar to that of freshwater marshes (Singleton 1965) with regard to the construction of dikes and emergency spillways, location of borrow pits and ditches, installation of mechanical spillways and artificial nesting and loafing sites, construction of dug and blasted potholes, etc., and, to a large extent, management of vegetation. Typical plant associations based on elevation, water level, and salinity for various wetland regions of the United States and Canada are shown in Figs. 1.1 to 1.4 (U.S. Army Engineer Waterways Experiment Station 1978).

Wetland habitat improvement involves alteration of wetlands and associated uplands to enhance wildlife populations. Wetland management can involve developing areas for manipulating water levels, establishing water on areas without the capability of manipulating water levels, controlling vegetation, controlling wildlife, and establishing artificial nesting and loafing habitat. Most wetland management is directed toward waterfowl, with the goal being to improve production, to aid during migration and winter, or to do both. Other objectives could be to aid in managing preferred, endangered, or threatened species; to prevent threatened destruction of high-quality habitat; to disperse large concentrations of birds to reduce mortality from diseases and parasites or to reduce crop depredations; and to provide public use of the wildlife resource through consumptive and nonconsumptive use.

Generally, other wetland wildlife species are managed incidental to waterfowl. Upland wildlife species benefit directly as herbivores from food and cover in uplands managed for duck nesting cover and goose pasture or indirectly through the food chain as predators (Schitoskey and Linder 1978, Mathisen 1985, Rakstad and Probst 1985, Weller 1986). Rakstad and Probst (1985) summarized the relationship of wetland habitat improvement to various species of wildlife in the lake states.

Wetlands are dynamic; they change with seasonal and annual precipitation and flooding events. Plant phenology has adapted to the water regime, and wildlife to the plant regime. Wetlands with constant water levels lose their marshy characteristics and tend to become lakelike, with detrimental effects on marsh wildlife. To complicate matters further, migratory waterbirds use a variety of wetland types in various stages of dynamics to satisfy annual needs. For example, ducks on breeding grounds eat mostly animal food; on wintering grounds they eat mostly vegetation (Fredrickson 1985).

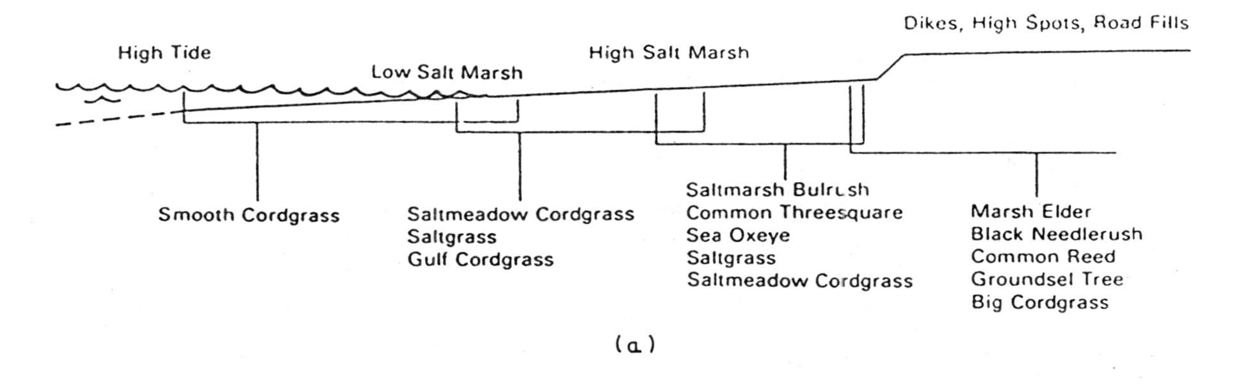

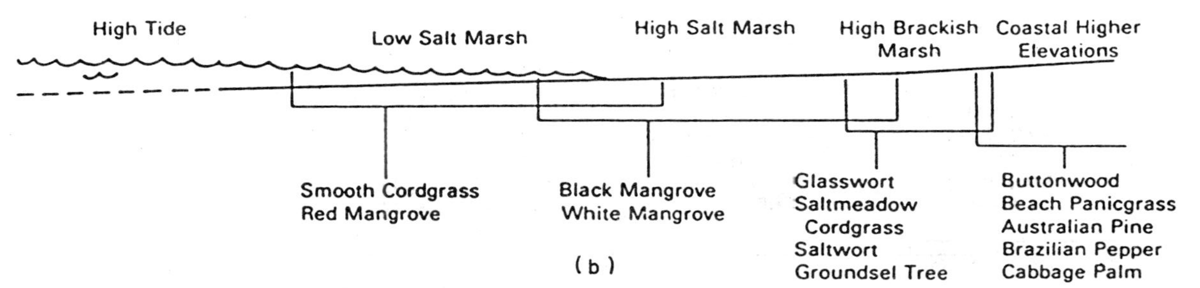

Figure 1.1 Typical tidal salt marshes of the east coast (*a*) and Florida (*b*), with plant associations and usual occurrence in the wetlands (*U.S. Army Engineer Waterways Experiment Station 1978*).

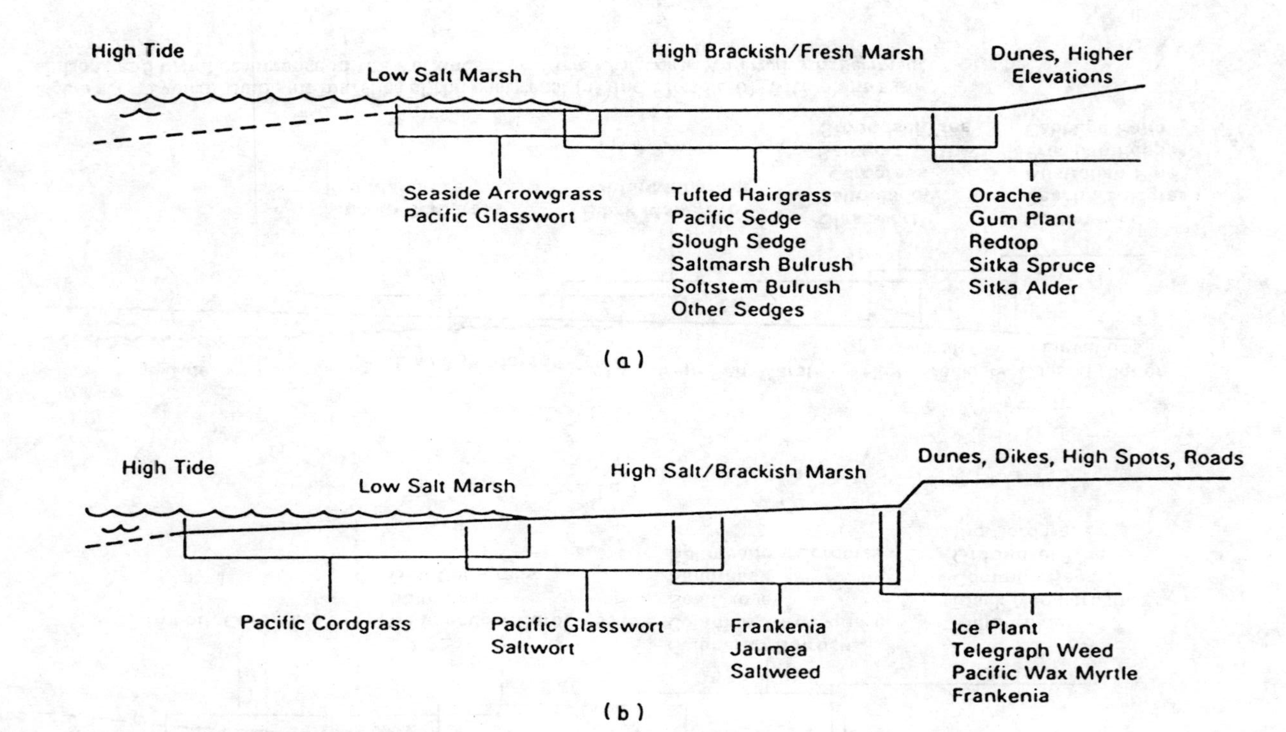

Figure 1.2 Typical tidal salt marshes of the Pacific Northwest (*a*) and California (*b*), with plant associations and usual occurrence in the wetlands (*U.S. Army Engineer Waterways Experiment Station 1978*).

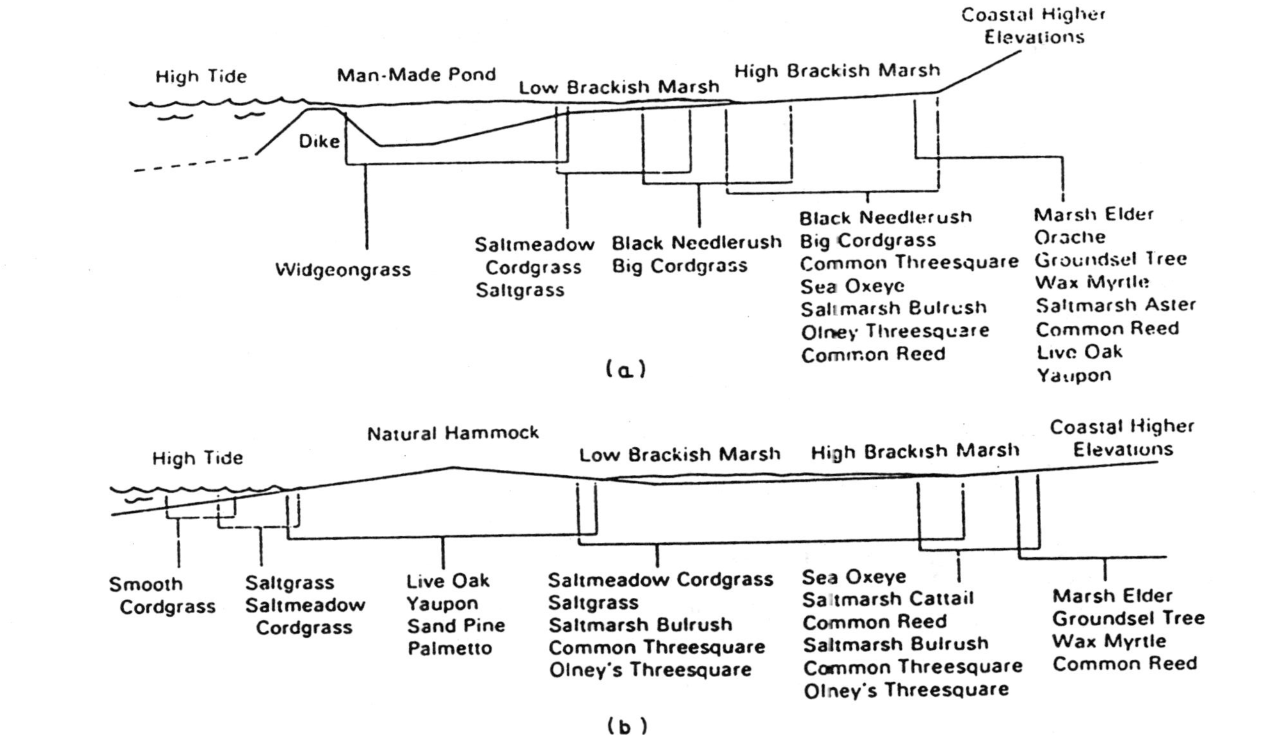

Figure 1.3 Typical brackish marshes with dike (*a*) and hammock island (*b*), with plant associations and usual occurrence in the wetlands (*U.S. Army Engineer Waterways Experiment Station 1978*).

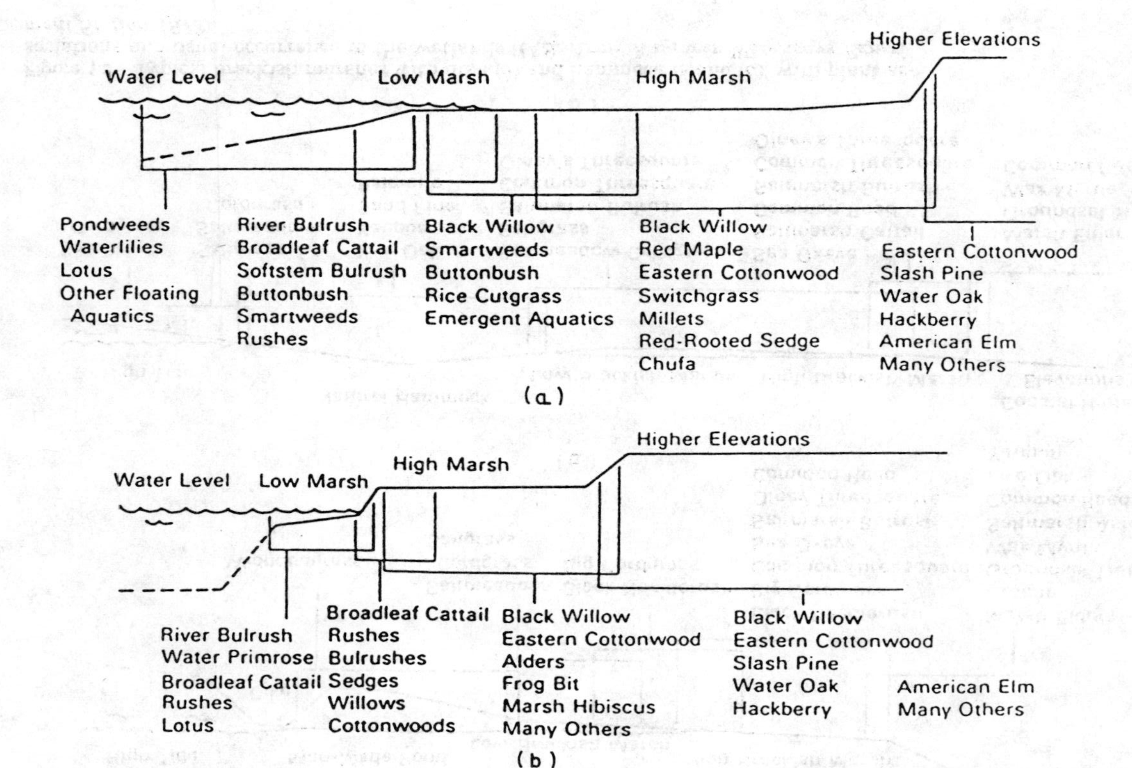

Figure 1.4 Typical lake or pond (*a*) and river (*b*) freshwater wetlands, with plant associations and usual occurrence in the wetlands (*U.S. Army Engineer Waterways Experiment Station 1978*).

Habitat complexes satisfy more objectives than individual types of habitat do. Forested wetlands provide good wildlife habitat with low management costs, but with relatively low food production for waterfowl (Reinecke et al. 1989). Moist-soil impoundments provide diverse habitat for wetland and upland wildlife, with intermediate management costs and food production. Cropland provides generally poor wildlife habitat with high management costs, but with high food production for waterfowl.

All methods to manage vegetation aim to improve the carrying capacity for the target wildlife, usually waterfowl. Carrying capacity of food = duck use-days of moist-soil areas + duck use-days of bottomland hardwoods (Reinecke et al. 1989).

$$\text{Duck use-days} = \frac{\text{food available [g(dry)]} \times \text{metabolizable energy [kcal/g(dry)]}}{\text{daily energy requirement (kcal/day)}}$$

To illustrate, 450 kg/ha is a conservative estimate of food production in moist soil impoundments for some areas, but about 50 kg/ha is assumed to be unused by ducks (Reinecke et al. 1989). The metabolizable energy of moist soil seeds is about 2.5 kcal/g, and the energy requirement of a mallard-size duck is about 292 kcal/day. In a 50-ha impoundment, the food available is

$$\frac{50\text{ ha}(450\text{ kg/ha} - 50\text{ kg/ha}) \times 1000\text{ g/kg} \times 2.5\text{ kcal/g}}{292\text{ kcal/day}} = 171{,}233\text{ duck use-days}$$

or about 1 month of use for about 6000 ducks. A better estimate would include production of invertebrates. Reid et al. (1989) estimated about 1344 kg/ha as reasonable for moist soil seed produced in intensively managed moist soil impoundments of the upper Mississippi Valley.

To evaluate cropland, the type of crop, production rate, and percentage harvested must be known. A 50-ha grain sorghum field might produce 2000 kg/ha after crop damage by deer and blackbirds, with a metabolizable energy of about 3.5 kcal/g, resulting in 1,168,664 duck use-days. (See "Goose Pasture" in Chap. 5.)

To evaluate bottomland hardwoods, tree species composition, mast production, and invertebrate production must be known. In a 50-ha greentree reservoir where red oaks comprise 80 percent of the basal area of large trees, acorns average 71 kg/ha, and invertebrates average 10 kg/ha, with a metabolizable energy of 3.5 kcal/g for acorns and invertebrates, the carrying capacity of food is 18,579 duck use-days (Reinecke et al. 1989).

Chapter

2

Wetlands With Water Control

Some wetlands can be developed so that the water level can be readily controlled. These are referred to as controlled marshes, greentree reservoirs, and dredge fill marshes. In controlled marshes and greentree reservoirs, water is added to expand areas of inundation. A dredge fill marsh is actually a form of controlled marsh, but water is removed through filling to reduce inundation. All three need an extensive system of dikes or levees to contain the water. Water is controlled by water control structures sometimes called mechanical spillways and occasionally by pumping.

Basically, three methods exist to control water levels, i.e., to put water into and take water out of an impoundment: tides, gravity, and pumps. These methods are used with four management categories (Wicker et al. 1983): passive estuarine management, controlled estuarine management, gravity drainage management, and forced drainage management.

With passive estuarine management, water levels are stabilized through the placement of low-level weirs and earthen plugs in the primary natural bayous and canals, with or without a system of dikes encircling the management unit. Water spills over the weir with high tides, but cannot drain out with low tides.

With controlled estuarine management, a system of dikes partitions the marsh into separate units. Various control structures inserted into the dikes permit control of water level and salinity mainly through tidal influence.

Gravity drainage management also consists of a system of dikes dividing the marsh into separate units, with water control structures connecting the impoundments through the dikes. But these impoundments use surface runoff from gravity flow rather than tidal influence to manipulate water levels. If the location is estuarine, tides are used to fill the impoundments only when runoff is inadequate. In such areas, gravity drainage can lower water levels only to the low-tide stage plus loss from evapotranspiration. With heavy spring rains, complete drawdown and subsequent germination of herbaceous annuals are impossible. Then the objective is to grow widgeongrass. During drought, brackish water is used to fill the impoundment, but the increased water salinity might be deleterious to emergent annuals, most of which thrive in salinities of 0 to 3 ppt. It might be best to wait for rain. In general, successful gravity drainage management depends on adequate rainfall, even when water is released from storage reservoirs or diverted from streams, the two most dependable sources of gravity flow. Stream diversion involves placing a gate or stoplog device in the streambed (Rudolph and Hunter 1964). Closing the structure diverts the streamflow into the impoundment. Usually a diversion ditch is needed.

Forced drainage management uses pumps with fresh water to manipulate water levels in impoundments, similar to gravity drainage management. Although the most expensive method, forced drainage management affords the best control of water levels.

CONTROLLED MARSHES AND GREENTREE RESERVOIRS

Design Features

Controlled marsh

The minimum management area is the minimum size of wetland that will maintain water levels to support shallow marsh vegetation for a seasonal wetland, assumed to be 0.04 ha, and deep marsh vegetation for a semipermanent wetland, assumed to be 0.8 ha (Stewart and Kantrud 1971, Sousa 1987). But wetlands less than 4 ha can be too costly to develop (Hoffman 1988). One large marsh would be cheaper to manage, but a cluster of smaller marshes increases habitat variety and duck territories, and reduces risk of disease or problems from introductions. Pond density for waterfowl production should be at least 2 to 3/2.6 km^2 (Proctor et al. 1983a, b, c).

The maximum size should be 405 ha (Beule 1979), although impoundments range to 1500 ha (Reid et al. 1989). In South Carolina, waterfowl will use managed wetlands up to 40 ha diurnally before and after the hunting season, but only nocturnally during the hunting season due to high hunter density. Waterfowl will use managed wetlands larger than 100 ha diurnally, with limited hunter density within the wetland and evenly spaced hunter density around the perimeter (Gordon et al. 1989). Generally, the larger the impoundment, the more attractive it is to waterfowl.

Breeding bird surveys indicate that the variety of species increases with the size of wetlands up to 4 ha, with clustered wetlands, located less than 0.4 km from at least two other wetlands, being most attractive (Williams 1985), although marshes of 20 to 30 ha preserve bird species better than do marshes of 30 to 180 ha or less than 20 ha (Brown and Dinsmore 1986). Diving ducks prefer large marshes for nesting within 0.4 km of large lakes (Low 1945). The number of bird species increases with the number of plant communities, leveling off to 15 in Minnesota wetlands having seven or fewer plant communities (Williams 1985).

In areas of high contour, subimpoundments are made by diking next to the main pool. Water is raised in the subimpoundments by diverting either part of or all the main water source into the impoundments through a control structure built into the dike and then letting it spill through other control structures from the subimpoundments into the main pool (Linde 1969). Pumps are used to supplement gravity and tides in dry years or to lift water into uphill impoundments.

A borrow pit is needed for fill material to build and maintain the dike. An emergency spillway allows excessive floodwater to escape. It is built into the dike or borrow pit.

Most areas developed for waterfowl contain several impoundments beside each other separated by dikes so that water is always available in some impoundments when others are drained. Such impoundments sometimes are referred to as embankment ponds. Typically, outer levees surround inner dikes that separate the impoundments. The outer levee helps keep floodwaters of nearby rivers or ocean from flooding the impoundments uncontrollably. Impoundments are connected by control structures through the dikes, with the outer (higher) impoundments connected to a tributary, or lake, upstream by a control structure through the main dike (outer levee). The innermost impoundment and perhaps some others are connected to the tributary, or lake, downstream by a control structure through the outer levee at that point. In this manner water can be let into and removed from the impoundment complex.

Flooding systems are of two types. A stair-step system has a floodgate connecting the first (highest-elevation) impoundment to the main water source, with other impoundments connected sequentially to the first impoundment so that water flows across several contours (Fig. 2.1) (Reid et al. 1989). A header ditch system involves a series of flood-

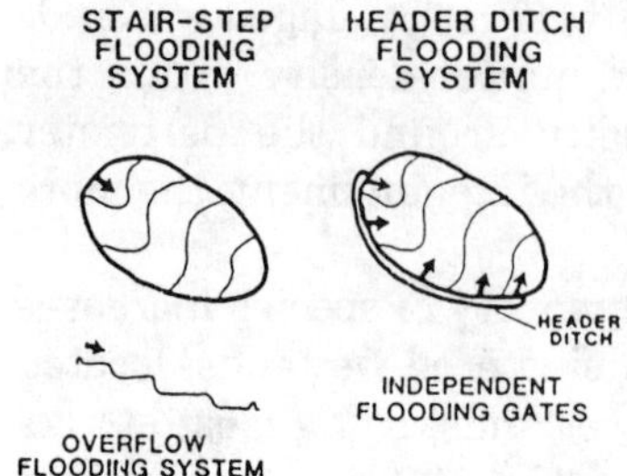

Figure 2.1 Design of stair-step and header ditch flooding systems in waterfowl impoundments (*Reid et al. 1989*).

gates connecting a series of impoundments independently to the main water source (Fig. 2.1). The stair-step flooding system might be cheaper to build, but it does not provide independent flooding control for impoundments.

Greentree reservoir

Greentree reservoirs are bottomland hardwood forests shallowly flooded and ice-free during the dormant growth period of fall and winter to attract migrating and wintering ducks (Mitchell and Newling 1986). Such flooding makes available mast (mainly acorns), benthic organisms, understory food plants such as wild millet and smartweed, and cover for resting or roosting habitat. Diking and flooding compensate for unreliable fall rains, creating habitat similar to bottomland hardwoods which flood annually from natural river overflow. Greentree reservoirs are used mostly in the lower Mississippi and Atlantic flyways.

Greentree (seasonally flooded) reservoirs differ from dead-tree (permanently flooded) reservoirs which usually contain deeper water and resting areas but little food. Fallen trees and branches from dead trees provide excellent loafing sites (Cowardin 1969), and if the water is shallow enough, the dead-tree reservoir can be managed for food production as a controlled marsh.

In the Mississippi Flyway, 75 to 90 percent of the ducks using greentree reservoirs are mallards; wood ducks are the second most numerous; and black ducks, green-winged teal, American widgeon, gadwalls, shovelers, hooded mergansers, and ringneck ducks comprise the rest (Rudolph and Hunter 1964, Hunter 1978). In the Atlantic Flyway, major species are mallard, wood duck, and black duck, with wood ducks the main user in the southern part of both flyways. Many other species of both wetland and upland wildlife benefit from management of greentree reservoirs.

A greentree reservoir can serve as a waterfowl refuge or a hunting area, two compatible functions generally. Hunting requires somewhat different design and operation of the reservoir and proper management of hunters as well as ducks (Mitchell and Newling 1986).

Unlike management of controlled marshes, which maintain their water several years between drainings, greentree reservoirs are drained every year (Mitchell and Newling 1986). Thus, a series of them for rotation of various water manipulation strategies is needless, and one large impoundment is generally as good as several smaller ones. It also is cheaper to build, without inner dikes and additional control structures. Terrain should be flat to gently sloping, isolated or near a source of ducks, such as a large river, refuge, or lake (Hall 1962), or near agricultural fields where ducks can feed on waste grain. The site should contain impervious clay soils close to a low-gradient stream, with a slope up to 2 dm/km to prevent excessive diking costs, or to other sources of water. If the slope is less than gentle, a large impoundment will result in water too shallow at the upstream end and too deep at the downstream end. In such cases, at least two impoundments are best. Multiple impoundments also are used if an impoundment is not flooded for winter because the soil needs additional drying. Size is a function of site capability and availability, personnel available for operation and maintenance, and expected hunting pressure. Potential hunting pressure probably will be the main determinant of reservoir size where hunting is allowed. Reservoirs with unrestricted public hunting should contain at least 600 ha; if hunting access is restricted, a reservoir should be more than 80 ha, although reservoirs of 41 ha attract ducks and are reasonably economical to operate (Hall 1962, Mitchell and Newling 1986). Smaller areas also will attract ducks but might not be cost-effective.

The shape of a greentree reservoir depends on the boundary, which usually follows land contours. To facilitate hunter access, boat trails can be made with a small bulldozer or similar bladed equipment to scrape out a trail slightly deeper than ground level (Hunter 1978). Such trails should be marked so that small flat-bottomed boats and canoes can be paddled or pushed through the shallow water.

Site Selection

Inventory

Areas must be inventoried (Cooperrider et al. 1986) to determine the extent and quality of wetlands (App. A) and their management potential for water control. Verry (1985a) described the steps.

1. Matching USGS 7.5-min topographic maps and standard resource vertical aerial photographs [23 × 23 cm (9 × 9 in)] at a scale of 1: 24,000 are obtained for the inventory unit. These permit mapping

of wetlands of at least 2.4 ha. Photographs at a scale of 1:15,840 also can be used, on which wetlands as small as 1 ha can be mapped. For small units, small-format (35- or 70-mm) aerial photographs can supplement or substitute for the 23 × 23 cm standard photograph.

2. Wetlands then can be delineated and identified by type on an acetate overlay of the aerial photograph, supplemented by the map. If possible, inflows and outflows to each wetland are indicated with arrows. The area of each wetland is then determined with a dot grid. For each wetland, three numbers on the overlay are used to indicate wetland number, area, and type (for example, 34-20-2 means wetland no. 34 has 20 ha of Type 2).
3. Ownership, lakes, beaver dams, and special areas such as wildrice are outlined with potential impoundment sites indicated.
4. In the top right corner of each overlay is the photograph number, and in the top left corner are the township and range, with section corners drawn in where they occur. All information is recorded separately, including comments on land acquisition and landowner cooperation needed.
5. Field-checking of questionable sites might be needed. Atlantic Waterfowl Council (1959, 1972) and Federal Interagency Committee for Wetland Delineation (1989) described the procedures.

Before consulting engineers are brought in, the state or provincial wildlife director and waterfowl biologist should collaborate with a flyway representative of the U.S. Fish and Wildlife Service or Canadian Wildlife Service to determine (1) the need for wetland development relative to local conditions, (2) the potential effect on density and distribution of the state or provincial waterfowl population, and (3) the relationship of the development to long-range flyway wetland programs (Atlantic Waterfowl Council 1972). Considerations for site selection include the area's location within the state, province, or flyway and therefore its relation to wintering grounds, breeding grounds, migration routes, and existing management areas within the state or province, as well as accessibility, ownership and acquisition, social and economic factors, vegetative cover, aspect, topography, geology, soils, water supply, water quality, and wildlife use. Refuges generally should be 10 to 30 km apart (Bellrose 1954, Delnicki and Reinecke 1986). After the wetland inventory is examined, two types of field reconnaissance are made: biological and engineering (Atlantic Waterfowl Council 1972). The engineering reconnaissance proceeds if the biological reconnaissance determines that development is feasible. If either reconnaissance recommends that the project is unfeasible, further consideration and expense cease.

Verry (1985a) listed conditions qualifying potential impoundment sites for construction:

1. Existing surface water with a specific conductance over 25 micromhos or high ratio of watershed area to surface water area.
2. No endangered plants or animals.
3. Streams not designated by state or provincial agencies as having potable water or viable trout populations. Trout streams should be below 21°C with a slope of 7 to 8 percent if 10 m wide, 6 to 7 percent if 10 to 23 m wide, 5 to 6 percent if 23 to 60 m wide, and 4.7 to 5 percent if 60 to 90 m wide. Impounding streams below 10°C improves fish growth.
4. No major spawning runs blocked.
5. No winter deer yards flooded.
6. No nearby lakes or roads flooded.
7. Low cost. Flood areas should be at least 10 ha.
8. Minimal construction of new roads to the site.
9. Existing roadways incorporated into the design of the dam.
10. Constricted terrain for a shorter dam.
11. New impoundments within 1.6 km of other wetlands including lakes, to encompass minimum home range for breeding pairs and broods of puddle ducks.
12. Few landowners.
13. Irregular shoreline and islands.
14. From 30 to 70 percent of the normal pool area less than 9 dm deep.
15. No potential for floating mats to clog spillways.
16. Food plants present.

Beule (1979) and Dobie (1986) described methods to calculate values to evaluate wetlands for development (Tables 2.1 and 2.2).

Engineers in the state, provincial, or federal agencies should be consulted before a water impoundment is built. Engineers will provide specifications for impoundment construction, aerial photographs, and soil maps which identify soil types and drainage systems. Such information, combined with that from a traverse survey, probably will be enough to develop specifications related to drainage and soil factors such as site location and dimensions of dikes and levees (U.S. Soil Conservation Service 1979, Mitchell and Newling 1986). A detailed plan then will include location and design of dikes and levees, diver-

TABLE 2.1 Water Supply Conditions for Rating the Management Potential of Wetlands

	Point value possible	Point value of an individual marsh*
Condition of water supply		
1. Ample at all seasons	50	
2. Ample through May or June, then uncertain	25	
3. Unreliable at any season	0	
Water-level control potential		
Condition of marsh basin		
1. Gradient allows good natural drainage	25	
2. Gradient allows slow natural drainage	13	
3. No gradient, no natural drainage	0	
Condition of the outlet structure		
1. Structure sufficient to release floodwaters rapidly and drain the marsh at any season	25	
2. Structure allows slow release of floodwaters, slow or incomplete drainage of marsh	12	
3. No structure	0	

*A fertile marsh with a rating of 62 or more usually is manageable for bird production.
SOURCE: Beule (1979).

TABLE 2.2 Evaluating Wetlands for Development for Waterfowl

Total points for each item are multiplied by the importance factor.

1. Proximity to any existing wetlands available for use by waterfowl—importance multiplier 5:
   a. <0.8 km = 5 × 5
   b. 0.8 to 1.6 km = 3 × 5
   c. >1.6 km = 1 × 5 Score: ________

Rationale: More waterfowl production will occur on basins which are in or near wetland complexes than in isolated basins.

2. Proximity to other state, provincial, and federal wildlife management units or legally protected private lands—importance multiplier 4:
   a. ≥1 unit within 0.8 km = 5 × 4
   b. ≥1 unit within 1.6 km = 3 × 4
   c. ≥1 unit 1.6 to 3.2 km away = 1 × 4 Score: ________

Rationale: As above, more production will occur where several blocks of wildlife lands, such as waterfowl production areas, easements, wildlife management areas, water bank tracts, etc., can be managed as a unit.

3. Restored wetland size in acres (maximum score 100)
   a. ≤10 acres (4 ha) = acres × 3
   b. 11 to 30 acres (4.5 to 12.1 ha) = acres × 2
   c. ≥31 acres (12.1 ha) = acres × 1 Score: ________

Rationale: While total area is important, small wetlands are generally more productive per unit area than large ones.

TABLE 2.2 Evaluating Wetlands for Development for Waterfowl (*Continued*)

4. Shape of restored wetland—importance multiplier 2:
   a. Very irregular = 5 × 2
   b. Somewhat irregular = 3 × 2
   c. Round or straight-edged = 1 × 2 Score: ______

Rationale: In many species of ducks, territory size is determined by the distance to the next visible pair. Thus irregularly shaped wetlands offer more territories for breeding ducks than regularly shaped wetlands.

5. Presence of existing marsh vegetation—importance multiplier 2:
   a. Desired marsh vegetation relatively plentiful = 5 × 2
   b. Some desired marsh vegetation present = 3 × 2
   c. No marsh vegetation present = 1 × 2 Score: ______

Rationale: Presence of marsh vegetation speeds the recovery of restored basin.

6. General soil fertility of basin and surrounding uplands—importance multiplier 2:
   a. High = 5 × 2
   b. Medium = 3 × 2
   c. Low = 1 × 2 Score: ______

Rationale: Soil fertility is related to the ability of the basin to produce food and cover.

7. Environmental condition of basin and uplands—importance multiplier 1:
   a. Basin and uplands clean = 5 × 1
   b. Some debris present = 3 × 1
   c. Extensive cleanup of trash, junk, fencing, stumps; containers, etc., needed = 1 × 1 Score: ______

Rationale: Debris might provide dens for nest predators; cleanup might disturb existing nest cover.

8. Ratio of secure upland nesting cover to wetland area (secure cover is permanent undisturbed grassland, no-till small grains, or alfalfa left unmowed until after July 31)—importance multiplier 5:
   a. 3:1 = 5 × 5
   b. 2:1 = 3 × 5
   c. 1:1 = 2 × 5
   d. <1:1 = 1 × 5 Score: ______

Rationale: Ideal upland/wetland ratios for waterfowl nesting range from 3:1 to 4:1; 1:1 might be acceptable if seasonal predator management is practiced.

9. Upstream-downstream landowner or local government negotiations needed—importance multiplier 4:
   a. None = 5 × 4
   b. Negotiations required but no problems foreseen = 3 × 4
   c. Difficult negotiations foreseen = 1 × 4 Score: ______

Rationale: With limited staff available, undue complications reduce the cost effectiveness of the project.

10. Water impoundment devices needed—importance multiplier 3:
    a. Single ditch plug or tile extension = 5 × 3
    b. Multiple plugs, extensions, or small dike = 3 × 3
    c. Extensive dike or control structure = 1 × 3 Score: ______

Rationale: With limited funds available, initial investments and long-term maintenance should be considered in maintaining cost effectiveness of the program.

TABLE 2.2 Evaluating Wetlands for Development for Waterfowl (*Continued*)

11. Ease of access and visibility—importance multiplier 1:
   a. Close to good roads = 5 × 1
   b. Moderately accessible = 3 × 1
   c. Access difficult = 1 × 1 Score: ______

Rationale: Access affects total project cost. Visibility affects public relations value of the project.

12. Ownership of wetland basin—importance multiplier 5:
   a. Public ownership, open to hunting = 7 × 5
   b. Public ownership, not open to hunting = 5 × 5
   c. Private ownership, protected by deed restriction = 5 × 5
   d. Private ownership, public or protected waters = 3 × 5
   e. Other private ownership = 1 × 5 Score: ______

Rationale: Public hunting lands provide both resource and recreational benefits. Legal protection of projects on private lands provides permanence to project benefits.

13. Percentage of basin <9 dm deep—importance multiplier 2:
   a. 75% = 5 × 2
   b. 50% = 3 × 2
   c. 25% = 1 × 2 Score: ______

Rationale: More waterfowl will use the larger percentage of shallow water area.

14. Adequate outlet capability to permit drawdown—importance multiplier 2:
   a. Yes, good outlet = 3 × 2
   b. Maybe, outlet is adequate in normal to drier years = 2 × 2
   c. Doubtful, very poor outlet = 1 × 2 Score: ______

Rationale: Feasibility of water-level control effectiveness is, in part, determined by this.

SOURCE: Dobie (1986).

sion channels for flooding the pool, spillways, and water control structures.

Accessibility

The project is impractical without suitable access to the project area for later users and for development and maintenance with heavy equipment (Atlantic Waterfowl Council 1972). Access rights-of-way might need to be bought or leased (at least 20 years), or an easement obtained, and developed into all-weather roads with parking facilities.

Ownership and acquisition

Ownership of the proposed project area and a surrounding buffer zone can be determined from a county plat book. Power lines, telephone lines, underground cables, pipelines, railroads, other public utilities, and rights-of-way might have to be moved—a costly venture. Water rights and riparian rights on the drainage system must be determined and coordinated, especially if impoundment, water manipulation, or excavation will raise or lower the water table on adjacent land. If the proposed project area is owned by a public agency, the agencies involved might enter into an interagency agreement detailing the rights of each (Atlantic Waterfowl Council 1972).

Several choices pertain to acquiring private land. If possible, the project area and buffer zone should be purchased outright from the landowners. Although undesirable and delicate, the use of eminent domain or condemnation might be needed, but the site will be difficult to develop without the landowner's cooperation and even active participation in the project. Agreeing to certain easements to the landowners, such as allowing them hunting or trapping rights, rights-of-way, or flowage rights, might facilitate purchase. Flowage easements should be discouraged because they do not include drawdown, hunting, or other activities needed to develop and manage the project area. But flowage easements on surrounding land prevent legal suits resulting from drainage or accidental flooding. Hunting easements on surrounding land facilitate managing species such as geese, establishing regulations, and preventing exploitation for private use.

Leasing the project area from the landowners, usually for a specified rent, is an alternative only for short-term projects of low expense, usually for at least 20 years with an option for renewal or purchase at specific cost. The lease period can be calculated by depreciating construction costs per hectare to ensure a valid return on the state's or province's investment.

Social and economic factors

Good public relations should be established with the surrounding community (Atlantic Waterfowl Council 1972). Information meetings should be held. If the proposed wetland development is incompatible with the surrounding community, it should not proceed even though biological and engineering considerations are suitable.

The positive impact of the project area on the local community should be determined, such as increased consumptive and nonconsumptive activities, environmental education, and increased employment and tourism and their economic impact. The positive impact on the local community of the proposed project should be determined, such as political support and revenue from user fees. The negative impact of human activities on the proposed project area should be determined, such as water contamination by rural or urban sources; noise pollution from highways, railways, airports, industry, military, etc.; nearby development of residential, commercial, or industrial complexes; and so on. The negative impact of the proposed project area on the local community should be determined, such as increased mosquitoes, nuisance wildlife (e.g., crop depredation by geese and ducks), hunting problems, offensive odors, changes in the local tax base and individual property taxes, and unemployment. Wetland management should be part of a more comprehensive management plan because wetlands are influenced by, and can influence, activities within the watershed including adjacent uplands, the local community, and even distant locations.

Waterfowl on marshes with ideal habitat will tolerate human activity nearby. Because waterfowl will use lower-quality marshes, choosing such sites for impoundment development will work only if they are distant from human disturbance (Atlantic Waterfowl Council 1959).

Vegetative cover

Based on the frequency of occurrence, wetland plants can be divided into four groups: (1) obligate plants that occur mostly (99 percent) in wetlands, (2) facultative wetland plants that occur mostly (67 to 99 percent) in wetlands but at times in uplands, (3) facultative plants that can occur equally (34 to 66 percent) in wetlands and uplands, and (4) facultative upland plants that can occur in wetlands (1 to 33 percent) but occur mostly in uplands (67 to 99 percent) (Federal Interagency Committee for Wetland Delineation 1989). Nearly 7000 vascular plants grow in wetlands of the United States and Canada, but only about 27 percent are obligate wetland species (Sipple 1987a, b, Reed 1988).

Controlled marsh. Plant indicators reveal information about not only bottom type (soft or hard) and elevation but also pH and fertility, salinity, marsh depth, and frequency of flooding (Linde 1969, Atlantic Waterfowl Council 1972, Gordon et al. 1989) (Table 2.3). Monotypic wetlands such as inland fresh meadows (Type 2), inland shallow fresh marshes (Type 3), and shrub swamps (Type 6) (Shaw and Fredine 1956) can be converted to inland deep fresh marshes (Type 4) to improve waterfowl potential and wildlife diversity (Table 2.4) (Verry 1985a, 1989). An improvement site usually is selected to convert relatively monotypic wetlands of shrub swamps, sedge meadows, and sometimes lowland deciduous forests to a complex of ponds, emergent persistent, emergent nonpersistent, and shrub swamp communities (Mathisen 1985). Otherwise, agricultural lands are the best choice for construction of shallow marshes, followed by grasslands and brushy areas. The potential host cover types for constructing seasonal and semipermanent wetlands include temporary and ephemeral wetlands, woodland, shrubland, native grassland, tame grassland, and cropland (Sousa 1987), although converting existing ephemeral and temporary wetlands to seasonal or semipermanent wetlands might be undesir-

TABLE 2.3 Coastal Wetland Types and Dominant Flora in the Santee River Estuary, South Carolina

Wetland type	Water salinity, ppt	Dominant plant species	
		Unmanaged marsh	Managed marsh
Fresh marsh	<1	*Zizaniopsis miliacea* *Alternanthera philoxeroides* *Polygonum* spp. *Pontederia cordata* *Sagittaria* spp. *Peltandra virginica* *Taxodium distichum* *Nyssa* spp.	*Polygonum punctatum* *Polygonum arifolium* *Panicum dichotomiflorum* *Aneilema keisak* *Cyperus* spp.
Intermediate marsh	1–5	*Zizaniopsis miliacea* *Spartina cynosuroides* *Typha angustifolia* *Scirpus validus* *Scirpus americanus* *Scirpus robustus*	*Panicum dichotomiflorum* *Echinochloa walteri* *Scirpus robustus* *Setaria magna* *Polygonum punctatum* *Cyperus* spp.
Brackish marsh	5–20	*Spartina cynosuroides* *Scirpus robustus* *Spartina alterniflora*	*Scirpus robustus* *Eleocharis parvula* *Ruppia maritima* *Leptochloa fascicularis*
Brackish/salt marsh	20–30	*Spartina alterniflora* *Spartina cynosuroides* *Juncus roemerianus*	*Ruppia maritima* *Eleocharis parvula* *Sesuvium maritimum*
Salt marsh	30–35	*Spartina alterniflora*	*Ruppia maritima* *Sesuvium maritimum*

SOURCE: Gordon et al. (1989).

TABLE 2.4 Most Useful Classification for Wetlands Suited to Water Impoundments for Wildlife

Circular 39 type*	Classification of wetlands and deepwater habitats†	
	Classes	Water regimes
Type 2—Inland fresh meadows These are largely sedge meadows in the lake states. They are without standing water during most of the growing season, but the soils are waterlogged within a few centimeters of the surface.	Emergent wetland	Saturated
Type 3—Inland shallow fresh marshes In most years they retain water until midsummer but often dry up before brood rearing is complete. Water is generally <15 cm deep.	Emergent wetland	Saturated, seasonally flooded, semipermanently flooded
Type 4—Inland deep fresh marshes These are covered with ≥15 to 90 cm of water during the growing season.	Emergent wetland	Permanently flooded, intermittently exposed
	Aquatic	Semipermanently flooded
Type 6—Shrub swamps These are usually waterlogged during the growing season and often covered with ≤15 cm of water.	Scrub-shrub wetland	All nontidal regimes except permanently flooded

*Shaw and Fredine (1956).
†Applied to fresh water only (Cowardin et al. 1979).
SOURCE: Verry (1989).

able. In the northern Great Plains, impoundments are designed to create new Type 3 and Type 4 wetland habitat for species that use such wetlands, such as waterfowl, muskrats, and associated wildlife, or to restore drained wetlands to Type 2 or Type 3 (Shaw and Fredine 1956, Stewart and Kantrud 1971, Sousa 1987).

The highest waterfowl productivity seems to occur in areas with a combination of woody cover, open water, and emergent vegetation (Atlantic Waterfowl Council 1959). Next in importance are marshes of open water and emergents. Open water devoid of woody and emergent plants might attract migrants, but not breeders. Marsh sites covered completely by woody plants are least valuable if the watershed is small. The first few years these are highly productive until the cover dies off. But as emergent cover develops and replaces woody cover, large watersheds seem to retain much of their productivity, apparently because greater volumes of runoff water purge the site of stained standing water, which indicates low pH and which reduces light penetration and inhibits aquatic plant growth. Potential cover from emergents is estimated in advance, based on soil associations. Sites with poor cover are rejected. Proper management of herbaceous and woody vegetation will minimize exposure of waterfowl to prevailing winds and drifting snow, thus reducing thermoregulatory costs (Ringelman et al. 1989).

The needles from coniferous trees produce acidic, unproductive water if flooded. Clearing of trees from the project area will increase the cost, with the decaying wood residue reducing productivity of the marsh if the watershed soils are generally infertile. The entire bog mat might float when flooded, with little or no change in habitat, or beneficial small floating nesting islands might result (Atlantic Waterfowl Council 1959).

If adjacent uplands in the watershed are barren, flooding and siltation will occur in the wetland after each storm. Well-vegetated watersheds produce constant, slower runoff of better quality. Sites near tall grass, shrubs, or woodlots attract wildlife by providing a protective corridor for animals seeking water (Kierstead undated). Sites without adequate upland nesting area surrounding the wetland should be rejected (Markell 1986). A 3:1 or 4:1 ratio of uplands to wetlands has been used in Minnesota (Piehl 1986).

Before any sites are flooded, plants especially in bogs should be examined adequately to determine if endangered or threatened species occur, which halts the project. Flooding alters plant communities and associations drastically. A diagram of the existing plant cover should be made, with a list of species. This aids site evaluation and might be useful in future ecological study or management of the marsh (Atlantic Waterfowl Council 1959). The Federal Interagency Committee for Wetland Delineation (1989) described the techniques.

Greentree reservoir. Tree species vary in response to water-level changes (Fig. 2.2) (Teskey and Hinckley 1977a, b, Fredrickson 1978, Larson et al. 1981, Reinecke et al. 1989). The site selected must have mast-bearing oaks (Table 2.5) that are adapted to flooding and can be flooded. This opportunity generally is limited strictly to broad, geologically old-age valleys of the south-central and southeastern United States (Rudolph and Hunter 1964). The red oak group (subgenus *Erythrobalanus*) should be targeted for management because it produces most of the acorns eaten by ducks (Reinecke et al. 1989): water oak, willow oak, Nuttall oak, cherrybark oak, pin oak, Shumard oak. In years of poor acorn production, mast producers such as blackgum, sweetgum, hickories, and baldcypress also might become valuable sources of food for ducks (Rudolph and Hunter 1964). When flooding or soil saturation is relatively permanent, water tupelo and baldcypress dominate the canopy. If the soil usually is saturated more than 25 per-

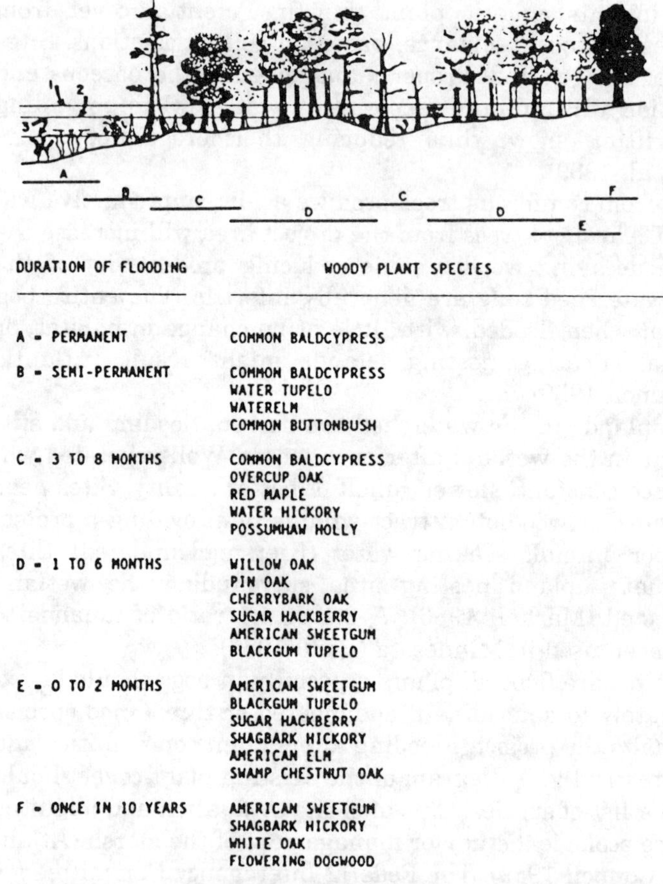

Figure 2.2 Cross section of forested wetlands in the Mississippi alluvial valley showing the distribution of woody plants relative to duration and depth of flooding. 1 = 10-year flood; 2 = mean annual high water; 3 = mean annual low water; A = permanently flooded; B = semipermanently flooded; C = flooded 3 to 8 months; D = flooded 1 to 6 months; E = flooded 0 to 2 months; and F = flooded once in 10 years (*after Fredrickson 1978* in *Reinecke et al. 1989*).

cent of the growing season, water hickory and overcup oak occur. When flooding is common during the dormant season but saturation is irregular and brief during the growing season, sugar hackberry, sweetgum, and several species of oak occur (Reinecke et al. 1989).

Aspect

Long, narrow wetlands oriented east-west, especially in cold climates, maximize southern shoreline exposure for wintering ducks and

TABLE 2.5 Bottomland Trees of Best Food Value to Ducks

Best	Good	Poor
Water oak *Quercus nigra*	Cow oak *Quercus michauxii*	Elm *Ulmus* spp.
Willow oak *Quercus phellos*	Swamp chestnut oak *Quercus muehlenbergii*	Ash *Fraxinus* spp.
Nuttall oak *Quercus nuttallii*	Swamp white oak *Quercus biocolor*	Sycamore *Platanus occidentalis*
Cherrybark oak *Quercus falcata pagodaefolia*	Overcup oak *Quercus lyrata*	Yellow poplar *Liriodendron tulipifera*
Pin oak *Quercus palustris*	Blackgum *Nyssa sylvatica*	Beech *Fagus grandifolia*
Shumard oak *Quercus shumardii*	Sweetgum *Liquidambar styraciflua*	Birch *Betula* spp.
	Hackberry *Celtis occidentalis*	Soft maple *Acer* spp.
	Black locust *Robinia pseudo-acacia*	Boxelder *Acer negundo*
	Honey locust *Gleditsia triancanthos*	Pine *Pinus* spp.
	Water locust *Gleditsia aquatica*	
	Tupelo *Nyssa aquatica*	
	Cypress *Taxodium distichum*	

SOURCE: Payne and Copes (1986).

the shoreline/surface-area ratio (Ringelman et al. 1989). A thermal refuge with good loafing sites results from wetlands with sheer or steep-sloped banks next to broad, gently sloping shoreline.

Topography

The impoundment basin should be examined to determine if it is flat enough for development with or without water control structures (Atlantic Waterfowl Council 1972). It should have a gradient of less than 1 percent, i.e., less than 1-m elevation in 100 m, so that up to 75 percent can be flooded with water less than 6 dm deep for puddle ducks. The construction height of the dike thus is reduced along with the cost. Ideally for impoundment, the flat area should have gentle, undulating surfaces for islands and side channels, diverse water depths, and an irregular shoreline for edge effect. Side slopes should be no steeper than 5:1 horizontal to vertical and of stable material (Proctor et al. 1983a, b, c).

Three important considerations in evaluating the watershed are its size, channel slope, and storage (Lejcher 1986). If the watershed is large and the slopes are steep, then the volume and speed of runoff will be high, thus causing impoundment water levels to increase faster and higher with longer flood levels, requiring a larger overflow (Atlantic Waterfowl Council 1972). The watershed is determined by using 7.5-min topographic maps from U.S. Geological Survey or Canada Department of Energy, Mines and Resources, or aerial photographs if contour intervals are too great, and essentially drawing a line along the high points from one side of the project site completely around all streams and tributaries leading into the project area until the line ends at the opposite side of the project site, as though one were trying to determine the path to walk around the entire river system from one bank to the other without getting wet feet. Contour lines on the map will help. Then a dot grid or planimeter is used to calculate the size of the watershed feeding into the project site. The size of large watersheds (over 1000 acres, or 405 ha) can be estimated from a quadsheet by the number of 40s (40 acres, or 16 ha) or sections (640 acres, or 259 ha) involved (Lejcher 1986). The area hydrologist can help.

If more than one channel occurs in the watershed, the length (in kilometers) of the longest channel is measured with a map wheel from the end to the start where it no longer exists; then the map wheel is used along the contour indentations until the top of the watershed is reached (Atlantic Waterfowl Council 1959, Lejcher 1986). Next the length of the major drainage is measured at 10 percent and 85 percent of the main channel length. The slope is the difference in elevation at the 10 and 85 percent points divided by the length. The storage of the watershed is determined by using a planimeter or dot grid to measure all the lakes, swamps, marshes, and depressions that hold water and adding all the measurements.

Ravines or natural catch basins indicate topography suitable for wetland development. The topography of the catchment basin determines the shape and extent of the resulting wetland.

Adjacent highlands should be bored to determine if suitable material exists for the core of the dike, preferably clay, to minimize transportation costs. Natural diking factors such as hard bottoms, ridges, and islands, relative to a suitable borrow pit, should be exploited. They can be determined by surveying the area with standard levels and taking soil borings. Plants can be used as indicators of bottom type, elevation change, etc.

Geology

Knowledge of the geology of the wetland site within the province or state can be gained from textbooks or journals. This will help to determine the regional water table, whether a 2- or 3-dm fluctuation in the regional water table will affect the site, how surface runoff will get to the site, whether there are porous materials that serve as a conduit to the site, whether bottom materials will hold water without leaking, and whether certain rock types will yield a unique water quality, such as specific conductance (Verry 1985a).

Soils

For the purpose of site selection, soils can be divided into two general types: organic and inorganic (mineral) (Atlantic Waterfowl Council 1959, U.S. Soil Conservation Service 1977, Mitsch and Gosselink 1986). A soil with less than 20 to 35 percent (dry weight) organic matter is a mineral soil. Organic soils usually are classified either as peat (if the organic remains are fresh enough for identification of plant forms) or as muck (Atlantic Waterfowl Council 1972, Federal Interagency Committee for Wetland Delineation 1989). In wetland (hydric) soils, organic soils often tend to have fewer total nutrients, with more minerals in organic forms unavailable to plants (Mitsch and Gosselink 1986). Thus, mineral soils are preferred for production of plants, compaction of the pond bottom, and construction of dikes (Atlantic Waterfowl Council 1959, U.S. Soil Conservation Service 1977). Soils high in clay content (at least 20 percent) tend to be best, for the small soil particles reduce leakage in the impoundment and through the dike. Yet clay can be a problem in the headwaters or basin of the area if it causes turbidity due to ready suspension of the small clay particles or in barren areas which do not vegetate well (Atlantic Waterfowl Council 1972).

Organic and mineral soils also differ in other ways (Table 2.6). Perhaps the best indicator of soil suitability for wetland development is pH (Atlantic Waterfowl Council 1959). Sites are rejected where organic soils comprise most of the basin. But if good soils do not exist in regions where the need for wetland development is great, the project should proceed anyway. Because soils vary in productivity with flooding depth, the poorest soils should be flooded up to 3 dm (Atlantic Waterfowl Council 1959), which is practical only in flat basins.

Highly organic soils might be very productive initially, but plant growth soon might decrease (Atlantic Waterfowl Council 1972). Such soils might have to be drained and dried completely to oxidize toxic substances and reverse other changes before reflooding. But soils such as the cat clays of the southeastern coastal states contain iron oxides.

TABLE 2.6 Comparison of Mineral and Organic Soils in Wetlands

Characteristic	Organic soil	Mineral soil
Typical wetland	Northern peatland	Some marshes; riparian forest
pH	Acid	Circumneutral usually
Organic matter	>20 to 35%	<20 to 35%
Nutrient availability	Often low	Generally high
Porosity	High (80%)	Low (45 to 55%)
Water-holding capacity	High	Low
Hydraulic conductivity	Low to high	High, except clays
Cation-exchange capacity	High, hydrogen ion dominant	Low, major cations dominant
Bulk density	Low	High

SOURCE: Adapted from Mitsch and Gosselink (1986).

If dried out, such soils become irreversibly unproductive when the pH drops drastically and the sulfur combines with the iron oxide to form sulfuric acid upon rewetting. Many desirable aquatic plants prefer soft bottoms of mucky, peaty, loamy, or organic soils, and not rocky bottoms or hard sands.

Soil should be analyzed at the impoundment site (Federal Interagency Committee for Wetland Delineation 1989) by augering four or more 2-m test holes around the proposed impoundment and dike location (Kierstead undated) and at least one hole 1.5 m or more deep above water near the stream channel (Verry 1985a). The texture, depth, and color of each horizon are then recorded. For field expediency, the texture of organic soils can be determined by squeezing the soil by hand. Fibric (fibrous) soils release almost clean water and no soil, sapric (rotten) soils ooze readily between the fingers, and hemic (in-between) soils release colored water, leaving fibers lighter in color (Knighton and Verry 1983).

Wetland vegetation tends to have shallow root systems. If the topsoil horizon is organic less than 0.3 m thick, underlain with mineral soil, roots will penetrate the mineral soil, anchoring the bog mat. Organic soils of 67 percent fibric material and 0.3 to 1 m thick produce floating bogs when flooded (Verry 1985a). Indicator plants are bog shrubs or herbaceous plants.

Flooded peat soils are usually hemic or sapric organic soils and thus will not float. But such soils in the shallow wildlife edge around multipurpose impoundments over 405 ha erode if spring runoff is over 28 m^3/s, the impoundment is held over winter at normal pool, and the ice freezes into the bottom. Then the spring flow will lift about 1.5 dm of soil and plant roots with the ice, which will be pulverized with wind-pushed ice and flushed out in overflow. This action deepens and reduces the value of the shallow wildlife zone. Erosion can be prevented by dropping the water level in these larger, deeper impoundments to 0.6 to 1 m in the winter after 7.6 cm of ice has formed. This accommodates spring runoff and leaves a collapsed layer of ice on the shallow areas of organic soil which melts before peak flow of snowmelt. The drawdown could begin after the waterfowl season, but muskrat populations are protected better if the drawdown begins after 7.6 cm of ice forms (Verry 1985a).

Coarse sands to pebbles within the top 3 to 4 dm usually indicate a good supply of groundwater and rapid accumulation of spring snowmelt through the porous soil, requiring an adequate emergency spillway in the dike. Such bottoms will leak unless the regional water table is above the bottom of the impoundment as in floodplains. In very dry years when the water table drops, heavy rains and melt from heavy snowpacks are needed to raise the regional water table to fill the impoundment. Impoundments with leaky soils are controllable only when the regional water table is high enough to fill the impoundment (Verry 1985a).

Sites with very fine sand, silt, or clay in the top 1.5 m will hold water. Horizons of medium to fine sands above horizons of clay, silt, or very fine sand indicate that a direct groundwater supply is possible. If not, the groundwater might supply the site indirectly through an artesian well upstream (Verry 1985a).

Each horizon should be tested with a 10 percent solution of hydrochloric acid (HCl) to determine how calcareous the soil is, i.e., how much calcium is present. Strong fizzing indicates highly calcareous soil. With such soils, the specific conductance cannot be used to interpret the potential for groundwater supply and thus the suitability of the site for impoundment.

The U.S. Soil Conservation Service (1982) classified soils into four hydrologic groups according to infiltration and transmission rates of water: (1) Soils with a high infiltration rate are mainly deep, well-drained sands or gravels with low runoff potential. (2) Soils with a moderate infiltration rate are mainly moderately deep and well drained, with moderately fine to moderately coarse texture. (3) Soils with a slow infiltration rate are moderately fine- to fine-textured with a layer that impedes downward movement of water. (4) Soils with a very slow infiltration rate are mainly clay, with a high swelling potential, a permanent high water table, a claypan at or near the surface, shallow soils over nearly impervious material, and a high runoff potential. Help in classifying soils into one of the four hydrologic groups for a pond site can be obtained from district soil conservationists. Dennis (1979) described basic concepts, soil classification, field identification, engineering behavior, site investigation, and uses of soils.

In coastal areas, subsidence can occur. Subsidence is the settling or sinking of the coastline from compaction of sediment below the land surface, loss of groundwater buoyancy, and biochemical activity (U.S. Soil Conservation Service 1977). Natural subsidence is accelerated by canal building, with resulting saltwater intrusion and blockage of natural marsh drainage by construction of spoil banks (Morgan 1973). Subsidence accelerates erosion and salt intrusion and must be considered in marsh planning. Wind and wave erosion of the shoreline, especially of sandy soils, also must be considered and controlled with vegetation, barrier, or design of impoundment. Gabions, riprap of rock or concrete pieces, vegetation with a good root system, or other methods are used above the waterline to retard erosion by wind, waves, or muskrat burrowing. Below the waterline a band of dense, emergent aquatic plants should be established to retard erosion by waves, or islands can be built to reduce fetch. (See "Erosion Control" and "Retention and Protective Structures" in this chapter.)

In all cases, soil survey maps from the federal government should be consulted. Such maps provide fairly accurate information on the characteristics of soils in most localities. Maps of wetlands and agricultural capability should be consulted. Soils can be tested by the appropriate government agency.

Water supply

Wetland sites might have potential for improvement with or without water-level control. Sites can be grouped into three hydrologic types that relate directly to management control (Knighton and Verry 1983): those with (1) little water inflow, with water levels controlled by weather; (2) moderate inflow, with best potential for management-controlled water levels; (3) much inflow, with potential for management-controlled water levels and nuisance beaver problems and expense. The three general sources of water are surface water only, mostly groundwater, and about equal mixtures of both. Surface water results from precipitation, tides, streams, and lakes or reservoirs. The water level will fluctuate widely if weather controls it. In dry years, shorebirds will benefit but waterfowl will not. Controllable water levels should favor waterfowl. If groundwater is the main source, furbearers will benefit more than waterfowl. But groundwater is cold and can be nutrient-deficient, with reduced plant growth (Reid et al. 1989). Groundwater is especially important in arid regions where evapotranspiration is high, although pumping is expensive.

The specific conductance indicates the size of the surface drainage area or the extent of groundwater in the wetland sites proposed for development (Verry 1985a, b). It is measured during summer and fall streamflow (June, July, August, September, or October) when the water supply must match open-water evaporation. Sampling periods should be avoided 2 or 3 days after a large storm (over 3 cm) or during snowmelt. In most cases, confident estimates can be obtained with two or three samples. With experience, one sample will suffice. The specific conductance can be measured in the field or laboratory with a type RQ battery-operated electrolytic recorder adjusted to 25°C by dialing the temperature correction knob to the ambient water temperature (Christiansen and Low 1970, Verry 1985a). Measurements are expressed as siemens (formerly mhos, or reciprocal ohms) or microsiemens (micromhos) per cubic centimeter. The interpretation is as follows (Verry 1985a):

1. *Readings below 26:* Such low readings indicate water has traveled only a short distance and dissolved only minor amounts of mineral salts. A limited amount of surface water occurs only during spring. An impoundment should not be built.
2. *Readings of 26 to 100:* The water source in the project site is mainly surface water from the watershed. If the ratio of watershed area to impoundment area at normal pool equals or exceeds that of Table 2.7 or Fig. 4.3, enough water for impoundment control exists in average to wet years. Otherwise water levels are controlled by weather.
3. *Readings of 101 to 150:* Surface water is still dominant, but groundwater mixing with it is increasing. Favorable ratios of watershed area to impoundment area can be used in decisions to control water by impounding.
4. *Readings over 150:* The water source in the project is mainly groundwater, inadequate only during years of the most severe drought. Ratios of watershed area to impoundment area can be ignored. If the ratio is less than the value of Table 2.7, the groundwater supplying the site exceeds the surface water from the watershed.

To ascertain that enough water is available to keep the pond full, the pond capacity is estimated (U.S. Soil Conservation Service 1982). At the pond-full elevation (normal pool) the waterline is staked, and the width of the valley at that level is measured at regular intervals in the field. Otherwise, if maps are large and detailed enough to reveal small changes in topography, the pond-full elevation can be plotted and measured with planimeter or dot grid. The surface area is multiplied by 0.4 times the maximum water depth at the dam to cal-

TABLE 2.7 Area of Watershed per Area of Impoundment Surface Needed to Ensure a Dependable Water Supply in the Northern Lake States*

		Inches of open-water evaporation				
		24	26	28	30	32
Annual streamflow, in	Summer and fall streamflow, in	Acres of watershed per acre of impoundment water at normal pool				
2	0.7	34	37	40	43	46
4	1.8	13	14	16	17	18
6	3.0	8	9	9	10	11
8	4.1	6	6	7	7	8
10	6.2	5	5	5	6	6
12	6.4	4	4	4	5	5
14	7.5	3	3	4	4	4
16	8.7	3	3	3	3	4

*Table derived from the equation

$$Y = \frac{\text{inches of open-water evaporation}}{-0.46 + 0.57(\text{inches of annual stream flow})}$$

= surface watershed acres needed per acre of normal pool

This equation contains a measure of annual streamflow because annual values are easiest to obtain. The ratios in fact are based on summer and fall streamflow because this is the time when the water supply must match open-water evaporation. Spring streamflow (April and May) fills the impoundment, which overflows. It is useless to maintain the water at normal pool during the summer when evaporation is high. The denominator is a regression of summer and fall streamflow against annual streamflow from a small surface water area over a range of annual precipitation. (1 acre = 0.405 ha; 1 in = 2.54 cm.)

SOURCE: Verry (1985a).

culate the capacity of the pond. For example, a pond with a surface area of 6.7 acres and a maximum depth at the dam of 3.2 ft has a capacity of about 8.6 acre · ft (0.4 × 6.7 × 3.2 = 8.6), or about 2.8 million gal.

The volume of surface runoff water collected in the basin from the watershed determines the size of the control structure in the impoundment (Atlantic Waterfowl Council 1959). This volume is influenced by the size, shape, and cover of the watershed and the slope and porosity of the ground (Kent and Styner 1979). Ideally, the ratio of drainage area (watershed) to flooded area should be 10:1 to 30:1, but it depends on factors such as groundwater flow (Anderson 1985). The watershed/impoundment ratio commonly used to design waterfowl impoundments is 15:1 (Poff 1985). Most coastal marshes in Texas have a watershed of less than 1000 ha (Stutzenbaker and Weller 1989).

As an example for modification, in New York the ratios of watershed size and cover relative to the size of the marsh impoundment basin are 40 ha of woodland per hectare of marsh, 30 ha of pastureland per hectare of marsh, and 20 ha of cropland per hectare of marsh (Atlantic Waterfowl Council 1959). A sample watershed might produce the following size impoundment in the marsh basin:

Sample watershed		
Cover	Size, ha	Size of basin, ha
Woodland	80	2
Pastureland	30	1
Cropland	60	3
Total	170	6

In areas where annual precipitation varies substantially from that in New York, adjustments to the ratios must be made. In Louisiana the watershed/impoundment ratio is 5:1 to 20:1 for pasture or cover cropland and 10:1 to 20:1 for timberland (Summers 1984).

In areas with little or no watershed, annual rainfall of over 100 cm provides enough water beyond evapotranspiration losses to fill compartmentalized marshes with good dikes and water control structures (Stutzenbaker and Weller 1989). In the eastern part of the northern Great Plains where annual precipitation is about 50 cm, 10 ha of grass-covered watershed normally provides enough water to flood 1 ha 6 to 18 dm deep. In the western part of the Great Plains where annual precipitation is 38 cm or less, the watershed/pond ratio is 20:1 (Lokemoen et al. 1984). Kent and Styner (1979) and the U.S. Soil Conservation Service (1982) presented a range of soil-slope-cover curves developed from a runoff equation, based on the rainfall frequency atlas published by the U.S. Weather Bureau, to estimate the volume of water entering the marsh basin from a 24-h storm.

The degree of seepage and the rate of evaporation or transpiration influence the amount of water flow needed to maintain adequate water levels in the pond (Christiansen and Low 1970). In eastern North America, e.g., when seepage is minimal, a flow of 75 to 140 L/min is generally enough to maintain a 1-ha pond (Addy and MacNamara 1948). In Colorado, a natural surface flow of at least 850 L/min will permit water-level manipulation (Rutherford and Snyder 1983). Otherwise little can be done to change water levels if the natural wetland is maintained by seep water only.

The transpiration rate of most marsh vegetation exceeds that of upland vegetation (Atlantic Waterfowl Council 1972). Evapotranspiration is highest in July and August when air temperatures are highest and plants have reached maximum growth and transpiration (Linde 1969). For example, in central Wisconsin, evapotranspiration losses are about 7529 m^3 of water per hectare per year, or 7,622,785 L/(ha · yr). In Utah, seasonal losses for April through October are about 10,400 m^3/ha (Christiansen and Low 1970). The higher the density of aquatic plants and phreatophytes, the greater the transpiration losses. Evapotranspiration losses must be considered in site selections for impoundment management. Water levels in the impoundment will decline continually if the water source is inadequate to compensate. Early fall recovery of a partial summer drawdown is doubtful if evapotranspiration losses are high. In contrast, evapotranspiration losses can aid a complete summer drawdown. The basic equation for determining evapotranspiration (Christiansen and Low 1970) is

$$E = I - O - \Delta S + P - \Delta \mathrm{SM} + \mathrm{GW}$$

where E = evapotranspiration
I = surface inflow to area
O = surface outflow from area
ΔS = difference in surface storage from beginning to end of period
P = precipitation during period
$\Delta\mathrm{SM}$ = change in soil moisture in root zone of unflooded area during period
GW = groundwater contribution or deep percolation loss

Determination of evapotranspiration can be made more applicable to various areas if the values are related to climatic and other parameters. Christiansen and Low (1970) discussed various equations for doing so.

Water losses for some marshes exceed that of normal or theoretical evapotranspiration (Atlantic Waterfowl Council 1972). Some of the loss is from the lateral movement of water leaking through the dike. Some of the loss is from vertical downward movement, detected after drawdown by unusual concentrations of essential minerals, such as calcium, in the mineral portion of flooded soils. The U.S. Geological Survey, U.S. Soil Conservation Service, U.S. Army Corps of Engineers, U.S. Weather Bureau, U.S. Coast Guard, Coast and Geodetic Survey of the U.S. Department of Commerce, the U.S. Department of Commerce National Oceanic and Atmospheric Administration, Canadian counterparts, state or provincial hydrologic bureaus, local community records, newspapers, and landowners can furnish detailed information about streamflow, precipitation, evaporation, seepage, and frequency and size of storms and tides.

Development of tidal marshes often is expensive. Management often is a compromise between waterfowl and mosquito control, commercial and sport fishing, spawning and rearing grounds for estuarine species of commercial value, or some combination thereof. Large volumes of salt water should never be introduced into fresh, intermediate, or brackish marshes (U.S. Soil Conservation Service 1977). Hurricanes, canals dug through the marsh connecting tidal channels, cutoff tributaries of river systems, and subsidence will increase salinity. Salinity can be measured with a salinity determination kit, which includes a thermometer and Vogel unimometer (U.S. Soil Conservation Service

1977). Although not as accurate as electronic devices, it is accurate enough for field use. Coastal salt marshes are mainly intertidal; they are not flooded during low tide, but sometimes are during high tide.

Sites likely to be polluted from groundwater or surface water contamination by agricultural, industrial, commercial, residential, or other sources are avoided. Unlike in most lakes, nutrients increase greatly in shallow marshes in winter, but usually remain within eutrophication standards and pose no threat to water quality downstream because outflow is limited then (Verry 1985b).

Before the proposed site is developed potential and actual drainage activities in the area should be examined, e.g., developments upstream, ditching and tiling, possible backwater, evidence of floods. These might render the site unsuitable. Legislation affecting riparian rights, water rights, control of malarial mosquitoes, and other potential public health problems must be consulted (App. A).

If an impoundment is desired, it should be controlled by spillway manipulation with enough water from reliable surface water and groundwater sources rather than from the vagaries of weather. But an impoundment can produce habitat during fall, winter, or early spring if enough water is available from precipitation, irrigation systems, or wells (Addy and MacNamara 1948). Impounding reliable fall rainfall is the most economical method of flooding (Yoakum et al. 1980). But the most effective methods do not depend on seasonal rainfall. Pumping from lakes, streams, or wells affords complete control of flooding, but expense limits the size of the impoundment. The ideal arrangement is a storage reservoir or stream from which water can be released or diverted to flow by gravity into the impoundment.

Site selection might be best in winter or early spring when standing water or ice is an indicator of wetland conditions. Other indicators are drainage ditches and water speed in small streams of the project basin (Atlantic Waterfowl Council 1972).

Sources of water for flooding and draining greentree reservoirs are pumping and gravity flow from rainfall and runoff, storage reservoirs, and stream diversion. In rice-producing regions, irrigation projects can furnish water, but pesticide and fertilizer loads might prevent it. Silt-laden waters and running water that overflows often are unsatisfactory because silt loads can kill entire stands of vegetation (Broadfoot and Williston 1973, Hunter 1978). Swift currents and deep water might cause ducks to leave for more accessible feeding and resting sites and are unsafe for hunting (Mitchell and Newling 1986).

Water quality

Before plans are developed for the site, the water quality should be studied comprehensively, to include tidal fluctuations, availability of

TABLE 2.8 Water Quality Parameters that Tend to Support Fish and Other Aquatic Organisms

Parameter	Range
pH	6.5–9.0
Alkalinity	≥20 mg/L
Hardness	20–150 mg/L
Dissolved oxygen	≥5 mg/L
Total dissolved solids	Productivity generally positively correlated
Temperature	≤20–30°C, depending on species and acclimation

SOURCE: Herricks (1982).

fresh water from groundwater and surface water sources, periods and amounts of runoff, turbidity, pollution, stain, salinity, pH, methyl orange alkalinity, and total alkalinity (Atlantic Waterfowl Council 1972). These factors must be examined relative to the tolerances and preferences of the plants concerned, and they will largely determine the type of management to use on the site. Water-quality parameters that generally support fish and other aquatic organisms fall within general ranges (Table 2.8) (Herricks 1982).

Water quality probably is the most important factor affecting production of submerged vegetation for waterfowl (Atlantic Waterfowl Council 1972). Most good food plants need reasonably hard, fertile water. Many pest plants thrive in soft, less fertile water, extremes of which produce no worthwhile growth of submerged spermatophytes.

The quality of water entering the marsh basin from the watershed depends on the concentration of nutrients in the soils of the drainage area (Atlantic Waterfowl Council 1972). Marsh basins are best if surrounded by agricultural soils which produce drainage water with high concentrations of essential minerals. But such land use can produce runoff with dangerous levels of nitrates from fertilizers, chemicals from pesticides, and sediment from erosion.

Water is unproductive if highly acid (pH 5.0 or lower), with total alkalinity below 10 ppm (Linde 1969). Water with pH 6 to 7 and alkalinity at least 50 ppm has fair to good potential in the north. But these acid waters tend to be stained from spruce and leatherleaf, and they must be held at low levels to ensure light penetration for adequate plant growth. A larger watershed might dilute the stain problem.

The level of salinity in the soil water and overlying water influences plant associations. Salt-marsh vegetation is replaced by freshwater plants if the water salinity is below 5 ppt (Mitsch and Gosselink 1986). Studies in Utah (Christiansen and Low 1970) suggest that a specific

TABLE 2.9 Tentative Classification of Water Quality for Waterfowl in Marshes of Utah

Rating*	Conductance, mmhos	Salinity, ppt
Excellent	<1	0.00–0.64
Good	1–2	0.64–1.28
Fair	2–4	1.28–2.56
Poor	4–8	2.56–5.12
Restrictive†	>8	>5.12

*Relative to growth of marsh plants.
†Only the most tolerant marsh plants are present.
SOURCE: Christiansen and Low (1970).

conductance of less than 1 mS is excellent for waterfowl and that over 8 mS is restrictive (Table 2.9). In coastal areas, more fluctuation often occurs at the mean high-water level.

The tidal marsh can be divided into two zones: the lower marsh (or intertidal marsh) and the upper marsh (or high marsh). Unlike the upper marsh, the lower marsh is flooded almost daily with less than 10 days of continuous exposure, resulting in generally higher salinity than in the upper marsh. Salinity in the marsh soil water is influenced by tidal inundation, rainfall, tidal creeks and drainage slope, soil texture, vegetation (evapotranspiration), depth of water table, and freshwater inflow.

Control of salinity is effected by mixing tidewaters of varying salinity via draining and flooding the impoundment. Water salinity influences the types of plant communities in the impoundment. The closer a tidal impoundment is to a river, the more fresh water it contains. Generally, freshwater impoundments have less than 1-ppt salinity. Freshwater to brackish, or intermediate zone, impoundments have 1 to 5 ppt, brackish impoundments have 5 to 20 ppt, brackish to saline impoundments have 20 to 30 ppt, and saline impoundments have 30 to 35 ppt (Gordon et al. 1989).

A conductance meter at each wetland management area allows the manager to check the specific conductance at inlets and outlets so that the water of highest salinity can be released (Christiansen and Low 1970). Otherwise, large flows of relatively good-quality water or small flows of low-quality water might be discharged at the outlets of the various impoundments.

Plans and specifications

Provincial, state, or federal agencies probably will require permits before wetlands are altered, resulting in impoundment, drainage, or filling (U.S. Fish and Wildlife Service 1976a, 1989, Office of Technology

Assessment 1984). Such agencies are consulted before proceeding, to determine if they will issue a permit and to learn their requirements for it. Eventually, application for the permit(s) might include plans and specifications prepared by a registered engineer (Farmes 1985). The final design incorporates low construction cost, ease of operation and maintenance, and visual attractiveness (Anderson 1985).

Without engineering skills or an engineering staff readily available, the supervisor in charge of the project might not be involved in the details of construction design (Farmes 1985). The supervisor must review all plans and specifications, taking nothing for granted, particularly relative to elevations and drawdown capability. Structures must accommodate the highest normal-pool elevations anticipated, for corrections later might be too costly to do.

Detailed plans and specifications are always desirable, always expensive, but not always needed. To save engineering costs on topographic work, the extent and depth of flooding from spring runoff and heavy rains should be determined on areas that flood naturally. Otherwise an engineer must be employed if the project supervisor and staff are unfamiliar with surveying instruments for the relatively simple surveys for most ponds (Mosby 1980, U.S. Soil Conservation Service 1982).

All planning information should be recorded on an engineering plan (Stanley 1979), which shows all elevations; dimensions of the dikes; dimensions of the cutoff trench and other areas needing backfill; location and dimensions of the emergency spillway, trickle tube, and other water control devices; other pertinent information such as use of wave action control, riprap, and artificial earth islands; and the kind and amount of building materials needed. The specifications supplement the plan by delineating the type and method of work, quality of material and workmanship, method of measuring progress, and unit of payment. Without a clear understanding between owner and contractor, the desired standard of work will not result. Local soil conservation specialists or private consultants can assist in preparing plans and specifications. During the planning stage, several types of plans, or work sheets, are needed to facilitate development and construction of the impoundment so that the specifications can be transferred readily from paper to field (Atlantic Waterfowl Council 1959).

The marsh site investigation sheet lists the important features of the proposed site that would qualify or disqualify it. Characteristics include the marsh basin, watershed, outlet, and ownership.

The design and estimate sheet refers to data about the design of the emergency and mechanical spillways and the cost estimates. Data include runoff characteristics, size of impoundment and spillways, and costs of structures, excavation, seeding, and riprap.

The preliminary work sheet is used in the field to plot the survey information and the contours and other features of the basin. The plan view sheet records topographic and geographic features of the basin, with alterations of terrain needed for construction. This sheet is developed from the work sheet. It is essentially a map, and it contains the scale; north arrow; soils data; location of benchmarks, contours, fences, roads, buildings, etc.; positions of dike (top and bottom widths), emergency spillway, and mechanical spillway; and centerlines of the marsh, dike, and emergency spillway (Fig. 2.3).

The marsh profile sheet is profile paper containing three profiles: (1) the profile along the centerline of the dam, across the emergency spillway; (2) the profile of the centerline of the emergency spillway; and (3) the profile along the centerline of the marsh. These three profiles present a different perspective of the items already shown on the plan view sheet. The profile of the dam contains the following information:

1. Original ground line
2. Ground line after removal of topsoil
3. Elevation of top of dike
4. Elevation of maximum water level
5. Elevation of spill crest
6. Elevation of normal water level
7. Cross section of emergency spillway, with side slopes labeled
8. Location of mechanical spillway
9. Vertical and horizontal scales

The profile of the emergency spillway contains the following information:

1. Original ground line
2. Proposed spill crest
3. Size and slope of outlet from spill crest
4. Size and slope of inlet to spill crest
5. Direction of flow
6. Vertical and horizontal scales

The profile of the marsh includes the same features shown on the profile of the dam, except for information about the emergency spillway, and includes upstream and downstream slopes of the dike.

The water control structure sheet depicts drawings and measurements of the mechanical spillway.

Unless the owner has the equipment and personnel needed for con-

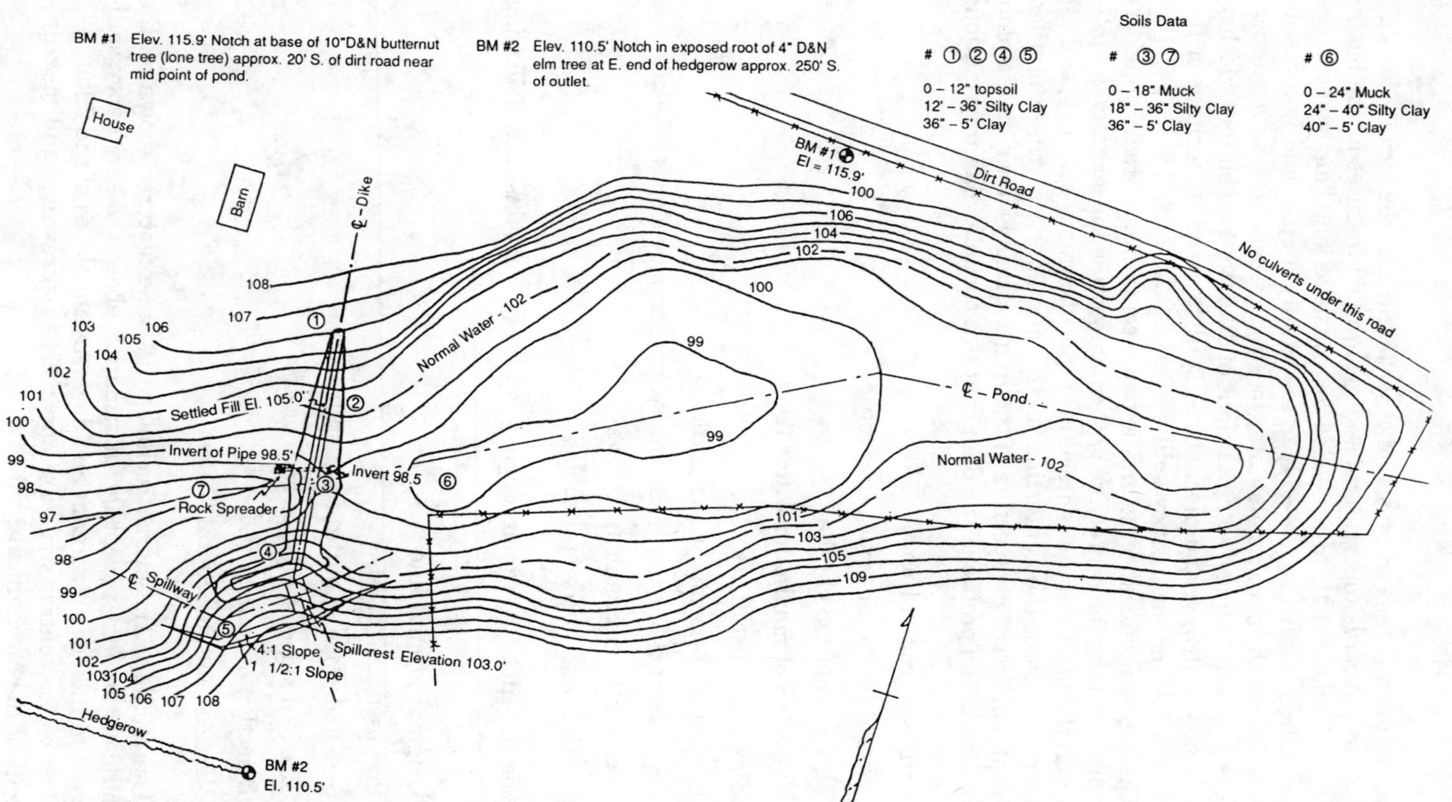

Figure 2.3 Example of plan view sheet (*Atlantic Waterfowl Council 1959*).

struction, a contractor is needed. Even when the owner has the equipment, cost comparisons for large projects often indicate that a private contractor will be cheaper, releasing the owner's labor force for other projects (Farmes 1985). Contracts usually are awarded by bidding to find the least expensive contractor. The plan and specifications provide a basis for contractors to bid on the project and compete fairly. Bid invitations should be prepared to maximize the number of qualified bidders. Fewer bids at higher cost usually result by specifying specialized equipment for construction. The project should be advertised to attract smaller contractors, who usually can bid lower than large companies having higher overhead costs.

Projects can be bid in one lump sum; a complete set of plans and specifications must be furnished to the bidders. If the earthwork is bid separately from the control structures, more bids will be received because many contractors do not have expertise in both areas. The job likely will be done at lower cost, too. The earthwork can be bid in one lump sum, by volume (cubic meter) or by the hour. If bid by the hour, good specifications must be furnished on the type of equipment desired. For payment based on hourly work, all machines should have an instrument called a *Servus Recorder* that records on a paper disk when the machine is operating. For hourly work, equipment and operator must meet specifications and operate well.

If the contract for the earthwork is awarded by volume or in a lump sum, the contractor determines the equipment to be used. If the work is paid on an hourly basis, the project supervisor (owner) determines the equipment. If soil and moisture conditions permit, dozers, dozers with scrapers, and draglines should be used in that order. A dragline often moves soil for twice the cost of dozers; costs of dozers with scrapers are in between. Draglines with mats usually are needed for wet, heavy soil. Low-ground-pressure (LGP) crawlers can operate unless the area is extremely wet (Green and Rula 1977, Willoughby 1978).

Estimating the volume of earth excavated for the impoundment helps to estimate costs, invite bids, and award the contract. Estimates of earth to be excavated include the dam, cutoff trench, holes in the foundation, dikes, stream channels to be backfilled, and any other earthmoving. The sum-of-end-area method, described by Renfro (1979) and U.S. Soil Conservation Service (1982), probably is the most efficient method to estimate the volume of earthfill. Instead of calculating the volume of earth in an irregularly shaped dam, the method involves reading the volume from a table for different side slopes, top widths, and heights.

Generally the larger equipment is more cost-effective. Usually, draglines, crawler tractors with scrapers, crawlers with dozers, front-end loaders, or a combination of these are used. At times a backhoe, clamshell, sheepsfoot roller, grader, and pile driver are used. Dump trucks are used with front-end loaders and clamshells. Crawler tractors (D8H-46A or larger) should have power shift with at least 200 flywheel horsepower. Draglines should carry 18.3 m of boom and usually a perforated bucket with at least a 0.76-m^3 (1-yd^3) rating. A larger backhoe can substitute for a dragline in some cases. Big jobs might require a backhoe with a 1.5- to 2.3-m^3 (2- to 3-yd^3) rating, carrying 24.4 to 30.5 m of boom (American 700 series or equivalent) (Farmes 1985). Work often is done during winter. Equipment can be moved more easily and prices usually are lower, for winter is the off season for construction work (Linde 1969). Excavation for fill is accomplished in winter by using a crane with a heavy steel ball to break the frost. During summer, where the ground is wet, use of log mats to support the equipment is time-consuming and costly, for a helper is needed to move the mats as the dragline advances (Farmes 1985).

Surveying and staking

Most biologists directing wetland projects have little or no training in topographic surveying. Thus, to survey and stake impoundment sites in small marshes, in-service training or hiring of an engineer is needed to survey the site and plot a topographic map with 3-dm contour intervals. Mosby (1980) and the U.S. Soil Conservation Service (1979) described the use of surveying instruments and mapping for small ponds. The two commonly used methods are centerline with cross sections and perimeter traverse with lateral extensions, both of which are detailed in Atlantic Waterfowl Council (1959). The latter method requires more skill and is used on densely wooded or brushy sites to eliminate the extensive brush cutting needed if the centerline method is used. The centerline method, also described in U.S. Soil Conservation Service (1982), is used mostly on open meadow sites where brush and trees do not interfere with the establishment of the centerline or on brushy and partially wooded sites where excessive clearing is not needed to establish lines of sight. The entire centerline survey can be done with a level. Pond surveys usually consist of a profile each of the dam's and spillway's centerline and enough measurements to estimate pond capacity (U.S. Soil Conservation Service 1982).

Before construction begins, each job is stated clearly and adequately. Staking transfers the information on the plan to the job site, thus locating the work and providing the lines, grades, and elevations needed for construction (U.S. Soil Conservation Service 1982). Much of the staking is done during surveying. Each area to be cleared, namely the sites for the dam, spillway, borrow pit, and impoundment, is marked with stakes. The proposed waterline is located with a level and rod to establish a baseline for clearing limits.

The dam is located by setting stakes along its centerline at intervals of 30 m or less. To mark the outer limits of construction and the points of intersection of the side slopes with the ground surface, the stakes for fill and slope are set upstream and downstream from the centerline stakes. These stakes also establish the height of the dam.

The earth emergency spillway is located by staking the centerline first. Then the cut and slope stakes are set along the lines of intersection of the natural ground surface with the side slopes of the spillway.

After the foundation is prepared, stakes are set to show the location of the centerline of the trickle tube or other control structure, including a permanent pump. The stakes are marked to show cuts from the tops of the stakes to the grade elevation of the tube. The locations of the riser, drainage gate, antiseep collars, outlet structures, and other structures are staked also.

To mark the limits of construction activity in the borrow pit and to allow proper drainage, cut stakes are set to indicate how deeply to excavate suitable material, as determined from soil borings.

Timber sales and clearing

Controlled marsh. In the north, where greentree reservoirs are not established, all merchantable timber (sawlogs and pulp) in the proposed impoundment and immediate environs should be sold to private contractors (Atlantic Waterfowl Council 1972). The immediate environs include upland areas as needed to provide interspersion of cover types beneficial to upland wildlife and nesting ducks. Stumps should be cut at least 3 dm above the proposed normal pool elevation to serve as nesting and loafing sites for waterfowl. They are removed later after a drawdown if they serve as perches for raptors and crows, which cause extensive undesirable predation of key wildlife species.

In semiwet woodlands, trees cannot always be cleared efficiently or economically with machines. Under such conditions, clearing is accomplished eventually by flooding the trees for 2 to 3 years to kill them (Green et al. 1964). A biennial summer drawdown for 8 to 10 years following flooding will speed decay. Where cold temperatures occur, trees can be cut in winter on ice.

All cutting and clearing is done in consideration for proper landscaping. A bulldozer is used to clear and convert wooded, brushy, or uneven ground to open ground capable of maintenance by standard farm equipment as grassland or cropland. Debris is piled in properly constructed brush piles to benefit wildlife or is burned, buried, or removed. A rotary shredder or brush crusher, pulled by a rubber-tired or crawler tractor, removes light brush invading open lands. To kill undesirable vegetation on open lands, approved chemicals can be used with caution at proper rates, with tractor-mounted spray equipment (Atlantic Waterfowl Council 1972).

Greentree reservoir. On a greentree reservoir, 70 percent of the timber can be retained in mast production under a 100-year rotation (Mitchell and Newling 1986). Regeneration cuts can vary from 0.4 to 6.1 ha. To allow time for regeneration, cuts should be made before the growing season. Intermediate cuts can be made in 20- and 30-year-old stands and should result in a residual stand of 5.6 to 6.5 m^2 of basal area, a density that produces crown development conducive to maximum mast production and volume growth. Only areas scheduled for regeneration cuts should be flooded. Such harvest cuts, especially for red oak, combine some features of shelterwood, clear-cutting, and coppice silvicultural methods to create an uneven-aged forest of even-aged patches (Lea 1988). But seedling red oak and other shade-intolerant hardwoods must be established before harvest. Then partial or shelterwood cuts should be used to create the openings needed for seedling development. Acorns also can be planted.

Oaks more than 25 cm diameter breast height (dbh) and in dominant or codominant positions in the forest canopy should be retained to produce acorns (Reinecke et al. 1989). Oaks less than 25 cm dbh should be thinned to promote growth and acorn production of survivors. The forest should be thinned about every 10 years, depending on age, density, and vigor. Trees removed are less desirable for timber or mast production. Cavity trees and dead snags should be retained. Openings of 0.8 to 2.0 ha should be created during regeneration cuts. To encourage reproduction of oaks, openings should be larger than 0.5 ha (Reinecke et al. 1989). To restore acorn production, at least 20 to 30 years are needed (McQuilkin and Musbach 1977).

Bottomland forests can be developed with two separate planting sequences (Herricks et al. 1982). Eastern cottonwoods, sycamores, and willows should be planted in clumps of 6 to 15 cuttings with clumps 6 dm apart, 25 to 50 clumps per hectare. Hardwood seedlings (see Table 2.5) are planted in a 2.1- by 2.4-m spacing (1925 trees per hectare). One recommendation is for 50 percent of the seedlings to be pin oak, silver maple, and boxelder and 50 percent to be bur oak, swamp white oak, bitternut hickory, green ash, hackberry, overcup oak, baldcypress, and river birch, all of which are dominants and of good food and cover value (Herricks et al. 1982). (See "Riparian habitats" in Chap. 5.)

Soil disturbance and sunlight penetration after logging operations will promote growth of various smartweeds, grasses, and sedges of value to ducks when flooded (Rudolph and Hunter 1964). Although costly, supplemental food plots of a high-energy, high-volume food, such as corn, help to offset mast failure. Clearcuts can be seeded with Japanese millet to supplement mast (Mitchell 1989), but this generally is costly and inefficient.

Landscaping

Principles of landscape architecture and design techniques can improve a pond's appearance. Good design includes shoreline configuration, relationship to surrounding use and landscape patterns, size, and site visibility (U.S. Soil Conservation Service 1982).

Existing landforms, vegetation, water, and structures should enhance the location and design of the pond. To reduce the expense and difficulty of reestablishing vegetation and to conserve the visual quality of the site, clearing of the site should be minimal. Interest in the configuration of the water's edge can be created by forming inlets, peninsulas, or islands. Strategic use of landforms to impound water often will minimize excavation. Irregular shape of a pond with smooth, flowing shorelines usually will complement the surrounding landscape. But usually the shape of the impoundment is dictated by water depth and topography. Clearing extended above the water's edge should leave an irregular, natural-appearing edge of open area and vegetation, achieved by selective clearing, new plantings, or both. The height and density of vegetation can be feathered progressively from the water's edge to the undisturbed vegetation. Costly clearing and grubbing can be reduced by leaving groups of trees and shrubs along the shoreline. These often tolerate 3 to 6 dm of graded fill over their root systems, or their root systems can be pruned in excavated areas, or they can be placed in tree wells and raised beds. Such vegetation will provide shade in summer.

Existing structures such as trails and stone walls can be retained to control vehicular and pedestrian traffic and to reduce disruption of existing use. Landform and vegetation will direct passage around the pond, frame the water for emphasis, blend the pond into the landscape, and guide attention to or from the water. Landform and vegetation also provide interesting reflections in the water, which help to create a contrast or focal point to the landscape.

Major viewpoints and sightlines should be identified. Where visible isolation is not desired, a pond visible from the main building, entrance road, or home improves the attractiveness of the landscape and often the land value to neighbors. If feasible, the pond should be located so that the main sight line crosses the longest dimension of the pond and so that the water is noticed before the dam, spillway, or pipe inlet is. Minor changes in alignment and location of these structures often can shift them from view or prominence.

A dike curved downstream, i.e., following the contour, is more aesthetic than a straight dike (personal communication, T. Meier, Wisconsin Department of Natural Resources 1991). The borrow pit for the dike could be in the proposed impoundment site if kept shallow by expanding the area of impoundment so that water covering the scar will not be hazardous to hunters. But even after impoundment, fill for dike maintenance requires a borrow pit somewhere, perhaps off the project site, but otherwise landscaped by minimal size and appropriate concealment.

All material cleared and grubbed from the area should be burned or buried under 6 dm of soil or in a disposal area such as a sanitary landfill. Some excess excavated material should be used for dike maintenance or road surfacing if suitable. Maybe county, state, or provincial highway maintenance crews will remove it if they can use it. Otherwise, removal is costly. Excavated material can be placed to improve the site's suitability for recreation, to screen undesirable views, or to buffer noise and wind. Waste material placed on the windward side of the pond serves as a snow fence for collecting drifts in the pond and as a wind break to reduce evaporation losses, especially in the Great Plains.

Waste material should not be placed so that it erodes into the pond or endangers the stability of the side slopes with its weight. It should be shaped, spread, and feathered to appear natural and to blend with the landscape. It should not protrude above the horizon. Proper placement of waste material improves the pond's appearance, prolongs its useful life, facilitates establishment of vegetation, and facilitates maintenance.

During construction of the dike, or after complete drawdown, the bottom should be altered 1 m below normal pool by excavating open-water patches of 0.1 ha to at least 0.4 ha for use by waterfowl broods and interconnecting channels for flight paths (Ambrose et al. 1983, Verry 1989). Channels should be dug 60 cm below normal pool. A road grader or similar equipment (Brown 1977) then should be used to develop low ridges and furrows along the contour in a washboard pattern 30 cm deep to provide microtopography (Kadlec and Smith 1984, Verry 1989).

A planting plan should be developed for the species, distribution, functions, and planting dates of the plants desired. Functions include wildlife habitat, climate control, erosion control, screening, and space definition. Planting should begin during or soon after construction. Local varieties should be used because they tolerate local conditions with minimum maintenance, look most natural, and are cheapest to establish. A vegetative buffer strip to reduce erosion should be at least 15 to 20 m wide to remove 50 to 75 percent of the sediments (Barfield and Albrecht 1982). To expedite erosion control, sodding is better than seeding, although more costly.

For additional information, a landscape architect should be consulted. Experienced wetland biologists can help with basic principles of landscape architecture relative to impoundments. Local representatives of the U.S. Soil Conservation Service or Agriculture Canada can recommend grass mixtures and other plants, planting and fertilization rates and dates, and mulching procedures.

Normal pool

The selection of the normal pool level determines the size of the pond, height of the dam and emergency spillway, and to some extent the size and type of control structure (Anderson 1985). An engineer should survey the proposed impoundment site and draw 3-dm (1-ft) contour lines on a map. After the area enclosed by each contour line is calculated with a planimeter or dot grid, the wildlife biologist should select the elevation producing the most area flooded to proper depth, for example, 3 to 9 dm (1 to 3 ft) for dabbling ducks. In Table 2.10, the desired range of depth (3 to 9 dm) is enclosed between parallel lines. The area flooded for each corresponding contour interval is then added to give the total area flooded to a depth of 3 to 9 dm. The example below shows that flooding to the 335-dm (110-ft) contour line will produce the largest area (12.2 ha, or 30 acres) 3 to 9 dm deep (Table 2.10). Often the elevations around the impoundment and at the ends of the dike actually will determine the maximum allowable pool elevation without constructing extensive and costly dikes and levees.

Normal pool elevation		Calculated area flooded	
dm	ft	ha	acres
329	108	2.4 + 3.2 = 5.6	6 + 8 = 14
332	109	3.2 + 6.5 = 9.7	8 + 16 = 24
335	110	6.5 + 5.7 = 12.2	16 + 14 = 30
338	111	5.7 + 2.8 = 8.5	14 + 7 = 21

The three main considerations for constructing the impoundment are to (1) maximize the flooded area to minimize the cost per hectare, (2) allow enough freeboard (distance from normal pool to top of dam) to store flood waters temporarily while most of it passes through the emergency spillway, and (3) keep the distribution of vegetation to wa-

TABLE 2.10 Sample Calculation to Determine the Normal Pool Elevation (ft) for the Most Area Flooded at Various Pool Elevations

Contour interval		Calculated area flooded*		Normal pool elevation							
				dm	ft	dm	ft	dm	ft	dm	ft
dm	ft	ha	acres	329	108	332	109	335	110	338	111
314–317	103–104	0.4	1	12–15	4–5	15–18	5–6	18–21	6–7	21–24	7–8
317–320	104–105	1.2	3	9–12	3–4	12–15	4–5	15–18	5–6	18–21	6–7
320–323	105–106	2.4	6	6–9	2–3	9–12	3–4	12–15	4–5	15–18	5–6
323–326	106–107	3.2	8	3–6	1–2	6–9	2–3	9–12	3–4	12–15	4–5
326–329	107–108	6.5	16	0–3	0–1	3–6	1–2	6–9	2–3	9–12	3–4
329–332	108–109	5.7	14			0–3	0–1	3–6	1–2	6–9	2–3
332–335	109–110	2.8	7					0–3	0–1	3–6	1–2
335–338	110–111	2.8	7							0–3	0–1

*Area enclosed by each 3-dm (1-ft) contour line on map, calculated with dot grid or planimeter.
SOURCE: Anderson (1985).

ter between 30 and 70 percent, with 50 percent optimum (Verry 1985a). Most emergent vegetation occupies the area 0 to 9 dm (0 to 3 ft) below normal pool. The area over 9 dm below normal pool will be mostly open water because emergent vegetation in stable water tends to die out below 6 dm (2 ft). Water levels in impoundments fluctuate naturally 3 dm (1 ft) below normal pool. Thus, a 9-dm increment typically reflects the emergent zone in most impoundments.

When an impoundment is planned, a profile of the site in 30-cm contour intervals can be used as a guide to establish the elevation of normal pool. Figure 2.4 shows that at an elevation of 300 cm for normal pool, the area impounded was best at 35.6 ha, the area of open water at 63 percent, and the area of emergent vegetation at 37 percent (Verry 1989). Normal pool would be established at 270 cm to allow enough freeboard and to protect adjacent roads. A 60-cm instead of a 90-cm increment is used in an area with a strong source of groundwater and thus a stable elevation of water. Water depth is not critical where floating mats occur. Interspersion is determined by the combined areas of impounded water and floating mats (Verry 1985a).

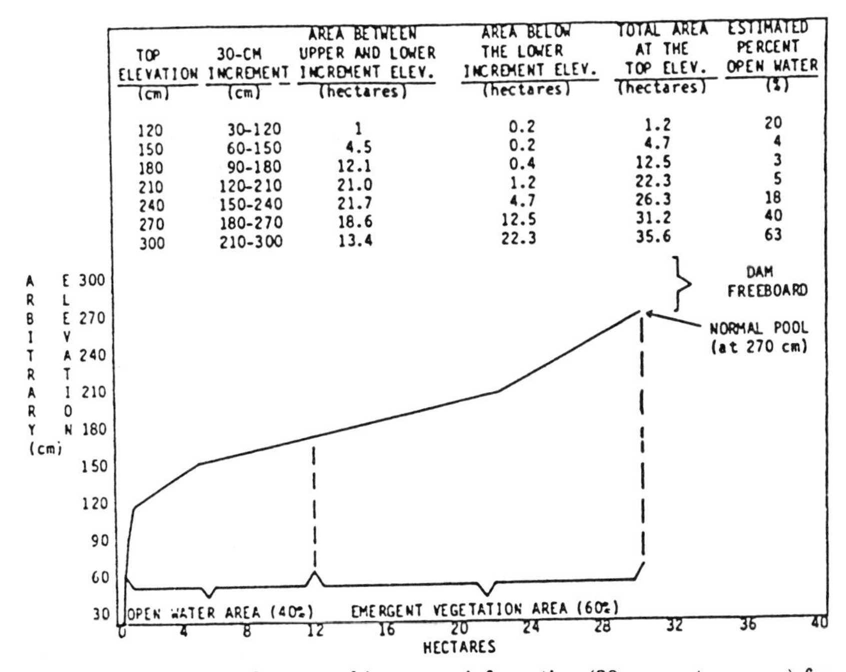

TOP ELEVATION (cm)	30-CM INCREMENT (cm)	AREA BETWEEN UPPER AND LOWER INCREMENT ELEV. (hectares)	AREA BELOW THE LOWER INCREMENT ELEV. (hectares)	TOTAL AREA AT THE TOP ELEV. (hectares)	ESTIMATED PERCENT OPEN WATER (%)
120	30-120	1	0.2	1.2	20
150	60-150	4.5	0.2	4.7	4
180	90-180	12.1	0.4	12.5	3
210	120-210	21.0	1.2	22.3	5
240	150-240	21.7	4.7	26.3	18
270	180-270	18.6	12.5	31.2	40
300	210-300	13.4	22.3	35.6	63

Figure 2.4 Summary of topographic survey information (30-cm contour map) for Bear Brook Impoundment on the Chippewa National Forest, Minnesota. Maximum depth of the open water area was 30 cm, but actual height of stoplogs was 270 cm to preserve an existing roadbed (*Verry 1989*).

Sealing the pond

If the site selected for the impoundment cannot be of impermeable soils, then the permeability of the soil must be reduced to tolerable limits or the site rejected. The method employed depends mainly on the ratio of fine-grained clay and silt in the soil to coarse-grained sand and gravel (Renfro 1979, U.S. Soil Conservation Service 1982). All methods are expensive.

Compaction. If the pond contains at least 10 percent clay mixed with silt, small gravel, or coarse to fine sands, soil compaction alone might suffice to seal the pond and is the cheapest method. Except for greentree reservoirs, all trees and other vegetation must be cleared within the pond perimeter; stump holes and other crevices filled with impervious material; the soil scarified 20 to 25 cm deep with disk, rototiller, pulverizer, or similar equipment; and all tree roots and rocks removed. Then with enough moisture to lubricate soil particles, four to six passes should be made with a sheepsfoot roller to compact the bottom of the pond to a depth of 20 cm. A sheepsfoot roller cannot be used with saturated soils.

Clay blanket. If the pond area contains coarse-grain soils with less than 10 percent clay, the entire area including upstream slope can be sealed by blanketing with well-graded soil mixtures of small gravel or coarse to fine sand and clay, each mixture containing at least 20 percent clay by weight. A borrow pit close to the pond permits hauling at reasonable cost. After all vegetation is removed from the pond area of the marsh to be controlled, and all holes and crevices are filled, the blanket material can be hauled from the borrow pit to the pond site in tractor-pulled wheeled scrapers or similar equipment and spread evenly in layers 15 to 20 cm thick. Each layer should be moistened and then compacted with four to six passes of a sheepsfoot roller. To prevent the clay blanket from cracking due to drying, freezing, and thawing, a cover of gravel 30 to 46 cm thick should be spread over the clay blanket below the anticipated high-water level. To protect areas where water flow into the pond is concentrated, suitable material such as rock riprap should be used.

Bentonite. Bentonite is a fine-texture colloidal clay that swells 8 to 20 times when moistened (Tunberg 1966). When mixed with well-graded coarse-grain soil, compacted, and moistened, bentonite will seal the pond. But it will crack if dried. Thus it is not recommended for ponds planned for complete drawdowns; and in any case it might be prohibitively expensive if the source of bentonite is far from the pond site. It is generally available in granules or powder from livestock supply, chemical, and cement companies.

The soil moisture level should be examined before bentonite is applied. It should not be too moist to compact. If it is too dry, moistening with a sprinkler truck will be needed for compaction.

In the marsh to be controlled, the pond area must be cleared of all vegetation, all holes and crevices filled, and exposed gravel covered with a suitable fill material. Bentonite should be spread evenly at the rate of 5 to 15 kg/m^2 usually, as determined from laboratory analysis, then mixed with 15 cm of surface soil with rototiller or disk, and compacted with four to six passes of a sheepsfoot roller. The pond should be filled immediately, or else to prevent drying, it must be covered with a mulch of hay or straw pressed into the soil with the last pass of the roller. Another method is to remove several centimeters of soil from the bottom of a dry impoundment, spread a solid membrane of bentonite throughout, cover with at least 10 cm of soil, moisten, and compact thoroughly (Tunberg 1966). To protect soil from concentrated water inflow, rock riprap or something similar should be used (Renfro 1979, U.S. Soil Conservation Service 1982).

If draining and drying the pond is impractical or impossible, powdered bentonite can be mixed with water to form a slurry, which is allowed to flow into the pond water. Wave action mixes the heavy slurry over the bottom. A harrow pulled over the bottom then mixes bentonite and soil (Tunberg 1966).

Granular bentonite can be sprinkled from a boat in a pattern to cover the entire bottom thoroughly. Then a harrow should be pulled along the entire bottom (Tunberg 1966).

In a large pond, complete coverage is impractical, so a series of holes can be drilled in the leakage area and liquid bentonite slurry injected. Ground pressure forces the slurry into the seepage area. Or a trench can be dug along the seepage area, filled simultaneously with bentonite slurry, and then backfilled (Tunberg 1966).

Bentonite seals can hold well after 10 years of use. Factors such as excessive vegetative growth, extreme drying, very saline water, and livestock use will reduce the effectiveness of bentonite (Tunberg 1966).

Chemicals. If the soils in the pond contain at least 15 percent clay and 50 percent silt and clay, chemical treatment might be needed to disperse porous aggregates of clay soil particles arranged end to end or end to plate (Renfro 1979, U.S. Soil Conservation Service 1982). These chemical dispersing agents usually are sodium chloride (common salt) and sodium polyphosphates of which tetrasodium pyrophosphate and sodium tripolyphosphate are best. Soda ash, technical grade, 99 to 100 percent sodium carbonate, also can be used. Laboratory analysis of the soil is needed to determine the best dispersing agent and the application rate. The application rate usually is 0.98 to 1.61 kg/m^2 for sodium chloride, 0.24 to 0.49 kg/m^2 for sodium phosphates, and 0.49 to 0.98 kg/m^2 for fine-grain soda ash, with at least 95 percent passing a no. 30 sieve and less than 5 percent passing a no. 100 sieve.

After the pond area of the marsh is cleared of all vegetation and trash, rock outcrops and other exposed areas of highly permeable material should be covered with 3 to 6 dm of fine-grain soil. The dispersing agent should be applied evenly over the pond area with hand broadcaster, fertilizer spreader, drill, or seeder; mixed into 15 cm of surface soil with disk, rototiller, or pulverizer operated in two directions for best results; and compacted with four to six passes of a sheepsfoot roller when moisture conditions are optimum. To protect the high waterline from erosion, the perimeter should be covered with 30 to 45 cm of gravel, and riprap where the water inflow is concentrated.

Waterproof lining. Waterproof linings of polyethylene, vinyl, butyl rubber membranes and asphalt-sealed fabric will reduce excessive seepage, but they are expensive and impractical for most wildlife impoundments due to the large size of the pond area. Moreover, side slopes must be chemically sterilized, or butyl rubber linings 20 to 30 mils thick installed, to prevent plants like quackgrass from penetrating the lining (Renfro 1979).

Dams, Dikes, Levees

Technically, *dikes* separate impoundments, and *levees* prevent water from flooding nonproject land such as along rivers. In practice, the two terms often are used interchangeably. Construction methods are essentially the same (Foreman 1979). Dikes and levees are low forms of dams, consisting of earthen embankments constructed across a drainage system. The term *dam* often is reserved for a concrete or steel embankment within the dike or connecting two ridges to impound water usually at the narrowest gap in the flowage (Fig. 2.5) (Anderson 1985). To allow or prevent flow of water, some type of control structure penetrates the dike to connect adjacent impoundments or is built into the dam to allow or prevent release of water from the flowage. Many dikes and levees serve also as roads for vehicles and maintenance equipment.

At least one dike per impoundment is needed. Outer levees might be needed on uniformly flat terrain to contain the controlled marsh or to prevent uncontrollable flooding from a nearby river or ocean. Land el-

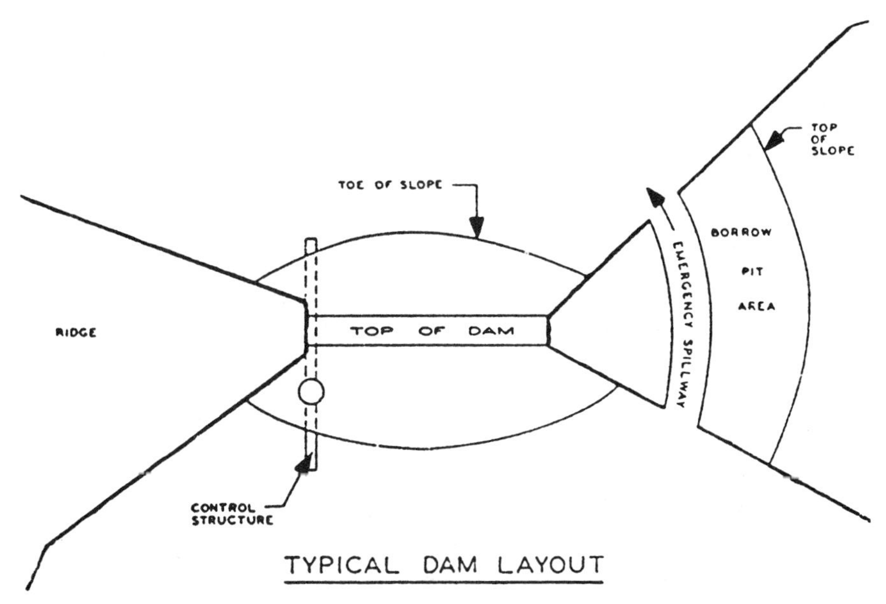

Figure 2.5 Typical dam, with emergency spillway and borrow pit (*Anderson 1985*).

evations are used as natural levees whenever possible, to reduce construction and maintenance costs and to promote drainage (Mitchell and Newling 1986). Inner dikes separating impoundments within the marsh are best constructed on contours at a contour interval of 15 cm when possible (Fredrickson and Taylor 1982). On fairly level terrain this interval will result in a series of fairly large impoundments which provide maximum water-level control. Relatively inexpensive construction of low dikes, adequate for up to 2 years, is achieved with laser technology that quickly and accurately locates contours, and a rice-levee plow. On intensively managed marshes, contour dikes improve waterfowl use of all impoundments and water-level control for vegetation management (Reid et al. 1989). Dikes not located on contours reduce bird use of traditional swales and sloughs. Depending on the slope, the amount of habitat that can be managed effectively for ducks is increased by at least 50 percent if contour dikes are used. The long-term benefits of contour dikes usually justify the high initial cost of construction. In brackish tidewater areas, constructing dikes on contours is less important because water is let into and removed from the impoundments through the same control structures by tidal fluctuations.

For integration with mosquito control in salt marshes, the dike is built along the natural division of low marsh and high marsh where mosquitoes breed (Provost 1968). In Florida the dike should be where

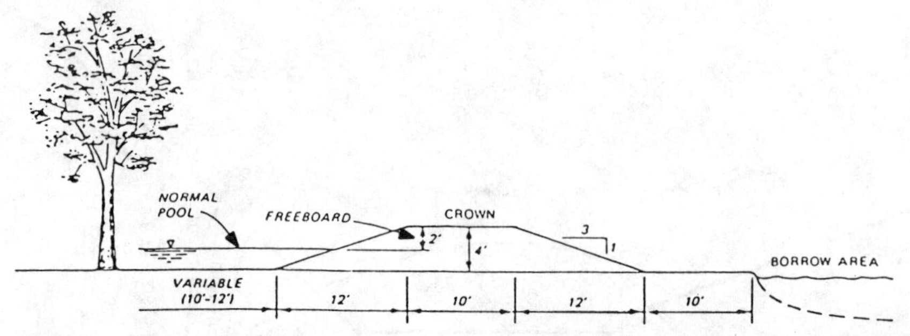

Figure 2.6 Levee dimensions recommended for a greentree reservoir (*Mitchell and Newling 1986*).

smooth cordgrass and black needlerush replace saltmeadow cordgrass and seashore saltgrass. Wave action can erode dikes, which should be located preferably so that prevailing winds blow across the impoundment, i.e., parallel to the dike (Atlantic Waterfowl Council 1972).

Dikes and levees for greentree reservoirs can be of the simple contour type used for ricefields (Rudolph and Hunter 1964). The heights of dikes and levees for greentree reservoirs vary with irregular terrain, but 1.2 m is average (Fig. 2.6) (Mitchell and Newling 1986). In bottoms subject to natural overflows, the dike and levee are low and wide to avoid or reduce damage from overtopping floodwaters (Rudolph and Hunter 1964). Higher dikes and levees need an emergency spillway.

Types

Sand dike. Sand dikes are used to prevent storm tides from moving salt water into marshes. They can be hydraulically placed, dozed or dumped in place (see Table 2.13 and Fig. 2.39), or sand-fenced (Atlantic Waterfowl Council 1972, Eckert et al. 1978, U.S. Army Engineer Waterways Experiment Station 1978, Woodhouse 1978).

Where existing dunes are too close to the ocean or too eroded to protect marshland effectively, bulldozers are used to develop dunes 18 to 24 dm high, sloped 1:10 to 1:20 on the ocean side, 46 to 76 m from mean high tide. Where few or no dunes exist and enough wind exists for drifting sand, an untreated snow fence (lath/space ratio of 3.8:2.5 cm) can be stretched 76 m from mean high tide, with 1.8-m support posts (5.2 by 10.4 cm) embedded 60 cm deep and spaced 3 m apart on alternate sides. Brush panels can be used as wings at right angles on each side of the snow fence, spaced regularly along its length. Parallel fencing might be needed in two or three rows spaced 4.9 to 6.1 m apart. When these fences achieve a broad, low, 1.2-m dune, another

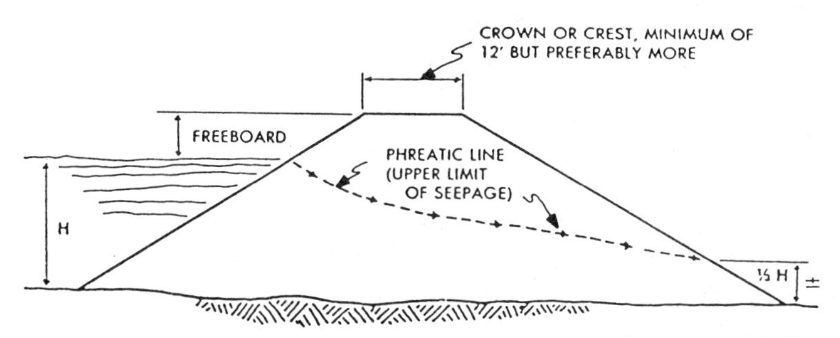

Figure 2.7 Typical dike of homogenous fill (*Atlantic Waterfowl Council 1972*).

row or two of snow fence should be erected on top, to build the final dune 2.4 m high. Buildup takes only a few months for fine drifting sand to 3 to 5 years for a two-tiered dune, during which time the buildup is susceptible to storm damage.

To stabilize the dune, native grasses such as beachgrass, sea oats, and cordgrasses should be planted as soon as bulldozing is completed and sand-fencing is 80 percent completed. A mechanical planter should space the plants 3 dm apart in rows 7 dm apart.

Simple embankment dike. Simple embankment dikes consist of uniform material throughout (Fig. 2.7). These dikes are used where onsite soils must be used, and therefore they are the least expensive (Atlantic Waterfowl Council 1972). Often they are built atop remnant ricefield dikes, at least 1 m above mean high water to withstand all waves except those from hurricanes and catastrophic floods (Gordon et al. 1989). For impoundments of at least 0.8 ha with or without a water control structure, the dike can be a simple earthen dam protected by sandbags or a blanket of coarse gravel riprap, or an earthen dam with a shallow V-shaped spillway of two or more layers of any size fieldstone grouted together (Poston and Schmidt 1981).

Core dike. Core dikes have a core and cutoff trench of the most impervious material available, usually clay, with the outer surface of onsite material (Fig. 2.8; also see Fig. 2.17). The core should extend above the waterline. The core base should be one-third the width of the dike base (Summers 1984). Low dikes usually are not built with a core unless the supply of impervious soil is readily available (Atlantic Waterfowl Council 1972).

Diaphragm dike. Diaphragm dikes have a thin section of wood, steel, or concrete extending the length of, and within, the dike, as a barrier

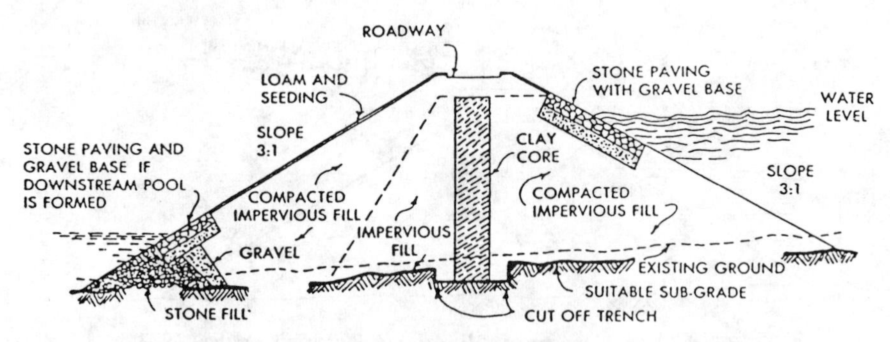

Figure 2.8 Typical dam with clay core (*Atlantic Waterfowl Council 1972*).

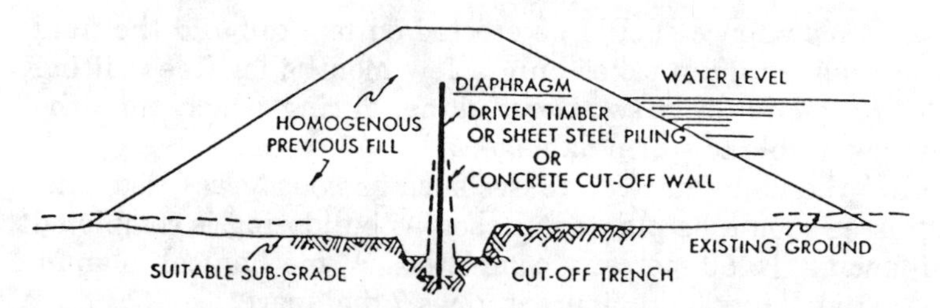

Figure 2.9 Typical dam with diaphragm (*Atlantic Waterfowl Council 1972*).

to percolating water (Fig. 2.9). In a full-diaphragm dike, the barrier extends from the level of the impounded water into an impervious foundation, unlike a partial-diaphragm dike.

Borrow pit

Often two types of borrow pit are used if one is not enough. One type is located on high ground off the end(s) of the dike, or else within reasonable travel distance, for dike construction and maintenance fill (see Fig. 2.5). This type of pit should be graded so that it is well drained and does not allow stagnant water to accumulate as breeding places for mosquitoes (U.S. Soil Conservation Service 1982).

The other type of borrow pit runs parallel to the dike on one or both sides of it, becoming a water-filled ditch on the downstream side or inundated by the pond on the upstream side (Fig. 2.10). The ditch is constructed by a dragline which pulls in soil from the side(s) of the dike (see Fig. 2.42). The soil is deposited on the dike. As the dike advances, so does the dragline on top, thus extending the ditch as soil from the ditch is pulled onto the dike to extend it. If enough material for mucking the dike can be obtained from one side of the dike, the borrow ditch can be located on the upstream side of the dike if a low

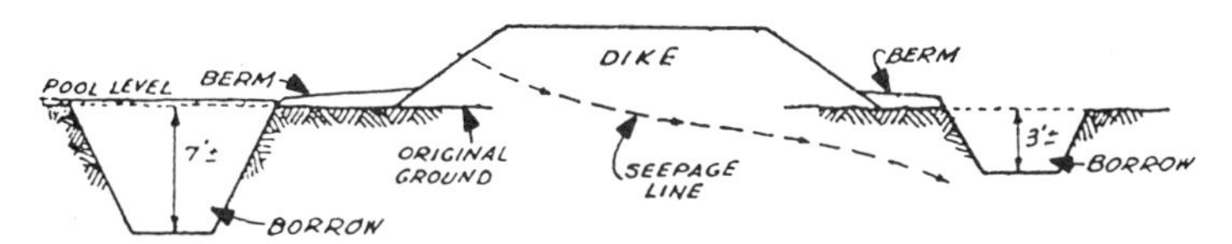

Figure 2.10 Dike constructed with dragline (*Atlantic Waterfowl Council 1972*).

head of water normally is maintained in the impoundment. Such a ditch allows personnel to travel by boat on a waterway around the management area, and it improves water movement in the impoundment. But an elevated access should be established across any borrow pit inside the dike to allow equipment to pass over the flooded borrow pit into the impoundment during drawdown, so that woody or other undesirable vegetation can be eliminated or a seedbed prepared as needed (Fredrickson and Taylor 1982). The borrow ditch is best located on the downstream side of the dike (outside the impoundment) if a high head of water normally is maintained in the impoundment, to reduce a potential hazard to hunters, unless borrow ditches inside the impoundment are marked as boat trails to provide hunter access (Mitchell and Newling 1986). Placement outside the impoundment also reduces pumping costs during initial flooding (Reid et al. 1989). In tidewater areas, dike fill is taken from only one side of the dike to reduce erosion and maintenance costs (Stutzenbaker and Weller 1989).

The *berm* is the distance between the toe of the dike and the top of the borrow pit ditch (Fig. 2.10). On stable soils the berm is at least 3 m wide or twice the depth of the borrow pit, whichever is greater. The berm should be at least 6 m wide on unstable soils (Atlantic Waterfowl Council 1972) and at least 4.5 m wide on stable soil (Farmes 1985). The berm should be wide enough for earth-moving equipment to operate during construction and later for maintenance rebuilding, without causing future sloughing of the near slope of the borrow pit.

The width and depth of the borrow pit are influenced by the amount of material needed for the dike, the amount and position of suitable and unsuitable soils available in the borrow pit, and the type of earth-moving equipment available. The maximum depth of a pit inside the dike should be about 2 m from berm to bottom. Outside the dike, the maximum depth should be about 1 m, with the width of the berm extended as needed so that the bottom of the pit will be above the anticipated seepage line through the dike, foundation, and berm (Fig. 2.10).

Foundation material

The foundation must prevent water seepage and support the control structure and traffic from vehicles and heavy equipment (U.S. Soil Conservation Service 1982). Soil borings under the proposed dam and at the abutments at the ends of the dam should be taken to determine porosity and strength. If rock, it must be thick, without fissures and seams, or it will need sealing. Coarse-texture materials such as gravel, sand, and gravel-sand mixtures provide good support but are porous and need sealing with a core of impervious soil under the dam or a blanket of it along the impoundment side of the dam and under the pond.

Fine-texture materials such as clays and silts have little porosity, but also little stability, although generally enough for the small dikes of waterfowl impoundments. If a dike must be built on organic soil such as muck and peat more than 6 dm deep, the soil is removed from the foundation and suitable mineral soil hauled in, or the dike is built in stages to permit at least 40 percent shrinkage and settling (Foreman 1979). Usually the site is rejected for the dike if the peat is more than 1.2 m deep (personal communication, T. Meier, Wisconsin Department of Natural Resources 1991). If the peat is 1.8 to 2.4 m deep in spots, but more than 1.2 m deep elsewhere, the site can be used because removal of peat is still cost-effective. (The impoundment bottom can consist of deep peat.) Peat thus removed is cast out on the pond side.

Good foundation materials that are both impervious and stable are mixtures of coarse- and fine-texture soils such as sand-silt, sand-clay, gravel-sand-silt, and gravel-sand-clay. Less desirable but still acceptable foundation soils for ordinary pond dams include gravelly clays, sandy clays, silty clays, silty and clayey fine sands, and clayey silts with slight plasticity (U.S. Soil Conservation Service 1982).

Fill material

Except for organic clays and silts, soils acceptable for the dike's foundation material usually are acceptable for the fill material (U.S. Soil Conservation Service 1982). The best fill material ranges from small gravel or coarse sand to fine sand and clay, with about 20 percent by weight of clay. Other materials can be used, but the greater the variance, the more precautions are needed. Dams built with soils having much coarse sand or gravel are pervious, requiring a core of clay unless an impervious diaphragm of concrete, steel, or wood is used. Fill that has a high clay content shrinks when dry and swells when wet, perhaps causing cracks and leakage. Dams built with soils mostly of silt, e.g., the loess areas of western Iowa and along the Mississippi

River in Tennessee, Arkansas, and Mississippi, need proper moisture for compaction during construction.

The proportion of clay, silt, and sand in the fill material can be estimated by sifting a sample through a 0.6-cm (¼-in) sieve or screen to remove the gravel and filling a large, straight-sided bottle or jar one-third full with the sifted material. Then the bottle is filled with water, shaken vigorously for several minutes, and set down for 24 h to allow the soil to settle. Then a ruler is used to measure the thickness of the soil layers, the coarsest layer (sand) settling to the bottom first and the finest layer (clay) last (U.S. Soil Conservation Service 1982).

Construction

Foundation. The marsh area should be scalped where the mineral soil core of the dike will be placed, to allow the core to settle quickly, thus reducing seepage. Such marsh soil is removed with a dragline preferably, or hydraulic trackhoe, and placed for reuse in mucking the slopes (Atlantic Waterfowl Council 1972). Removing 15 to 60 cm of soil from the wetland basin can improve water retention, even changing the basin from a Type 2 (inland fresh meadow) to a Type 3 (inland shallow fresh meadow) wetland, e.g. (Shaw and Fredine 1956, Petersen et al. 1982). Removal of too much bottom soil can break the wetland seal, resulting in inadvertent draining. Soil test cores determine how deeply to dig. Dikes might have to be built entirely of organic soils in areas of low tide and at some coastal and inland sites where inorganic (mineral) soil is unavailable (Atlantic Waterfowl Council 1972).

In areas that are less marshy, sod, topsoil, and rocks must be removed from the area over which the dike will be built, preferably with a tractor-pulled wheeled scraper (U.S. Soil Conservation Service 1982). The topsoil should be stockpiled for later use. All holes in the foundation should be filled with fill from the borrow pit and compacted with hand or power tampers in areas not readily accessible to other compacting equipment. Then the ground surface should be turned 15 cm deep throughout, leveled with a disk harrow, and compacted by running the equipment over it six or more times with a bulldozer (Atlantic Waterfowl Council 1959).

At this point a cutoff is installed, if needed; this is a trench cut along the centerline of the dam and filled with clayey material. If the dam's foundation contains alluvial deposits of porous gravels and sands above an impervious layer of rock or clay, the cutoff is needed to join the base of the dam with the impervious layer to reduce seepage. Seepage is a function of the imperviousness and compaction of the soil used. If sand or sand-gravel soil in the foundation is too deep over the impervious layer to dig a cutoff economically, then a trench is cut 12 to 15 dm deep to the width of a dragline bucket and backfilled with puddled inorganic soil (Atlantic Waterfowl Council 1972), in consultation with an engineer.

The cutoff usually is dug with a dragline or backhoe at least 3 dm into the impervious layer (U.S. Soil Conservation Service 1982), at least 2.4 m wide, with sides no steeper than 1:1 (see Figs. 2.8, 2.9, 2.17, 2.19, and 2.20). It should extend into and up the abutments of the dam if needed, until no seepage is anticipated. If the cutoff is free of organic matter, roots, boulders, and trash, material removed from the cutoff can be stockpiled for later use on the downstream third of the dam and compacted.

All water must be pumped from the cutoff before it is filled with clayey material. When dry, the cutoff should be filled with successive thin layers of clay or sandy clay material. Each layer should be compacted at proper moisture conditions, preferably with a sheepsfoot roller, otherwise by running the equipment over it repeatedly. Stream channels crossing the foundation should be made deeper and wider to remove all organic matter, stumps, roots, rocks, gravel, sand, and sediment. Then the channels should be filled with clay or sandy clay and compacted, as with the cutoff.

Foundation soils usually will be stable enough to support the dike, control structure, and maintenance equipment (Atlantic Waterfowl Council 1972). An engineer should be consulted if the foundation consists of or is underlaid by a highly plastic clay or unconsolidated soil (U.S. Soil Conservation Service 1982). Such soils can be removed and replaced with more stable soil. Otherwise, rows of sheet piling or round piling can be used to confine the unstable soils, but it is costly. Double rows of first rooted dragline buckets have cored a dike adequately in some areas, but extensive areas of highly plastic soil probably should be abandoned (Atlantic Waterfowl Council 1972).

Dike body. The basic difference between simple embankment dikes and core or diaphragm dikes is that the simple embankment does not have a core or diaphragm extending into the cutoff trench of the foundation (see Figs. 2.7, 2.8, and 2.9). Otherwise, construction of these dikes is similar, whether they are used as outer levees or inner dikes.

Dikes for waterfowl impoundments should be no higher than 3 m unless detailed soil studies are conducted (Addy and MacNamara 1948), although 1.8 m is best for mineral soils and 1.2 m for organic soils (Foreman 1979). With or without such studies, the services of an engineer are preferred. Dikes in areas of deep frost must be somewhat higher to allow for frost damage and settling. In fact, all dikes should be built 10 percent higher than desired to allow for settling. Muck or other soil of high organic content is least suitable for dike construction, but if it must be used, the dike should be 50 percent higher than desired to allow for settling (Addy and MacNamara 1948). If the dike is not tightly compacted as the construction progresses, 15 to 20 percent more fill is added to compensate for shrinkage and settling (Summers 1984). Use of a dragline requires 25 to 30 percent more fill (Farmes 1985). Dikes constructed by dragline usually contain wet fill. If muck clay is involved, 1 year of drying might be needed before dozers can level the fill. A dragline is used because the fill excavated for the dike is wet, which renders compaction impractical after each lift. But with a low head, meeting rigid compaction standards should not be needed, unless loss of life or property could result from dike failure (Farmes 1985). Where seepage might cause the dike to fail, strengthening or enlarging the dike might be the cheapest way to treat anticipated seepage problems. The proper width of dikes built on soils preventing proper shaping is 5 times the settled height plus the top width (Foreman 1979).

A stoppage is a strengthened location on the dike needing bulkheading to hold excavated material in place (Williams 1987). Stoppages are constructed to close off small tributaries and canals where they intersect the dike.

Freeboard—the additional height of the dike above the pool's surface at maximum flood stage—should be sufficient to keep waves from a maximum-velocity wind from overtopping the dike from wave height plus wave run up the slope (Linde 1969, Atlantic Waterfowl Council 1972). Where flooding from storms occurs regularly, as along large rivers, large protective levees are damaged more by flooding than low levees that submerge quickly and uniformly (Fredrickson and Taylor 1982). A freeboard of 3 dm is needed for ponds less than 0.2 km long, 4.5 dm between 0.2 and 0.4 km long, 6 dm between 0.4 and 0.8 km long (U.S. Soil Conservation Service 1982), and 9 dm if more than 6 ha (Addy and MacNamara 1948). An engineer should determine the freeboard for longer ponds (U.S. Soil Conservation Service 1982). Farmes (1985) recommended that the dike be 9 dm above maximum normal pool elevation to include 6 dm of freeboard during flooding and 3 dm of bounce. The height and width of the dike depend on the area and depth of the pond.

In tidewater areas, embankments are constructed to exclude stormwaters if conditions warrant, i.e., type of management desired, type of soil available, and cost of construction. After shrinkage and settling, dikes should be 1 m or more above mean high water to withstand all but catastrophic floods and storm surges during hurricanes (Williams 1987, Gordon et al. 1989, Stutzenbaker and Weller 1989).

Sizes of spillway and control structures should admit stormwater so that water rises in the impoundment about as fast as outside. Then when overtopping occurs, dike erosion is minimal. Low-level dikes, about 46 cm high and well sodded, are designed to be overtopped by high water. They use small, simple water control structures with flapgates or stoplogs, with wide wings that extend into the dike to prevent erosion. Such impoundments usually are maintained at a relatively constant water depth of 15 to 25 cm, rather than on a drawdown system (Green et al. 1964). Tidal ranges are available annually from U.S. Department of Commerce (e.g., U.S. Department of Commerce 1986). Average dimensions of dikes in South Carolina tidal areas are 2.1 m high, 3.7 m wide on top, and 9.1 m wide at bottom (Williams 1987). Alluvial soils with a high clay content usually occur in coastal wetlands located at the terminus of rivers rising in the mountains or piedmont. Such soils can be used to construct good dikes that settle little.

Generally, the dike will cut across several contour lines between the two elevations to be connected to impound the water. Once the height of the dike is determined, with allowance for settling, and the width is established, the upstream and downstream toes of the dike (i.e., the bottom edge of the slopes) are staked. After the topsoil and porous soil are stripped for the dike's foundation, the toe of the dike's slope may be located at any particular contour line that the dike crosses by first determining the height of the dike at each different contour it crosses, multiplying the height by the slope ratio (by 3 for a 3:1 slope or 4 for a 4:1 slope), and adding one-half the dike's width (for example, 1.5 m for a dike 3 m wide). This distance then is measured from the dike's centerline. Stakes should be located regularly along the dike's toes, especially where the dike crosses different contour lines. Depositing and compacting the fill to the dike's established height, width, and toes, with allowance for settling, produces the desired slope (Atlantic Waterfowl Council 1959). On the land side of any dike crossing an old channel, the base should be widened to form a banquette at least the height of the dike above normal ground, sloping away, with the top at least 3 dm above normal ground (Foreman 1979). To prevent seepage through a dike of permeable soil, the landside base is extended to increase stability.

The entire dike normally can be built from impervious soil taken from the borrow pit or ditch so that the core is not needed if the impervious soil is well tamped and puddled during construction (Addy and MacNamara 1948, Atlantic Waterfowl Council 1959). If such soil is in short supply, it should be placed on the dike in layers up to 15 cm thick and compacted by traveling over it with at least six passes of the equipment, so as to continue the core wall upward from the cutoff trench to above-normal pool. The impervious soil is extended to one-third the dike's width; the lighter fill material is extended to the full width of the dike's bottom (Addy and MacNamara 1948). These applications in the ratio ⅓:⅔ are continued in layers up to 15 cm deep and compacted with at least six passes of the equipment. The lighter fill material must be moist for proper compaction (Atlantic Waterfowl Council 1959). Wetting the fill, as on large construction jobs, usually is impractical. Moisture conditions are proper if a handful of fill retains its shape after being squeezed. If the fill is too dry, construction is delayed until after a suitable rainfall. Otherwise, water must be sprinkled onto the soil by machine. Although ideal for compaction, sheepsfoot rollers usually are impractical for controlled marsh construction. To achieve best compaction, the equipment operator should not operate the equipment always along the same path on the dike (Atlantic Waterfowl Council 1972). If the dike is constructed with a dragline pulling on-site material from a borrow ditch paralleling the dike, normally two passes are needed, and perhaps a third pass after 2 years to bring the dike up to grade (Williams 1987).

A costly diaphragm embankment might have to be constructed where no impervious material or clay for a core is available. Then steel or wood sheet piling usually is driven down to an impervious foundation, or a concrete cutoff wall is installed (Atlantic Waterfowl Council 1972, Farmes 1985).

To avoid costly revamping of the earthwork after the desired height of the dike has been attained, the required slopes are maintained as construction proceeds. To avoid unnecessary earth moving, frequent rod readings are taken as the dike is built up to its desired height. The top is crowned to conform to the height of unsettled fill, as shown on the plan, and is slightly wider than shown on the plan so that final grading and seedbed preparation do not render the width substandard.

Heavy equipment and motor transport used to construct the dike usually involve a front-end loader, bulldozer, mobile or tow scraper, dragline, mobile or tow grader, disk, dump truck, low-bed tractor-trailer to transport tracked equipment especially, and sometimes a backhoe (trackhoe), clamshell on a mobile or crawler crane, sheepsfoot roller, and farm (rubber-tired) tractor (see Table 2.14). The dozer removes rocks and stumps from the dike area. The scraper removes topsoil. The dragline, backhoe, or blasting digs out the core (key) area. The impervious fill is bulldozed in the borrow pit, loaded by front-end loader or clamshell into a dump truck, dumped onto the dike area, and spread and compacted with the dozer. A sheepsfoot roller will improve compaction but usually is not needed. A scraper, perhaps pulled or pushed by a dozer, could be used to scrape up the impervious fill from the borrow pit and then deposit it onto the dike area.

The Atlantic Waterfowl Council (1972) reported the dimensions of a dike with muck topping built in large expanses of tidal salt marsh to be as follows:

Width at top = 4.3 m

Width of solid fill at base = 14.6 m

Width of dike topping at base = 21.9 m

Elevation = 2.7 m

Berm = 6.1 m each side of dike

Solid fill (sand and gravel) on 2.67:1 slope

Muck (mud and sod from borrow pit) on 5:1 slope

Topsoil on top of 4.3-m-wide dike

The solid fill was hauled into the marsh from selected borrow pits on the mainland by 9.9-m^3 (13-yd^3) Tournapull scrapers, loaded by a D-8 dozer used as a pusher. After the scrapers unloaded the material on the centerline of the developing dike, a D-7 dozer pushed the material out ahead. When the solid fill was to dimension and properly sloped, a dragline with 24.4-m boom was moved onto the dike to pull muck from a linear borrow ditch on each side of the dike, where they formed two long ditches as the dragline advanced. (Sometimes the entire dike is built with a dragline.) The muck was placed on the slopes of the solid fill where it eliminated most of the severe tide and wind actions. The berm, between the toe of the dike and the top of the borrow pit ditch, was 6.1 m wide. After all the muck was placed on the slopes of the dike, it was allowed to dry, shrink, and settle before the slopes were smoothed with a 3- by 3- by 12-dm wooden drag. After about 6 months, a small tractor with wide treads disked the slopes to prepare them for planting. An alternative to the dragline is a dump truck bringing in fill from the borrow pit and a dozer to spread it on the dike, which is then smoothed by the grader.

Slopes should be at least 3:1 or 4:1; a 3.5:1 slope is considered minimum for the safe operation of brush mowing equipment with a tractor. A 4:1 slope, with its wider base, deters muskrat burrowing. Construction is better with gentler slopes, but the cost is higher because more fill is needed (Atlantic Waterfowl Council 1959). If the fill is not rich enough to support fast plant growth, topsoil is spread on the dike's slopes and top, if the top is not used for frequent travel. On a dike with light traffic, no surfacing is needed. The core can be covered with topsoil if needed, arched for drainage, and planted to develop a heavy sod. Where traffic will be heavy, tops of dikes are surfaced with 15 to 30 cm of gravel, stone, cinders, or shell. More surfacing is needed with heavy freezing or excessive rainfall. Gravel should be applied only in late fall or winter after the dike has frozen adequately, especially on relatively soft new dikes. Dikes should be at least 3 m wide on top to accommodate vehicles and maintenance equipment, with at least one turnout point every 274 m, or dikes should be 6.1 m wide for two-lane travel (Atlantic Waterfowl Council 1972, U.S. Soil Conservation Service 1982, Farmes 1985).

Maintenance

The dike should be examined for repair regularly throughout the year and especially after heavy rains. All rills and washes must be filled with suitable material, compacted, and reseeded or resodded and fertilized as needed. Cave-ins from burrowing muskrats and nutria will need filling. If wave action has caused serious washing or sloughing, booms or riprap should be installed to prevent or reduce the problem. Evidence of seepage through or under the dam requires an engineer's advice for proper corrective measures before major damage occurs.

The protective plant cover on dike and emergency spillway should be mowed and bushhogged often and fertilized when needed. Cutting helps develop a cover and root system more resistant to runoff, controls undesirable woody growth, and renders the dike less attractive to nesting ducks and associated predators. Levees and dikes should be mowed twice a year. Mowing or burning in fall removes residual vegetation which would attract nesting ducks in spring and predators. In some areas, fertilizing vegetation on dikes and levees is best done every 3 years, with reseeding or overseeding every 4 to 6 years (Mitchell and Newling 1986). To reduce erosion, overseeding, as with renovation of pastureland, is better than disking and replanting.

In malaria areas, aquatic and shoreline vegetation should be controlled to reduce mosquito habitat. Top-feeding fish such as Gambusia minnows also should be stocked to feed on mosquito larvae (U.S. Soil Conservation Service 1982).

Emergency Spillway

Design

A basin that impounds runoff water or streamflow needs an emergency spillway to dispose of floodwaters, or else the dike will be damaged severely. Impoundments need emergency spillways that can discharge the peak flow expected in a 24-h storm for a 50-year flood (U.S.

Soil Conservation Service 1982), or better, a 100-year flood (Hoffman 1988), less any reduction for detention storage and discharge through control structures. A 100-, 500-, or 1000-year-frequency storm might occur during operation of the marsh and cause severe damage or complete loss of the structures, but that is a normal calculated risk. A marsh with a watershed area less than 101 ha should have an emergency spillway with a capacity of at least 0.7 $m^3/(s \cdot ha)$ of watershed (Atlantic Waterfowl Council 1959). The spillway also must convey the excess water safely through the outlet channel below the dike without damaging the downstream slope of the dike. No emergency spillway is needed if the dike surrounds the impoundment with no entering runoff (Foreman 1979).

The peak flow can be determined from streamflow records or calculated (Atlantic Waterfowl Council 1959, U.S. Soil Conservation Service 1982). Peak flow is influenced by the amount of precipitation and the size, steepness, soil, and plant cover of the watershed feeding the impoundment. Information on streamflow beds is available from the local Soil Conservation Service office, the U.S. Army Corps of Engineers, the U.S. Geological Survey, and/or the Federal Emergency Management Administration and their Canadian counterparts, depending on which agency has collected data on a particular stream.

Various formulas and methods are used for calculating the maximum rate of runoff for a given watershed and hence the size of the emergency spillway needed (Atlantic Waterfowl Council 1959, U.S. Soil Conservation Service 1982). Emergency spillways made of earth generally are considered for impoundments draining less than 2.6 km^2 (Atlantic Waterfowl Council 1972). With larger watersheds, reinforced-concrete curbing and riprap of stone or rubble large enough to prevent washout should be placed on the downstream side of the earth spillway. Maintenance of this modified earth spillway might be high relative to initial costs. It should be considered only in special situations when concrete overfall structures or other conventional spillways are not applicable.

Generally, the best layout for the emergency spillway is to construct it as part of the borrow pit at one end of the main dike (see Fig. 2.5), where the natural terrain abuts the end slope of the dike (Linde 1969, Anderson 1985). An area of natural drainage or one with the proper level can be used to form a natural spillway with minor or no further improvement. Such a situation is highly desirable because of reduced construction costs and resistance to washing due to the natural sod cover. Otherwise an area higher than the desired spillway must be cut down to the desired level.

Emergency spillways are built on natural undisturbed soil if possible. Often the emergency spillway is built as part of the dike during dike construction, with fill from the spillway used for the dike. The topsoil from the spillway also is used to construct the berm of the spillway and to refine the downstream slope of the dike adjoining the spillway.

A dike along a river can have a reversible spillway (personal communication, T. Meier, Wisconsin Department of Natural Resources 1991). When the river floods, water flows through the spillway and floods the pond until the dike is overtopped and the pond's water level is even with the river's water level. When the river drops, water flows out through the spillway, thus reducing maintenance costs to the dike. Reversible spillways prevent gullies from developing in the dikes on the pond side if the dike is overtopped by the river. Spillways are at least 30 m wide, depending on the river's length and flooding volume, and are dug to about 1.5 dm above normal (full) pool. Both sides of the spillway must be covered with a mixture of small (10- to 13-cm) to large (20- to 25-cm) rock so that it locks together. The spillway should be surfaced with 8- to 10-cm chip rock and clay mixture. Grass is no good with heavy flows. A plastic cellular confinement system called GEOWEB, laid on the side slopes and on top of the spillway and filled with gravel or crushed granite, has been successful in Wisconsin. (See "Riprap" in this chapter.)

Emergency spillways are of two types: natural and excavated (Renfro 1979, U.S. Soil Conservation Service 1982). If a natural site with good plant cover is available, soil borings generally are needed. If the spillway is excavated, soils must withstand reasonable water velocity without serious erosion. Loose sands and other highly erodible soils are avoided.

Natural spillway. The natural spillway is located between the end of the dike and the natural ground. The slope of the end of the dike extending to the natural ground is fixed usually at 3:1, but the slope of the natural ground abutting the dike can be anything.

After the peak discharge rate from the watershed is determined, which is the required discharge capacity of the spillway, then the velocity of the water to be discharged over the spillway and the maximum depth of water above the spillway crest (where excess water begins to flow over the spillway) are determined. The length of the spillway is not determined because it varies with the variable slope of the natural ground abutting the spillway. The velocity to be determined is influenced by the height of the uncut herbaceous plant cover on the spillway (Renfro 1979, U.S. Soil Conservation Service 1982). The velocity to be determined must be within the range of permissible velocities relative to the type of vegetation covering the spillway, the soil's erosion resistance, and the slope of the exit channel from the spillway (Table 2.11). The size of spillway needed often is estimated from experience, but Renfro (1979), Urquhart (1979), and the U.S. Soil Conservation Service (1982) provided details.

TABLE 2.11 Permissible Velocity (m/s)* of Impounded Water Spilled over Vegetated Emergency Spillways

	Slope of exit channel, percent			
	Erosion-resistant soils†		Easily eroded soils‡	
Vegetation	0–5	5–10	0–5	5–10
Bermudagrass (*Cynodon dactylon*)	2.4	2.1	1.8	1.5
Bahiagrass (*Paspalum notatum*)	2.4	2.1	1.8	1.5
Buffalo grass (*Buchloa dactyloides*)	2.1	1.8	1.5	1.2
Kentucky bluegrass (*Poa pratensis*)	2.1	1.8	1.5	1.2
Smooth brome (*Bromus tectorum*)	2.1	1.8	1.5	1.2
Tall fescue (*Festuca elatior*)	2.1	1.8	1.5	1.2
Reed canarygrass (*Phalaris arundinacea*)	2.1	1.8	1.5	1.2
Sod-forming grass-legume mixtures	1.5	1.4	1.2	0.9
Sericea lespedeza (*Lespedeza cuneata*)	1.1	1.1	0.8	0.8
Weeping lovegrass (*Eragrostis curvula*)	1.1	1.1	0.8	0.8
Yellow bluestem (*Andropogon* sp.)	1.1	1.1	0.8	0.8
Native grass mixtures	1.1	1.1	0.8	0.8

*Values are increased 10 percent when the anticipated average use of the spillway is once every 5 years or less, or 25 percent if once every 10 years.

†Soils with a higher clay content and higher plasticity. Typical soil textures are clay, silty clay, and sandy clay.

‡Soils with a high content of fine sand or silt and lower plasticity, or nonplastic. Typical soil textures are silt, fine sand, sandy loam, and silty loam.

SOURCE: Renfro (1979).

Excavated spillway. The spillway must be excavated to the proper length, width, and depth if the required discharge capacity of the spillway (peak discharge rate of the watershed) is too large (Renfro 1979, Urquhart 1979). The sides of the excavated emergency spillway should have 4:1 slopes from the base of the spillway to the elevation at the top of the dike (Atlantic Waterfowl Council 1959), although 3:1 is permissible (U.S. Soil Conservation Service 1982). The original ground on the side away from the dike above the dike's elevation can be cut on the natural slope of the soil's repose, usually 1.5:1 (Fig. 2.11) (Atlantic Waterfowl Council 1959).

Where practicable, the spillway's entire depth is cut in undisturbed ground, i.e., no part of the spillway is cut into the dike's construction fill (Fig. 2.11*a*). Cutting into the fill area of the dike is permissible if that area remains above the maximum water level in the impoundment (Fig. 2.11*b*), but not if it is below (Fig. 2.11*c*).

The entrance to the inlet channel of the emergency spillway should

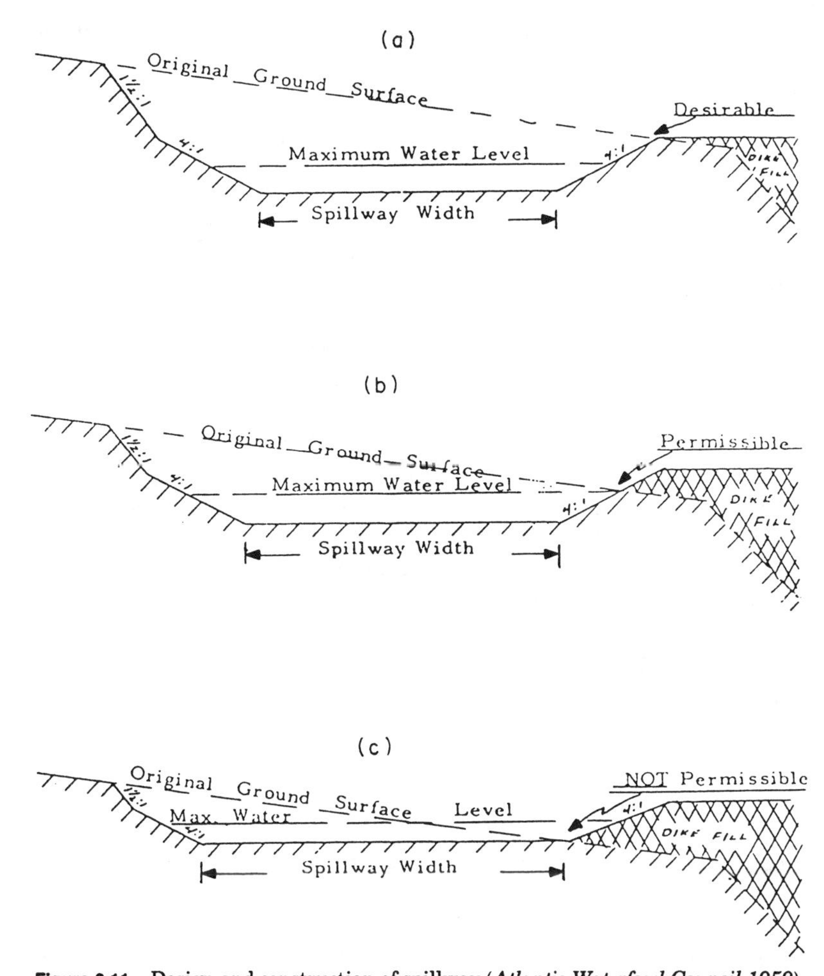

Figure 2.11 Design and construction of spillway (*Atlantic Waterfowl Council 1959*).

be at least 50 percent wider than the width of the rest (level part) of the spillway to which the inlet channel leads. The inlet channel should have smooth, gentle curves for alignment and should be reasonably short, with a slope of at least 2 percent to ensure drainage and low water loss at the inlet (U.S. Soil Conservation Service 1982).

As a general rule, the outlet channel of the emergency spillway should be at least 15 m from the downstream toe of the dike, preferably at about a 1 percent slope to convey the discharged water back to the original outlet channel without causing erosion (Atlantic Waterfowl Council 1959). In most cases topography will govern the location of the outlet channel. The discharged water must be spread to reduce velocity, usually by widening the spillway at the outlet. Outside edges of the outlet channel should be as wide as the outlet side of the emergency spillway's crest and should extend at least 15 m at right angles to the spillway before bending toward the original channel. Special legal and engineering approvals are needed before construction if floodwaters from the emergency spillway are not routed to flow in the original channel.

The crest (spill crest or level portion) of the emergency spillway should be 3 dm above normal pool. The maximum head on the emergency spillway crest is the depth of water expected to flow through the emergency spillway when an intense storm produces the maximum peak discharge from the watershed for a 50-year flood. The maximum permitted on a standard plan is 3 dm. The head can be increased only if soil conditions, plant cover, and economy of construction warrant. The difference between the maximum expected head (in the emergency spillway) and on the top of the dike is termed *freeboard*. This is the extent to which the top of the dike will project above the maximum flood stage. Freeboard should be 3 dm after settlement.

The maximum depth of flow is controlled by the crest of the spillway and the slopes of the inlet and exit channels leading into and out of the crest. The natural slope of the exit channel should be altered as little as possible. Once built, the spillway is planted with herbaceous cover which will slow the speed of excess water spilling over the spillway (retardance factor). Renfro (1979), Urquhart (1979), and the U.S. Soil Conservation Service (1982) provided details. The spillway also can be covered with crushed granite, riprapped with flat quartz stones (Linde 1969), or covered with GEOWEB. The free overfall (straight drop, broad crested) spillway (weir) and the ogee spillway (weir) are concrete or wooden structures used to prevent erosion where overflow is frequent.

The free overfall spillway or weir, sometimes called a straight-drop or broad-crested spillway, is used most often with large, shallow impoundments where discharge efficiency is relatively unimportant (Fig. 2.12). Scouring and structural damage probably will occur without a plank or concrete apron extending below the spill crest (Atlantic Waterfowl Council 1972). Wooden straight-drop spillways have square notches; concrete spillways have square or V notches. The design can be modified to permit water control, often with stoplogs.

The concrete free overfall spillway contains reinforcement rod and is more expensive initially and more satisfactory than wood. Ultimately it probably will be cheaper than wood due to decreased maintenance and increased longevity. Often the concrete spillway is an integral part of a more complete dam structure.

A wooden free overfall spillway used in Wisconsin has proved satisfactory (Linde 1969), and it obviates retardance factors in calculating

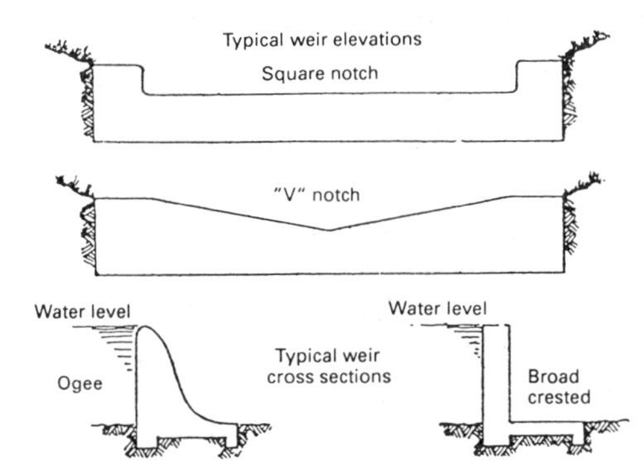

Figure 2.12 Concrete weir spillways (*Atlantic Waterfowl Council 1972*).

the size of the spillway. It consists of a heavy, flat-plank barrier bolted together securely with cross bracing, somewhat longer than the desired spillway length and about as wide as the dike is high. The long edge is buried in the dike on the downstream side so that the top edge is at spillway height. The ends extend into the dike far enough that water does not wash around them. The dike level is reduced to the level of the wooden spillway. The downstream side of the wooden spillway is heavily riprapped with stones to prevent erosion. The earth on the upstream side is seeded. The plank edge prevents the sides of the spillway from eroding.

The ogee spillway or weir is built of reinforced concrete and has an ogee or S-shaped weir for maximum discharge efficiency (Fig. 2.12) (Atlantic Waterfowl Council 1972). The water encounters minimum interference as it flows over the crest and along the profile of the spillway. The ogee spillway is expensive. Drawdown features should be incorporated with the design.

Construction

When the earth fill in the dike reaches the elevation of normal pool, construction shifts to the emergency spillway (Atlantic Waterfowl Council 1972). If the emergency spillway can be built as part of the borrow pit (see Fig. 2.5), the topsoil should be scalped from the entire area of the proposed cut, which has been staked previously, and piled nearby for later use. Then the cut at the outer top edge is started at a 1.5:1 slope. The dozer pushes the fill directly onto the dike. The spillway excavation is used as a borrow pit; the depth of the cut is controlled by frequent rod readings and computations. The cut is continued until the approximate elevation and width of the crest are attained. The spillway should be dug to the full depth of the pond (U.S. Soil Conservation Service 1982). If that is impractical, vegetation or riprap is used to protect the end of the dam and any earthfill constructed to confine the flow. Excavation of inlet and exit channels is omitted if the natural slope suffices. The exit channel is aligned so that discharge will not touch the dam. Otherwise the outflow is directed to a safe point of release with wing dikes, also called kicker levees or training levees.

At this point the dike is completed with fill from the original borrow pit unless enough soil from the spillway remains for the dike. Then the spillway is completed. The inlet and outlet of the spillway are flared. The flow should approach the spillway's entrance from the center or upper portion of the marsh. The flared outlet is constructed so that water passing through the spillway will be spread on sod.

The crest of the spillway is crowned slightly and is usually a continuation of the centerline of the dike. Final grading eliminates small depressions on the crest which would store water and thus hinder establishment of good sod in this critical area (Atlantic Waterfowl Council 1959).

Mechanical Spillway (Water Control Structure)

Mechanical spillways are used with tides and gravity, and they vary from simple tubes to large radial gates mounted on concrete structures. Permanent water control structures should be installed on all major inner and outer dikes, low enough that the impoundment and borrow ditches can be drained completely. Mechanical spillways can be controlled or uncontrolled. The uncontrolled spillways allow water to escape the impoundment over a fixed dam or through an uncapped pipe built into the dike. Uncontrolled spillways will prevent the impoundment from exceeding its maximum water level, but will not permit the impoundment to be drained or held at any controlled level except maximum.

Most control structures can be used with tidewater marshes or inland streams. The structure should be designed to allow water-level manipulation as well as complete drawdown (drainage).

Structures tend to fall into two categories or modifications thereof (Fig. 2.13): drop inlet and whistle tube (Anderson 1985). The drop inlet releases surface water from the impoundment. The whistle tube releases bottom water. Most dissolved substances and nutrients are located near the bottom of the impoundment. Water discharged through low tubes or whistle tubes tends to remove more nutrients from the impoundment than water discharged through high tubes, drop inlets, or over weirs (Linde 1969, Atlantic Waterfowl Council 1972). Also surface water tends to be warmer than bottom water.

Often the final wetland management area will consist of a series of

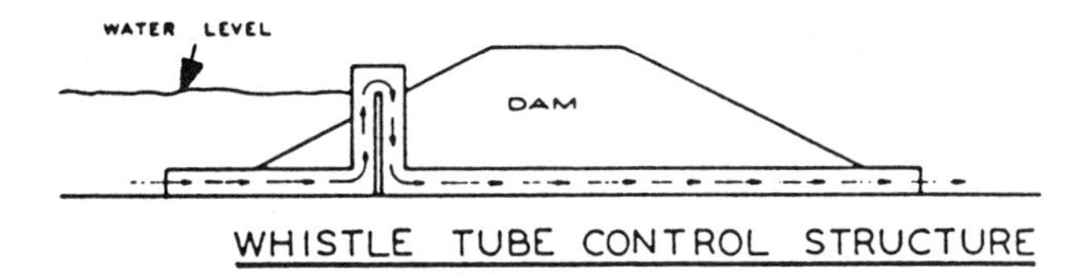

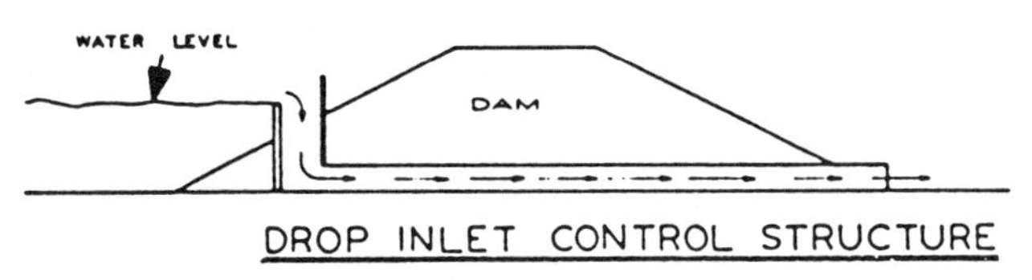

Figure 2.13 General design of water-level control structures (*Anderson 1985*).

impoundments separated by dikes, but connected by water control structures. Water exchange between impoundments occurs via tides or gravity through the control structures, or pumps if no control structure exists or if supplemental control is needed. With tides, water enters and leaves the complex of impoundments through the control structure in the main dike as the tide rises and falls in the bay or river outside the main dike. With gravity, water could enter the complex via a control structure in a dike separating the impoundment complex from a stream or river, returning to the same stream or river downstream via a control structure in the dike of the farthest (lowest) impoundment in the complex.

Less control is afforded with gravity-filled impoundments from runoff or direct streamflow, especially when combined with groundwater seepage. With such impoundments, water discharge is controllable, but water input is not.

Marsh impoundments need control structures that can discharge the peak flow expected in a 24-h storm for a 10-year flood (Hoffman 1988). In tidal areas, one control structure per 61 ha of impoundment is recommended (Williams 1987, Gordon et al. 1989). Fresh water distributed over the salt marsh usually creates better waterfowl habitat (Green et al. 1964). This can be accomplished with check dams installed permanently in freshwater streams flowing through the salt marsh.

Features

Size. Mechanical spillways should be large enough to discharge the estimated flow from snowmelt, seepage, springs, or storms and to drain the impoundment quickly. Calculations from the estimated flow are used to determine the size of the structure, in consultation with an engineer (Atlantic Waterfowl Council 1959, Beauchamp 1979, U.S. Soil Conservation Service 1982). A corrugated pipe must be larger than a smooth pipe to discharge the same amount of water. Generally, the largest structure anticipated for the maximum discharge is used, with capability to regulate the size of the opening. Ponds of 0.4 to 1.2 ha generally need a 10-cm-diameter pipe, ponds of 1.6 to 3.2 ha need a 15-cm pipe, ponds of 3.6 to 4.9 ha need a 20-cm pipe, ponds larger than 4.9 ha need a 30-cm or larger pipe (Summers 1984). The U.S. Soil Conservation Service (1982) recommended drainage tubes at least 15 cm in diameter. Generally, a pipe 46 cm in diameter and long enough to extend through the bottom of the levee will drain ponds up to 16.2 ha (Fredrickson and Taylor 1982). The emergency spillway helps remove excess water from flash floods.

The pipe should extend out the downstream side of the dike at least 1.2 m (Atlantic Waterfowl Council 1959), but preferably 2.4 to 3.0 m (U.S. Soil Conservation Service 1982), to prevent damaging the dike from the flow of water. Large pipes might need the downstream extension supported by a timber brace. Most marshes need about 14.6 m of pipe. The control structure should be designed so that water can flow uninterrupted from the marsh and not collect in or over the barrel of the pipe (Atlantic Waterfowl Council 1959).

Metal, concrete, wood, and plastic tubes. Most discharge tubes are corrugated, galvanized, asphalt-coated steel (Linde 1969) similar to highway culvert which tends to come in 6-dm lengths (Atlantic Waterfowl Council 1959). To improve watertightness, double-riveted pipe should be bought for marsh use. Care should be taken not to chip or crack the asphalt coating during installation. Peeled asphalt and subsequent rusting vary with water chemistry. Aluminum pipes will not rust, but might corrode in acid water, and they tend to bend more easily than steel tubes, making installation harder during backfilling and soil compaction. But aluminum tubes are lightweight relative to steel tubes and often can be handled manually, simplifying installation and reducing transportation and equipment costs. Where rusting and corrosion are a problem, reinforced-concrete tubes are used, although they are more expensive and harder to install than metal tubes (see Table 6.1).

The cost of creosoted or pressure-treated wooden structures might nearly equal that of concrete, and concrete is preferable, especially if fire damage and the relatively short life of timber alternately exposed to water are considered (Atlantic Waterfowl Council 1972). Wakefield piling should be used to build cutoff walls in impervious material and to build wooden spillway structures. Wakefield piling consists of usually three courses of creosoted timber planking, lapped to prevent leakage, fastened together with galvanized bolts, and driven vertically to impervious material as support for the control structure. In salt-marsh impoundments, 5-cm tongue-and-groove cypress has been used successfully without Wakefield piling. In salt water, wooden tubes (wooden trunks) are better than metal tubes because the pressure-treated creosote wood lasts longer.

Plastic polyvinyl chloride (PVC) pipe will not deteriorate and thus has a long life if it does not crack (Fig. 2.14). It is used mainly on small ponds (Wellborn et al. 1984).

Antiseep collar. A properly installed antiseep collar on a horizontal tube prevents water from seeping along the tube, eventually causing severe leaks or a costly washout (Linde 1969). These collars are flat, square metal plates which encircle the tube and are clamped or welded to it to serve as a vertical barrier to the flow of any water (see Figs. 2.17 and 2.19 to 2.22) (Beauchamp 1979, U.S. Soil Conservation Service 1982, Summers 1984, Kierstead undated). Overtightening installation bolts might shear off the rivets that secure the flanges. With concrete pipe, the collar is a concrete slab at least 15 cm thick poured around the pipe (U.S. Soil Conservation Service 1982). Concrete or steel collars suit smooth steel pipe. Special metal diaphragms can be used with corrugated metal pipe. Collars should project at least 6 dm perpendicular to the pipe on all sides. (See also p. 79.)

All lengths of metal corrugated pipe should be joined with the usual watertight bands manufactured for that purpose. Metal antiseep collars are installed at a joint connecting two pipe sections in the core fill at the centerline of the dike preferably (U.S. Soil Conservation Service 1982), although it could be 12 dm upstream from centerline (Atlantic Waterfowl Council 1959). Dikes at least 4.6 m high or corrugated tubes of at least 18.3 m need two or more collars at points equidistant between the dike's centerline and upstream side (Linde 1969, U.S. Soil Conservation Service 1982). Concrete antiseep collars should be installed at 3.7- to 4.6-m intervals to prevent leaks and to add strength to the span of pipe (Summers 1984).

Placement

To facilitate complete drainage of the impoundment, the mechanical spillway (control structure) should be located in the dike at the lowest elevation of the impoundment, usually not far from the emergency

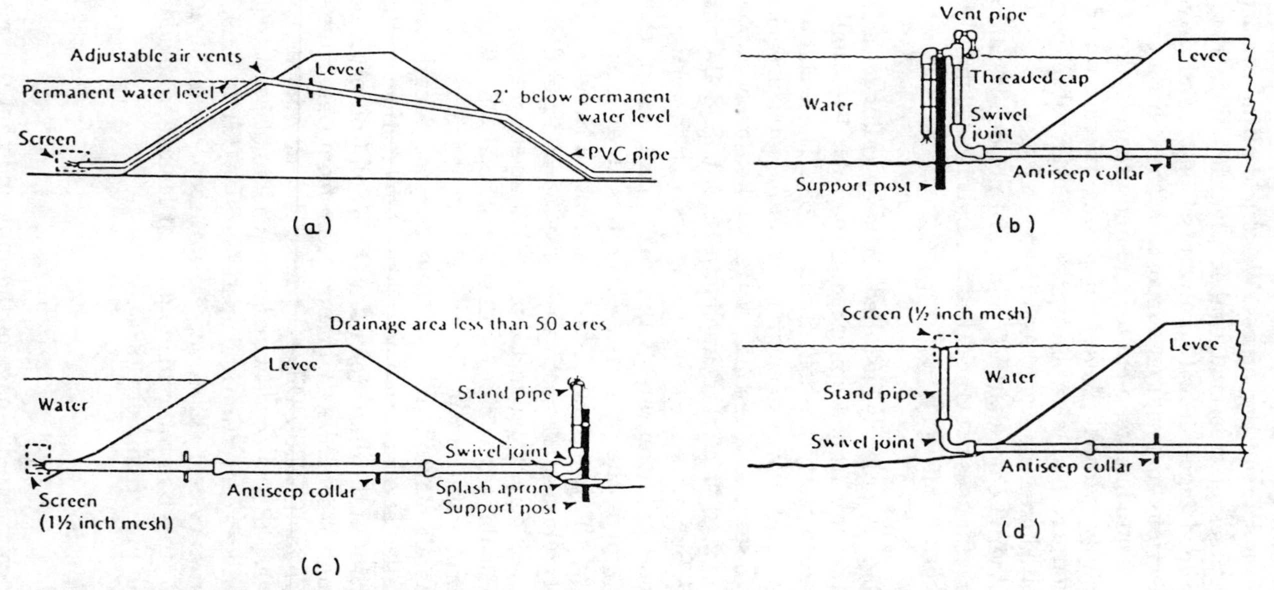

Figure 2.14 Typical design for drain pipes of PVC plastic pipe (*Wellborn et al. 1984*). [For metric equivalence, see metric conversion table in the front of the book.]

spillway if built into the dike but at a higher elevation. Position of the control structure is influenced by the approach route that equipment must use to get to the site for installation. Control structures which impound streamflow are located in firm ground next to the streambed. Muck in the streambed often is too deep to remove to firm soil. A control structure built on such muck would shift and rise, resulting in leakage. A bulldozer rather than a dragline then can be used for excavation, a faster process (Linde 1969). Generally, a structure built on a stream outlet or drainage is constructed by use of a coffer dam so that the stream is bypassed around the dam site, and the entire dam with control structure can be constructed simultaneously. Otherwise if half the structure is built, a stoplog section or gate must be built to bypass water while the second half of the dam is built (Atlantic Waterfowl Council 1972). Pumping might be needed to remove excess or accumulating water that seeps into the construction area.

Soil borings will indicate how much topsoil and shaky subsoil must be removed to place the control structure on firm ground. Unfirm ground might result in rupture of the control structure, costly washout of the dike, and repair or replacement of the control structure.

After the core trench of the dike is dug, a ditch at least 3 to 6 dm wide for the control pipe is dug at right angles to the core trench. Then the core trench is filled with clay up to the intended position of the pipe, so that the fill below the pipe is completely compacted. Next the pipe is laid, and all joints are thoroughly sealed and braced. The rest of the ditch is filled with heavy clay around the pipe in layers at least 30 cm thick, tightly packed, and then the core of the dike is constructed (Summers 1984).

With prior engineering approval, antiseep collars can be made from sheet steel welded in place, if heavy steel pipe is used. Concrete collars are poured in forms built in place. Other collars are prefabricated before placement. All collars need careful backfilling and compaction by hand tamping to prevent subsequent settling and leakage (Atlantic Waterfowl Council 1959).

Impoundments manipulated with tidewater (brackish or saline) need control structures with pipes perfectly level so that water can flow in either direction. Impoundments manipulated with gravity flow of water usually need control structures with pipes angled higher at the inlet end.

For a 14.6-m pipe, which most marshes need, the inlet end should be 46 cm above the outlet end (Atlantic Waterfowl Council 1959). Once firm ground is exposed, a simple means to accomplish this uneven placement is to nail a 15-cm piece of board at a right angle to one end of each of three 4.9-m 2 × 4's (5- by 10-cm boards), and to place them along the path of the pipe starting at the end at the control structure

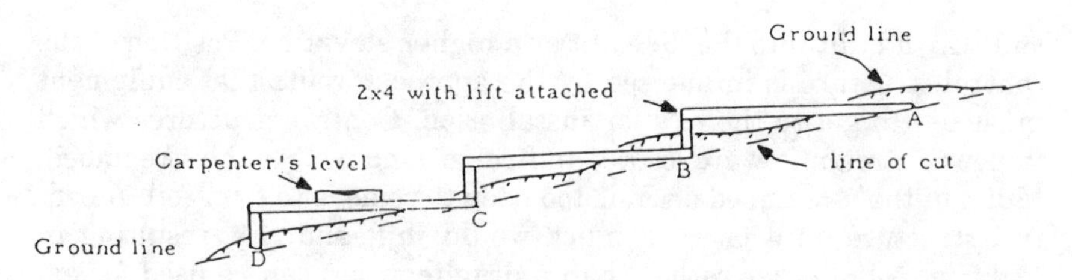

Figure 2.15 Method for placing pipe unevenly through a dike (*Atlantic Waterfowl Council 1959*).

(Fig. 2.15). The 2 × 4's are leveled with a carpenter's level, and soil is removed beneath the 15-cm boards if needed. Then the high spots along the route are removed with a pick and shovel to minimize adjustments that might be needed after the assembled pipe is rolled into the shallow trench thus dug (Atlantic Waterfowl Council 1959).

For best support, the pipe should lie on cradles or bents located at intervals on firm ground beneath the pipe. For concrete pipe, concrete cradles are poured at joints to provide support, seal the joint, and act as an antiseep collar if poured large enough (Linde 1969).

A concrete inlet box with flashboards (see Fig. 2.25) (Fredrickson and Taylor 1982) can be built into the upstream slope of the dike such that the bottom front edge of the box is even with the dike's toe (Atlantic Waterfowl Council 1959). The tube runs through the dike from the box to the downstream side. Where frost action is a problem, the metal, concrete, or wooden riser (control box) supporting the stoplogs should be reinforced with a 15-cm-thick reinforced-concrete slab poured around the riser footing at ground level.

Concrete water control structures (boxes or risers) should be constructed with reinforcement rods in properly constructed forms. The box and footing are completed in one continuous pour for good bondage (Atlantic Waterfowl Council 1959). The concrete around and especially beneath the pipe is tamped or vibrated completely to ensure bondage, and then it is covered with a wet blanket, straw, or other material to prevent premature drying, especially during hot, dry weather. Forms are left in place 3 to 5 days and then saved for reuse. No concrete work is attempted if temperatures approach freezing.

Types

High tube overflow. Small impoundments often do not warrant the expense of installing sophisticated control structures. In such cases, a simple overflow tube will suffice. The tube should be large enough to accommodate the expected spill. The tube is set horizontally, with the lower edge at the height of the desired water level of the impoundment. Water rising above that level will flow out the tube. If the tube becomes plugged with debris or an exceptional storm occurs, excess water escapes through the emergency spillway. No water-level manipulation is possible with this type of control structure. If a low-draining tube also is installed (Fig. 2.16), the water level can be lowered or the pond drained—a highly desirable feature for management (Linde 1969).

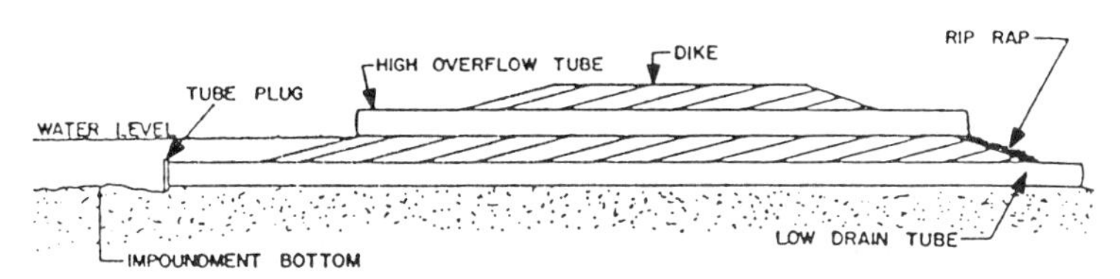

Figure 2.16 Horizontal overflow and drain tubes *(Linde 1969)*.

Low tube with cap, gate, or stoplogs. In small impoundments where heads are low, a single low tube, capped on the impoundment side, can be installed (Fig. 2.16). The cap is a flanged wooden plug which fits snugly inside the tube and has a shoulder that fits against the end of the tube. The cap is held in place by water pressure. Water levels must be shallow, or the cap will be hard to remove to drain the impoundment.

A better arrangement is a header plate welded to the top of a low tube on the impoundment side, with a sliding gate installed. Angle iron is welded to each side of the header, to serve as slots for sliding a steel gate up and down. The plate should be thick enough to withstand distortion from water pressure. The slots are made large enough to facilitate sliding the gate against water pressure or any distortion the gate might eventually acquire. The gate can be closed tightly with wooden wedges if needed. A plywood plate can be substituted for steel, but plywood distorts easily, is harder to slide, and is subject to ice damage.

Header plates can be added to culvert drain tubes, with control effected by either sliding gates or stoplogs. Such tubes equipped with stoplogs are set so that normal pool is below the top of the tube. The water level in the impoundment is set by adding or removing stoplogs. Stoplogs have no advantage over a sliding gate if the water level is to be above the top of the tube, because all stoplogs would need to be in place (Linde 1969).

If caps and gates are too costly, an elbow or tee joint can be installed on the outlet end of the control pipe, with a standpipe the same height as normal pool (Fig. 2.17). It functions as a trickle tube when upright, and when it is lowered, the pond can be drained (Summers 1984). The

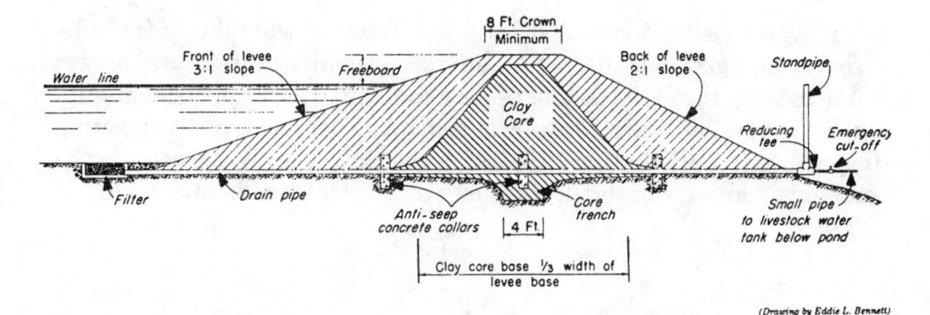

Figure 2.17 Cross section of the levee or dike (*Summers 1984*).

inlet end of the drain pipe should extend into a sump filled with gravel or should have screening or a trash rack to prevent clogging.

Trickle tube. The trickle tube is similar to the high tube overflow in that the opening is set at normal pool to remove excess water, and no water manipulation is possible. Hence its use, too, is limited to small impoundments. The trickle tube is not set horizontally though. It must be large enough to discharge excess water from seepage springs, snowmelt, estimated prolonged surface flow from intense storms, or any combination of flows. The crest elevation of the entrance should be at least 30 cm below the spill crest of the emergency spillway. The tube through the dike should be at least 15 cm in diameter. Trash racks (Fig. 2.18) are recommended to reduce clogging. Two types of trickle tube are in common use: drop inlet and hood inlet (U.S. Soil Conservation Service 1982).

A drop-inlet trickle tube consists of a pipe barrel through the dike and a vertical tube or riser pipe connected to the upstream end of the barrel (Fig. 2.19). If a suitable valve or gate is attached to the upstream end also, the drop inlet can be used to drain the pond. The ratio of barrel diameter to riser diameter is a function of the required discharge capacity determined, and it varies with smooth or corrugated pipe. For the tube to flow full, the diameter must be larger for the riser than for the barrel. The U.S. Soil Conservation Service (1982) listed discharge values for drop-inlet trickle tubes for various ratios of barrel and riser diameter for both smooth and corrugated metal pipe for a particular size dike.

The hood-inlet trickle tube is a straight pipe penetrating the dike, having the inlet end cut at an angle to form a hood (Fig. 2.20). To increase the hydraulic efficiency of the tube, an antivortex device, usually metal, is attached to the pipe's inlet (see Fig. 2.18). The hood inlet cannot drain the pond, but it often is less costly than the drop inlet.

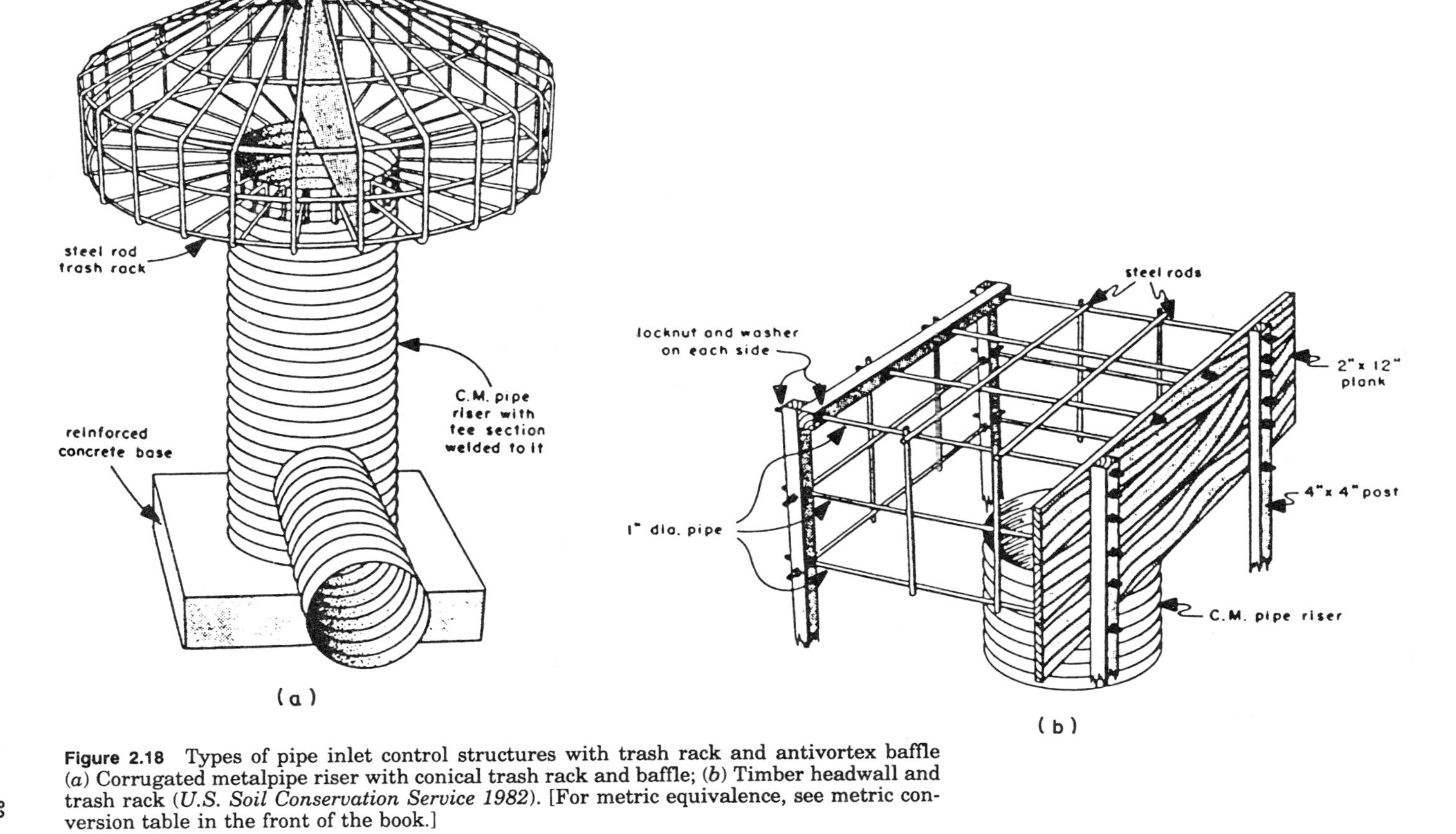

Figure 2.18 Types of pipe inlet control structures with trash rack and antivortex baffle (*a*) Corrugated metalpipe riser with conical trash rack and baffle; (*b*) Timber headwall and trash rack (*U.S. Soil Conservation Service 1982*). [For metric equivalence, see metric conversion table in the front of the book.]

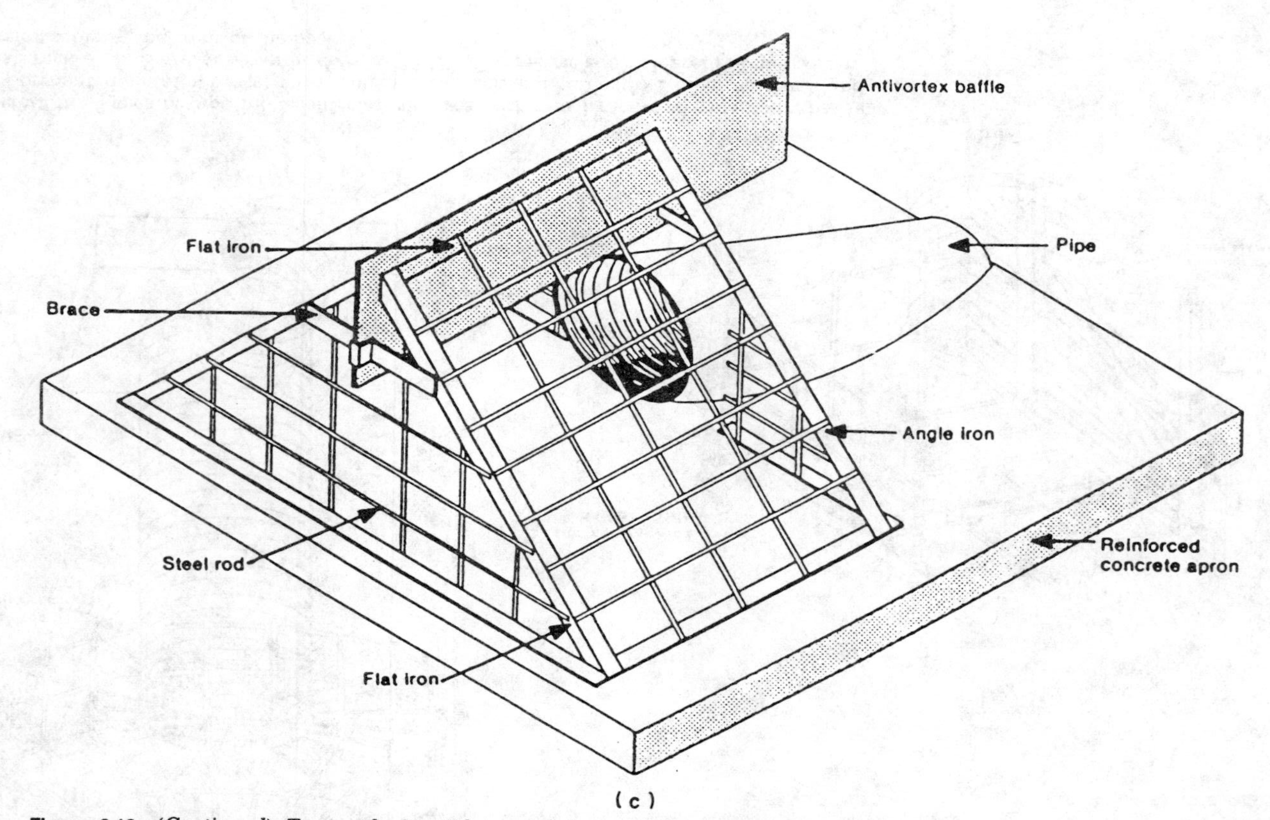

Figure 2.18 (*Continued*) Types of pipe inlet control structures with trash rack and antivortex baffle (*c*) Hooded inlet with trash and baffle (*U.S. Soil Conservation Service 1982*). [For metric equivalence, see metric conversion table in the front of the book.]

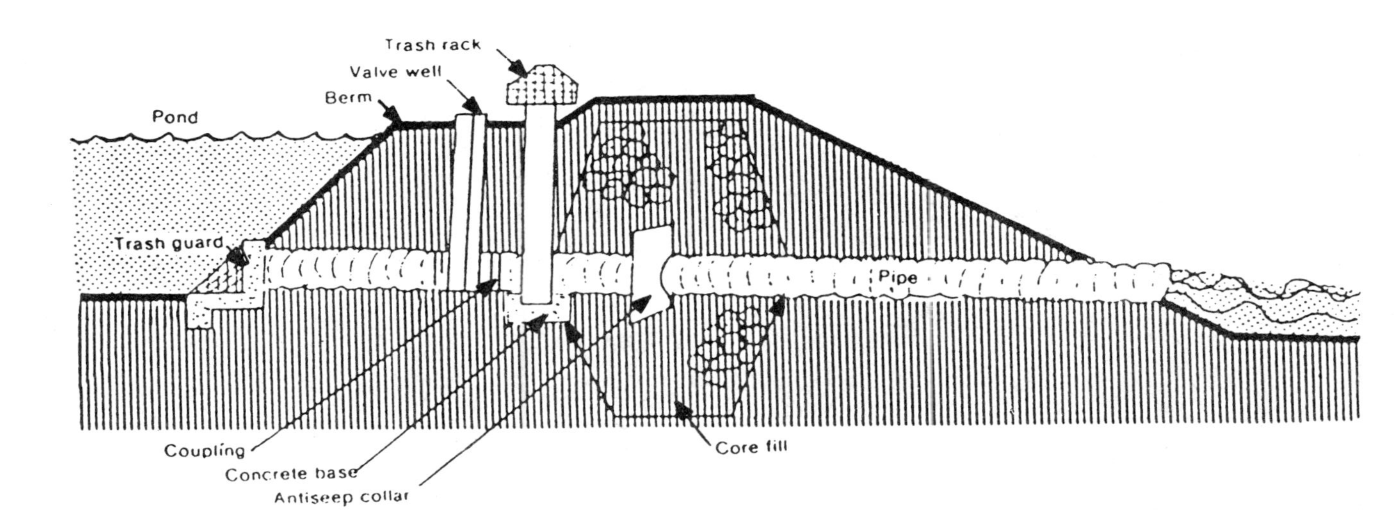

Figure 2.19 Dam with drop inlet pipe control structure (*U.S. Soil Conservation Service 1982*).

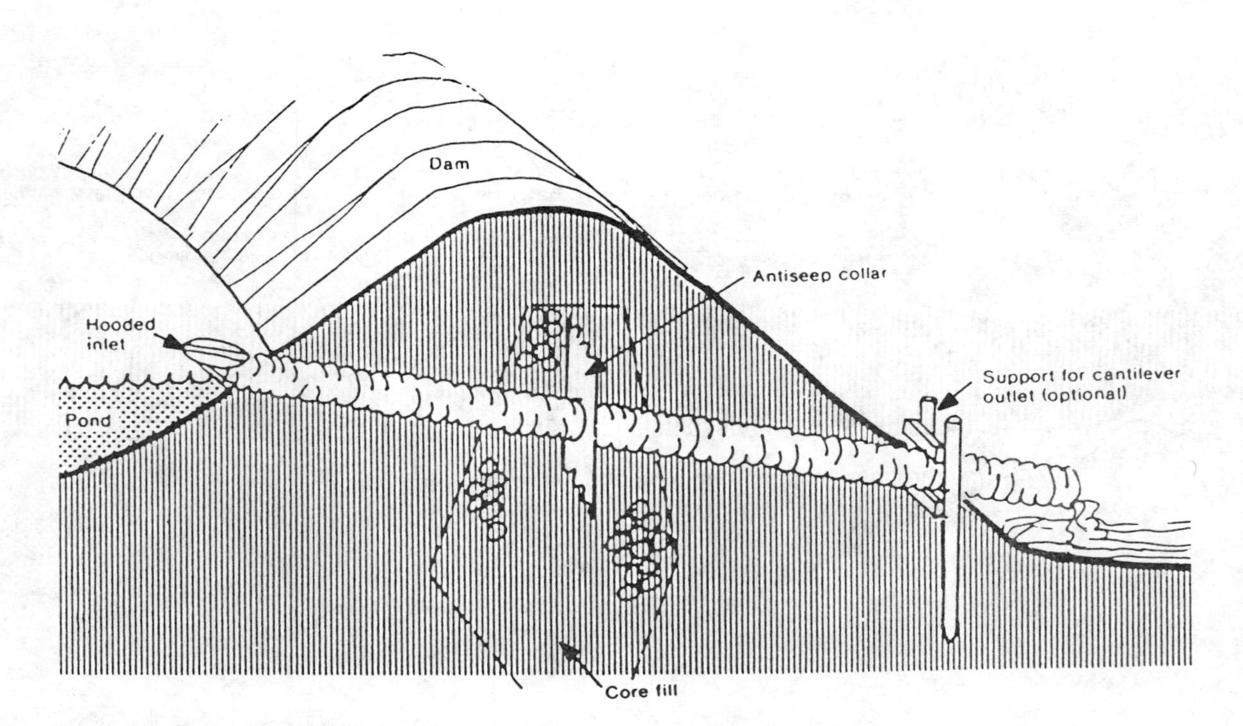

Figure 2.20 Dam with hooded inlet pipe control structure (*U.S. Soil Conservation Service 1982*).

The U.S. Soil Conservation Service (1982) listed the minimum head required above the invert (crest elevation) of the tube entrance of hood inlets to provide full flow for various sizes of smooth and corrugated metal pipe for a particular size dike.

Whistle tube (tin whistle). The whistle tube or tin whistle probably is the most common type of water control structure for inland marshes. It comes prefabricated in a variety of sizes, and it is easy and inexpensive to install and maintain (Anderson 1985). But in soils where frost heaving is a problem, gated tubes and drop inlet are preferred (Linde 1969).

The whistle tube consists of a vertical full-circle riser tube (conduit) located on the impoundment side or the middle of the dike, or a half-circle riser often with a grill-like beaver baffle on the impoundment side (Figs. 2.21 to 2.23) (Beauchamp 1979, Kierstead undated). The bottom end is connected to a horizontal intake barrel tube extending into the impoundment and a horizontal discharge barrel tube extending through the dike, essentially a culvert (Linde 1969). The horizontal tube is set with the low edge into the bottom of the impoundment. The half-circle riser allows marsh gases (hydrogen sulfide and methane) to escape, reducing potential hazard to persons cleaning clogged control structures. The riser has channels along each side to accommodate stoplogs. When inserted, the stoplogs form a wall through the riser tube perpendicular to the barrel tube, forcing the impounded water to rise up and over the stoplogs before entering the discharge tube. The water level of the impoundment is regulated by the height of the stoplogs. Removing all stoplogs allows the impoundment to drain. A locking device prevents tampering (Linde 1969).

A more expensive modification of the conventional whistle tube, which offers easier and more precise water-level control, is the replacement of stoplogs with a sliding steel gate raised or lowered with a handwheel that turns a long screw shaft (Fig. 2.24). The gate has two sections which slide across each other such that water can flow from above or below it. More water is released faster if it flows below the gate. By turning the handwheel, ease of operation is unaffected by heavy flows or flood conditions as it would be with stoplogs.

Whistle tubes vary in diameter, being usually 33 to 122 cm for the barrel and 18 dm or less for the riser (Linde 1969). The riser is made of metal usually, but can be made of wood. Wooden risers are square tubes made of 5-cm creosoted or pressure-treated planks. Slots for the stoplogs are made by bolting two wooden rails on each interior side. Metal plates welded to the horizontal metal tubes are bolted to the lower end of the riser. The construction cost is low, the tube is rigid

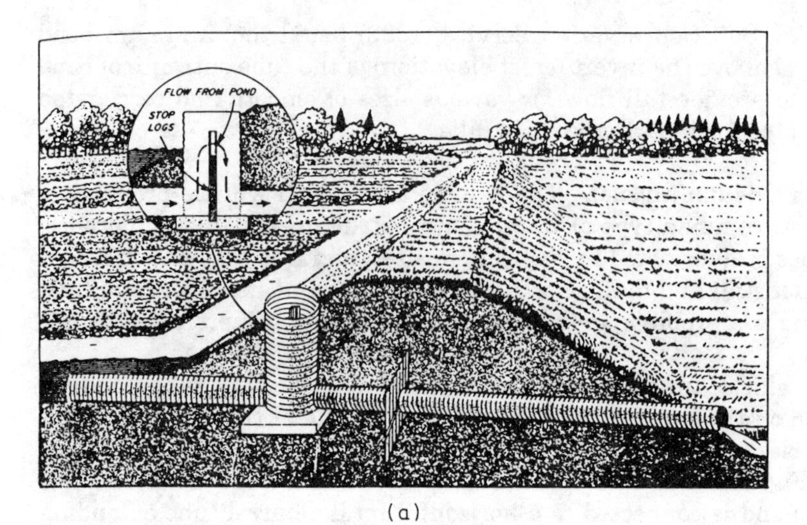

(a)

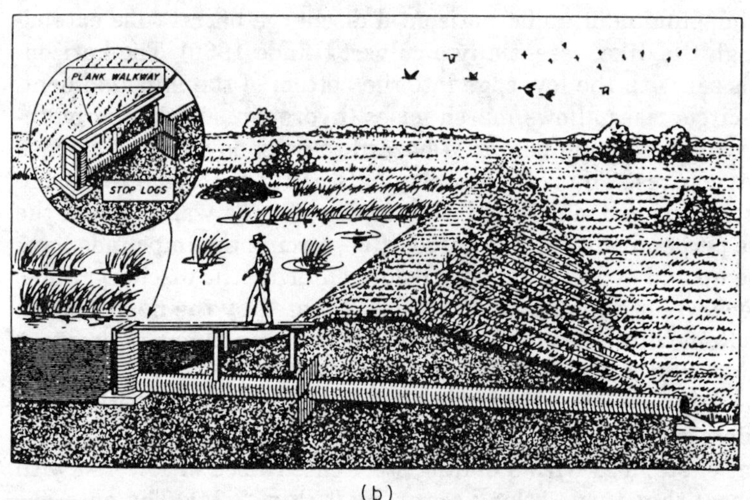

(b)

Figure 2.21 Dams with corrugated metal pipe drop inlet control structures for water level control by use of stoplogs in the riser (*Beauchamp 1979*).

and not easily distorted during installation or by frost action, and it is relatively durable.

Unconventional, budget-saving whistle tubes can be made by substituting 208-L (55-gal) drums for the corrugated metal barrel. The ends are cut out; the resulting tubes are spot-welded together to the desired length and painted with Rustoleum primer to minimize rusting. Pressure-treated yellow pine or Douglas-fir attached to six 208-L drums has been used in Wisconsin (Linde 1969). This is reasonably durable. So are smaller structures made from 33-cm grease drums for

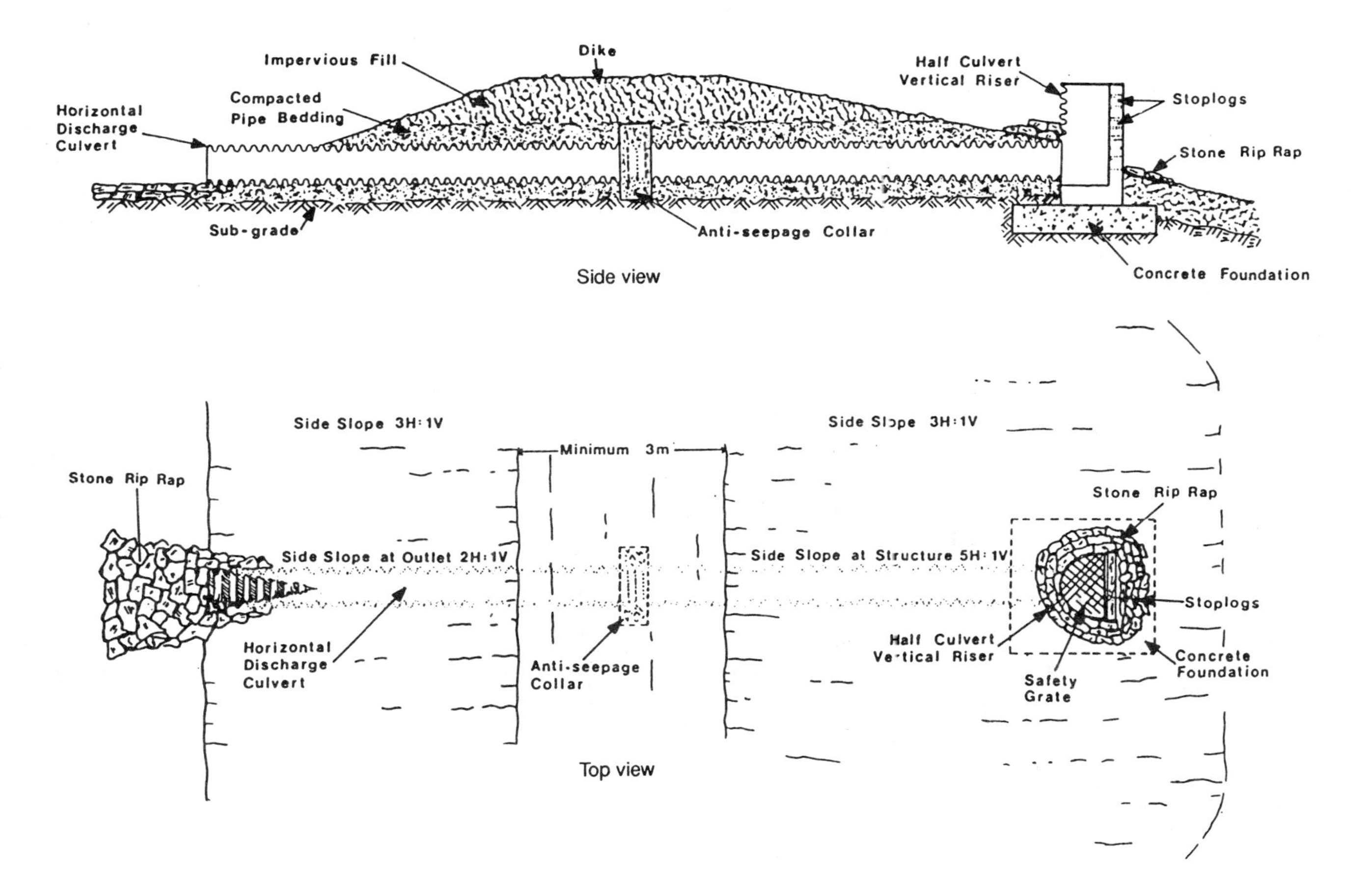

Figure 2.22 Half culvert drop inlet control structure (*Kierstead undated*).

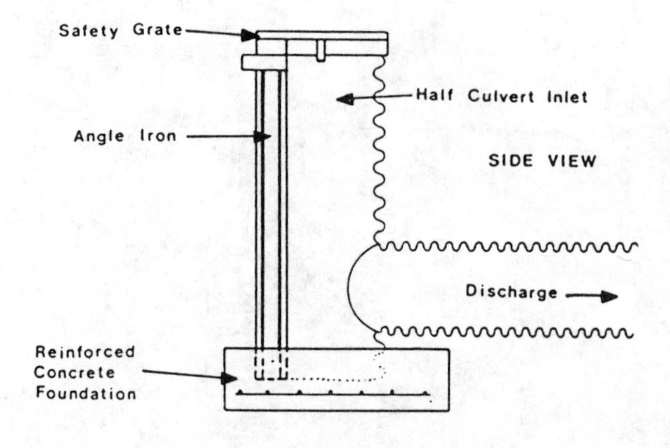

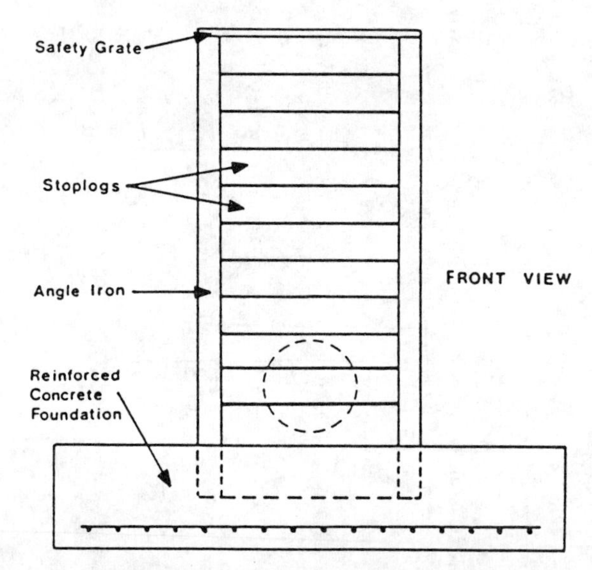

Figure 2.23 Detail of half culvert drop inlet control structure (*Kierstead undated*).

tubes and 51- by 51-cm wooden risers, some of which used 5- by 20-cm salvaged lumber thoroughly creosoted.

One of the best mechanical spillways in common use, for both tidal and inland marshes, is a stoplog (stop plank or flashboard) structure used with a concrete riser or whistle tube as a form of drop inlet (Atlantic Waterfowl Council 1972, Fredrickson and Taylor 1982). In tidal areas, the riser usually is attached to both ends of the barrel tube penetrating the dike, especially if limited runoff is available from the sur-

Figure 2.24 Full-circle whistle tube with safety grate and handwheel-controlled sliding gate in middle of dike.

rounding watershed. Thus, most water entering such impoundments is salt water, from high tide, controlled by the height of the stoplogs.

Fredrickson and Taylor (1982) considered the best control structure to be a boxlike concrete riser without top and front sides, attached at the bottom of the back wall to a corrugated, galvanized steel drain pipe (Fig. 2.25). The walls and bottom are reinforced concrete 13 to 15 cm or thicker. A perpendicular groove large enough to slide boards (stoplogs) 5 cm thick extends from top to bottom along each side of the box toward the inside front edge. The inside front-to-back distance with boards installed is 46 cm. To reduce soil erosion and prevent water seepage, the bottom of the box should extend 15 to 20 cm beyond the front and be as flat and level as possible. Normal pool and the depth of the internal borrow ditch dictate the height of the riser. The riser should be at least 3 dm higher than normal pool. Extension of the back wall as an antiseep collar reduces rodent burrowing and consequent water seepage.

Stoplogs are harder to remove and install if the guide channels are asphalt-coated. Metal stoplogs work well and prevent swelling and warping. The best wooden stoplogs are of rough-cut redwood or creosoted or pressure-treated boards (Fredrickson and Taylor 1982). Cypress stoplogs are durable in salt water and economical (Atlantic Waterfowl Council 1972). Tongue-and-groove edges should not be used

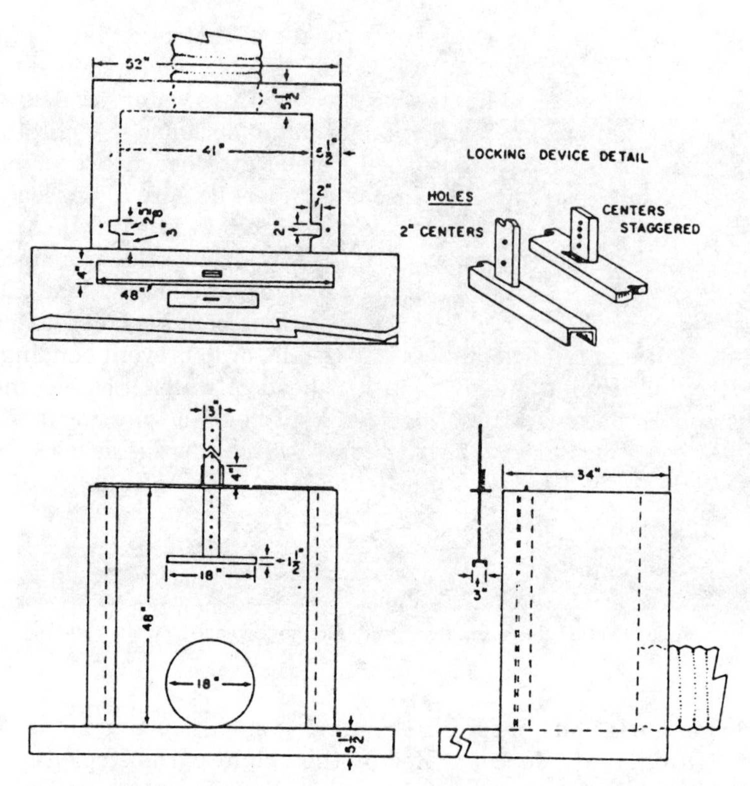

Figure 2.25 Specifications for a box-type water control structure (*Fredrickson and Taylor 1982*).

because some warping always occurs and the edges will not fit together. Small leaks between boards can be sealed by placing dry cinder coal ashes a handful at a time immediately upstream over the leak so that the ashes float into and clog the leak (Atlantic Waterfowl Council 1972). Plastic sheeting also controls leaks if placed over the pool side of the boards and fastened with thumb tacks or bulletin board pushpins for fast removal (Fredrickson and Taylor 1982). An oil-base caulk will seal lower boards not removed for minor water-level manipulations. To prevent leaks at the sill, the bottom edge of the bottom stoplog can be fitted with a rubber gasket, which can be a length of rubber hose (Linde 1969).

Boards are about 5 cm thick. They should vary in width (height) to facilitate minor water-level changes as small as 1 cm. For accurate and fast installation, stoplogs should be numbered according to size and fit (Fredrickson and Taylor 1982). Thin wooden wedges can be used to hold stoplogs tightly to the channels to prevent them from lifting (Linde 1969).

Stoplogs are undesirable as the only means of control on large impoundments (Linde 1969). Control is not precise, and removal of stoplogs difficult and often impossible against water pressure. Stoplogs can be equipped with a screw eye bolt at each end to facilitate removal with hooks, or a strong wire can be attached between the screw eyes so that a board can be inserted under the wire to pry loose the stoplog. A sheet of 2-cm plywood can be placed in front of the horizontal discharge tube on the upstream (impoundment) side to reduce water pressure on the stoplogs during removal. Stoplogs at least 18 dm long should be 7.6 cm thick to prevent their bending from water pressure, rendering removal almost impossible, or to prevent bending, deflecting, and popping out under a full head of water. Some dams have more than one bay (space between abutments or piers or abutment and pier) which use stoplogs to control water depths at least 18 dm deep (Atlantic Waterfowl Council 1972).

Flapgate and wooden trunk. Used exclusively in tidal areas, cast iron or aluminum flapgates are hinged on the top of a perfectly level horizontal steel or, preferably, heavy-gauge noncorrosive aluminum corrugated pipe so that the gates hang perpendicular over the ends of the pipe (Atlantic Waterfowl Council 1972, Rollins 1981, Wicker et al. 1983, Williams 1987). The gate is raised or lowered by screwing it up or down on a threaded rod (Fig. 2.26) (Wicker et al. 1983) or by hand winching with a cable from a creosoted beam extending from the dam out over the gate (Atlantic Waterfowl Council 1972). The gate covers the opening of a pipe that is usually 91 cm in diameter and 6.1 to 14.6 m long (Williams 1987). If runoff water from the surrounding watershed is adequate, the gate is installed only on the impoundment side of the tube. If water manipulation is mainly from tidewater, gates are installed on both ends of the pipe penetrating the dike. To fill the impoundment, the tidal gate is raised, and the high tide enters the pipe

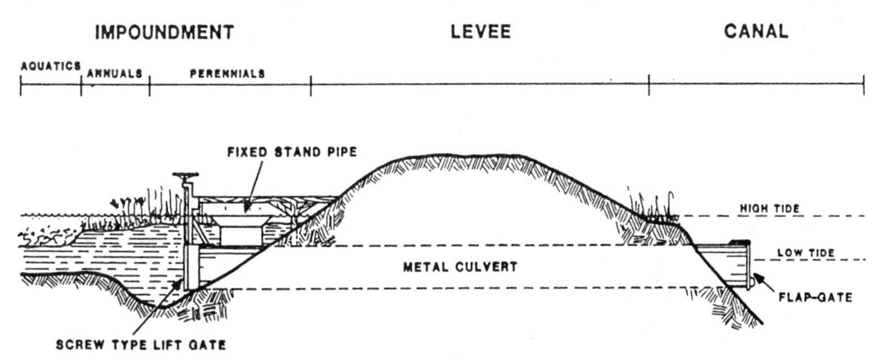

Figure 2.26 Longitudinal section of a 91-cm flapgate metal control structure (*Wicker et al. 1983*).

to push the impoundment gate open to enter the impoundment. As the tide recedes to a level below that in the impoundment, water in the impoundment is prevented from exiting when the impoundment flap is forced against the tube by water pressure from the impoundment. To drain the impoundment, the tidal gate is lowered and the impoundment gate raised. The impounded water enters the tube, and during low tide, forces open the tidal flapgate to exit. As the tide rises, it forces shut the tidal gate, which prevents tidewater from reentering the impoundment.

Modifications of the flapgate include using a flapgate on the outside (tidal) end of the pipe, and a screw-type slide (lift) gate on the inside (impoundment) end (Fig. 2.26), mainly to drain the impoundment, but not in areas with much runoff, unless supplemented with another control structure (Atlantic Waterfowl Council 1972). Or the outside (tidal) flapgate can be used with a stoplog riser (Williams 1987) such as a whistle tube on the inside (impoundment side) (Fig. 2.27), perhaps with a screw gate (Wicker et al. 1983). Still another modification in areas with little runoff is a combination slide and flapgate (Fig. 2.28). A handwheel lifts the slide to allow outflow. When lowered, the gate functions as a flap which allows inflow and prevents outflow.

Wooden trunks are costly and bulky to handle during initial instal-

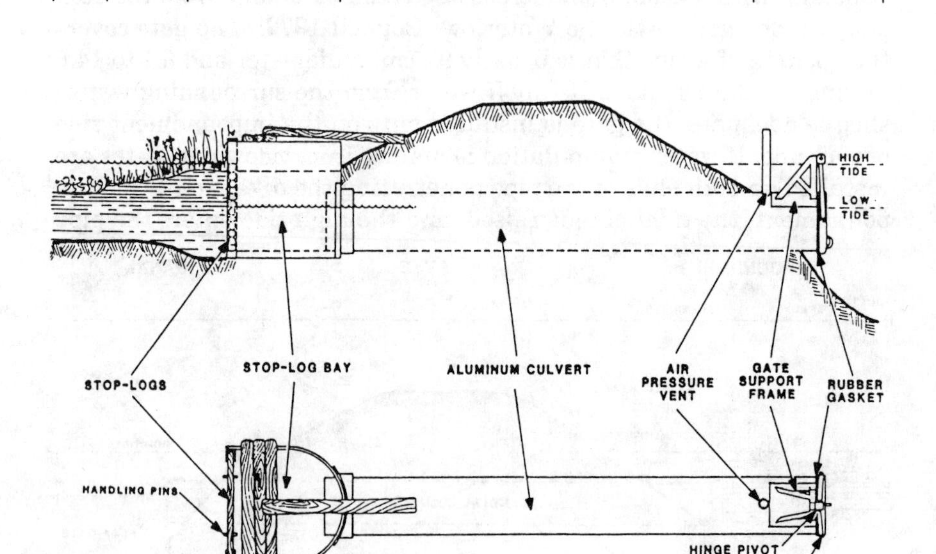

Figure 2.27 Longitudinal and top views of a 122-cm aluminum flapgate control structure (*Wicker et al. 1983*).

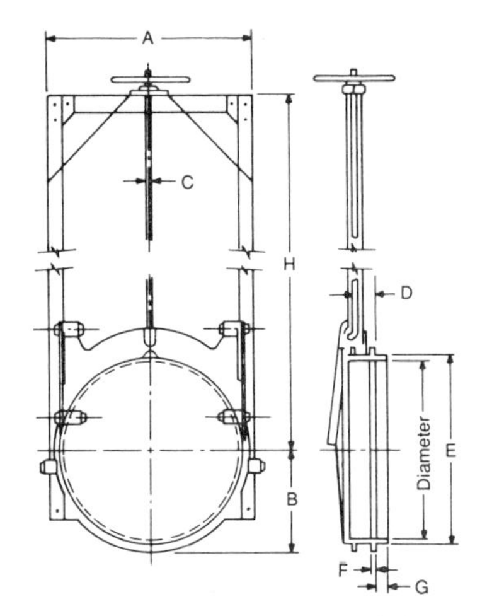

(all dimensions in inches)

Gate size (diameter)	A	B	C	D
18	22 3/4	10 1/2	25 3/4	2 5/16
24	28 3/4	13 1/2	32 1/4	2 5/16
30	35 1/2	17 3/8	39 1/4	3
36	41 3/4	20 3/8	45 1/4	3
42	47 3/4	23 5/8	51 1/4	3
48	53 3/4	26 3/4	57 1/4	3
60	68	33 1/2	73 1/8	4 1/8

Gate size (diameter)	E	F	G	H
18	18 3/4	7/16	1 1/4	48
24	24 3/4	7/16	1 1/4	60
30	32 1/4	5/8	2 1/8	72
36	38 3/4	5/8	2 1/8	84
42	44 1/4	5/8	2 1/2	96
48	50 1/2	5/8	2 1/2	120
60	63	7/8	2 5/8	144

Figure 2.28 Combination slide and flapgate for a control structure (*Atlantic Waterfowl Council 1972*).

lation. But they are preferable to metal covers because they can handle greater volumes of water in a shorter time, and they last longer in salt water, at least 50 years in some cases, with minimum maintenance (Atlantic Waterfowl Council 1972). They allow a variety or combination of incremental settings of flapgates and flashboards to circulate water and control levels within or between impoundments automatically at a preset level as the tide rises and ebbs (Williams 1987). For best results, the density of trunks should be one or more per 61 ha of impoundment (Gordon et al. 1989).

Wooden trunks typically consist of a wooden box 0.6 m high, 1.5 m wide, and 8.5 or 11 m long, usually with a wooden flapgate attached to each end and a wooden drop-inlet flashboard riser attached to the box (Figs. 2.29 and 2.30) (Carlton and Jackson 1984, Williams 1987). Other designs have two sliding flapgates and no riser, or a sliding flapgate on one end and a riser on the other. Other common dimensions include trunks 0.3 by 1.2 by 8.5 or 11 m and 0.6 by 1.2 by 8.5 or 11 m, with appropriate-size flapgates and risers. Williams (1987) provided illustrations of the wooden trunk and listed materials for the three different sizes most common.

Wooden trunks are built of precut pine pressure-treated before construction with creosote or copper chromate arsenic applied at 33.3 kg/m^3. Stainless steel bolts and nails are used for trunks installed in brackish and saline water. In freshwater areas, hot-dipped galvanized bolts and nails will suffice.

Wooden and metal flapgates function similarly. Such a design also can be used to take in fresh water resulting from tidal influence. If the

Figure 2.29 Flapgate on wooden trunk used for tidal creeks (*Carlton and Jackson 1984*).

dike is installed far enough upstream to be beyond the reach of salt water but still under tidal influence, the rising tide will push fresh water into the impoundment to permit flooding at high tide. It might be desirable to leave both gates open so that tidal changes produce a flushing action, especially where accumulated organic sludge is to be removed from the channels within an old field. To facilitate complete drainage and maximum reflooding potential, trunks should be installed 10 to 15 cm below the mean water mark (Gordon et al. 1989).

The best design has a flapgate at each end of a trunk and a flashboard riser. Trunks designed with a flapgate on the outside and only a flashboard riser on the inside cannot manipulate water levels as readily. The flashboards (stoplogs) require individual removal or replacement. Water adjustment is facilitated if a sliding gate is used in the riser slots instead of stoplogs, but then bottom water with its nutrients is removed instead of surface water.

Amphibious draglines and backhoes, or standard types perhaps brought by barge to the site, are used to install trunks by excavating a level ditch through the dike (Williams 1987). Flapgates are removed, the machinery lifts the trunk in place and tamps it down until level, and the flapgates are replaced. The trunk then is partially covered in the middle with a clay-based soil to hold it in place while the bulkhead is constructed.

The bulkhead is a wall on each side of the dike and across the width of the ditch for the trunk. The bulkheads must be well designed for the control structure to be installed and to function properly. The bulkhead is constructed on both sides of a linear path of a tidal rivulet in

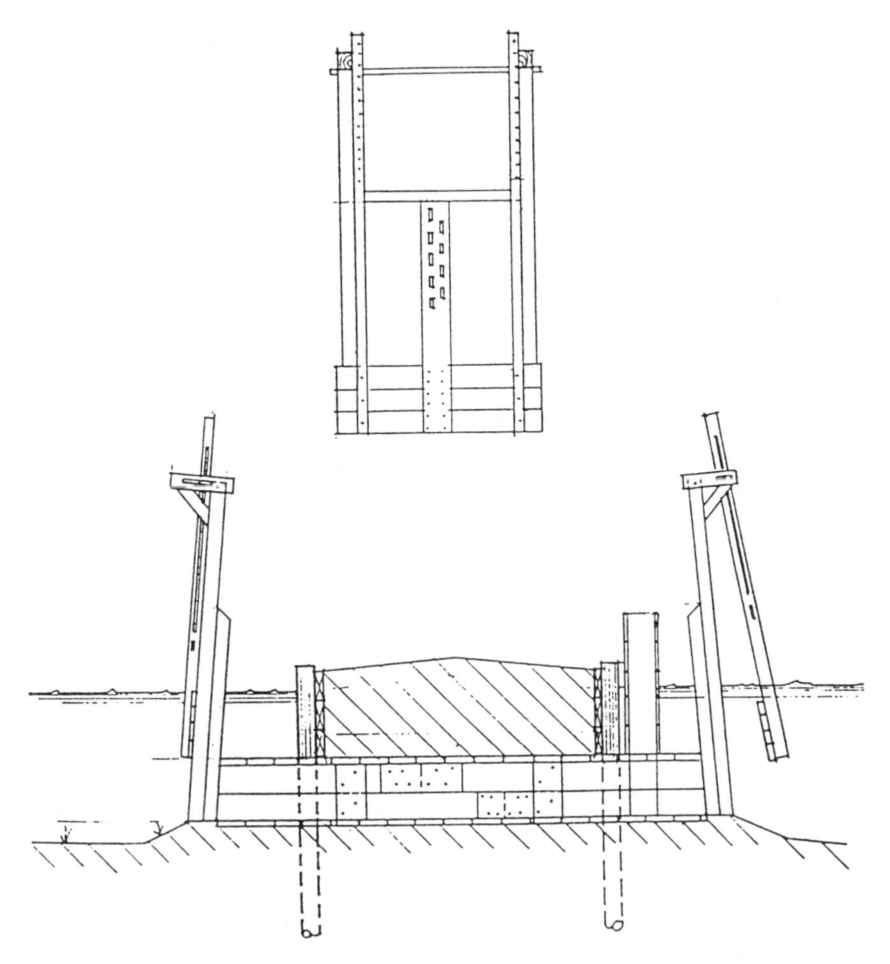

Figure 2.30 Side view of 6 × 15 × 84-dm wooden trunk with 18-dm flashboard riser and flapgates (*Williams 1987*).

the tidal marsh. About four to eight pressure-treated copper chromate arsenic or creosote 9-m pilings are driven to rejection every 1.5 m on both sides of the path on the inside (impoundment side) and on the outside (tidal side). These form two parallel bulkheads 6 to 9 m apart, perpendicular to the rivulet. Stringers are then toenailed in place behind and perpendicular to the pilings, one positioned at mean low water and one at mean high water (personal communication, R. K. Williams 1989). Then pressure-treated tongue-and-groove or rough pine (7.5 cm by 25 cm by 5.5 m or 5 cm by 20 cm by 5.5 m) called sheeting is driven vertically side by side behind the stringers so that two walls are formed perpendicular to the dike. Sheeting is placed carefully to prevent backwash and erosion of the backfill when the trunk is operating. To prevent buckling from the weight of the backfill, pilings are cabled together across the dike. Finally, impervious backfill is placed inside over the trunk and allowed to dry 4 to 6 months as the trunk settles before it is used.

Large gates. Gates are used in large impoundments that need larger concrete dams designed and constructed by engineers. Lift gates must be heavy, usually stainless steel or cast iron, and slide perpendicular in guide grooves of supporting piers by an overhead hoist device such as a handwheel which raises a threaded control rod and gate. Roller gate dams have sliding gates equipped with rollers in tracks, to ease the lifting action from the high water pressure endured by ordinary sliding gates.

Radial gates are large prefabricated steel gates mounted on control arms which pivot the gate up or down in an arc to discharge or hold water. Heavy rubber flaps seal the sides and bottom of the gate channel by water pressure. A winch with pull chains moves the heavy gate (Linde 1969). Sometimes the structure will contain three stainless steel radial gates about 2.4 by 2.4 m, which are raised and lowered by cables connected to a windlass and gearboxes powered by a removable motor or hand cranks (Fig. 2.31) (Wicker et al. 1983). Fully raised, such gates can pass small boats.

A series of stainless steel flapgates in a single concrete control structure can be used to manage water levels (Fig. 2.32) (Wicker et al. 1983). To prevent erosion around the structure in tidal areas, steel sheetpile wing walls extend from the structure into the dike. Articulated concrete mats (Gobi-mat) should cover the dike's berm and the outflow channel to prevent erosion. The main portion of the structure consists of a series of vertical concrete walls with a concrete walkway on top. To form stoplog bays, stainless steel U channels are embedded in the sides of 12- by 24-dm holes within the vertical concrete walls.

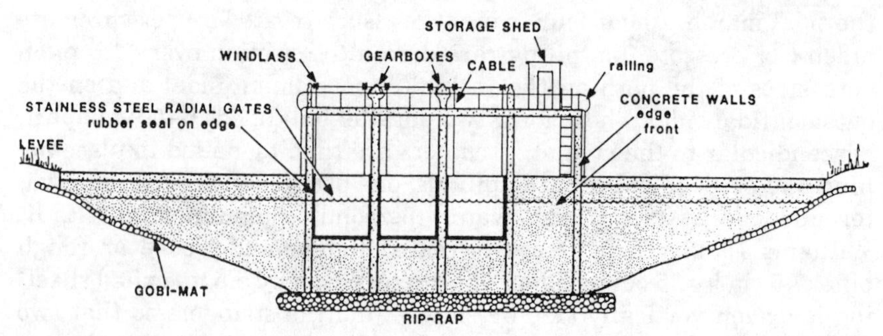

Figure 2.31 Front view of a stainless steel and concrete radial lift gate control structure (*Wicker et al. 1983*).

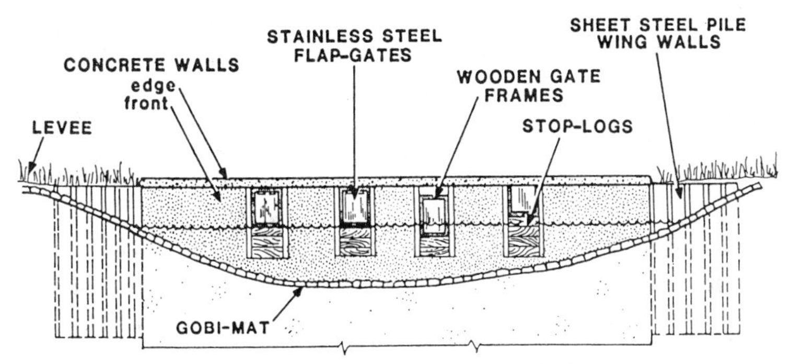

Figure 2.32 Front view of a variable crest, reversible flapgate concrete control structure (*Wicker et al. 1983*).

Equipped with two protruding handling pins, each stoplog is seated in the bays through slots in the walkway up to the desired water level in the impoundment. Hinged flapgates, seated in the bays above the stoplogs, are attached to one side of a wooden frame the same width as the U channels and stoplogs. Thus, the impoundment can be drained or flooded by adjusting the number of stoplogs, relative to rainfall and tidal conditions. With stoplogs at the desired elevation, the flapgates can discharge excess precipitation from the impoundment while preventing inflow of brackish water and estuarine organisms (Wicker et al. 1983).

Weir and earthen (gut) plug. Weirs are damlike structures placed in the drainage system of tidal marshes about 15 cm below the marsh level (Chabreck 1968, Wicker et al. 1983). Tidal water moves in and out of the drainage system over the weir (Fig. 2.33), creating an impoundment by preventing excessive drainage of the marsh during low tides

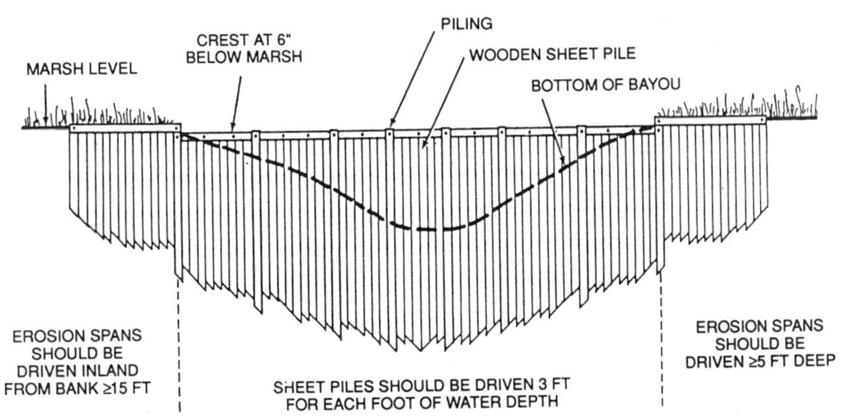

Figure 2.33 Typical Wakefield weir (*Wicker et al. 1983*).

and recharging the impoundment with saline or brackish water during high tides. Weirs are used for water control on unstable coastal marshes that will not support continuous levees (Chabreck and Hoffpauer 1962). They have less effect on salinity and turbidity than on water-level stability and production of aquatic vegetation, and they improve conditions for waterfowl and furbearers, although production of marine animals might be reduced (Chabreck and Hoffpauer 1962, Spiller and Chabreck 1975, Larrick and Chabreck 1976, Rogers and Herke 1985). A vertical slot in the weir enhances passage of aquatic organisms (Rogers et al. 1987). Some water-level manipulation can be achieved with a flashboard riser built behind the weir (Chabreck 1968).

Weirs can be used with earthen plugs or dikes that partially impound high-phase tidal wetlands (Williams 1987). The earthen plug or dike is constructed from two opposing points above mean high water. The weir is installed where the two sections of earthen plug or dike meet. Some earthen plugs have a corrugated metal pipe with a riser for stoplogs on the upstream end and a flapgate on the downstream end, and they function as a weir. The riser with stoplogs allows the impounded water level to be set as desired. The flapgate allows excess water to drain from the impoundment, but prevents tidewater from entering unless locked open. The structure functions as a weir when the flapgate is open and the stoplogs are set in the riser about 15 cm below marsh level. A drainage area of 405 ha needs a 91-cm culvert about 8.5 m long, which lasts at least 15 years. Earthen plugs with a control structure typically are used to block bayous 4.5 to 6 m wide (Chabreck 1968).

Weirs cost a lot initially, but need little maintenance if built properly. Weirs built of interlocking steel sheet piling last 25 years, wooden sheeting pressure-treated with creosote lasts 20 years, and aluminum sheeting lasts 15 years (Chabreck 1968). Metal piling is more costly than wood, but not subject to marine wood borers or fire damage. Metal piling is used usually in deep channels or for large structures. One type of wooden sheet piling used is a single row of 8- by 25-cm center-matched piling called Wakefield piling (Fig. 2.33). The other type of Wakefield piling is either 5 by 20 cm or 5 by 25 cm, driven in two or three rows, each row spliced over the seams of the previous row by nailing several rows of piling together and driving them all at once. Material is handled and driven with a dragline.

The site for the weir should have well-defined banks along a straight reach up to 18 dm deep. The length of the sheet piling is determined by multiplying the deepest depth by 3, so that two-thirds of the sheet piling will be in the ground and one-third in the water.

Weirs are designed to hold water in bayous and ponds, but not over the marsh itself. A level line should be located through the marsh to determine the mean elevation. The sill of the weir is set about 15 cm below the surrounding marsh, even lower on small streams with large watersheds. Setting the weir too high will force water around the ends, which will quickly erode the marsh soil and form a new channel. Planks and sheet piles extend at least 4.6 m into the banks on each side of the weir. Splashboards built into weirs and riprap prevent eroding and deepening of the bottom as water spills over the weir. Water from large watersheds eventually can erode the soil to the bottom of the sheet piling and thus flow under the weir (Chabreck and Hoffpauer 1968). Weirs are hard to correct if they lean downstream in areas with soft subsoil or with side tidal fluctuations, resulting in little support during low tides. Wedge-shaped weirs can be used, or diagonal bracing and tie-back piling where trouble is anticipated (Chabreck 1968).

Conveyance Channels and Ditches

Properly designed ditches to move water between impoundments can provide additional wetland habitat and edge effect (Snyder and Snyder 1984). They should be at least 3 dm deep, at least 12 dm wide at the bottom with 3:1 slopes, over 30 m apart if parallel (60 m is best), 90 degrees to the prevailing wind if possible, and zigzagged at least 20 degrees every 90 m.

Main ditches, also known as supply or circulation ditches, have been used in California, especially in tidal areas, to connect managed wetlands with a major water source and to facilitate flooding and drainage (Rollins 1981). Main ditches eliminate the fragmentation of large pond areas with unnecessary dikes which can disrupt flight patterns of waterfowl.

Main ditches should be at least 6 dm wide and at least 6 dm below the ground surface, but they must be large enough to accommodate the control gate to which they are connected. They should be large and numerous enough to flood the area in 10 days and drain in 20 days; this will allow the necessary two full leaching cycles each spring to maximize waterfowl food production.

Main ditches usually are dug with a dragline. Ditch spoil can be used to repair nearby dikes, to fill undesirable low spots in the pond, or side cast parallel to the ditch. Side-cast material should be placed at least 12 dm away from the ditch and below the anticipated water level unless an island is desired. Ideally, it should be disked evenly for establishment of marsh plants.

Spreader ditches, also called bleeder, spud, and lateral ditches, speed the drainage of isolated low spots in the pond, enhance spring leaching especially of organic pond soils distant from main ditches, and improve circulation. Spreader ditches are connected to main ditches; they should be at least 46 cm wide and 46 cm below the ground surface in order to remove salt from the soil comprising the root zone of most marsh plants. Gradual slopes fill in faster than steeper slopes, but they are less hazardous to unwary waders. Enlarged spreader ditches can serve as boat channels. Spacing is experimental, influenced by such factors as location and prevalence of low spots, level of soil salinity, soil type, and economics.

Spreaders are dug by backhoe and various types of plow blades attached to tractors or tracklayers. Some equipment can cut several thousand dm of spreader ditch per day. Side-cast spoil material should be spread so that it will be covered by inundation.

Ditches not completely drained during summer will fill in with undesired vegetation which can be removed by fall burning, reexcavation, or systemic herbicides. Systemic herbicides kill the entire plant and thus do not need annual application; dead plants should be burned before fall flooding. Ditches should be maintained on a rotation schedule (Rollins 1981) (see Chap. 4).

Pumps

With a gravity diversion system, water levels in the stream above the impoundment must be raised to gain the necessary advantage in head. If this adversely affects other interests along the floodplain, pumping must be used to flood such impoundments (Linde 1969). Pumps also can be used to drain low impoundments or supplement gravity or tidal drainage and flooding. In some cases groundwater can be pumped from wells to provide the necessary water (Wentz 1981). Low-head turbine pumps used for flood irrigation of ricefields in the south are suitable for flooding impoundments (Neely and Davison 1971). Windmills can be used in some situations (Beintema 1982).

A double divergent pumping unit can be used to manage water levels in some impoundments (Wicker et al. 1983). Concrete support walls form two separate boxes connected to the dikes or levees on each side (Fig. 2.34). A diesel-powered pump is mounted in the center of a concrete platform on top of the walls. The intake pipe is in one of the boxes, the outfall pipe in the other. To pump water out of the impoundment, stoplogs are removed in the intake box from the impoundment side and added to the exit side, and removed in the outfall box from the exit side and added to the impoundment side. Then the pump is started. Water can be pumped into the impoundment by reversing the position of the stoplogs. Pumping is not needed if enough water exists to flood the impoundment to the desired level by simply removing enough stoplogs from each bay.

The size of the power unit for the pump determines the volume of

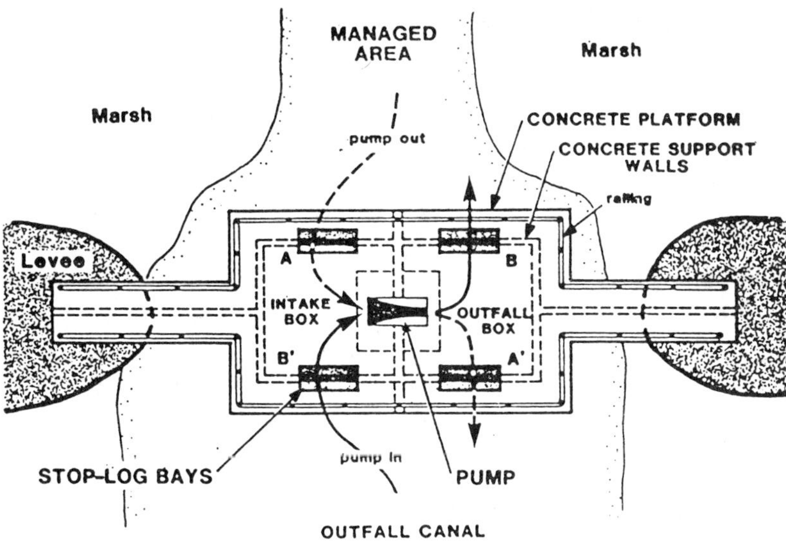

Figure 2.34 Top view of a double divergent pumping unit (*Wicker et al. 1983*).

water which can be lifted a given height or distance (Atlantic Waterfowl Council 1972). A 61-cm Crisafulli pump can lower a 40-ha pond 3 dm/day if pumping 94,625 L/min (Pierce 1970). Electricity is the most economical power source, followed by diesel, bottled gas, and gasoline, but perhaps more important is the availability of replacement parts. Electric pumps also are quietest and need less monitoring and maintenance (Reid et al. 1989). Pumps can be stationary or portable, mounted as a self-contained unit on a pickup truck, or off the power takeoff of a tractor or in a boat, operating off the boat motor (Linde 1969). Stationary pumps 15 to 91 cm in diameter and portable PTO-drive pumps 20 to 41 cm in diameter are used commonly (Bookhout et al. 1989). Portable pumps are preferred for impoundments less than 100 ha. PVC pipe 38 cm in diameter will transport pumped water while maintaining stability in the water against the pumping action.

Most pumps are of three types: propeller, mixed-flow, and centrifugal. Competent engineers should be consulted in advance. Sources of information include local soil service offices, university extension services, pump sales companies, hydraulic engineering concerns, and state, provincial, and federal soil agencies.

Propeller pump

The propeller pump is especially adapted for low-head, high-volume operations such as wetland work, where the lift usually is 30 dm or less. It can pump 75,700 to 113,550 L/min (Atlantic Waterfowl Council 1972). This pump has an impeller at the intake end of a large pipe. A shaft runs up the pipe to a power connection. A discharge pipe runs off at an angle. The pump can be designed to reverse flow, needed for quick drainage and flooding. The propeller pump has low operating costs because horsepower increases with lift.

Centrifugal pump

The centrifugal pump is used commonly where lifts are under 6 dm. It has high head and pressure, and it can force more water through smaller pipes than other pumps can, but at higher operating cost. Horsepower generally peaks at a lower lift (Atlantic Waterfowl Council 1972).

Mixed-flow pump

The mixed-flow pump is a hybrid between the propeller and centrifugal pumps, with an open-vaned, curved-blade impeller combination from both (Linde 1969, Atlantic Waterfowl Council 1972). Used mainly for irrigation where moderately high pumping heads are needed, it can handle an 8-m head.

Siphons

Impoundments can be drained or flooded from higher to lower elevations of adjacent impoundments with suitably sized siphons (Green et al. 1964). As with any siphon, the tube must be filled with water to begin the flow from the water source at the higher elevation. To reduce salinity, increase nutrients, and add sediments to help offset subsidence, fresh water can be introduced to a tidal wetland by connecting a siphon to a river and a tidal canal in the wetland (Davis et al. 1983, Chabreck et al. 1989). To form an impoundment, existing spoil banks are raised along the canal, and other connecting dikes are built. A double flapgate is installed in the dike to connect the impoundment to the canal near the siphon connecting the canal to the river. Other double flapgates are installed in the dike some distance away. All structures are connected to the tidal drainage system. The canal water is essentially fresh when the siphon is operating. When the tide is rising, the flapgate on the canal side of the control structure near the siphon is locked open, and the tide holds the fresh water in the canal and pushes it through the control structure into the impoundment. When the tide falls, the hinged flapgate on the impoundment side of the structure near the siphon prevents escape of the fresh water.

Impoundment-side flapgates on the distant control structures then are locked open. The rising tide cannot enter the distant control structures because it pushes against the down flapgates. But the fresh water entering the impoundment can flow through the impoundment and leave by escaping through the distant control structures via the raised flapgates when the tide falls.

The siphon might be operational for 4 to 6 months when water levels in the river are high enough. During this time the daily rise and fall of the tide pumps fresh water, sediments, and nutrients across the wetland. From about June through November when the siphon might be inoperable due to low water level in the river, the double flapgate structures are used (1) to lower water levels to allow germination of annual sedges and grasses as wildlife food and cover, (2) to retain and conserve rainfall to moderate salinity during late summer and fall, or (3) to raise the water level through tidal flow to allow entry of postlarval estuarine finfish and shellfish, to facilitate access for trapping, and to encourage use by migratory waterfowl.

Fishways

Recreational fishing can be a major activity on certain waterfowl management areas. A fishway for passage of desirable species of fish should be incorporated into the design of the dam unless the passage of undesirable species or public pressure for fishing becomes a major problem. Three basic types of fishway—pool-and-weir, denil, and vertical-slot—are widely used (Payne and Copes 1986), but only the pool-and-weir or denil and their modifications are used for the relatively low-head, low-velocity water characteristics of wetland impoundments. Both types can operate through periods of low water if the upper baffle is kept below the impounded water level (Atlantic Waterfowl Council 1972). Damage from floodwater is minimized by sodding or riprapping around the fishway or by installing a concrete spillway. A trash rack in the upstream end prevents entry of debris. Culverts also can pass fish if adjusted properly. Payne and Copes (1986) provide details of requirements and design of fishways, including swimming speeds and stamina of various species and sizes of fish.

Pool-and-weir fishway

The pool-and-weir fishway (Fig. 2.35) is a channel or flume constructed around the dam or beside it as part of it, with vertical partitions or baffles at intervals along its incline (Payne and Copes 1986). Baffles are progressively lowered by about 15 to 30 cm and function as

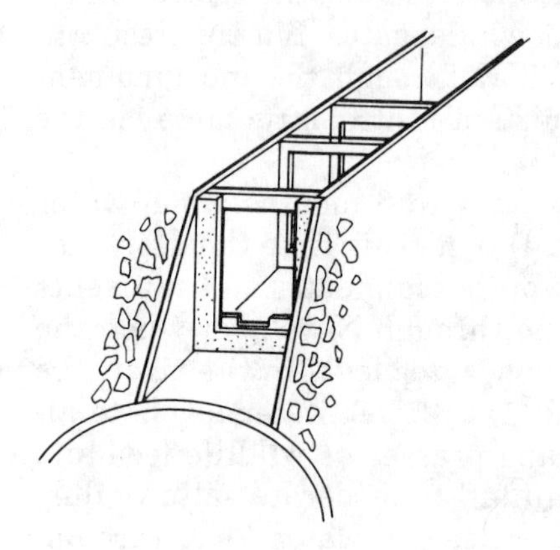

Figure 2.35 Typical pool-and-weir fishway (*Atlantic Waterfowl Council 1972*).

a series of weirs forming stepwise pools as water flows over each baffle. The pools must be large enough to dissipate the energy of the flow. Fish swim up over the baffles from pool to pool. Fluctuating water levels need adjustment at the upstream end to maintain optimum flow down the fishway (Payne and Copes 1986).

The pool-and-orifice ladder is a modification of the pool-and-weir ladder for substantially fluctuating flows. Water flowing too fast prevents fish from ascending the ladder. The baffle has a notch or orifice of about 0.37 m^2 cut out at the bottom, permitting water and fish to pass. Fish swim through the baffles more easily than over them.

Too large an orifice will not reduce the water velocity effectively. In some designs, water also passes over the baffles. In other designs, the pool and orifice are constructed at the top of the ladder, with the pool and weir at the bottom. But then the top pools with orifices must be about twice as high as the maximum depth of flow entering the ladder. Orifice ladders need less adjustment of flashboards than weir ladders do, but the orifices plug with debris during high flows and are hard to clean.

Denil fishway

The denil fishway (Fig. 2.36) has regularly spaced lateral baffles projecting at an angle from each side and usually the bottom of a straight chute that slows the water and allows fish to pass up the middle (Payne and Copes 1986). Each denil unit should be less than 2.4 m high. Otherwise a resting pool is needed between units, usually unnecessary with waterfowl impoundments. Denils pass debris well and

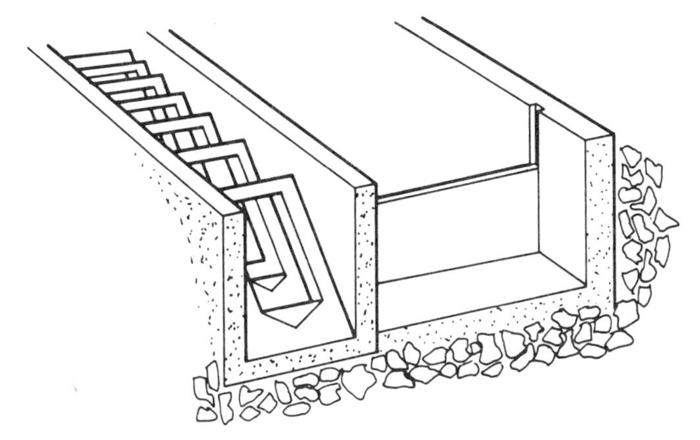

Figure 2.36 Typical denil fishway (*Atlantic Waterfowl Council 1972*).

need minimum maintenance and adjustment. They usually are built over low barriers, and they are ideal for streamflows up to 28 m^3/s.

The Alaskan steep-pass fishway, an improved modification of the denil fishway, is cheaper, prefabricated, portable, smaller, and easily installed. Both versions operate in varying water levels, which allows fish to swim at various depths. Designed to pass fish over low-head barriers, sections of the steep pass can be connected to pass fish over relatively steep natural and artificial obstructions. They last over 25 years as permanent structures or can be installed temporarily to assess the potential of a site for a larger, more expensive fishway.

Culvert

For culverts less than 24 m long, water velocity should average under 1.2 m/s for warmwater fish and less than 23 cm deep (Payne and Copes 1986). Angled baffles can be installed in culverts to reduce water speed, but baffles collect debris. Culvert overfall of 61- to 91-cm vertical drop is recommended as a barrier for unwanted fish (Dillon et al. 1971). For desired fish, culvert overfall can be corrected with at least 1 gabian, rocks, logs, or concrete weirs, cribs, or low head dams of 3-dm rise (Fig. 2.37) (British Columbia Ministry of Environment 1980, Payne and Copes 1986). Only the arch culvert effectively maintains natural water flow (Nelson et al. 1978, Proctor et al. 1983a, b, c). Culverts should be set below the streambed level to allow fish to pass.

Erosion Control

Erosion of dikes and shorelines can occur from wind, rain, waves, tides, and muskrat and nutria burrowing. Wave action can prevent

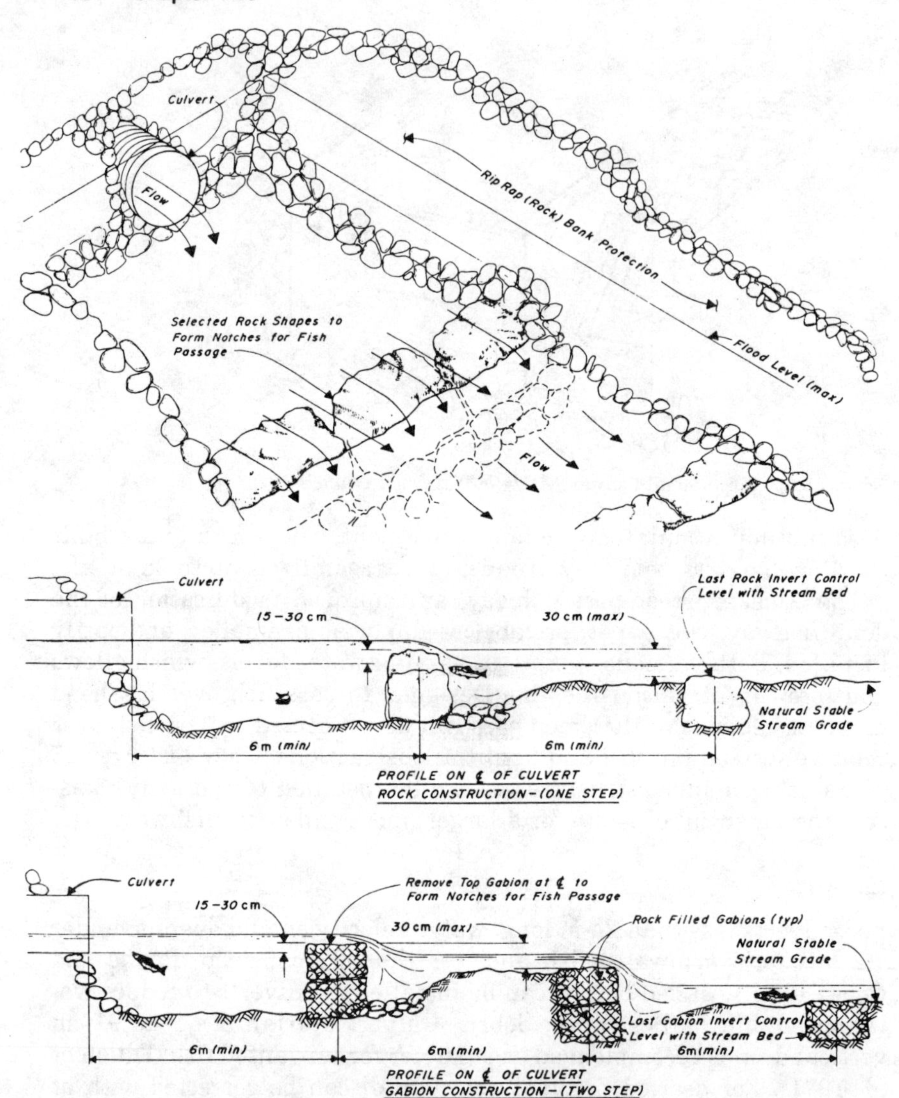

Back flooding a culvert to eliminate the downstream drop to improve fish passage. The banks and stream bed must be stabilized with large rock (rip-rap) or rock filled gabions.

Figure 2.37 Fishway from backflooding culverts (*British Columbia Ministry of Environment 1980*).

growth of aquatic vegetation either through turbidity from soft pond bottoms or from physically breaking the vegetation. Methods for preventing erosion depend on whether the water level in the impoundment is always kept at normal pool or fluctuates with drawdown (U.S. Soil Conservation Service 1982). The U.S. Army Engineer Waterways

Experiment Station (1978) described protective and retention structures for coastal wetlands (see Table 2.13 and Fig. 2.39).

Vegetation

The construction of a dike is incomplete until it is well vegetated to prevent erosion. Seeding is the usual method. Sod can be used for immediate protection of steeper slopes, but it is expensive. A dense strip of emergent plants along the shoreline will reduce wave action and erosion. Among the best are the hardstemmed bulrushes and common threesquare bulrush (Addy and MacNamara 1948). Once they establish, reduced wave action might allow the desirable pondweeds to grow. In the southern half of the United States, water primrose can be used to reduce wave action and turbidity in small ponds. Dike soils arc stabilized with vegetation having a good root structure. Reed canarygrass often is used for this purpose, but it develops into dense nesting cover that attracts ducks and exposes them to high mortality from predators that use the narrow dikes as travel lanes. To eliminate nesting cover, dikes are mowed or burned in late fall or early spring. In coastal fresh marshes, dikes can be seeded with bahiagrass and/or bermudagrass (Baldwin 1968). The dike's toe and berm should retain plants like the original maidencane and giant cutgrass. In coastal areas of regularly flooded salt marshes, new dikes can be vegetated rapidly by sprigging with saltmeadow cordgrass, with black needlerush planted on the dike's toe and berm (Baldwin 1968). Black needlerush is unattractive to feeding waterfowl and muskrats. Standard procedures are used for site preparation, fertilization, and seeding. (See "Plant Propagation" in Chap. 5.)

A good cover of sod-forming grasses and forbs will protect from erosion the exposed surfaces of the dike, spillway, borrow pit, and other disturbed surfaces. Muck requires about 6 months to settle and dry before smoothing and seeding are attempted. Recovery of vegetation is faster on dikes started by bucket clumps of plants with muck or peat between. Dikes made entirely of mineral soil should be seeded immediately (Atlantic Waterfowl Council 1972). As soon as possible, slopes of all dikes are sodded, planted with a legume-perennial grass mixture adapted to local soil and climatic conditions, or mulched for later planting (Addy and MacNamara 1948, U.S. Soil Conservation Service 1982).

The dike might need 5 to 8 cm of topsoil (Farmes 1985), but the soil used for most dikes is suitable for seeding with the proper amount of fertilizer and lime, which depends on soil quality and type of vegetation to be planted (Atlantic Waterfowl Council 1972). Subsoil, if derived from calcareous material, and topsoil might possess enough lime to establish a good sod. But if the pH is low, lime must be added. Soil samples can be analyzed by the local soil office for advice on liming and fertilization (Atlantic Waterfowl Council 1959). It usually pays to apply 112 to 224 kg/ha of fertilizer in a formula recommended for the local soils (Farmes 1985).

As soon after construction as practical, the area is limed and fertilized and the seedbed prepared by disking or harrowing. The surface area to be seeded is calculated, the recommended seeding rate is obtained from the local soil office, and the total amount of seed required is determined. Seed usually is applied with a cyclone seeder for economy and thorough, uniform seeding (Atlantic Waterfowl Council 1959). To ensure prompt germination and growth of seeds, soils must be irrigated if they are too dry after construction is completed (U.S. Soil Conservation Service 1982).

If the slope is too steep for equipment, the seedbed is prepared manually. On such areas, or where severe wind erosion occurs or the dike's slope is too rough, cuttings or shoots of plants are planted by hand usually, or by machine to control beach erosion (Atlantic Waterfowl Council 1972).

A good root structure helps prevent soil erosion, but use of deep-rooted grasses and legumes such as alfalfa should be avoided on the dike because deep penetration of the roots might cause a porous condition in the dike, especially after the roots die. Tall growing species are not always desirable. They shade the surface and prevent thick sod from developing, and such species provide good residual overwinter cover, which attracts nesting hens in waterfowl production areas, and serve as travel lanes for predators, resulting in excessive nest destruction. Such plants need frequent mowing to develop a good sod cover. Short fescues form a compact sod, gain dominance over taller species, and seem to maintain dominance for a long time (Atlantic Waterfowl Council 1959). Panic grass also provides a good cover crop, but it does not burn easily (Linde 1969). Generally, tall, heavy-bodied plants are planted along the toe of the dike and the berm to break wave action; grasses should be planted higher on the dike's slopes (Atlantic Waterfowl Council 1972).

Water-level manipulation and salinity determine which plants to use on and along the dike. In brackish to salt areas, plants such as smooth cordgrass, black needlerush, and common threesquare could be used on the toe of the dike and on the berm. On the higher slopes, saltmeadow cordgrass, bermudagrass, and saltgrass could be used. In fresh to brackish tidal areas, plants such as big cordgrass and common threesquare can be planted along the toe of the dike and on the berm, with bermudagrass, bahiagrass, fescue, and other agricultural erosion-retarding grasses on the upper slopes (Atlantic Waterfowl Council 1972). Bermudagrass and bahiagrass tolerate wide variation in soil conditions (Baldwin 1968).

To stabilize embankments, especially in areas subject to natural flooding, perennial warm-season grasses establish fast (Mitchell and Newling 1986). In wintering areas, grasses such as dallis grass or bermudagrass overseeded with legumes will provide winter cover and supplemental food for wildlife. For example, deer heavily use white clover and cool-season grasses such as ryegrass; turkeys use wheat during nesting and brood rearing. But planting of dikes and levees should be planned carefully. Poaching can be a problem on some dikes and levees planted to attract wildlife.

Mulching

To stabilize fill slopes and other disturbed areas temporarily directly after construction, various materials such as wood chips, bark, and straw can be used (Hynson et al. 1982). In most cases to establish stable slopes, mulching is used in combination with seeding and planting. Mulches are spread by hand or machine. Application rates vary with erosion potential and available materials and should be determined through a local soil or extension office. (See "Fertilizing and Mulching" in Chap. 5.)

Mulching is recommended on new dikes to protect the newly prepared seedbed, seeds, or small plants from rainfall damage and to conserve moisture and provide favorable conditions for germination and growth. A thin layer of old hay, straw, fodder, or a commercially manufactured material will suffice. Mulching is less important in early summer because a nurse crop included in the seed mixture normally will accomplish the same result more economically. If hay to be used as mulch contains undesirable seed, it should be cut late in the season so that most seed will have dispersed. All pads formed within bales of any mulch should be loosened completely for spreading so that some soil is visible after application. Then the mulch is anchored against strong winds. Scattered heavy brush might suffice as anchorage. Most of the mulch will decompose before mowing is needed (Atlantic Waterfowl Council 1959, U.S. Soil Conservation Service 1982). Sandy dikes might need mulch of 3-year-old hay applied 3 dm deep, crisscrossed with binder twine tied to stakes (Linde 1969). On most dikes, mulch produces good results when 5 to 8 cm deep so that 20 to 30 percent of the soil shows through the spaces (personal communication, T. Meier, Wisconsin Department of Natural Resources 1991). Mulching procedures can be obtained from the local soil office.

Berms

With fairly constant water levels, a berm 2.4 to 3 m wide, located at normal pool level and sloped 15 to 30 cm down toward the pond, provides additional protection from waves (see Fig. 2.10). Vegetation should cover the slope above the berm (U.S. Soil Conservation Service 1982).

Booms

Single, double, or triple log booms anchored in a line floating offshore break the force of waves and allow plants to grow along the shoreline and in the shallow-water zone near shore (Addy and MacNamara 1948). The boom consists of units of two or three logs attached end to end by chain or cable, with each log about 6 dm apart or closer if practical. Enough slack between logs allows the boom to adjust to fluctuating water levels. The boom is held about 1.8 m offshore by logs anchored horizontally between boom and shore.

Riprap

If the water level fluctuates widely and much protection is needed, rock riprap is used (U.S. Forest Service 1974b, Maynord 1978, U.S. Army Engineer Waterways Experiment Station 1978, Ohlsson et al. 1982, U.S. Soil Conservation Service 1982). It is generally readily available and relatively inexpensive. If colored similarly to stones in the immediate area, riprap has a natural appearance. Riprap is dumped directly from trucks or placed by hand. Dumping requires less labor but more stone. A bed of gravel or crushed stone 25 cm thick is placed before the layer of larger stone 30 cm thick is dumped over the gravel. Grass grows through the riprap to blend with surrounding vegetation.

The entrance and exit of the tube for the control structure are riprapped to prevent sloughing and erosion, which eventually might cause a washout. Stone usually is used for riprap, but if it is difficult to obtain, cement or burlap bags can be used (Linde 1969). Bags are filled usually with 4 parts gravel, 3 parts sand, and 1 part cement; mixed dry; and then laid in place, overlapped as with shingles. Each layer is soaked with water so that the mixture sets. Such riprapping is more durable than stone.

The area beneath the pipe's exit should be excavated 3 dm deep and riprapped 46 cm thick. The pipe should extend 12 dm beyond the toe of the dike. The riprap (spreader) should begin about 6 dm beyond the toe of the dike. At that point it should be 12 dm wide, extending 30 dm, at which point it should be 18 dm wide (Atlantic Waterfowl Council 1959).

Ultraviolet-resistant plastic material called GEOWEP (Presto Products Co., Appleton, WI) can be used on tundra, peat, dikes, and emergency spillways to support vehicles and reduce erosion (personal communication, T. Meier, Wisconsin Department of Natural Resources 1991). It contains honeycomb cells 6.7 to 20.3 cm deep and is assembled by attaching strips that are 20.3 m by 6.1 m and filling them with crushed granite or gravel. (See "Design" of "Emergency Spillway" in this chapter.)

If no other material is readily available, a willow mattress can substitute. This consists of 2.5- to 5-cm-diameter willow saplings wired together in bundles 30 to 46 cm thick, up to 6.1 m long, placed on the area to be protected.

Gabions

Gabions are rectangular rock-filled partitioned baskets of galvanized or plastic-coated wire mesh, available in a variety of sizes (Table 2.12) (U.S. Forest Service 1974b, Anderson and Cameron 1980, Ohlsson et al. 1982, Ambrose et al. 1983). Rounded rocks 15 to 30 cm in diameter are contained by 11-gauge wire mesh with 10-cm openings. The cage is anchored firmly to the bottom along the shoreline, rocks are added by hand, and the lid is wired shut (Hynson et al. 1982). Several gabions can be wired together to improve stability. It might be desirable or necessary to install two rows of gabions lashed together, with a third row on top. Gabions can provide habitat for invertebrates. Gabions are labor-intensive and therefore expensive to construct. They are used only when erosion cannot be corrected by more natural means.

TABLE 2.12 Standard Sizes of Gabions

Dimensions, m	No. of partitions	Capacity, m^3
2 × 1 × 1	1	2.0
3 × 1 × 1	2	3.0
4 × 1 × 1	3	4.0
2 × 1 × 0.5	1	1.0
3 × 1 × 0.5	2	1.5
4 × 1 × 0.5	3	2.0
2 × 1 × 0.3	1	0.6
3 × 1 × 0.3	2	0.9
4 × 1 × 0.3	3	1.2

SOURCE: Ambrose et al. (1983).

Beach prisms

Modular, precast, preassembled, detached breakwaters called *beach prisms* can stabilize sand spits in flowing water (Ailstock 1987). Suspended sediment settles behind the prism first, then in front, but plants such as cordgrass must be planted to reduce loss of soil during storms.

Islands

Properly placed islands, bulldozed in the impoundment before flooding or dredged with a dragline after flooding, can reduce wave action on the dike and serve as nesting areas. Islands are constructed to serve as relatively predator-free nesting sites for waterfowl and other water birds. The size of the islands depends on the size of the impoundment; the bigger the impoundment, the bigger the island. For erosion control of the dike or other shorelines of the mainland, the island should be relatively close. Yet islands farthest from the mainland are the most predator-free, at least from mammalian predators. Thus, a compromise is needed.

For erosion control of the mainland, islands should be long and narrow, closer to the windward than the leeward side of the mainland and perpendicular to the prevailing wind. For maximum use by waterfowl, islands should be closer to the leeward than the windward side of the mainland and parallel to the prevailing wind (Giroux 1981). A compromise would be to build a series of islands in clusters or parallel to the prevailing wind close to the windward shore to serve as wave (and wind) breakers. Islands should be heavily vegetated or riprapped on windward sides, with flat, open shorelines for loafing and access on the leeward sides (Giroux 1981, Duebbert 1982). (See "Islands" in Chap. 6.)

Ice Control

Some dredged or diked areas can be kept ice-free by pumping into them from shallow wells. Aerating ponds with compressed-air systems will provide ice-free areas, but the expense rarely is justified, unless electrical service is readily available (Rutherford and Snyder 1983).

DREDGE FILL MARSHES

Design Features

Instead of placing water on the land or removing soil to expose water, marshes can be created by reducing the amount of water through filling deepwater areas (U.S. Army Engineer Waterways Experiment Station 1978). Such opportunities generally occur along large rivers and coastal areas including the Great Lakes, where the dredging of shallow areas requires deposition of the dredge spoil. Dredge material can be used to create or improve marsh habitat by changing a deepwater habitat into a shallow-water wetland. (See "Dredging" in Chap. 4.)

In areas of high hydraulic energy, structural protection and containment in the form of a retaining wall or dike might have to be established in the water as an extension of the shoreline or around a proposed island site. As the area enclosed by the retaining wall is filled with dredge material, the water escapes through a weir built into the wall (Fig. 2.38). The area is then vegetated by natural invasion or artificial propagation (Smith 1978, U.S. Army Engineer Waterways Experiment Station 1978).

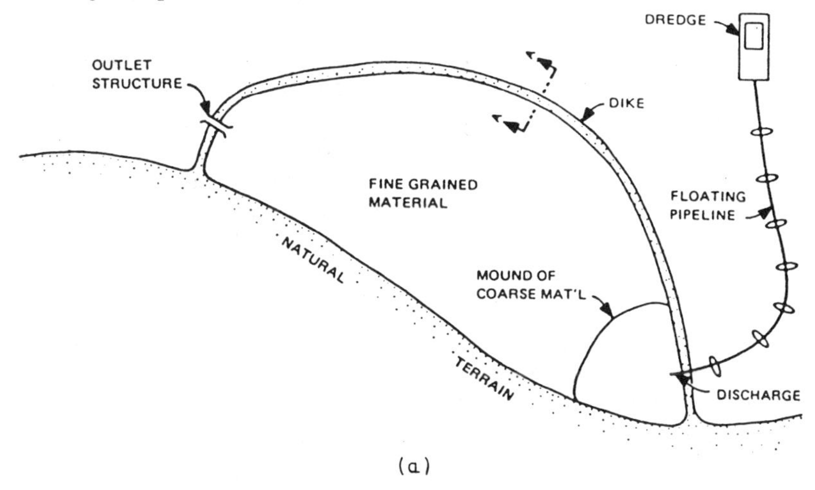

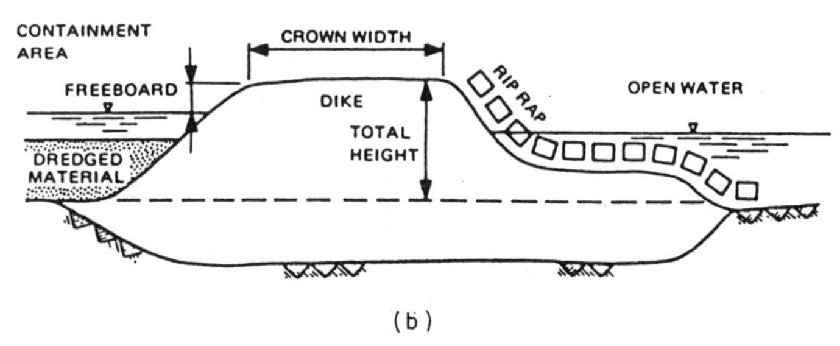

Figure 2.38 Dredged material containment area (*Allen and Hardy 1980*).

Site Selection

Creation of wetland habitat with dredge material is feasible if engineering, biological, and social aspects are suitable. Because of the nature of dredge material disposal and the formation of drainage patterns, most marsh development projects will contain elements of shallow and deep marsh in freshwater areas or high and low marsh in saltwater areas. The design of the marsh habitat is influenced by the location, elevation, orientation, shape, and size (U.S. Army Engineer Waterways Experiment Station 1978).

Location

Sites are best located in areas of low energy, e.g., in the lee of beaches, islands, and shoals; in shallow water where wave energy is dissipated; on the convex side of river bends; in embayments where marshes presently exist; and away from tidal channels, inlets, headlands, and exposure to fetches over 4 or 5 km if the new marsh is to be planted with sprigs or plugs and over 1 km if it is to be planted with seed (U.S. Army Engineer Waterways Experiment Station 1978, Woodhouse 1979). But such sites tend to be biologically productive. Productive areas such as seagrass meadows, clam flats, and oyster beds are avoided.

The shorter the transport distance, the more cost-efficient is the project. Sandy material is the ideal dredge material for the substrate of the new marsh. Sand might need some containment protection under moderate wave energy. Hydraulically placed clay usually needs containment under any situation. Silt might need no containment under very low energy situations.

With a topographic map, aerial photograph, or a diagram, the site's location is recorded relative to the dredging project and surrounding aquatic areas, wetlands, uplands, topographic features, cultural points (housing, industry, recreation, agriculture, water treatment, etc.), and land and water access routes. Aerial reconnaissance is helpful. Details of the site are recorded, e.g., dimensions, shape, topography, elevation, bathymetry, dikes, debris, human activity.

Plant and animal species existing at the site are documented, especially those of nearby wetlands which might invade the new marsh. The regional endangered species coordinator of the U.S. Fish and Wildlife Service or Canadian Wildlife Service should be contacted for updated information on federally protected species and their habitats. Wildlife departments of the states and provinces also should be contacted for locally protected species. The potential presence of wildlife species such as Canada geese or deer should be determined because they might feed on newly planted marsh plants.

Species composition, abundance, and distribution of invertebrates on the water bottom should be described, for they might be covered by dredge fill (Hirsch et al. 1978). Pollution potential and other ecological considerations should be examined (Environmental Effects Laboratory 1976, Barnard and Hand 1978, Brannon 1978, Lunz et al. 1978, Wright 1978). The dredge fill should be analyzed to determine suitability for plant growth (Palermo et al. 1978). Analysis includes particle size, available nutrients, pH, salinity, organic matter, sulfates, and contaminants (heavy metals, pesticides, oil and grease, etc.). Analysis for contaminants requires that samples be taken in an oxygen-free atmosphere and frozen to prevent chemical changes. The substrate of the site must be analyzed to determine its compaction potential for supporting the dredge fill.

Characteristics of the water energy and the water that will be flooding the marsh must be described. Such characteristics include wind and ship waves, flooding, tides, currents, salinity, freshwater flows, drainage patterns, and altering flow conditions (in rivers). The U.S. Army Coastal Engineering Research Center (1977) described techniques for observing and measuring the water energy regime.

Federal, state, and provincial laws relating to fish and wildlife must be examined (App. A). Even local laws might include lists of protected plants and animals, restrictions on collecting or propagating certain species, and zoning regulations. State and provincial laws on ownership of intertidal lands vary.

Elevation

Saltwater marshes are most productive within the upper third of the tidal range, and freshwater marshes when flooded 0.1 to 1.0 m deep (Smith 1978). Determining the final elevation of the marsh is based on knowledge of the elevational requirements of the plant community desired. Variation in topography produces desirable habitat diversity. After the site is filled, the dredged material settles, due to its self-weight consolidation and/or consolidation of foundation soils. Largely determined by settlement and consolidation (Palermo et al. 1978), final elevation of the marsh substrate is critical, as it dictates the amount of material needed and the biological productivity of the habitat established. Enough dredge fill is needed so that after settlement the elevation is about 6 dm below mean high water.

Orientation and shape

The marsh is placed and shaped to minimize high-energy exposure and adverse effects on drainage or flow patterns and to blend into the surrounding environment. Coves, islands, and breakwaters can serve as low-cost protection, minimizing the length of costly containing or protective structures. For maximum cost efficiency, protective structures are located in shallow water, and the fill area is located in deep water. The shape is determined by the budget. The smaller the budget, the more nearly the shape should resemble a circle, which requires the minimum length of containment structure. But irregular shapes are best for wildlife.

Size

The size of the new marsh depends on the amount of dredge material available. Filling options that affect size are one-time, incremental, and cellular. With one-time filling, the marsh site is filled from one discrete operation, and the site is not used again for disposal. With incremental filling, the site will be used during more than one season or dredging operation, and the site is considered full when the desired marsh elevation is attained. With cellular filling, a compartment or cell of the marsh site is filled to the desired elevation during each disposal project. Cellular filling creates a marsh at the end of each season, compared to the years required to create marsh with incremental filling.

Construction

Retention and protective structures

Sediment properties and site hydraulics determine if retention and protective structures are needed at marsh development sites. A structure might be needed to reduce erosion caused by currents, waves, or tides; to retain the dredge material until it settles and consolidates; and to control migration of suspended fine material. Mostly fine-grain dredge material generally will need a retention structure to reduce turbidity and to prevent excessive loss of fines during filling. The plan for the proposed structure might need alteration if the structure constricts water flow and increases local current velocities or reflects wave action, all of which might increase erosion (Eckert et al. 1978).

Selection of the containment structure includes consideration for the type of dredged material to be retained, maximum height of the dredged material above firm bottom, required degree of protection from waves and currents, permanence of the structure, foundation conditions at the site, and availability of structure material (U.S. Army Engineer Waterways Experiment Station 1978). The feasibility of a structure relative to the project goal, useful life, and total cost requires engineering site data. Although several types of retention and protective structures are feasible for use in marsh habitat development (Fig. 2.39, Table 2.13), two types are most likely to be used: sand dikes and fabric bags.

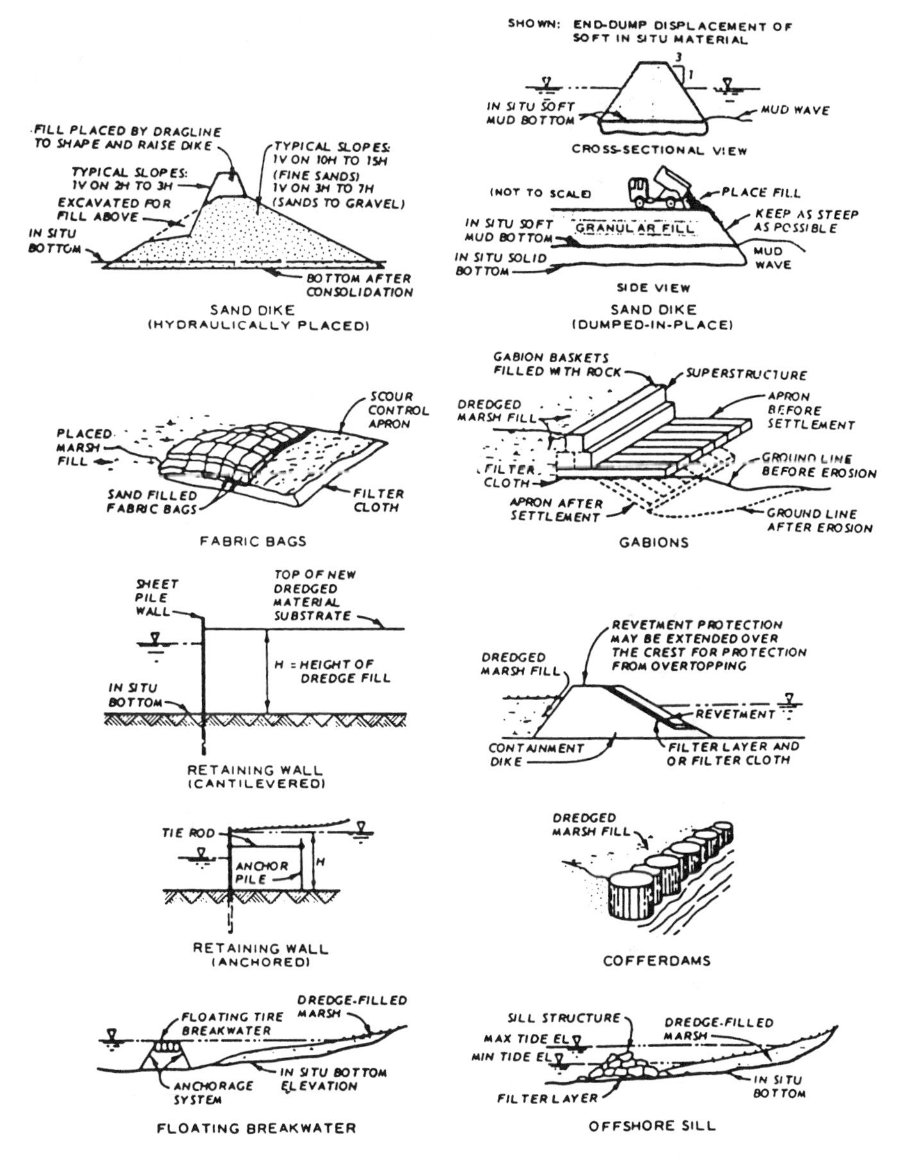

Figure 2.39 Retention and protective structures (*Eckert et al. 1978, U.S. Army Engineer Waterways Experiment Station 1978*).

Sand dikes are the most common type of containment structure. They are overbuilt to compensate for consolidation of foundation soils and settlement. Elevation is estimated from total expected settlement. Hydraulically placed dikes have flat slopes which are suitable for weak foundations, but which can be reworked by a dragline for a

TABLE 2.13 Protection and Retention Structures and Their Use

Structure	Function	Maximum feasible height, m	Special foundation requirements	Erosion resistance	Duration	Relative cost	Remarks
Sand dike (hydraulically placed)	Protection and retention	Foundation-dependent	None	Depends on material used	Long	Low	Built from coarsest material available
Sand dike (end-dumped)	Protection and retention	Foundation-dependent	None	Depends on material used	Long	Low	Land borrow might be available
Retaining wall (cantilevered)	Protection and retention	4.5	Firm bottom	Good	Long	Moderate to low	Wall usually built with sheet pile; reclamation of piling recommended
Retaining wall (anchored)	Protection and retention	12.0	Select backfill	Good	Long	Moderate to high	Construction usually performed by floating plant; adequate operating depth required
Coffer dam	Protection and retention	6.0	None	Good	Long	High	Limited applicability in habitat
Gabion	Protection and retention	3.0	None	Susceptible to scour	Intermediate	Moderate	Requires availability of small rock

TABLE 2.13 Protection and Retention Structures and Their Use (*Continued*)

Structure	Function	Maximum feasible height, m	Special foundation requirements	Erosion resistance	Duration	Relative cost	Remarks
Fabric bags	Protection and retention	Varies	None	Good	Long if concrete-filled, short if sand-filled	Low	Susceptible to vandalism; degrades in 2 to 3 years
Revetment	Protection	—	None	Good	Long	Low to high	Used in conjunction with dikes
Offshore sill	Protection	—	None	Moderate	Long	Low to medium	Causes waves to break before reaching substrate
Floating breakwater	Protection	—	None	—	Intermediate	Low	Reduces wave heights
Groin	Protection	—	None	Good	Long	Low to high	Causes waves to break before reaching substrate

SOURCE: U.S. Army Engineer Waterways Experiment Station (1978).

higher dike with steeper slopes. Dikes built with end-dumped construction have steeper slopes. Clean, free-draining, cohesionless sand is desirable for construction because dikes built in water cannot be compacted readily. The dike's exposed surface is revetted or riprapped if waves and current will erode it. Filter gravel or a filter cloth behind the revetment reduces erosion further. But the integrity of the revetment or riprap depends on the stability of the dike.

Where only short-term protection is needed, fabric bags can be used. These are large sandbags filled with saturated sand. Fabric bags last 2 to 3 years due to deterioration of the fabric from ultraviolet rays of sunlight, energy forces exerted against the dike, and susceptibility of the bags to vandalism if accessible to humans. Dikes of fabric bags should have filter cloth beneath, to prevent scouring and reduce loss of dredged material through the openings between bags.

To determine the height of the retention structure, the first step is to establish the desired elevation of the proposed marsh (Fig. 2.40*a*). Then the maximum fill level is determined (Fig. 2.40*b*), based on the volume of the area to be dredged relative to the volume of the area to be filled, the substrate configuration, and anticipated settlements. Then the maximum ponding (water) level in the containment (filled) area is established to retain suspended solids (Fig. 2.40*c*). The minimum water depth is 6 dm. Finally the maximum height of the dike (Fig. 2.40*c*) is established relative to settlement of the dike and desired freeboard. Freeboard is determined from the susceptibility of the site to storm erosion and/or overtopping by waves or high tides (Eckert et al. 1978, U.S. Army Engineer Waterways Experiment Station 1978). Freeboard for dikes is computed for open-water reaches less than 16 km (10 mi) and wind less than 97 km/h (60 mi/h) as 3 dm (1 ft) plus

$$\text{Wave height} = 0.75\left[1.5\sqrt{f} + \left(2.5 - \sqrt[4]{f}\right)\right]$$

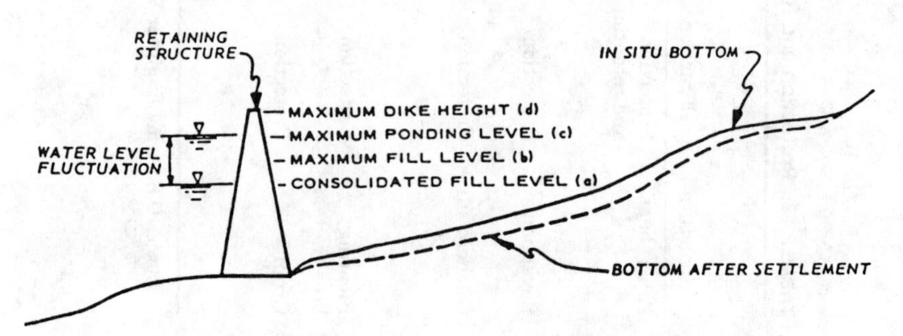

Figure 2.40 Retaining structure for a dredge fill marsh, with elevations defined (*Eckert et al. 1978, U.S. Army Engineer Waterways Experiment Station 1978*).

where f = fetch in nautical miles (6076 ft) and wave height is measured trough to crest (Foreman 1979).

Freeboard is determined by wave characteristics such as height, period, direction, and probability of occurrence. The erosive effects of wave run-up and overtopping of a retention structure depend on the slope of the seaward side, the bottom slope and water depth in front, the surface roughness of the structure, and wave characteristics (U.S. Army Engineer Waterways Experiment Station 1978). Ship- and wind-generated waves can erode marsh edges.

Equal hydrostatic pressure acts on both sides of the retention structure during its construction (Eckert et al. 1978). But placement of the dredged material exerts an unbalanced force against the inside of the structure. This force is maximized after completion of filling during low tide or low river stage. Pressure also varies with the rate, duration, and stoppages of filling; location and direction of pipeline discharge; and variations in grain size of the fill.

The retention structure is located in a low-energy environment, with flat outer slopes, streamlined upstream face, vegetation, and protection of inner and outer surfaces with filter cloth, revetment, or antiscour blankets of rubble if needed. Breakwaters or floating wave-attenuating devices also help, but are costly (Eckert et al. 1978). The U.S. Army Coastal Engineering Research Center (1977) presented information on predicting wave characteristics, wave-structure interaction, and tidal and littoral currents. Johnson and McGuinness (1975) and Hammer and Blackburn (1977) discussed site location, shape, and erosion control.

Most failures of retention structures result from overstress on foundation materials (U.S. Army Engineer Waterways Experiment Station 1978). Clays and silts typically found in marsh habitats have little shear strength. Slope stability, bearing capacity of the foundation soil, stress distribution caused by the retention structure, and the expected settlement of the structure must be evaluated, and allowances made in design if needed (Office, Chief of Engineers 1953, 1958, Hammer and Blackburn 1977).

The head difference (water level) between inside and outside the retention structure rarely will be more than 3.3 m (U.S. Army Engineer Waterways Experiment Station 1978). Thus, seepage will be minimal, with no control measures needed. Should they be needed, methods include placing an impermeable plastic membrane on the inner dike surface or a filter cloth on the outer surface under a protective layer of riprap or other revetment (Fig. 2.41). The quantity and velocity of seepage also can be reduced by increasing the length of flow lines by widening the dike crest, flattening outer dike slopes, or other means (Office, Chief of Engineers 1952, Eckert et al. 1978).

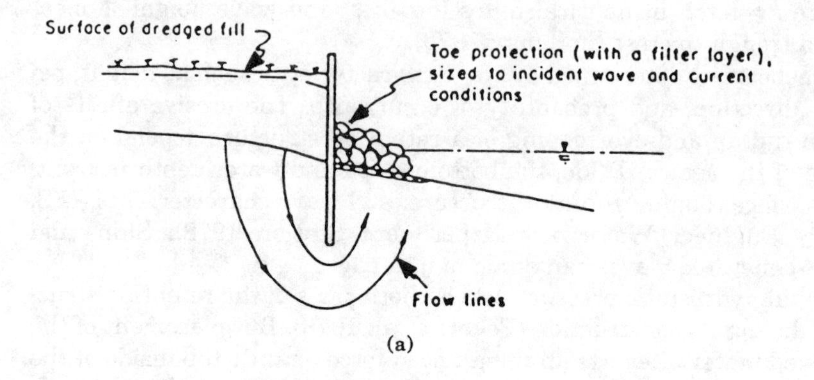

(a)

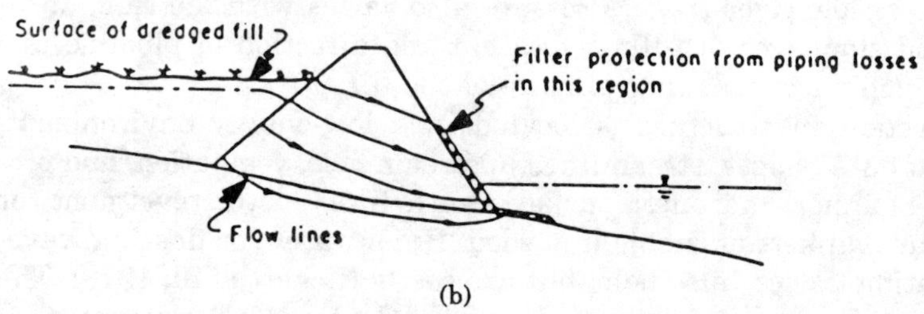

(b)

Figure 2.41 Paths of seepage flowing under or through a retaining structure for a dredge fill marsh (*Eckert et al. 1978, U.S. Army Engineer Waterways Experiment Station 1978*).

In-water retention dikes are built with hydraulic pumping, end-dumping, or draglines or clamshells (Murphy and Zeigler 1974, U.S. Army Engineer Waterways Experiment Station 1978) (Table 2.14). Large volumes are most economical with hydraulic pumping. The dredge fill is hydraulically pumped into the water along the site of the dike until the fill builds above the height of the water. The exposed portion of the dike, which is wide and flat from hydraulic placement, is then shaped with a dragline or other equipment.

With end-dumping, embankment construction begins next to land, with suitable borrow material transported by truck, progressing outward as a haul road is established. Material dumped from trucks is pushed into the water and shaped by bulldozer.

Dikes can be constructed with a dragline or clamshell from material taken within the containment area. In time the dike settles and might need raising (Fig. 2.42) (Bartos 1977). Eckert et al. (1978) recommended construction techniques for retaining walls, sills, breakwaters, gabions, and other site-specific structures (see Table 2.13 and Fig. 2.39). Throughout construction operations, thorough inspection

TABLE 2.14 Use of Various Construction Equipment for Habitat Development

Operation	Equipment used		
	On land	In shallow water	Off shore
Clearing foundation	Bulldozer Dragline	Dragline on timber mats	Floating dragline
Obtaining material	Bulldozer Dragline Truck transport from borrow area	Clamshell Dragline on pontoons Dragline on timber mats Hydraulic dredge and pipeline Truck transport from borrow area	Barged dragline Clamshell Hydraulic dredge and pipeline Barged transport from borrow area
Placing material	Dragline Bulldozer Hydraulic fill* End-dumping from trucks	Dragline on pontoons End-dumping from trucks Hydraulic fill	Bottom-dump scows Barge with conveyor Hydraulic fill* Barged dragline
Shaping and compacting†	Bulldozer Scrapers Haul traffic	Bulldozer Haul traffic Dragline	Bulldozer Dragline
Placing riprap	—	Clamshell	Barged clamshell

*Various hydraulic fill procedures have been used, including whooping crane (on land), bleeder pipe (on land, shallow water), spill barge (on water), and floating swing discharge line.

†Compaction normally is conducted on 3-dm added layers of fill on emergent portions of the dike.

SOURCE: U.S. Army Engineer Waterways Experiment Station (1978).

ensures that work is accomplished in compliance with plans and specifications and will reveal unanticipated details.

Weirs

Weirs are built into the containment structure to release water during and after the filling operation (Palermo et al. 1978, Walski and Schroeder 1978). The drop inlet and the box are two basic types of weirs, with the drop-inlet weir used more (U.S. Army Engineer Waterways Experiment Station 1978).

Similar in design to those used in controlled marsh management, the drop inlet consists of a half-cylinder corrugated metal pipe riser equipped with a gate of several stoplogs (flashboards) that can be added or removed as needed to control flow into and out of the disposal

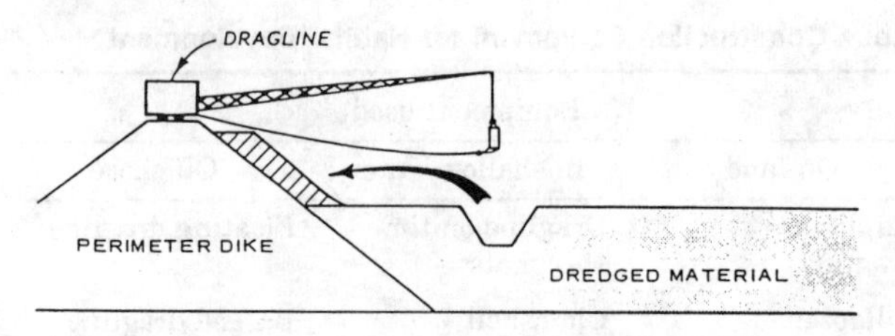

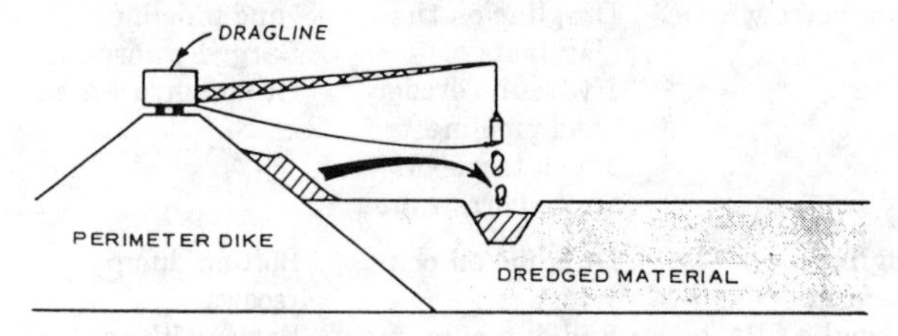

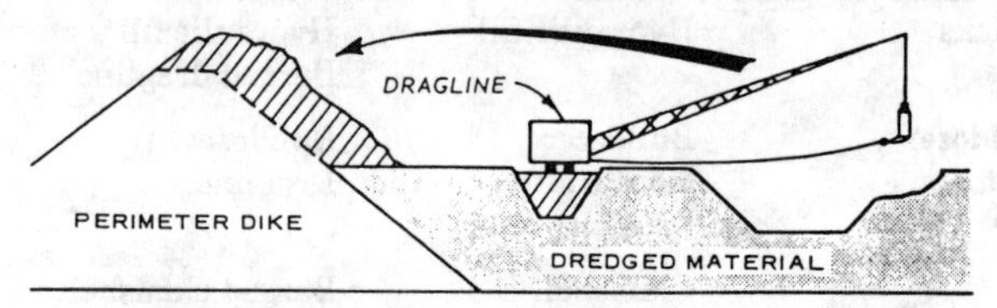

Figure 2.42 Use of dragline for perimeter trenching to dry dredged material for raising dikes (*Bartos 1977*).

(containment) area. A discharge pipe extends through the dike from the base of the riser inside the disposal area to the outside.

The box weir consists of an open cut through the entire dike, lined usually with wood, although steel or concrete can be used with stoplogs for water control. Box sluices can discharge large volumes of water rapidly, advantageous during development and operation of the new marsh. Box sluices are not used often because of high construction cost and susceptibility to failure by seepage.

Filling the marsh

The new marsh can be filled hydraulically or mechanically. The hydraulic pipeline dredge is most common (U.S. Army Engineer Waterways Experiment Station 1978). By adding intermediate booster pumps, pipeline length can be extended to several kilometers (Johnson and McGuinness 1975). The pipeline dredge can fill shallow-water areas by direct pump-out from bucket-loaded scows through

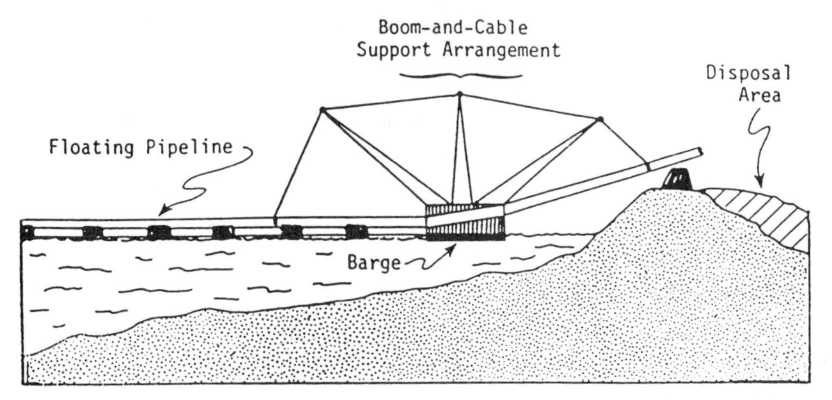

Figure 2.43 Use of spillbarge to deposit material in a dredge fill marsh (*Johnson and McGuinness 1975, U.S. Army Engineer Waterways Experiment Station 1978*).

shallow-draft floating pipelines or pipelines on shore. For long-term stability of the marsh, coarse-grain material is placed on the side of the marsh facing the direction of maximum erosive energy.

Floating pipelines can be moved by moving the pipe anchors. By use of elbows and swivels, the dredge can advance without moving the pipeline. A spillbarge can be used when the proposed marsh is next to the channel being dredged (Fig. 2.43). By use of booms up to 77 m long, dredged material can be directed over a dike, linear island, or jetty located parallel to the channel being dredged.

Pipeline on shore cannot be moved easily, so 30°, 45°, and 90° elbows are used for directional change. When discharged material mounds near the end of one pipe, the flow can be diverted to another pipe without interrupting the dredge operation. Then the mounded area can be leveled with heavy equipment on stable soils or by hand labor otherwise, or the pipeline can be extended. Timber mats or marsh buggies can extend the range of equipment capability.

Energy dissipaters, mainly baffle plates, wye joints, and bleeder pipes, are used to reduce the velocity of the slurry of dredged material at the discharge pipe. The energy from the slurry flow can damage the interior slope of the dike and can scour and resuspend material already deposited.

When placement of the dredged material begins in the proposed marsh area, flashboards of the outlet weir(s) are set at a predetermined elevation so that the ponded water will be deep enough to allow the soil to settle as the containment area is filled with slurry (U.S. Army Engineer Waterways Experiment Station 1978). Slurry pumped into the containment area raises the water level until water spills out the weir(s), decreasing the ponding depth as the thickness of the dredged-material deposit increases. When the elevation of deposited dredged material is completed in the new marsh, water is allowed to fluctuate with the tides through the weir(s). After the deposited dredged material has stabilized and potential for erosion is minimal, the weir(s) or retention structure can be breached to allow natural water circulation throughout the new marsh. Bartos (1977) and Palermo et al. (1978) provided details regarding management of containment areas of dredged materials.

Surface shaping

When the disposal operation is complete, the surface of the fill in the new marsh might need shaping to meet final elevation requirements, to improve water circulation, or to facilitate boat or vehicular traffic (Green and Rula 1977, U.S. Army Engineer Waterways Experiment Station 1978, Willoughby 1978). Coarse-grain material often can be graded with heavy equipment. Building finger roads or using matting will extend the operational range of tracked vehicles and draglines, but low-ground-pressure vehicles such as marsh buggies might suffice and will reduce cost.

Access channels dug into the fill area will extend the operational range of floating draglines or clamshell dredges, and these channels could be left open to improve water circulation in the new marsh, provide access, and support waterfowl and other wildlife. The distance to the center of the new marsh might preclude use of a conventional floating dragline. The narrower an area of marsh is, the easier it is to shape for removing or creating interior mounds and holes, providing a network of access channels or level ditches and improving water circulation. In extensive (up to 500 m) or other difficult areas, a highline arrangement can be used, consisting of a winch, a deadman at the opposite side of the disposal area, and a dragline bucket of 3-, 6-, or 10-m capacity (Fig. 2.44 and 4.25). The deadman consists of a stationary object such as a small tree or a dozer which is connected to the winch by a cable and which can support a bucket dragged along the bottom to excavate fill (U.S. Army Engineer Waterways Experiment Station 1978).

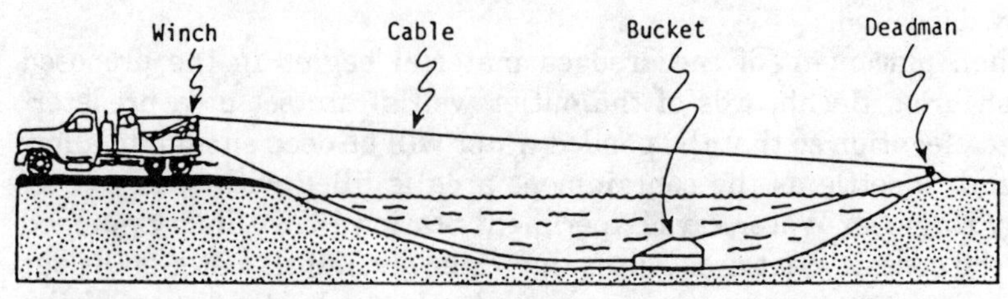

Figure 2.44 Portable dragline with deadman (*U.S. Army Engineer Waterways Experiment Station 1978*).

Chapter

3

Wetlands Without Water Control

Flooding by means of dikes or dam is not always feasible because of physical conditions (soil types or water supply), financial limitations, or legal restrictions (rights of adjoining landowners or the public). Or if a dam or plug is constructed, financial limitations might prevent use of a control structure. Thus, water-level control can be severely limited or impossible. Yet the water area will benefit wildlife, especially if built and maintained properly. A variety of ponds with limited or no water control, built in the same general area and varying in depth, vegetation, and open water, will attract a large variety of wetland and upland wildlife. Best density of wetlands of various sizes and shapes is 12 to 40 per square kilometer (Lokemoen et al. 1984). Where water manipulation is too expensive to produce duck food, methods can be used such as depositing cereal grains by permit in the marsh, as in Ontario, where no hunting is allowed within 400 m of the grain (Bookhout et al. 1989).

LEVEL DITCHES

Level ditches are ungraded ditches that fill with groundwater. They are dug in areas where the water table is 41 cm or less below the ground (Johnson 1984). Each end is closed so that the wetland will not drain. Generally, level ditches are installed about at right angles to natural channels to avoid intercepting natural channels or leaving unopened blocks between ditches and channels (Atlantic Waterfowl Council 1972). Water control structures can be used at the connecting point to allow floodwaters in or to regulate flow (Schnick et al. 1982), in which case level ditches are managed as controlled marshes. Level ditches usually are dug with a dragline in a dry marsh to benefit ducks and muskrats, but can be dug with marsh buggy plows, marsh cutters, backhoe, or blasting (Atlantic Waterfowl Council 1972). In Louisiana, ditches 12 dm wide and 6 dm deep are dug with a plow pulled by a marsh buggy (Chabreck 1968).

Level ditches are installed for the following purposes (Mathiak and Linde 1956, Atlantic Waterfowl Council 1972, Broschart and Linder 1986):

1. To improve the distribution of water in marshes with dense, unbroken stands of vegetation
2. To increase production of muskrats, mink, other furbearers, invertebrates, waterfowl, and amphibians
3. To provide open water for courtship and brood rearing of waterfowl
4. To increase production of aquatic plants
5. To provide access for management or harvest

Muskrat harvests will be 4 to 10 times higher in ditched marshes, which concentrate muskrats and also facilitate boat travel for trappers.

Soils suitable for ditching include peat, muck, clay, and silt. Sand, sandy loam, and clay high in salt content generally are unsuitable (Atlantic Waterfowl Council 1972, Yoakum et al. 1980). Scattered cattails can be used to indicate the proximity of the water table where level ditching will be satisfactory (Mathiak and Linde 1956). Suitable marshes for level ditches generally are larger than 0.8 ha.

Where slopes are more than 0.5 percent, ditches are built on contour. On flat marshes over 4 ha, ditches should be located 90° to prevailing winds and zigzagged 10° to 30° about every 30 to 90 m (Fig. 3.1) to improve cover for ducks and to impede wind and wave erosion but not boat travel. Smooth curves might be better than sharp angles where the ditch changes direction. Ditches should be spaced 60 to 120 m apart and dug 0.9 to 1.8 m deep and 3 to 9 m wide (Atlantic Waterfowl Council 1972). For muskrats, ditches can be 4 m wide at the surface, 1.5 m wide at the bottom, and at least 1.5 m deep to prevent freeze-out (Mathiak and Linde 1956). To reduce exposed vertical edges where the water table is subject to large drops in summer, ditches should be at least 9 m wide (Linde 1969). Wider ditches probably provide greater security for broods. Emergent aquatics will provide escape cover if occasional depressions about 3 m in diameter and 3 dm deep are dug in the bog.

Limited ditching for large marshes is best constructed 60 ha apart in one section, to concentrate muskrats, stabilize production, repopulate trapped-out areas quickly, and allow orderly development

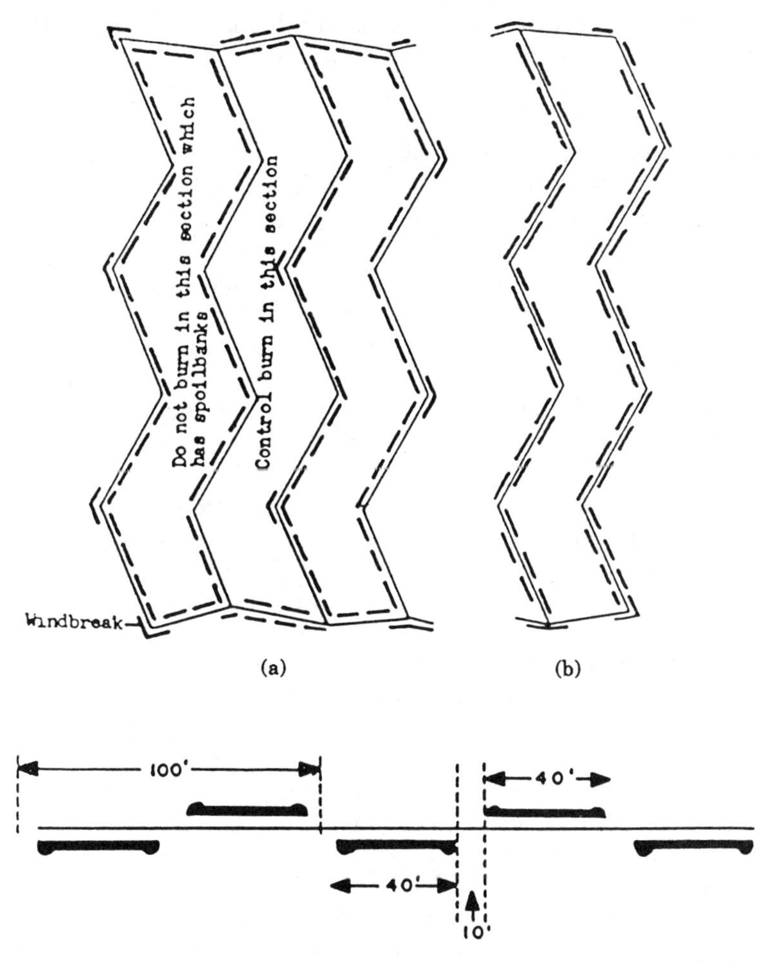

Figure 3.1 Proposed layout of level ditches and placement of spoilbanks (*Mathiak and Linde 1956*).

of the remaining marsh in the future (Mathiak and Linde 1956). But a 122-m spacing probably is better for territorial nesting ducks.

Sides should be sloped 5:1 or more on one side in all soil types and on the other side 3:1 in sand, 1:1 in peat, and 2:1 for all other soil types. Spoilbanks should be piled higher at the ends of each straight section of ditch as a windbreak. Where controlled burning is not needed, spoilbanks are piled 0.9 m from the ditch edge, up to 1.5 m high, flat on top, 12 m long on alternate sides of the ditch, with 3-m gaps between the ends of the banks (Fig. 3.1), and the ends of each bank are enlarged. Spoilbanks so constructed will improve conditions for duck nesting, discourage walking on the banks, and reduce the chance of fire's sweeping down the length of the bank. Such construction also reduces mosquito breeding in tidal areas by allowing tides to enter and leave the level ditch (Atlantic Waterfowl Council 1972). In peat marshes subject to drying, spoilbanks are placed so that alternate areas between ditches can be control-burned without allowing peat fires to endanger the banks (Fig. 3.1). Volunteer growth of aquatics might be encouraged by repeated burns of dry marsh which might lower the marsh level enough. Muskrats can be induced to use the marsh on the side away from the spoilbanks, and humans can be discouraged from walking on the banks, if short channels are dug into the marsh at breaks in the spoilbanks. To encourage growth of muskrat foods such as cattail, bulrush, and burreed in semidry marshes of sedges and grasses, a strip 1.8 m wide and 0.3 m deep can be dug next to the ditch.

Ditches mainly for ducks can be short and wide with low spoilbanks to enhance shooting chances with camera or gun. Construction is best accomplished with a bulldozer during dry soil conditions. Excavated materials should be leveled out around the ditch. Tall vegetation (e.g., trees, shrubs, and cattails) should be eliminated around the ditch.

Spoilbanks should be seeded to grasses and legumes as soon as possible after dredging to reduce erosion and provide dense nesting cover. The dense root structure and ground cover of reed canarygrass are effective. Seeding some white sweetclover with it will provide early cover while the slower-growing canarygrass develops. Although time-consuming and expensive, grading of soil and sod is recommended to reduce water pockets where mosquitoes can breed (Atlantic Waterfowl Council 1972).

Trespassing is reduced by digging a continuous ditch around the marsh with the spoilbank on the inside, because the water will be too deep to wade. Such a ditch will serve as a firebreak and as a starting point for back firing or controlled burning. Level ditching can be used as a firebreak if dug parallel to the upland edge of the area to be protected. If the spoil is peat, it is placed on the outside of the ditch away from the burn to prevent the banks from burning during a hot fire (Linde 1985).

POTHOLES

Potholes are shallow depressions, usually less than 1.6 ha, containing water that provides breeding territories and brood-rearing and feeding habitat for ducks in spring (Atlantic Waterfowl Council 1972). They can be naturally occurring or artificially dug or blasted. Management of naturally occurring potholes consists mainly of protecting them from filling, drainage, agricultural pollutants in runoff, and grazing. Artificial potholes cost about half as much as level ditches to build (Linde 1969).

The U.S. Soil Conservation Service recommends an average density of 2.5 potholes per hectare, but clustered in groups of 5 or more, within 1 km of larger bodies of water. A duck uses several potholes, so clustering encourages concentrations of waterfowl. Potholes should be 61 to 91 m apart (Mathiak 1965) or in clusters of 5 to 15 within a radius of 61 m, the clusters 152 to 304 m apart (Warren and Bandel 1968). Optimum size of a pothole is 6.1 to 7.6 m wide, 12.2 to 22.9 m long, and 1.2 m deep (Linde 1969). One or two sides are sloped gently to encourage dabbling ducks to use bottom foods. The other sides are steep to reduce encroachment of emergent vegetation. Spoilbanks need not be leveled, but a berm separating it from the water reduces silting. In pastured areas, potholes should be fenced at least 7.6 m and preferably at least 21.2 m back from the water (Atlantic Waterfowl Council 1972).

Where many natural potholes occur, improvement for ducks consists of creating a pothole 9 or 12 dm deep. Adjacent shallower potholes are connected by ditch to the main pothole to channel water there so that young ducks have permanent water when the shallow potholes dry up (Gavin 1964). Potholes over 0.8 ha can be partitioned to increase courting and brood-rearing territories (Atlantic Waterfowl Council 1972). Partitions are made with rocks and boulders placed on the ice to sink later, or during dry periods by constructing ridges or chains of small islands with a bulldozer or dragline.

Canvasbacks prefer potholes with open shorelines, but most species of ducks in the Canadian parklands prefer about one-third of the shoreline rimmed with trees (aspen or willow) or brush (Stoudt 1971). All species prefer potholes and other ponds with less than 50 percent of the surface covered with emergents and with uncultivated or ungrazed shorelines.

In constructing artificial potholes, digging is used more than blasting (Holsapple and Lott 1979). Although more expensive than blasting, mechanical construction is safer and more predictable, but its use is restricted to marshes dry enough to support heavy equipment. Ducks use excavated potholes more than blasted ones (Linde 1985).

Dug Potholes

Dug potholes have bottom contours with wedge or trapezoidal shapes (Linde 1969). Such bottoms provide shallow edge for puddle ducks and shorebirds, despite weather-controlled fluctuations in water levels, and a maximum surface area of water during droughts. But the shallow edge might need deepening to reduce rapid encroachment by cattail. Construction can proceed with bulldozer or dragline. Amphibious draglines are especially useful in soft, spongy marshes (Chabreck 1968).

In marshes that are dry enough, the bulldozer or tractor and scraper is the faster method, costing half as much as a dragline. Potholes can be accurately contoured, with well-shaped spoilbanks, if bulldozed in dry to moist mineral soils (Linde 1985).

The dragline is more useful if the water table is on the surface of the ground. If needed, log mats are used as support, but they must be moved when the dragline moves, slowing the operation and increasing the cost. The dragline can operate without mats and can move faster on frozen ground. But the ice and frost layer must be broken by a steel ball before excavation. A dragline cannot shape spoilbanks as well as a bulldozer can, but birds seem to use the potholes equally well. Because moving a dragline is costly, the cost relative to the benefit requires that many potholes be constructed. A U-shaped or circular course of operation will return the dragline to its starting point efficiently.

Blasted Potholes

Boggy conditions or expense might prohibit the use of heavy equipment for potholes. Blasted potholes do not last as long as dug potholes, but are much cheaper. For best results, the water table should be 0 to 20 cm below the surface (Hopper 1971). Marshes of heavier-mineral soils (clays and loams) are best. Acceptable sites include marshes with a layer of 3 dm or more of heavy surface soil over sandy soil, typical of floodplains and stream bottoms (Hopper 1971). Unless a mineral soil is within 9 dm of the surface, blasting in peat soils is not recommended because the bottom will be loosened by the blast and might float within 1 year to fill the pothole (Warren and Bandel 1968). Blasting in floating bogs of leatherleaf and sedge forms only shallow pockets. Floating bog can be improved by placing charges in the mineral soil a few decimeters from the bog's edge and blasting deep holes (Mathiak 1965). Blasting in sedge meadows often results in potholes covered with undesirable floating sections of sedge mat (Martin and Marcy 1989).

Soil borings are needed to determine that underlying substrates are suitable and safe for blasting, i.e., not of gravel and rocks. The blast site must be cleared of rank vegetation, charge lines marked and staked off, warning signs posted, safety distances determined, unauthorized people expelled, and test shots taken. Blasting is done on

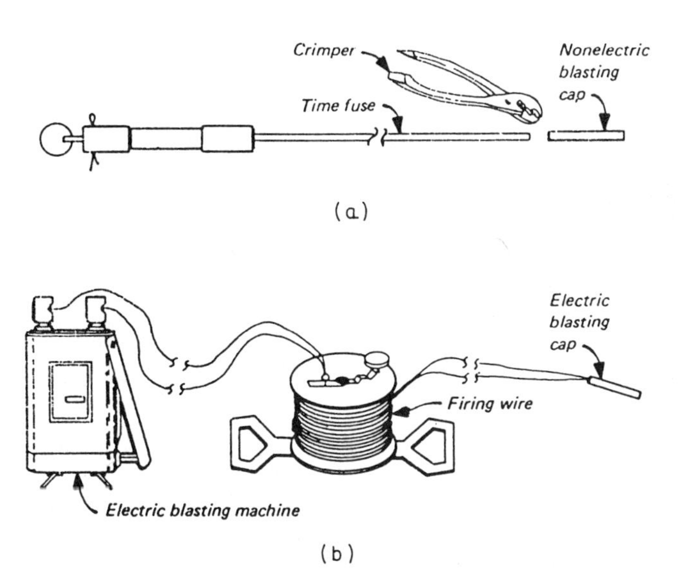

Figure 3.2 Major components and diagram of (*a*) nonelectric and (*b*) electric firing systems (*Martin and Marcy 1989, from U.S. Army 1986*).

windy days to dissipate the poisonous fumes, smoke, debris, and sound, but never when an electrical storm is approaching (Martin and Marcy 1989).

Blasting can be done with electric or nonelectric firing systems (Fig. 3.2) or with detonating cord to detonate the explosive, which can be dynamite, ANFO, or waterproof gel. The nonelectric system is ignited usually by a pull-type igniter that strikes a spark that begins a flame that is transmitted through a time fuse to the blasting cap which has a mild explosion strong enough to set off the charge. Care is taken to avoid detonating electric blasting caps by radio frequency, and electric and nonelectric caps by lightning.

Detonating cord is the most versatile (e.g., waterproof), least sensitive, and easily installed firing system (Martin and Marcy 1989). It looks like rope, but is an explosive. Detonating cord is looped around the main explosive, or a nonelectric blasting cap is crimped onto the detonating cord and inserted in the main explosive. It is ignited by an electric or nonelectric blasting cap crimped to the other end. Detonating cord allows use of multiple charges by attaching branch lines at 90° or greater to the main line with a detonating cord clip, a girth hitch with one extra turn, or a ring main (Fig. 3.3). Square knots are used for splicing.

All blasting is done a safe distance from the site by a person licensed to use explosives. The distance is determined from the equation

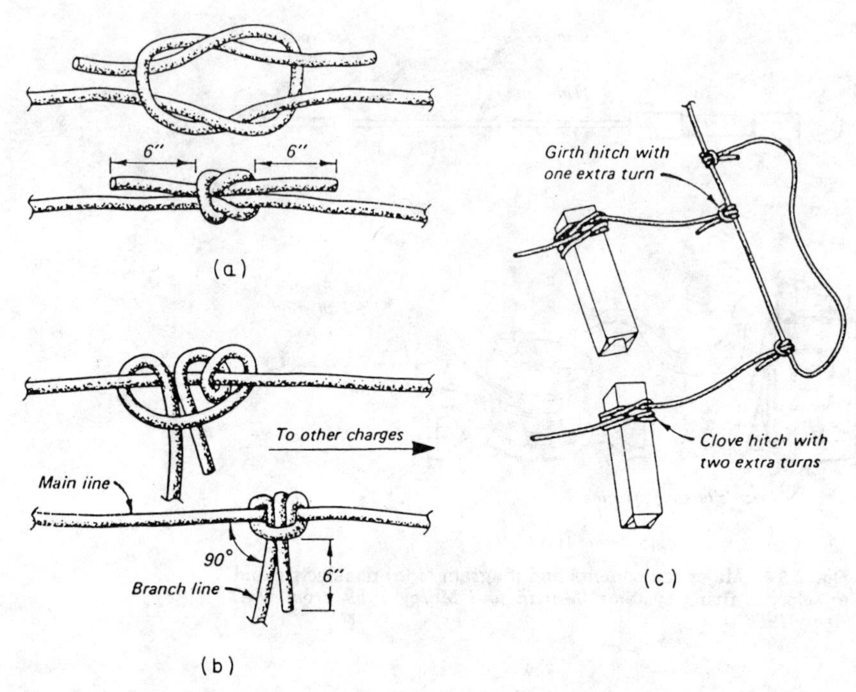

Figure 3.3 Det-cord connections: (*a*) square knot for splicing ends of det-cord, (*b*) girth hitch for fastening branch line to main line, (*c*) ring main with branch lines for detonating multiple charges (*Martin and Marcy 1989, from U.S. Army 1986*).

$D = 100\sqrt[3]{p}$, where D = safe distance in meters and p = pounds of explosive (Martin and Marcy 1989). The safe distance is 370 m for 23 kg (50 lb) of explosive and 500 m for 57 kg (125 lb).

ANFO, a mixture of ammonium nitrate (NH_4NO_3) and fuel oil, is a safe, inexpensive explosive widely used for potholes. Mixing can occur at the blasting site at the rate of 45.4 kg of NH_4NO_3 soaked in 3.8 to 4.7 L of no. 2 fuel oil for at least 30 to 45 min in a heavy plastic bag (about 10-mil strength) to prevent puncturing (Mathisen et al. 1964, Mathiak 1965, Hopper 1971). Diesel fuel can substitute for fuel oil (Warren and Bandel 1968). Dry ammonium nitrate (about 33.5 percent nitrogen) in prill (bead) form is commonly used for agricultural lands. It can be bought in plastic (polyethylene) sacks. An 11.4-kg (25-lb) bag should be tested before a given brand is chosen. No unabsorbed fuel oil should remain in the bottom of the bag after it is mixed and allowed to stand 30 min. The charge should blast a pothole about 4.3 by 5.5 m in most wetland soils.

To set off the charge, a stick (0.23 kg) of dynamite (50 to 60 percent nitroglycerin) is primed with a 152-cm length of safety fuse crimped

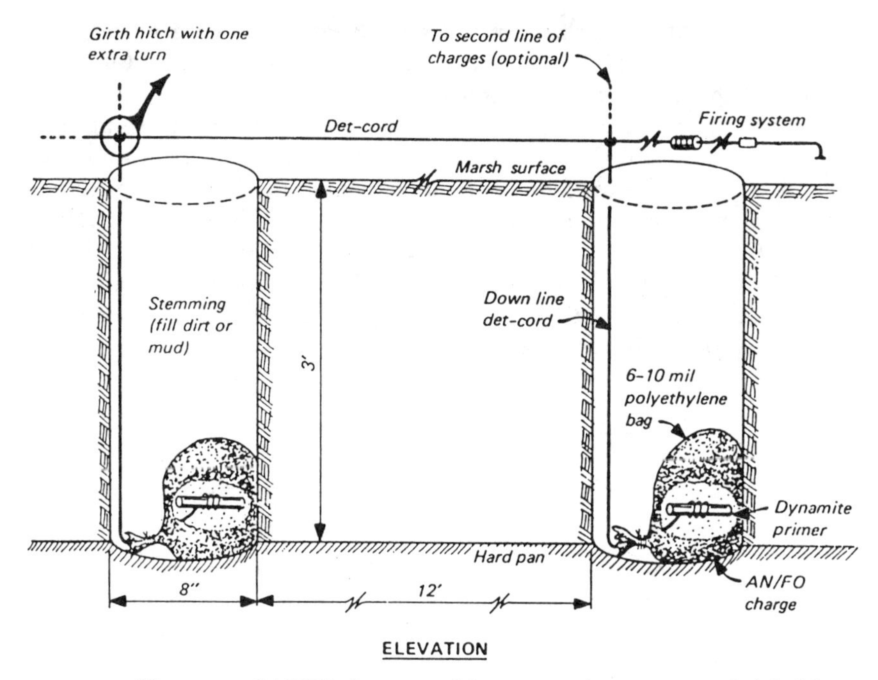

Figure 3.4 Placement of AN/FO charges and fuse connections recommended for blasting potholes or ditches (*Martin and Marcy 1989, modified from Mathisen et al. 1964, Mathiak 1965*).

into an electric blasting cap with a cap crimper and then inserted into the middle of a 23-kg bag of ANFO. The fuse extends out the top of the bag, and the bag is tied tightly. A hole just large enough for the charge (about 20 cm wide) is dug about 9 to 12 dm deep with sharpshooter shovels, posthole diggers, hand augers, or power augers. The top of the charge must be 3 dm beneath the surface in heavy soils and at least 3 dm deeper in light soils (Fig. 3.4). The water is baled out of the hole, the charge inserted at once, and the soil and mud packed (stemmed) tightly over the charge. Before excessive water can enter, the charge is detonated immediately with a blasting machine connected to the fuse by 305 m of wire (Hopper 1971). In marshes of the South where soils commonly are 0.9 to 21 m deep overlying limestone, 25-cm holes are dug to bedrock and anticave sleeves are used to prevent seepage (Atlantic Waterfowl Council 1972).

With multiple charges, holes are dug about 4.5 m apart in a triangle or circle. Three 11.3-kg charges produce a pothole of 52 m^2, three 23-kg charges one of 80 m^2 (Hopper 1978), and six to eight 23-kg charges one of 743 m^2 (24 m × 30 m) (Whitman 1982). Hopper (1971, 1972, 1978) found that three 23-kg charges produced the best results, but

Whitman (1982) found that six to eight 23-kg charges were best. Ideally, spoilbanks and islands are 9 to 12 dm above the water level which is 9 to 18 dm deep the first few months (Whitman 1982). Measurements are more meaningful after 2 years when initial sloughing should be completed (Mathiak 1965). Potholes about 0.1 ha and 3 to 3.7 m deep, with sloping sides, are made by placing five 13.6-kg ANFO charges 6 m apart in a pentagon shape, with five 6.8-kg ANFO charges 9 m apart in a pentagon shape outside, and a time-delayed detonation of 15 s for the outside charges (Fig. 3.5) (Martin and Marcy 1989). Other blasting patterns and charge sizes also can be used (Fig. 3.6) (Martin and Marcy 1989).

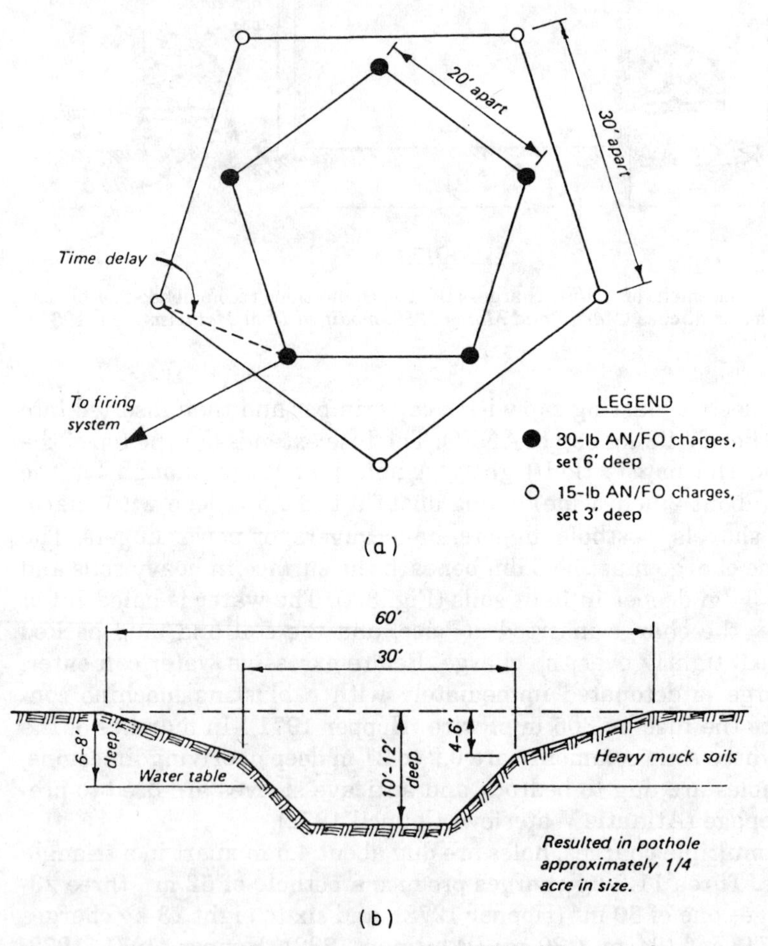

Figure 3.5 Blasting pattern (*a*) and sizes of AN/FO charges used to create large potholes (*b*) (*Martin and Marcy 1989*). [For metric equivalence, see metric conversion table in the front of the book.]

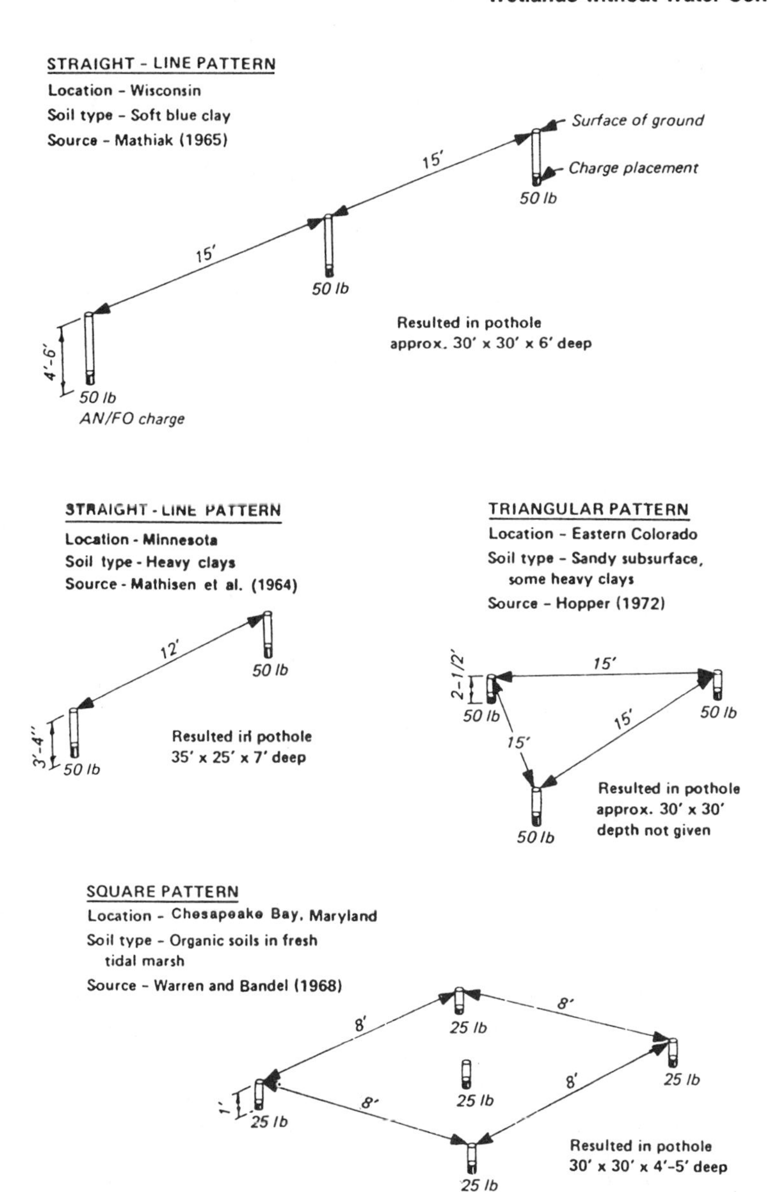

Figure 3.6 Examples of blasting patterns and sizes of AN/FO charges used to create potholes (*Martin and Marcy 1989*).

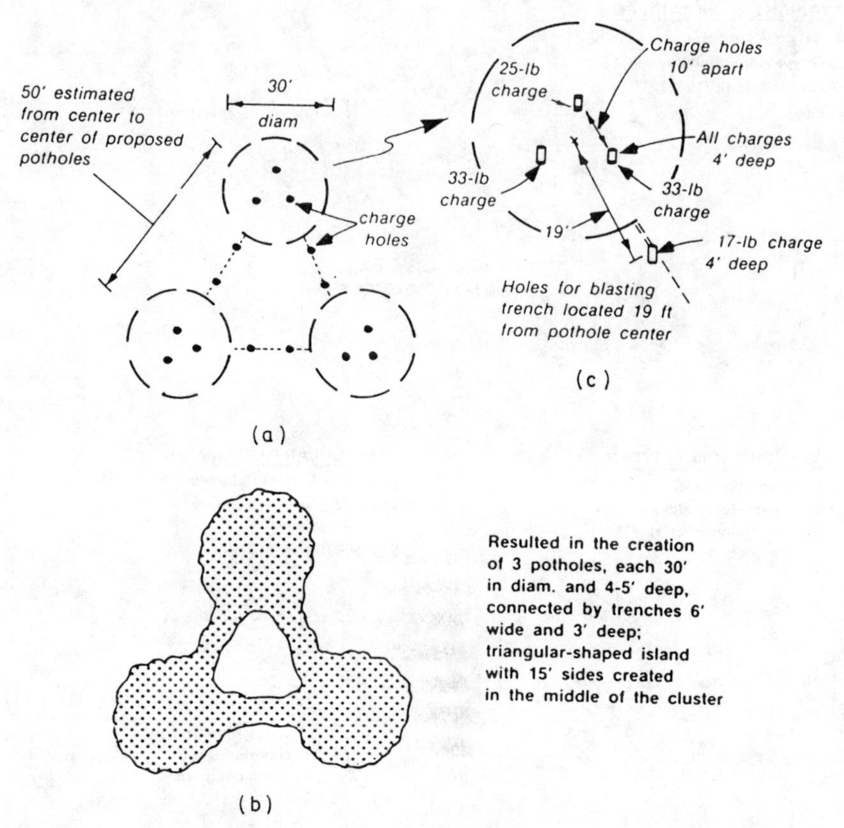

Figure 3.7 Design specifications for using waterproof gel to create clusters of three potholes (*Martin and Marcy 1989, after Skinner 1982*).

Waterproof gel (e.g., DuPont Tovex 800, Gulf Detagel, Gulf Iremite 60) will sink in water and is safer than ANFO because it is more stable and produces no toxic fumes (Martin and Marcy 1989). The gel is packaged in cartridges weighing 3.2 kg (DuPont Tovex 800). Clusters of three potholes connected with trenches, with an island in each cluster, can be designed (Fig. 3.7) (Skinner 1982). Dynamite can be used (Martin and Marcy 1989). It is more compact than the other explosives, but less safe.

In some parts of the country, local, state or provincial, and federal laws regulate the use and storage of explosives. These laws should be consulted and the necessary permits obtained. Various manuals about explosives are available which provide technical details and safety precautions (e.g., Lott 1977, U.S. Forest Service 1980, U.S. Army 1986, Cranney and Bachman 1987, Martin and Marcy 1989).

OPEN MARSH WATER MANAGEMENT

Properly designed *open marsh water management* (OMWM) for mosquito control results in increased populations of dabbling ducks and other waterbirds (Daiber 1987). OMWM is a combination of level ditches and potholes dug in tidal areas, except they are connected to tidal ditches, preferably with 15.2- or 30.5-m earthen plugs or sills placed in nontidal ditches to remove sheet water without lowering the subsurface water table, which is kept within 15 cm of the marsh surface. A sill system allows tidal exchange, passage of larvivorous fish, and a stable environment for growing submerged aquatic plants (Hindman and Stotts 1989). Tidal circulation, not drainage, is important (Atlantic Waterfowl Council 1972).

Where the density of mosquito-breeding depressions is high, as in saltmeadow cordgrass, smooth cordgrass, and seashore saltgrass, ponds are dug with a rotary ditcher (Condeletti 1979), to include several such depressions (Fig. 3.8) (Meredith et al. 1983, 1985, Hindman

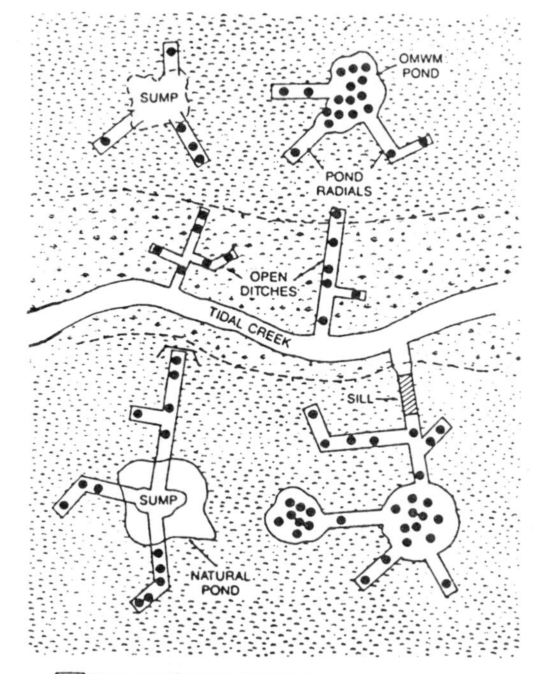

Figure 3.8 Alterations for open marsh water management to benefit wildlife (*from Meredith et al. 1983 in Hindman and Stotts 1989*).

and Stotts 1989). Ponds average 0.4 ha and are dug 30 cm deep to support widgeongrass, with a narrow fish reservoir 76 to 91 cm deep on one or two sides of the pond edge. Radial ditches 46 to 91 cm deep and 61 to 91 cm wide connect other mosquito-breeding depressions to the ponds. Ditches should not be dug with a gradual decrease in elevation because they revegetate and become isolated from tidal flow (Atlantic Waterfowl Council 1972). Spoil, spread 15 cm wide and 2 to 10 cm high, does not form pockets of temporary water and breeding mosquitoes and vegetates naturally in about 2 years.

The undesirable parallel-grid ditching method, which causes excessive drying, is improved by blocking old ditches with flashboard structures to maintain late fall and winter water levels at the marsh surface. These can be partially drawn down to maintain water at 20 cm below marsh level between April 15 and October 1 for production of submerged aquatic plants as food for duck broods. When used with flashboard structures, near the outlet of old mosquito ditches, potholes can be blasted at the head of the ditch, especially at the marsh-upland interphase, to increase water surface, growth of submerged aquatic plants, drawdown capability, mosquito control, and habitat for waterfowl and other wildlife (Stotts 1971). Water levels of wetlands managed as shallow impoundments (controlled marshes) should be kept 10 to 25 cm deep during the growing season for mosquito control and growth of submerged aquatic plants (Hindman and Stotts 1989).

PLAYAS

Playas are similar to potholes. They occur throughout the southern Great Plains in northern Texas and eastern New Mexico as round, oval, or elliptical depressions with a clay or fine sandy loam hydric soil (Mitsch and Gosselink 1986). Many support moist-soil plants, but most are dry most of the time (Guthery and Stormer 1984a). Typically, a playa has a watershed of about 55 ha and a wetland area of about 7 ha (Mitsch and Gosselink 1986). To enhance consistent use by ducks, management should focus on playas larger than 8 ha for maximum wildlife benefits (Guthery and Stormer 1984a), preferably 12 to 16 ha (Guthery and Stormer 1984b). Management options involve soil disturbance, excavation, burning, grazing, herbicides, planting, water manipulation, or some combination thereof. With water manipulation, management of playas can be a form of controlled marsh management.

Playas can be modified for wildlife and agriculture by excavating the shoreline on one side into two to five steplike terraces to make shallow, littoral habitat available to waterfowl and shorebirds at any

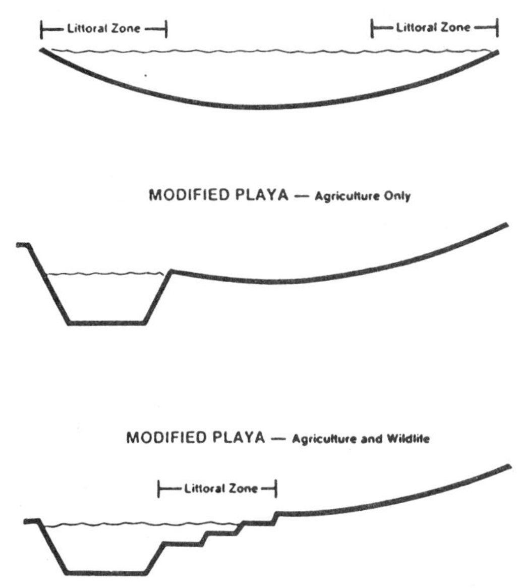

Figure 3.9 Cross section of a playa basin unmodified (top), modified for the storage of water for agricultural uses (middle), and modified experimentally for agriculture and wildlife (bottom) (*Bolen et al. 1989*).

water depth (Fig. 3.9) (Bolen et al. 1989). Spoil is feathered into surrounding uplands to prevent hunters from easy and unseen access. Recommendations are to buy or lease at least four to five large (over 40 ha) relatively permanent core playas per county, build terraces on pits dug in playa basins within 5 km of the core playas, and intensify postharvest management of waste corn on private and leased lands within 5 km of the core playas (Bolen et al. 1989). Ducks fly up to 5 km from playas to cornfields (Baldassarre and Bolen 1987).

Geese can be induced to use playas of at least 8 ha if water levels are maintained, usually by pumping from a well probably used also for crop irrigation, and if no hunting occurs except in nearby winter wheat fields (Guthery and Stormer 1984b). Playas in corn-growing regions attract more ducks (Guthery and Stormer 1984b). When the playa is dry, a bulldozer can push up 69 to 115 m^2 of low, flat island per hectare of playa, each island 6 to 12 m wide and free of vegetation in winter. Removing the fill from the bottom for the islands will cause the deeper areas to hold water longer, resulting in growth of cattails and bulrushes in the wet soils. Such emergents are desirable on one side of an island only. Smartweeds and barnyardgrass, excellent duck foods, will colonize the playa's bottom, especially if it is irrigated by shallow, temporary flooding one to three times during the growing season, depending on rainfall. After October 1 the playa is flooded less than 46 cm deep in most areas. Earlier flooding increases the risk of botulism. If supplemental water cannot be introduced to the playa, natural runoff must be concentrated and held longer by deepening the playa. Spoilbanks from the excavation are developed along the playa's perimeter, but up to 9 dm above the water level, with less than 1:5 slopes on at least one side.

Playas provide important winter and nesting cover to ringneck pheasants (Guthery and Stormer 1984b). Such playas benefit pheasants within a 1.2-km radius (452 ha) in irrigated areas with at least 121 ha of corn and at least 81 ha of small grains, ideally with a wheat field of at least 40 ha and similar-size cornfield bordering the playa. Best winter cover consists of dense, tall, robust vegetation, usually a cattail-bulrush community. Such plants occur mainly with shallow irrigation from supplemental water during the growing season, accomplished most economically with tailwater from crop irrigation. The tailwater is allowed to irrigate the playa periodically before it is circulated onto the crops again. For pheasants, the playa must be dry in winter. A central trench leading to a peripheral pit in the playa allows drainage of the playa in summer and winter.

Best nesting cover occurs on level spoilbanks vegetated by grasses and summercypress and on the adjacent watershed containing moderately dense residual cover at least 25 cm tall by March 1. Passive management of these areas is best, i.e., no grazing or other manipulation of vegetation.

A management plan (Guthery and Stormer 1984a) designed especially for ducks and pheasants in areas of extensive corn agriculture consists of an 8-ha playa, with 6 ha as a cattail-bulrush marsh separated by a dike from 2 ha of impounded water (Fig. 3.10). The impoundment side provides fill for the dike. A shallow pit dug next to the center of the dike can drain the impoundment via a pump on the dike, if fowl cholera or botulism occurs. Loafing sites for ducks are provided by scraping fill from the bottom of the playa to build one to three islands about 14 m in diameter, while creating variation in depth and aeration of water. Winter tailwater (runoff from agricultural irrigation) is channeled into the impoundment. Cattails and bulrushes usually will not occur in playas without supplemental water.

On the marsh side of the dike, a high volume of summer tailwater must enter the playa to develop a dense stand of bulrush and cattail. The marsh is drained during summer and winter from a shallow

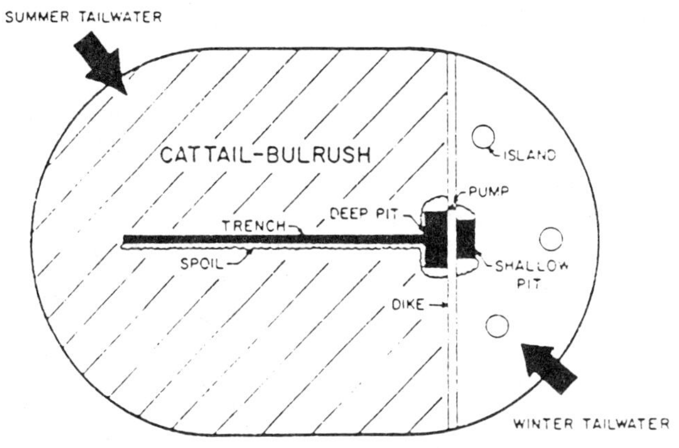

Figure 3.10 Playa management plan to provide quality hunting for ducks and pheasants in a region with large areas devoted to corn production (*Guthery and Stormer 1984a*).

trench dug through the center of the marsh and a deep pit dug next to the dike opposite the pit on the impoundment side. The summer tailwater is recycled onto crops. Water remaining in the marsh after October 1 is pumped into the impoundment. Deep flooding in summer kills emergent plants which provide winter cover. Flooding in winter renders the cover useless for terrestrial wildlife. This plan supports about 500 ducks per day during November if 202 ha or more of corn occurs within 4.8 km, a lower but still high density of ducks during other months, more than 90 pheasants during winter, shorebirds, coots, herons, mourning doves, red-winged blackbirds, yellow-headed blackbirds, harriers, barn owls, cottontails, coyotes, and raccoons. Willows established along the spoilbanks bordering the trench will attract songbirds and in time perhaps a black-crowned night heron rookery. Light to moderate grazing of the cattails and bulrushes improves structural diversity.

Stormer et al. (1981) mentioned other potential but untested habitat alterations for playas. Because most playas are privately owned, management of playas for wildlife is accomplished best through incentives to landowners, such as reduced property tax, fee-lease hunting, or federal cost sharing, e.g., the Great Plains Conservation Program (Guthery and Bryant 1982). In addition, problems must be solved relative to waterfowl depredations on sprouting wheat, mosquitoes breeding in playas, dispersal of crop weeds from playas, and introduction of pollutants into playas from tailwater containing pesticide and fertilizer.

PANNES

Pannes are natural depressions in an intertidal marsh that retain tidewater during low tide (Mitsch and Gosselink 1986). Pannes can form when shifting sediment or vegetation dams a natural depression or the outlet of a tidal creek. Because of the continued standing water and increased salinity with evaporation, pannes support different vegetation than the surrounding marsh. For example, widgeongrass, a desirable duck food, is tolerant of high concentrations of salt in the soil water. At higher elevations in the marsh, infrequent flooding by tides produces relatively permanent ponds. Ducks and shorebirds use pannes heavily due to the shallow depth and submerged vegetation. Pannes generally are not managed actively, but like potholes, artificially constructed or enlarged pannes are potentially feasible.

VERNAL POOLS

Vernal pools occur in uplands mostly of southern California as isolated shallow depressions with hardpan clay soils that retain water seasonally (Zedler 1987, Josselyn et al. 1989). Plant species tolerant of these seasonal pools are mainly annuals with seeds that survive the long dry period. Invertebrates survive the long dry period by mechanisms such as cyst formation. Vernal pools must be fenced against off-road vehicles and other disturbance. Artificial pools are designed and constructed similar to potholes, placed preferably within a large catchment to increase water depth and to prolong flooding.

STOCK PONDS

Stock ponds are beneficial to waterfowl and other wildlife and can be classified into three types (Lokemoen 1973, Eng et al. 1979): (1) retention reservoirs constructed by building short dams across intermittent streams usually in upland areas or across large gullies draining off upland slopes (Linde 1969) to catch spring runoff and rainwater; (2) dugouts or pit reservoirs similar to potholes but with steep sides, filled by groundwater or surface (runoff) water if in temporary or semipermanent natural wetland areas or by surface water if in intermittent waterways or on level ground in upland areas (Bue et al. 1964); (3) diked dugouts or pit retention reservoirs, built like regular dugouts but with the spoil placed on the downstream side as a dam to flood the shallow area around the dugout. Runoff ponds (Linde 1969), also called *paddies* (Atlantic Waterfowl Council 1972), are similar to retention reservoirs, except their primary function is for waterfowl rather than for livestock.

Stock ponds are most beneficial to waterfowl during spring and summer as breeding and brood-rearing habitat (Eng et al. 1979). Major breeders using the 88,000 stock ponds in South Dakota and 120,000 in Montana are mallards, blue-winged teal, pintails, and coot. Shovelers, gadwalls, widgeon, and ruddy ducks also nest there. Primary users during migration are green-winged teal, canvasbacks, redheads, lesser scaup, buffleheads, common mergansers, and Canada geese (Bue et al. 1964). New stock ponds and runoff ponds tend to be invaded first by the more adaptable puddle ducks such as mallards, blue-winged teal, and pintails. Some of the diving ducks, such as scaup and redheads, will use more mature ponds (Eng et al. 1979.)

To encourage seed-producing annuals on mudflats, ponds near reliable water sources (e.g., streams, reservoirs, wells) are dried naturally during summer and then flooded shallowly in fall or spring (Hamor et al. 1968). Building islands and rafts, leveling spoil piles, and intensively grazing selected shorelines provide loafing areas for waterfowl (Hamor et al. 1968, Poston and Schmidt 1981, Kantrud 1986b).

Shorelines can be steep with spoil piles for thermal coves (wind protection, south-facing shoreline) or sloping (≤ 1:5) to encourage native plants. Stock ponds are improved with irregular shorelines (e.g., oakleaf, kidney, or L shapes) and basins created where water depths generally are 0.5 m or less (Poston and Schmidt 1981).

Generally, retention reservoirs (and runoff ponds) are far more beneficial to waterfowl than either regular dugouts or diked dugouts. Diked dugouts are better than regular dugouts (Lokemoen 1973). Retention reservoirs are larger, with more diverse shoreline, depth, and vegetation.

For maximum use by waterfowl, stock ponds and runoff ponds should be built in gentle to rolling terrain away from major sources of siltation and pollution, within 1.6 km of natural wetlands, 0.5 to 1.5 ha in size, with more than 40 percent of the area being less than 61 cm deep, emergent vegetation (mostly smartweed and spikerush) covering 30 to 50 percent of the area, submerged vegetation covering more than 20 percent of the area, less than 10 percent of the shoreline bare, about 33 percent of the shoreline rimmed with brush, uncultivated and ungrazed shorelines, and a shoreline irregularity index greater than 1.5 (Stoudt 1971, Lokemoen 1973, Hudson 1983, Rumble and Flake 1983). The *shoreline irregularity index* is the shoreline length, as determined with a cartometer, divided by the circumference of a circle with an area equal that of the pond (Wetzel 1975). Best development of vegetation needs at least 5 years (Hudson 1983). Areas less than 61 cm deep generally produce more plants which provide both food and cover for duck and goose broods (Eng et al. 1979). Evans and Kerbs (1977) developed a form to evaluate the potential of existing ponds, past management programs, or future habitat improvement for brooding waterfowl (Table 3.1).

TABLE 3.1 Evaluation of Past Management Programs, Potential, and Habitat Improvement of Stock Ponds for Brooding Waterfowl

Habitat component	Rating
1. Size	
a. Temporary water	Unsuitable
b. Permanent water, <0.4 surface ha	Inadequate
c. Permanent water, >0.4 surface ha	Good
2. Average shoreline slope, measured from existing water level	
a. > 9 dm/15 horizontal dm	Unsuitable
b. 6–9 dm/15 horizontal dm	Poor
c. 3–6 dm/15 horizontal dm	Fair
d. 0–3 dm/15 horizontal dm	Good
3. Shoreline vegetation within 3 m of existing water level	
a. 0–25% of the shoreline vegetation covered, or shoreline completely covered with tall rank vegetation with no open shoreline for brood resting sites	Unsuitable
b. 25–50% vegetation covered	Poor
c. 50–75% vegetation covered	Fair
d. > 75% vegetation covered, except as in *a*	Good
4. Existing water conditions	
a. Water level low with shoreline vegetation either absent or excessively trampled by livestock	Poor
b. Water level low with a good cover of shoreline vegetation, or pond about half full	Fair
c. Pond full or nearly full	Good
5. Food and cover plants (circle selected rank as follows: 0 = absent, 1 = rare, 2 = occasional, and 3 = common)	
a. Pondweed	0 1 2 3
b. Smartweed	0 1 2 3
c. Spike rush	0 1 2 3
d. Bulrush	0 1 2 3
e. Cattail	0 1 2 3
f. Naiad	0 1 2 3
g. Buttercup	0 1 2 3
h. Water milfoil	0 1 2 3
i. Coontail	0 1 2 3
j. Chara (stonewort)	0 1 2 3

TABLE 3.1 Evaluation of Past Management Programs, Potential, and Habitat Improvement of Stock Ponds for Brooding Waterfowl *(Continued)*

Habitat component	Rating
6. Emergent and aquatic vegetation (refer to ranking of plants in item 5)	
a. No plant listed is ranked above the rare (1) category.	Unsuitable
b. No plant listed in *a* through *d* is ranked as common, but enough emergent and aquatic vegetation exists to rank some of the 10 listed species above the rare category. Listed plant species occupy < 25 percent of the area of water < 6 dm deep.	Poor
c. Of the plants listed in *a* through *d,* one or more, and preferably two, are common on the pond. Aquatic and emergent vegetation occupies 25 to 50 percent of the water area < 6 dm deep.	Fair
d. Of the plants listed, five or more occupy > 50 percent of the water area < 6 dm deep (except as in *e*). Of the species listed in *a* through *d,* two or more are common on the pond.	Good
e. Pond is completely or nearly completely covered with emergent and aquatic vegetation (choked).	Unsuitable

SOURCE: Evans and Kerbs (1977).

Good range management can be good waterfowl management (Bue et al. 1964). But cattle tend to congregate around water sources; overgrazing around stock ponds often is a problem (Eng et al. 1979). A rest-rotation grazing system is best, especially if no spring grazing occurs for 2 years and there was no late-season grazing the previous year (Mundinger 1976, Evans and Kerbs 1977). Early nesting ducks use residual vegetation from the previous growing season. Grassy shorelines preferred by pairs, and brushy and emergent shorelines preferred by broods, develop with grazing rates of 0.8 ha or less to 1.2 ha/AUM—Animal Unit Month (Lokemoen 1973). A stocking rate of 1 cow per 11 ha/yr or 37 cattle-days per year increases the duck population on stock ponds (Bue et al. 1952). Otherwise, fencing placed at least 12.2 m from the margin, except for access points for cattle, protects cover for ducks (Hamor et al. 1968). But fencing is expensive, and the results are questionable (Eng et al. 1979). Islands also partially offset the loss of shoreline cover from grazing and offer protection from mammalian predators. Because management for ducks and shorebirds is not always compatible, grazing systems which encourage mudflats for shorebirds might be used on ponds less than 0.4 ha (Evans and Kerbs 1977). (See "Controlled Grazing" in Chap. 5.)

Ducks also benefit from stock ponds near fields of small grain, which will be used for food and by hens nesting in residual and new growth. Water in stock ponds is more reliable if natural pond basins in adjacent areas are not drained (Rumble and Flake 1983).

Rock piles of riprap placed along the north shore of ponds 0.4 ha or larger are used as sunning sites by turtles and snakes and as shelter by bullfrogs and salamanders (Johnson 1983). To encourage amphibians, no fish are stocked. Brush piles and tree branches should be placed in water 6 dm or less deep for egg laying. Some downed tree branches should be placed along at least 25 percent of the total pond bank. In most ponds, 5 to 10 cedar logs (preferably), 1.5 to 2.4 m long and 15 cm wide, should be placed with part of the log in the water, preferably with the entire underside touching the bottom.

Retention Reservoirs

Retention reservoirs are essentially small impoundments with a dam sloped 3:1 that is 1.8 m wide at the top, 6 dm higher than the emergency spillway, and with a water control structure in watersheds larger than 10 ha (Atlantic Waterfowl Council 1972, Eng et al. 1979). In watersheds smaller than 10 ha, a vegetated spillway is used for normal runoff to escape from the pond (Fig. 3.11). If a natural spillway cannot be used, a one-blade cut on small watersheds at the desired flow level somewhere around the pond periphery usually will suffice to release excess water (Atlantic Waterfowl Council 1972). Spillways and dikes are paved with topsoil and seeded with a good grass mixture for stability.

Gentle rolling land is best for retention reservoirs because of the numerous small watersheds and basins, diversified drainage patterns, and gentle shoreline slopes (Atlantic Waterfowl Council 1972). Best sites contain side channels which can be flooded to increase the shoreline/pond-area ratio (Fig. 3.12) and to provide relatively inexpensive development of cutoff islands (Eng et al. 1979). (See "Islands" in Chap. 6.) The borrow pit for the dike is on the impoundment side of the dike and should be deep enough to reduce the chance of a complete dry-up in late summer.

Generally, aquatic plants pioneer into newly created ponds. But some ponds, especially isolated ones, might need transplants of native aquatic plants to accelerate the sequence toward marsh development (Eng et al. 1979). Planting rootstock might be best in late summer when low water levels expose suitable sites.

Many retention reservoirs are in small (less than 10-ha) watersheds. Thus, expensive water control structures are not used, although a means of draining the pond is desirable to control vegetation, undesirable fish and turtles and to make repairs (Linde 1969). In large watersheds where water control structures are used, retention reservoirs are managed as controlled marshes. Ponds built in upland watersheds that are too small will fail to fill or might dry up in sum-

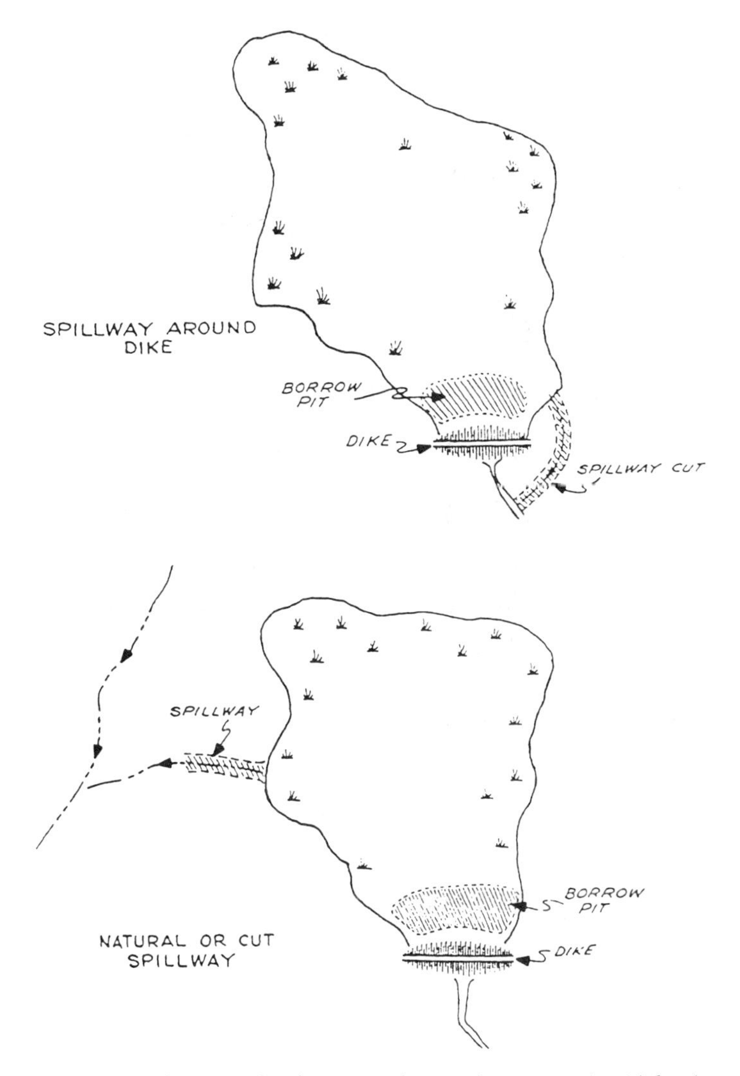

Figure 3.11 Common development of retention reservoirs (*Atlantic Waterfowl Council 1972*).

mer because only runoff (surface) water is used, rather than groundwater, too. A pond built on a watershed that is too large will need expensive dikes and control structures unless part of the flow can be diverted (Linde 1969). A watershed ratio of 20 ha/ha of impoundment seems about right (Atlantic Waterfowl Council 1972), but soil and vegetation types are influential, i.e., smaller watersheds for pasture and cropland, larger for woodland or brushland (Addy and MacNamara 1948). The retention pond will develop into a good

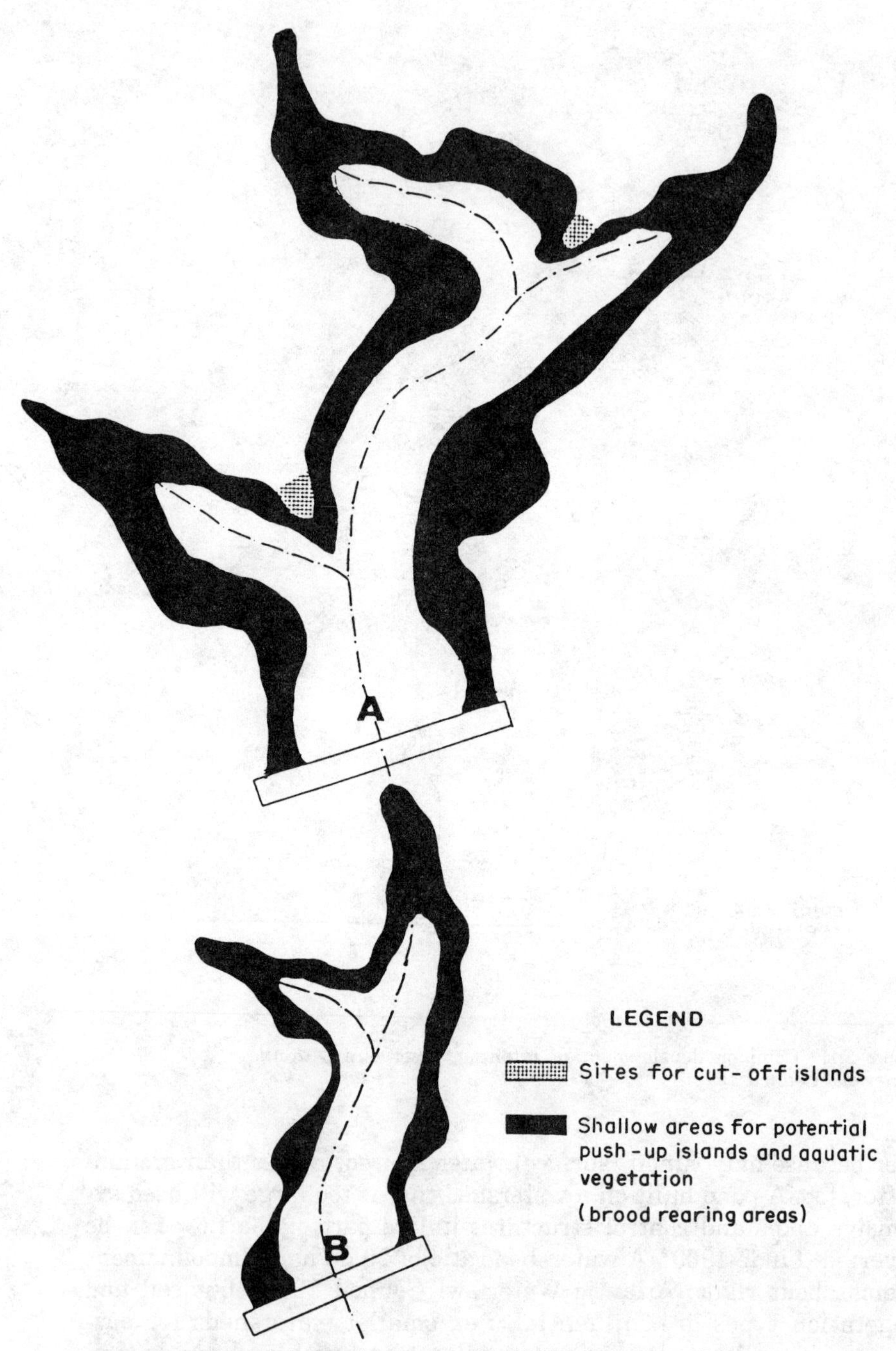

Figure 3.12 Potential sites for construction of islands in retention reservoirs (*Eng et al. 1979*). Site A is better than B because A has more shallow water and potential sites for islands with about the same size dike.

waterfowl nesting and brooding area if it is surrounded by grass cover and scattered small potholes are added (Linde 1969).

Other considerations for site selection are similar to the development of controlled marshes, including good water-holding capacity of the soil. Qualified soil scientists should be consulted if needed.

Dugouts (Pit Reservoirs)

Dugouts (pit reservoirs) are holes dug 1.8 to 3.7 m deep about 50 m long and 20 m wide with a surface area of 0.05 to 0.10 ha (Bue et al. 1964, Eng et al. 1979). Sides are generally steep, but one or both ends slope gently to allow livestock to drink. Ducks use dugouts in temporary or semipermanent wetlands more than those on intermittent waterways and level terrain. Pits designed to catch runoff often are dug in coulee bottoms or on the edge of large temporary potholes. Pits dug in areas with a high water table fill with groundwater. Ducks and geese prefer not to use dugouts because of the steep banks and consequent lack of emergent vegetation, unless the dugout is full and water flows into surrounding vegetation. But such a situation is more typical of diked dugouts, although it happens with dugouts constructed in large temporary potholes with ample water.

Dugouts serve ducks mainly as breeding-pair territories in spring, but contain little animal food for laying hens. Broods seldom use dugouts. Essentially, dugouts are unattractive to waterfowl. Other watering facilities more attractive to waterfowl should be used to provide water to a grazing program.

Improvements to dugouts include a 7:1 slope on one or both ends, grasses planted on the banks, a rest-rotation grazing system or fencing placed at least 12.2 m from the dugout's edge except for access points for cattle, islands, and 1.2-m rafts of logs or boards anchored in the center of the pond as loafing sites. Gentle slopes encourage growth of aquatic plants that supply food and cover. Dugouts are constructed mechanically as with dug potholes or level ditches. Dugouts can be designed more for waterfowl if dug 1.5 m or less deep in areas with a high water table and about 56 m^2 large, with an island for loafing and nesting sites (Vaughn 1976).

The dugout is begun with a sketch and notes on the depth, location of spoil, and bottom slopes and contours (Poston and Schmidt 1981). Irregular shorelines (crescent, kidney, rounded-L, dog-leg, oak-leaf) produce more edge than square or rectangular dugouts (Figs. 3.13 and 3.14). After the topsoil is removed and piled nearby, the subsoil is bulldozed around the edge and piled into at least one island and then spread and landscaped around the edge of the pothole. Next the topsoil is spread over the subsoil on island(s) and edge and seeded to a suitable grass-legume mixture.

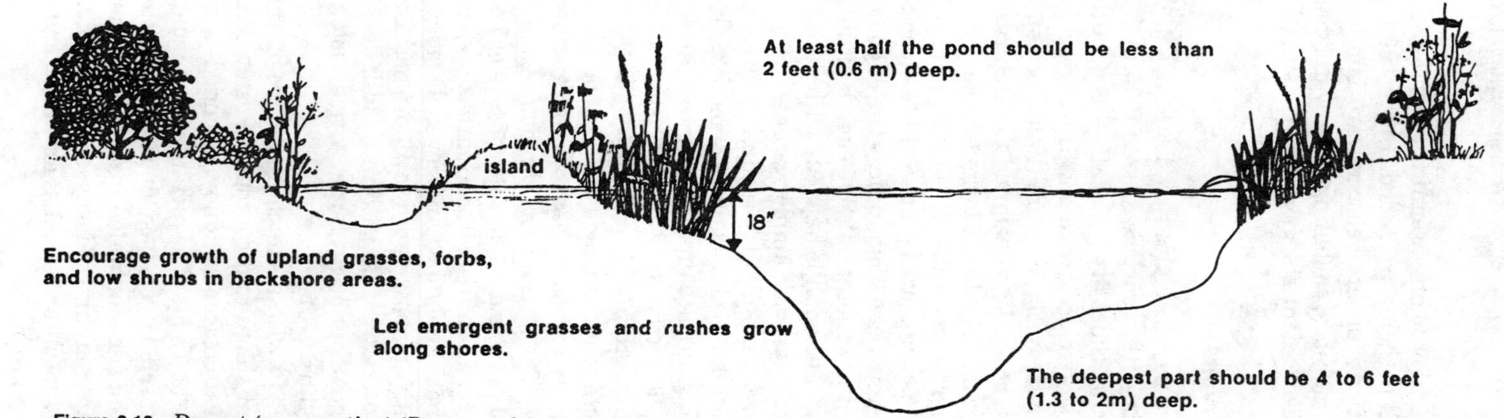

Figure 3.13 Dugout (cross section) (*Poston and Schmidt 1981*).

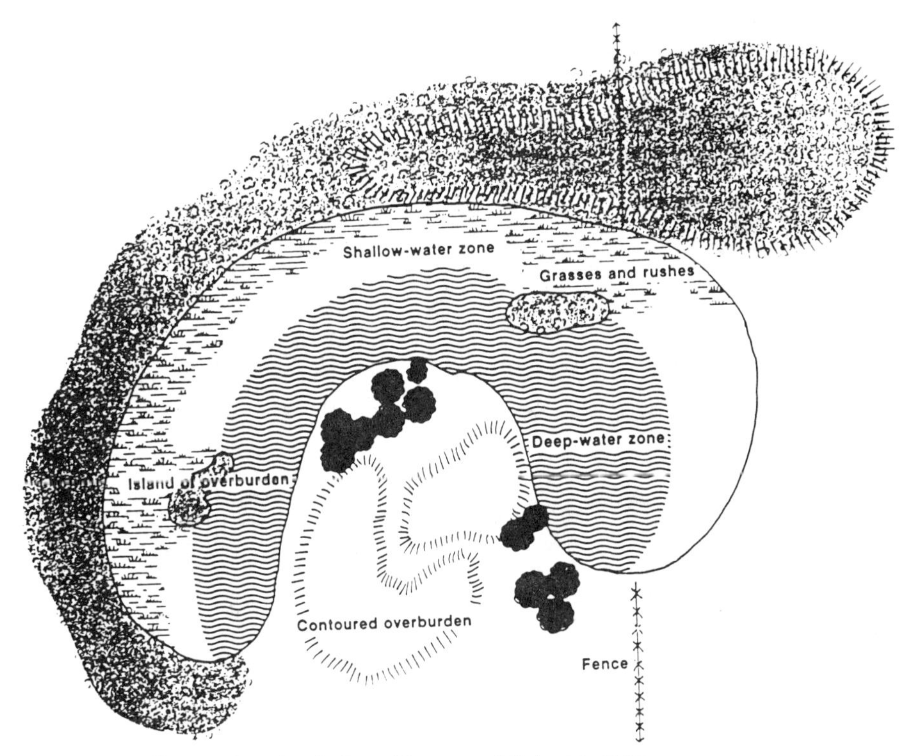

Figure 3.14 Dugout (overhead view) (*Poston and Schmidt 1981*).

Larger dugouts (1.6 ha or less) of irregular shape and bottom contour and depth, with islands (Fig. 3.15), can be readily surveyed and then dug in sedge meadow with a high flotation dozer (personal communication, T. Meier, Wisconsin Department of Natural Resources 1991). They are expensive but require no maintenance.

Diked Dugouts (Pit Retention Reservoirs)

Diked dugouts (pit retention reservoirs) collect runoff water if material excavated from the pit is deposited in a dike on the downstream side. In years of high water, the dike backs the water into the surrounding area, usually causing a dense growth of emergent vegetation to develop, which provides excellent feeding and brood-rearing areas for waterfowl (Eng et al. 1979). During years of low rainfall and runoff, the flooded area dries up, but the pit retains water for survival of broods nearing flight stage.

Dugouts are constructed mechanically as with dug potholes. The dike is developed across the low area as a simplified embankment dike.

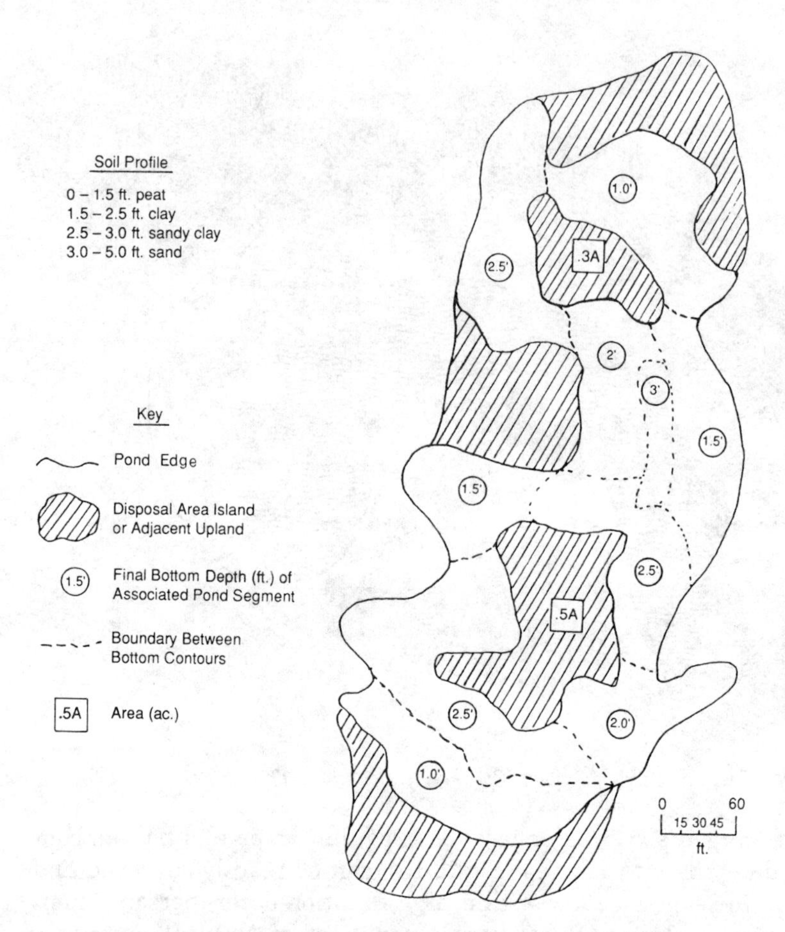

Figure 3.15 Example of a 1.5-ha (3.8-acre) dugout built in a sedge meadow community (*Courtesy, T. Meier, Wisconsin Department of Natural Resources, 1991*).

FARM PONDS

Farm ponds can be excavated ponds or embankment ponds (U.S. Soil Conservation Service 1982). Excavated ponds are similar to dugouts or diked dugouts. Embankment ponds are similar to retention ponds or runoff ponds or even controlled marshes. The considerations for construction are the same. Farm ponds are used for a variety of purposes such as stock water, irrigation, fishing, or other recreation, but usually the main reason for their construction is for fishing. The size, depth, water flow, vegetation, and location (close to buildings) render farm ponds less attractive to waterfowl than many other water areas, particularly for breeding (Edminster 1964). Ducks use them more for rest during migration, especially during spring. Improvement of farm ponds for waterfowl is similar to that for retention reservoirs or dugouts and diked dugouts.

MINES AND GRAVEL PITS

Mining companies must reclaim mined land to its original productivity. Wetlands can be developed from final cut areas (the last area mined in a surface mine) and from areas of original wetlands destroyed by mining (Rumble 1989). Sand and gravel pits, too, can be modified as wetlands during or after the mining operation (Matter and Mannan 1988).

Habitat improvement techniques are similar to those for dug potholes, playas, stock ponds, and even controlled marshes. Surface areas should be 0.4 to 4.0 ha, with the shoreline irregularity index at least 2.2 (Proctor et al. 1983b, Uresk and Severson 1988). Pits should have at least 493 m of shoreline for 0.4 ha and at least 1559 m for 4.0 ha. Overall bottom contours should approximate the shape of a shallow saucer, with gently sloping shores of 11 to 22 percent and local irregularities on the bottom to increase interspersion of shoreline and shallow- and open-water areas (Green and Salter 1987). Islands should be built in large wetlands.

Bottom materials should be covered with at least a 20-cm layer of topsoil originally removed during excavation, or muck brought into the new pit from nearby wetlands if possible, and plants should be established (Crawford and Rossiter 1982, Morrison 1982, Sanders et al. 1982, Ross et al. 1985, Green and Salter 1987). A 2-cm layer of hay or straw can be spread on the bottom of pits to expedite development of aquatic invertebrates and food chains (Street 1982, Green and Salter 1987). Pits should be managed to maintain 1500 stems per square meter of aquatic vegetation in shallow areas (Uresk and Severson 1988) and at least 150 benthic organisms per square meter (Belanger and Couture 1988). For dabbling ducks, pits should be 0.3 to 2.0 m deep, with 30 to 70 percent of the pit less than 60 cm deep; for diving ducks, pits should be 0.6 to 2.4 m deep and average 1.0 m (Proctor et al. 1983b, Lokemoen et al. 1984, Monda and Ratti 1988, Uresk and Severson 1988). Newly reclaimed areas often have no natural perches for birds. Small birds will perch on a cable strung across the pond (Allaire 1979). Management of sewage ponds is similar, but they should have a shoreline irregularity index of at least 1.5, at least 30 percent emergent cover, and a stem density of at least 30 per square meter (Belanger and Couture 1988). Snyder and Snyder (1984) presented feasibility diagrams and requirements for using oil shale wastewater in developing a closed-system waterfowl wetland, which

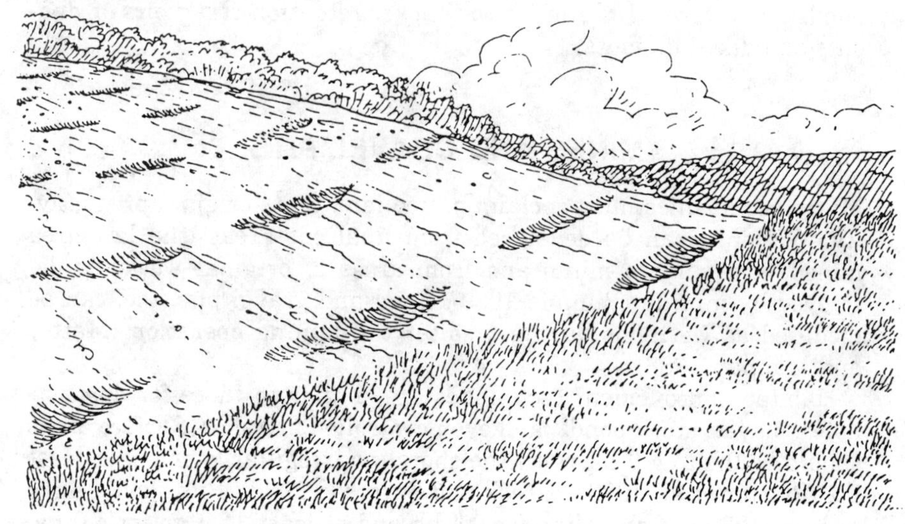

Figure 3.16 Slope with basins (*Ohlsson et al. 1982*).

also might have potential for sewage treatment and other wastewater management.

Basins and gouges can be used in restoring surface-mined lands to benefit wildlife by increasing topographic diversity and providing small seasonal ponds while reducing water runoff and increasing infiltration (Grim and Hill 1974, Ohlsson et al. 1982). They are not used on unstable or toxic fill materials or slopes steeper than 20 percent.

Basins should be 3 to 6 m long, 0.6 to 1.2 m deep, and 1.5 to 1.8 m apart, built on the contour in rows 6 to 9 m apart and aligned so that no uninterrupted runoff path occurs (Fig. 3.16). Basins are dug with the blade of a bulldozer dropped at an angle or with a basin-former attachment to a dozer (Brown 1977). Gouges should be 0.9 to 1.2 m long, 36 to 81 cm wide, and 15 to 20 cm deep dug along the contour with a gouger pulled by a tractor (Brown 1977). Herbaceous cover, with mulching, is established with broadcast seeding.

FLOODWATER-RETARDING STRUCTURES

Floodwater-retarding structures (FWRSs), also known as flood detention structures, flood prevention lakes, PL-566 impoundments, small watershed lakes, and pilot watershed structures, provide valuable habitat to migrant and wintering waterfowl in the South (Bates et al. 1988) and can be managed in limited fashion as a sort of controlled marsh. Typically 11 to 18 ha in size, they are designed to control floods mainly. They have dams with a mechanical spillway consisting of a vertical concrete inlet and a conduit through the dam through which water flows when levels exceed the elevation of the inlet. An FWRS usually contains drain valves to allow water-level manipulation. The emergency spillway seldom is used because of the large storage capacity. Impoundments built for grade stabilization, erosion control, and other limited uses, typically about 2 ha in size, contain some of or all the features of FWRSs. Many FWRSs have extensive littoral zones. FWRSs can be improved for waterfowl by

1. Designing the impoundment with 25 to 50 percent of the surface area less than 0.9 m deep
2. Installing a drawdown slot in the mechanical spillway to facilitate water-level control for food plantings and natural foods
3. Seeding annual grasses for waterfowl food and turbidity control on areas disturbed during construction
4. Planting corn, grain sorghum, or other foods near impoundments and submergent, floating-leaved, and emergent aquatic plants in impoundments for food and cover
5. Using the FWRS as a source of water to flood downstream bottomland hardwoods or food plantings
6. Fencing sections of the shoreline to exclude livestock and encourage waterfowl food plants

RESERVOIRS AND LAKES

Large reservoirs generally are built by impounding water behind a massive dam for purposes such as hydroelectric power, irrigation, and flood control. As with large lakes, such reservoirs provide resting sites for waterfowl. White-winged scoters, and probably many other diving ducks, prefer lakes over 45 ha with extensive areas 1 to 4 m deep and dense submergent vegetation for feeding (Brown and Brown 1981). Broods use shallow open-water areas with emergent vegetation and some protection from wave action. Islands are used for nesting if they have low spreading shrubs such as gooseberry, snowberry, rose, and raspberry and are about 50 to 150 m from water. Open shorelines are used for loafing.

Management practices include (1) applying proper soil and water conservation practices throughout the watershed, (2) controlling agricultural production on nearby lands, (3) seeding all disturbed areas around the reservoir or lake immediately with annual grasses, (4) establishing desirable natural food and cover plants, (5) establishing some areas of gentle slope for shallow water, (6) establishing some open shoreline for loafing, (7) planting windbreaks for thermal protection for loafing and roosting, (8) fencing selected areas around the lake or reservoir to exclude cattle partially, (9) controlling undesirable vegetation surrounding the lake or reservoir, (10) creating shallow subimpoundments, (11) manipulating water levels somewhat on smaller reservoirs [2 to 3 years of drawdown, then 2 years without drawdown (Cooke et al. 1986)], (12) providing artificial islands for loafing, and (13) protecting from disturbance by establishing boating restrictions or refuges (White and Malaher 1964, Hobaugh and Teer 1981, Johnson and Montalbano 1989, Ringelman et al. 1989). Kahl (1991) discussed shallow lake management.

Structural measures to reduce erosion include floating-tire breakwaters, plant rolls (see "Plant Propagation" in Chap. 5), erosion-control fabrics, willow-fence combination, wattling bundles, brush layering, brush mattress (matting), revetment, crib structures (Allen and Klimas 1986) as well as berms, booms, riprap, gabions, islands, and other measures (see Table 2.13 and Fig. 2.39). Federal, state or provincial, and private (e.g., Ducks Unlimited) wildlife agencies can establish waterfowl refuge and management areas along reservoirs when and where appropriate.

RIVERS AND SLOUGHS

Rivers generally are the first water habitats open to migrating waterfowl in spring, if enough current exists (Pederson et al. 1989). Spring floods and flowing water scour islands and shorelines to reduce overgrowth of woody vegetation. An open river channel of at least 150 m maintains roost sites for migrating waterfowl and cranes (Currier et al. 1985). To maintain river channels, techniques include maintaining river flow, mechanized clearing and burning, herbicide spraying, and bulldozing vegetation off islands and leveling them to near the base flow of the river (Aronson and Ellis 1979). In late summer when flow is low enough to accommodate farm machinery in some rivers, brush control is accomplished with large disks equipped with 92-cm notched disk blades, followed by repeated treatments (disking and brush hogging) along with scouring floods.

Streams with a gradient of less than 11 percent and a sinuous channel, with bends separated by distances five to seven times the stream width, receive maximum use by wildlife (Green and Salter 1987). During low water or when the channel is dry, pools on bends should be excavated 0.5 to 1.0 m deeper than the channel bottom. For fish and aquatic mammals, overhanging banks can be constructed on the outside bends of stream channels by using log or plank platforms. These are placed on the bank to extend somewhat over the channel, then are covered with rock and then sod. Grass and sedge are planted. Shrubs and trees (e.g., alder and willow) should be planted 2 to 5 m back from the stream bank to prevent excessive shading. In slow-moving areas, clumps of emergent plants (e.g., bulrushes, cattails, sedges) can be planted, with root clumps of 0.25 m^3 anchored with rocks or metal staples until established. To improve production of macroinvertebrates, at least 75 percent of a stream's length should be planted to woody vegetation with periodic gaps left unplanted along riffles (McCluskey et al. 1983, Melton et al. 1987) (See Fig. 5.1). Stream banks can be stabilized by reintroducing beaver to build dams.

Correcting or preventing damage to channels for fish management is similar for wildlife management and includes check dams, sills, flumes, revetments, gabions, jetties, lining, and stabilized plantings (U.S. Forest Service 1974b, Nelson et al. 1978, Seehorn 1985, Payne and Copes 1986). Large rivers with heavy navigational use need extensive bank stabilization and other protective measures (Schnick et al. 1982).

Warmwater sloughs fed by warmwater springs year round never or seldom freeze. Thus they provide winter habitat to waterfowl, including foods such as duckweed, watercress, and snails. Warmwater sloughs occur usually with sandy alluvial soils near major rivers (Ringelman et al. 1989). Beaver dams, fallen trees and other debris, and cattail and other emergents might require periodic removal if the water slows and the slough freezes.

HIGHWAY RIGHTS-OF-WAY

Where highways bisect wetlands, loss is minimized by expanding the wetlands into the right-of-way along each side of the highway (Fig. 3.17) or by creating 4 to 12 shallow, linear ponds per kilometer on each side of the highway, with ditch blocks (plugs) or excavation, each pond about 60 m long (Fig. 3.18), especially if near permanent or semipermanent natural wetlands (Fig. 3.19) (North Dakota State Highway Department 1978, Oetting 1982). Portions of wetlands next to and extending into the right-of-way should be protected from tillage and remain undrained during dry years (Fig. 3.20). Such ponds provide resting and feeding areas for ducks and invertebrates as food for broods. Waterfowl seem to prefer semipermanent ponds 30 m long and over 6 m wide with good cover along roads (U.S. Fish and Wildlife Service 1985).

Borrow pits for highway construction should be shaped to benefit

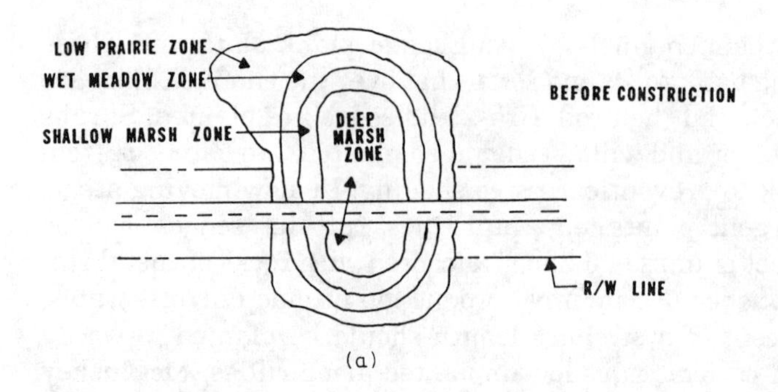

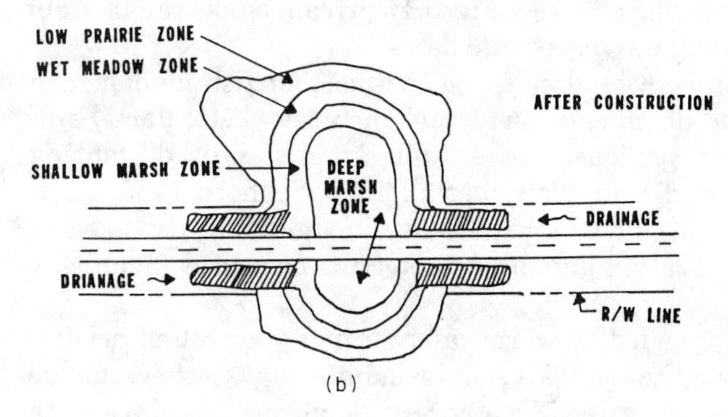

Figure 3.17 Expansion of existing natural wetlands into highway right-of-way (R/W) to minimize loss of wetland area from construction (*North Dakota State Highway Department 1978*).

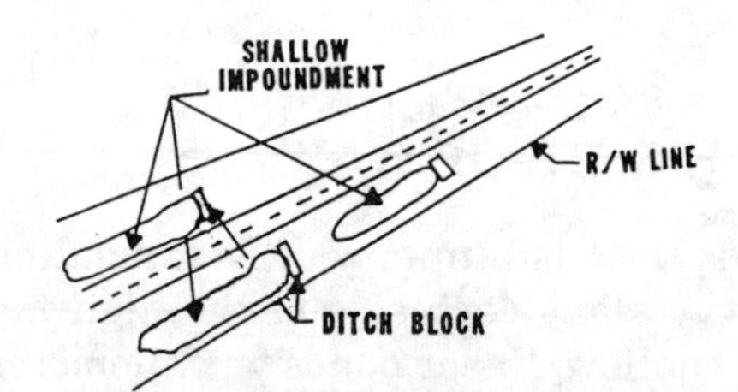

Figure 3.18 Highway right-of-way (R/W) with shallow ponds created by blocking ditches with plugs (*North Dakota State Highway Department 1978*).

waterfowl and other wildlife (Fig. 3.21) (Leedy and Adams 1982). A dam or partial dam in the borrow pit or right-of-way can be built of fill from the borrow pit to impound small streams or runoff, perhaps with a simple water control structure installed. Topsoil from a nearby donor marsh with a viable seed bank might be useful; otherwise, trans-

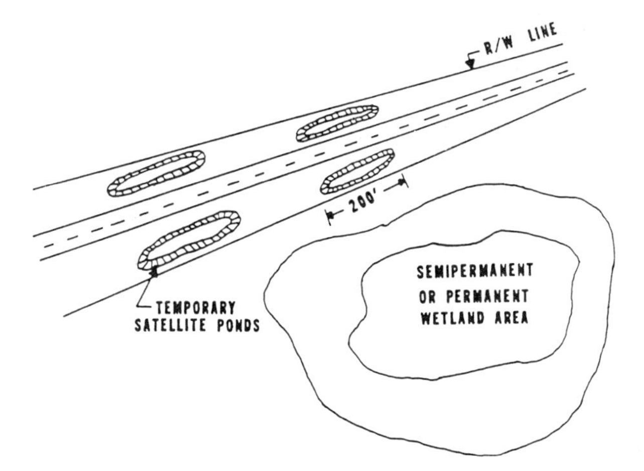

Figure 3.19 Shallow ponds dug in highway right-of-way (R/W) near permanent or semipermanent wetland (*North Dakota State Highway Department 1978*).

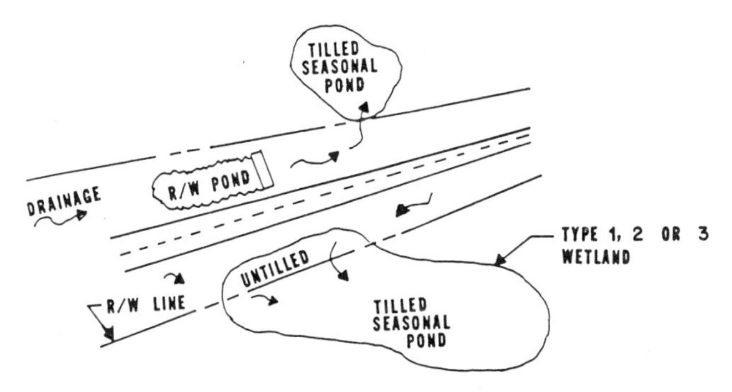

Figure 3.20 Parts of wetlands protected from tillage next to highway right-of-way (R/W) which remain undrained in dry years (*North Dakota State Highway Department 1978*).

planting might be needed (Garbisch 1986). Legal and engineering requirements must be met.

DRAINAGE DITCHES

Small seasonal wetlands drained with open ditches, for farming or mosquito control, are the simplest wetlands to restore (Piehl 1986). Restoration of drainage ditches usually is accomplished by plugging the ditch. Plugs, usually called *gut plugs,* hold water in the ditch, thus

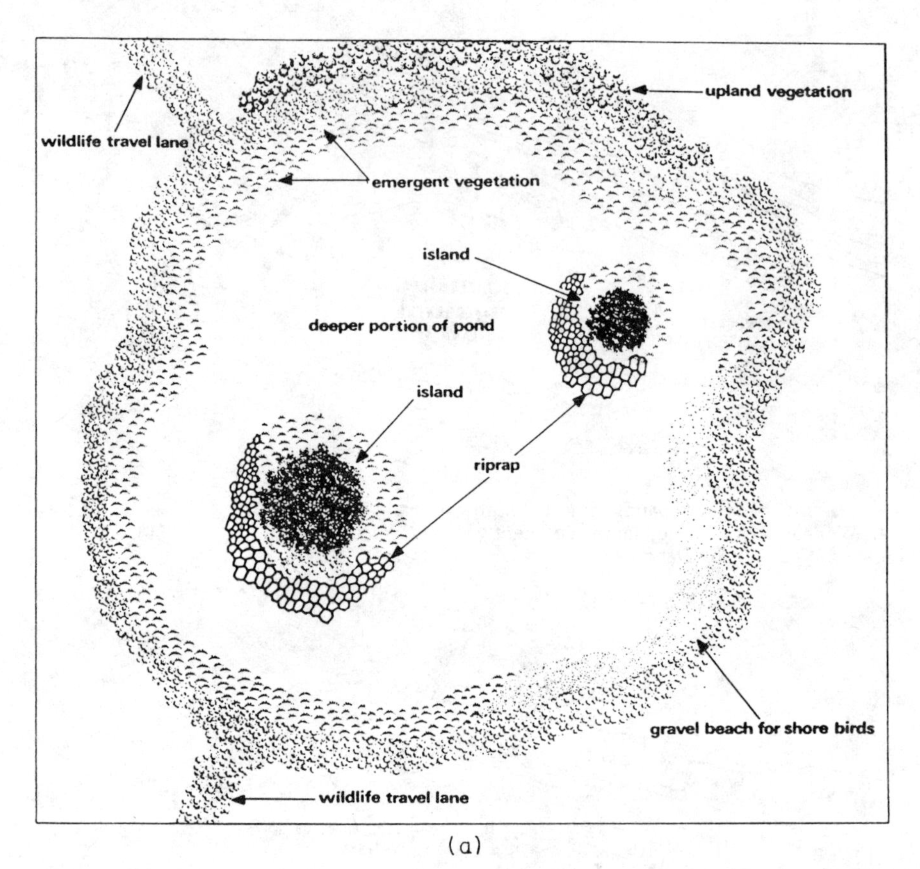

(a)

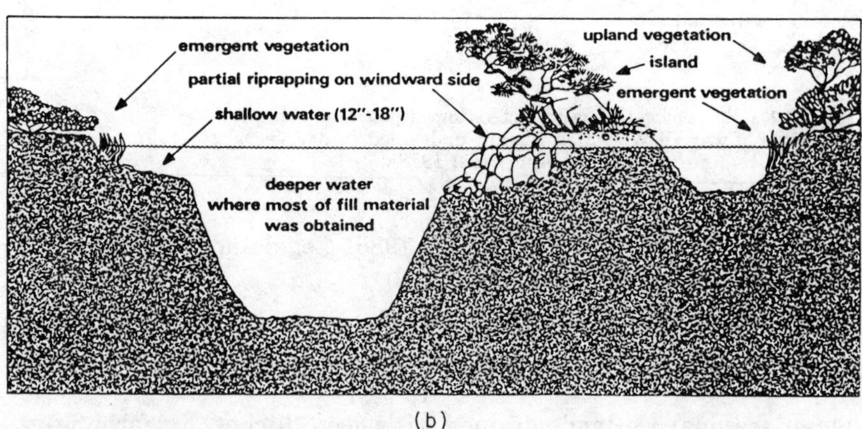

(b)

Figure 3.21 Hypothetical borrow pit developed into pond designed for waterfowl and other wildlife (*Leedy and Adams 1982*).

raising the water table in the wetlands. In tidal areas, high tides flow over the plug to fill the ditch and surrounding marsh. The area of stable water is increased when the tide falls, for the gut plug retains the water in the ditch at marsh level (Atlantic Waterfowl Council 1972). Gut plugs can be of a spilling or nonspilling type. Spilling gut plugs allow excess water to escape through the plug.

Nonspilling Earth Plugs

A good bond must be made between the plug and the bottom and sides of the ditch. A dozer cores organic material out of the ditch. Then the plug, built preferably with clay from a nearby borrow site, is dumped into the ditch, compacted by the dozer to a height several decimeters above the water level, topped with black dirt, and seeded (Piehl 1986). A spillway is cut at one side of the plug. This plug is fairly simple to construct. It also can be used to stabilize water levels in bayous of tidal areas to provide permanent water for ducks and access by humans (Chabreck 1968). As with weirs, small boats must be pulled over the plug to reach the ponds behind regardless of the tide. Major drainage bayous are impractical or undesirable to plug; shallow laterals less than 6.1 m wide are best. Plugs on bayous can be modified by installing a corrugated metal culvert with an automatic flapgate on the downstream end and a stoplog device on the upstream end. Such ponds are managed as controlled marshes.

Nonspilling Wooden Plugs

Instead of earth, Wakefield piling of creosoted lumber can be used to plug the ditch (Atlantic Waterfowl Council 1972). The piling must be long enough to prevent undercutting. Wing walls must be long enough to prevent water from cutting around the end of the plug. Nonspilling wooden plugs are used mainly in tidal areas of moving water.

Spilling Plugs

Wetlands with larger and faster watersheds need culverts installed in the ditch plugs to release excess water (Piehl 1986). A tough blanket of filter cloth reduces erosion and washout if placed below the culvert and covered with a layer of rock. A spillway also is constructed at one end of the plug at the same elevation as the top of the culvert.

Control of the spilling plug and management of the marsh can be achieved by installing in the ditch a standard water control structure with stoplogs and then plugging the ditch. Construction and management are similar to those for controlled marshes.

TILED WETLANDS AND CULVERTS

The first step to restore a tiled wetland is to locate and plug or remove all the buried tiles (Piehl 1986). This usually is done with a probe or backhoe. Once the tile is located, 6.1 to 9.1 m of it is removed, and the hole is filled in.

To leave the tile intact or to control the maximum pool elevation, a tile riser can be installed on the existing tile. First a backhoe locates the tile, removes 6.1 to 9.1 m of it, then exposes an undamaged tile. Depending on the size of the tile, a riser of plastic sewer pipe or corrugated metal pipe is tightly cemented onto the exposed tile so that head pressure will not blow the first downstream tile. An antiseep diaphragm is attached to the new horizontal pipe to prevent water from seeping back into the old tile. The elbow must rest on a good foundation of a small slab of concrete or even two cement blocks. Then the vertical riser pipe is installed, the hole filled in, and the debris cover set over the riser. The debris cover consists of a dome made of metal rod attached to about 0.9 m of corrugated metal pipe. The only maintenance is to keep the debris cover cleared of vegetation and other debris. To restore wetlands with both ditches and tiles, the ditches must be plugged and the tiles destroyed.

Road culverts can be raised so that less water escapes from some wetlands (Piehl 1986). Once the necessary permits are obtained from local town boards and other governing bodies, the culvert often is raised in a few hours with a dozer or backhoe and is reset higher at the same site with appropriate support structures and fill beneath.

DIKED HISTORIC BAYLANDS

Wetlands in many saltwater bays have been damaged or eliminated mainly through diking and filling. Harvey et al. (1983) described techniques to restore diked historic baylands to fresh and tidal marshes.

The tidal range and the period of inundation of the flat, former marsh plain are the most important design parameters to consider when diked historic baylands are reopened to tidal action. The tidal range should be measured at or near the site and compared to the local tide gauge from the U.S. Department of Commerce National Oceanic and Atmospheric Administration (1980). The two approaches to designing appropriate marsh elevations relative to the tidal range involve controlling the inundation and modifying the topography.

Where extensive subsidence has occurred, or where maximum tidal elevation would cause undesirable flooding, tidal elevations can be controlled by an automatic tide gate. This needs continued maintenance. It is used where the original marsh drainage network produces proper water exchange.

Three types of topographic modification usually are considered: marsh plain, islands and dikes, and channels. After an accurate topographic survey, grading proceeds carefully because errors of only 14 cm will alter the success of plantings.

Filling the marsh plain is not necessary if it has remained undisturbed, subsidence has not occurred, or tidal flooding will be controlled mechanically. Otherwise, soft estuarine mud must be used as fill material, the marsh plain must be graded, and a tidal drainage network should be graded. Usually, islands are built, and portions of dikes are preserved after breaching (Harvey et al. 1983). During high tide, wetlands designed with channels, ditches (narrow channels dug in the high marsh for mosquito control), and mudflats should provide roosting habitat and protected areas such as upland edges or curved or elongated islands for shorebirds, waterfowl, and other waterbirds (Josselyn and Buchholz 1984) (Tables 3.2, 3.3).

Once the levee is breached, tidal flows will establish a new drainage system after much time (Harvey et al. 1983). Channels can be graded to develop a new drainage system before the levee is breached. The design should be modeled on the natural pattern of nearby channels and sloughs. Relatively narrow channels with vegetated edges are preferred (Josselyn and Buchholz 1984). Two types of channels occur naturally (Harvey et al. 1983). Slough channels that are wider than about 1.5 m have shallow gradients and conduct ebb and flood tides equally. Smaller tributary channels have steeper gradients and flow appreciably only during ebb tide. Wider channels should be dug at least 6 m wide with no slope and with the bottom about as deep as the main outflow channel. In large marshes, side slopes can be at the angle of repose or cut back to 1:10 to produce mudflats and diverse vegetation. Tributary channels should be dug at least 1.8 m wide, 9 to 15 dm deep but not deeper than the main slough channel, with side slopes at the angle of repose and graded preferably toward the main slough channel.

No point on the marsh plain should be farther than about 30 m from a channel. Channels should meander in a pattern similar to the natural system, with junctions of slough channels at about 120° and junctions of tributary channels with slough channels at about 90°. Such a

TABLE 3.2 Habitat Design for Salt Marsh Channels and Mudflat

Species objective	Habitat	Elevation*	Design criteria: Channel shape	Design criteria: Critical periods	Remarks
Deposit feeding invertebrates (shorebird prey)	Mudflats	MLLW-MTL	Wide, flat bottoms	Spring and fall bird migrations	Slow currents
Filter feeding invertebrates (clapper rail prey)	Channel banks	MLW-MTL	Channels with rapid tidal flow	Periods of high sedimentation	Ribbed mussel in cordgrass
Reduction in burrowing isopod damage	Channel banks	MTL-MHW	Create low slopes to encourage sedimentation in burrows	Periods of strong wave action which undercuts banks	
Clapper rail nesting habitat	Channel edges	MTL-MHHW	Cordgrass/pickleweed next to narrow (1 m) wide channels	Nesting period in spring	Rats considered main predator
Heron/egret feeding habitat	Channels	MLW-MTL	Open channels with water present during tidal cycle	Predators on young birds in nests in spring	Tall, nesting vegetation at edge of marsh
Diverse juvenile fish populations	Channels and ditches	MLLW-MHW	Narrow, vegetated channels with moderate current	Highest abundance of juveniles in spring and early summer	Least use in fall
Mosquito fish dispersal to high marsh pools	Ditches	MHW-MHHW	Narrow ditches with deep sections to hold water during low tide	Periods of long exposure and drying	
Waterfowl	Open water	Below MHW	Broad expansive areas with brackish or freshwater plants	Winter storms	Often needs some protected refuge during storms

*MLLW = mean lower low water, MTL = mean tide level, MLW = mean low water, MHW = mean high water, MHHW = mean higher high water.

SOURCE: Josselyn and Buchholz (1984).

TABLE 3.3 Habitat Design Criteria for Upland Habitats Associated with Salt Marsh

		Design criteria		
Design objective	Habitat	Habitat configuration	Critical period	Remarks
Shorebird and waterfowl refuge during high tide	Bare ground	Island habitats with areas protected from wind	Spring and fall migrations	Birds easily disturbed by humans or feral animals
Shorebird and waterfowl refuge during high tide	Vegetated upland	Islands or upland edge above EHW* isolated from human disturbance	EHW* tides	Continuous cover from upland, resolve pH problems early
Shorebird and waterfowl refuge during high tide	Upland pathways near public access	Isolated from human and pet disturbance through channels or vegetated berms next to access points	High tides associated with frequent human use	Education on control of pets important

*EHW = extreme high water.
SOURCE: Josselyn and Buchholz (1984).

pattern should cover the largest drainage area in the shortest length. Preferably, islands should be developed between the channels. Channels can prevent intrusion of humans and feral animals into the marsh.

Net dilution of over 50 percent should provide adequate circulation in a typical tidal cycle to ensure adequate water quality. Marshes located at the upstream end of a long slough channel might contain low-quality water because some of the water leaving the marsh during the ebb tide might return at high tide. If the channel system is complex and pollution dispersion is a concern, computer modeling can be used to predict the transport of conservative (i.e., nondecaying) pollutants such as salt and the decay of nonconservative pollutants during successive tidal cycles, or the dispersion/exchange ratio can be estimated (Dyer 1973). Circulation and oxygen content are improved with areas of open water exposed to wind, a complex channel system allowing ebb and flow in different directions around marsh islands, and depressions on the marsh plain drained adequately to reduce stagnant water mosquito habitat and excessive soil salinity.

Once filling and grading are complete, the levee is breached. The size of the breach or culvert opening affects the circulation and tidal range in the marsh. Generally, the opening should be as wide as possible. To develop a tidal cycle inside the marsh for a typical tidal cycle in the bay, the difference between the tidal range (water height) on each side of the levee can be calculated (French 1985).

If the water level in the marsh will be managed, an automatic slide gate controlled by a water-level sensor is the most flexible where electric power is available. Otherwise, a gravity-controlled flapgate can be designed. A backup flap tide gate must be provided in case the control gate jams. Trash barriers must be included. Control gates constrict water flow and thus alter tidal hydraulics of the marsh. Control structures should be installed slightly higher than the bottom to avoid sedimentation and facilitate cleaning. When tidal action is restored, substantial sedimentation should be anticipated in areas where the water is deep and the velocity is low.

The following checklist specifies the information needed to plan restoration or enhancement of diked historic baylands (Harvey et al. 1983):

1. A topographic map of the site is prepared in 3-dm contours, showing storm drains, elevation (relative to national geodetic vertical datum) of adjacent surrounding properties, and the limit of the 100-year tide.
2. A topographic map is prepared in 3-dm contours, showing proposed modifications to the site: (*a*) typical cross sections showing proposed elevation of the marsh plain, any channels, and any high areas; (*b*) the estimated tidal range relative to mean higher high water, mean high water, mean lower low water, mean sea level, maximum predicted tide, and 100-year tide; (*c*) ratios of typical horizontal to vertical slopes for existing and proposed levees and channels or sloughs; (*d*) proposed plant species along the cross sections of their expected zone of growth.
3. Sizes of levee breaches or pipe installations are determined: (*a*) amount of cut and fill as well as material needed to strengthen the levee; (*b*) expected tidal exchange and range inside and outside the levee breach; (*c*) plant species, riprap, or other erosion control measures to moderate tidal forces at the breach; (*d*) detailed drawings of inlet-outlet structures used.
4. Soils are identified at the site and for fill with regard to type, salinity, pH, organic content, and bulk density.
5. Water quality is analyzed with regard to salinity, pH, biochemical oxygen demand (BOD), dissolved oxygen (DO), and pollutants.
6. Schedules are prepared, indicating occurrence of (*a*) fill, dredging, or grading; (*b*) soil settlement (include 10-year estimate); (*c*) operation of levee breaches or inlet structures; (*d*) planting.
7. Monitoring programs lasting 5 years are designed to measure the water quality, soil characteristics, plant survival and growth rates, and wildlife use.

Harvey et al. (1983) also described the hydraulic design of diked historic baylands used for storm water and wastewater treatment and flood protection.

BEAVER PONDS

A high beaver population results in many ponds and diverse wildlife habitat, creates or improves riparian habitat, and stabilizes stream banks (Hair et al. 1978). Management of beaver ponds for ducks and beaver requires a water level with little capability for manipulation. Beaver ponds usually are about 1.5 m deep (Buech 1985). They deepen when the water seeps over the top and around the ends of the dam, which beaver then raise and extend. Extended flooding kills trees, which provide nesting, perching, and feeding opportunities for many species of bird, especially great blue herons, cormorants, yellow-crowned night herons, ospreys, kingfishers, woodpeckers, wood ducks, goldeneyes, and mergansers (Rakstad and Probst 1985).

Management for resting and roosting ducks is passive, requiring adequate water in the beaver pond and minimal disturbance. Management for feeding ducks requires draining the pond, planting duck food, then reflooding, perhaps with a trickle-tube type of drain device installed in the dam to prevent the pond from deepening.

Beaver ponds must be evaluated to determine capability for active or passive management. For both types of management, beaver are relied upon to build and maintain the dam. Beaver ponds can be maintained as inland fresh meadows (Type 2), shallow marshes (Type 3), or deep marshes (Type 4) (Knighton and Verry 1983, Knighton 1985). Old beaver ponds should be regenerated if they have lost their capacity to support diverse wildlife due to age, acidity, or high water levels, resulting in little emergent vegetation and mostly open water (Kierstead undated).

Ponds selected for active management are at least 0.4 ha which will be exposed when drained, are flat enough for reflooding to 5 to 76 cm, have live streams to ensure water for flooding, and are suitable for planting millet (Arner 1963) unless suitable native vegetation volunteers (Teaford 1986). Most trees in the pond should be dead to allow sunlight to penetrate for proper growth of the desired plants. The pond should have few submergent and emergent plants that would cover the mud after draining and retard germination of desired plants. Reasonable access is needed to facilitate work and inspections. Beaver often build several dams below and sometimes above the main dam. A pond above the one being managed actively can furnish water for reflooding in dry years.

The pond is drained in June or July by notching the dam at the main channel about halfway through from the base to the top of the dam's downstream side without causing water flow (Teaford 1986). The adz side of a mattock works well for cutting through sticks and mud. Then a deep, narrow V-shaped cut is made at the top of the dam to start the water flowing. The force of the flowing water and continued chopping or pulling with a curved four-prong garden fork will clear away the debris as it is loosened. A 2- to 4-ha pond needs 4 to 10 h to drain. Work begins in the morning to be completed by nightfall before beaver, which are nocturnal, can repair the dam and undo the day's work.

Where the outflow must be controlled, especially on large ponds, and if enough drop exists on the downstream side of the dam, a siphon can be used to drain the pond by attaching two or three 3-m lengths of 10-cm PVC pipe, a 90° elbow on each end, and suitable lengths of drop pipe to each elbow (Kierstead undated). Laid in a 3-dm notch on top the dam, the siphon is filled with a pail of water to begin the draining process. Beaver probably will not block the siphon if it extends 5 to 6 m into the pond. The dam can be cut through when the water level is low enough.

So that emergent plants will grow, the siphon pipe can be used to maintain a drawdown for one growing season by placing it in the bottom of the cut in the dam with at least 6 m extending along the pond basin. A short riser placed on the downstream end of the pipe will raise the pond 5 to 10 cm after the dam is repaired, depending on the height of the riser. Beaver will abandon the area if the drawdown is extensive, especially over winter, and might have to be reintroduced after the dam is repaired and the pond reflooded.

To prevent beaver from reflooding the pond, a drainage device of perforated and solid PVC pipe, in 3-m sections of 10-cm diameter, is installed when the water flow slows enough and the area to be planted shows (Teaford 1986, Frentress 1989). The pipe is fastened with sheet-metal screws so that the solid sections lie through the slot in the dam and the perforated sections extend into the pond. The pipe is leveled and wired securely to 1.8-m fence posts driven into the pond bottom at intervals appropriate for supporting the entire length. To carry the flow and maintain the water level, more than one pipe might be needed in the same slot in the dam. PVC end caps placed on the outlets will cause water to flow around the exposed pipe, stimulating beaver to fill the breach. End caps are removed when beaver have restored the dam and are replaced or removed to raise or lower water levels. Such control pipes are most useful in watersheds less than 26 km^2 (Laramie 1963). If more than one pipe is needed, they are separated as much as possible or else are placed in a fan shape with the outlets together.

A variation includes removing enough dam so that the PVC pipe can be placed level with the pond basin and extended 8 to 9 m into it (Kierstead undated, Stabb 1989). A short length of 5- by 25- or 5- by 30-cm cedar, with a precut hole the exact size of the pipe, is fitted around the pipe on the upper edge of the dam to help support the pipe and serve as an antiseep collar (Fig. 3.22).

Ponds over 4 ha and those on the downstream end of a series of beaver ponds need 15-cm pipe instead of 10-cm pipe. A 90° elbow and short riser are attached to the inlet (upstream) end of the pipe, over which is placed a perforated section of culvert or old hot-water tank with the bottom removed (Fig. 3.22) and plates welded to the base covered with rock to keep it in place. A slot for the pipe is cut at the bottom edge, and another hole is cut to allow the inlet pipe to be covered or adjusted. The downstream end of the pipe extends beyond the dam, connects with a T fitting, a short piece of pipe, and a collar that accepts a threaded plug to allow periodic drawdown and a hole to clean out debris.

When the mudflats are exposed and the mud is ankle-deep, Japanese millet is broadcast if native plants such as smartweed and other heavy seed producers are unlikely to volunteer. (See "Planting" in "Impoundments" in Chap. 5.) Even if they do, seed production from native plants is generally far less than that from millet (Arner 1963).

The new water level for the pond is maintained at the height of the upper downstream pipe (pipe A, Fig. 3.22), which should be less than 6 dm deep in 50 percent of the pond so that the shallow areas grow emergent plants and the deeper areas suit beaver (Kierstead undated). The height is determined by extending a level line from the T fitting out into the pond to estimate the water depth relative to the dam at various heights above the T fitting. (A transit can be used.) Then the upper pipe (pipe A, Fig. 3.22) is installed with the 90° elbow and appropriate length riser, the dam is repaired with materials similar to those removed, and beaver are reintroduced to maintain the dam if they do not return the first year after completion. The outlet plug is removed once or twice per year to clear mud, leaves, or other debris. The inlet is blocked to reduce water pressure to replace the plug.

Because beaver are sensitive to water current and can dam it even if the inflow pipe extends well beyond the main dam, maybe the simplest method is best, with a short pipe that can be cleaned readily. Other methods to prevent reflooding by beaver or to limit the extent of flooding include a three-log drain (Arner 1963), modification of it with perforated PVC pipe (Teaford 1986) and unperforated PVC pipe (Wiley 1988), wooden beaver pipes (Laramie 1963, 1978), and others (Buech 1985, Almeida 1987).

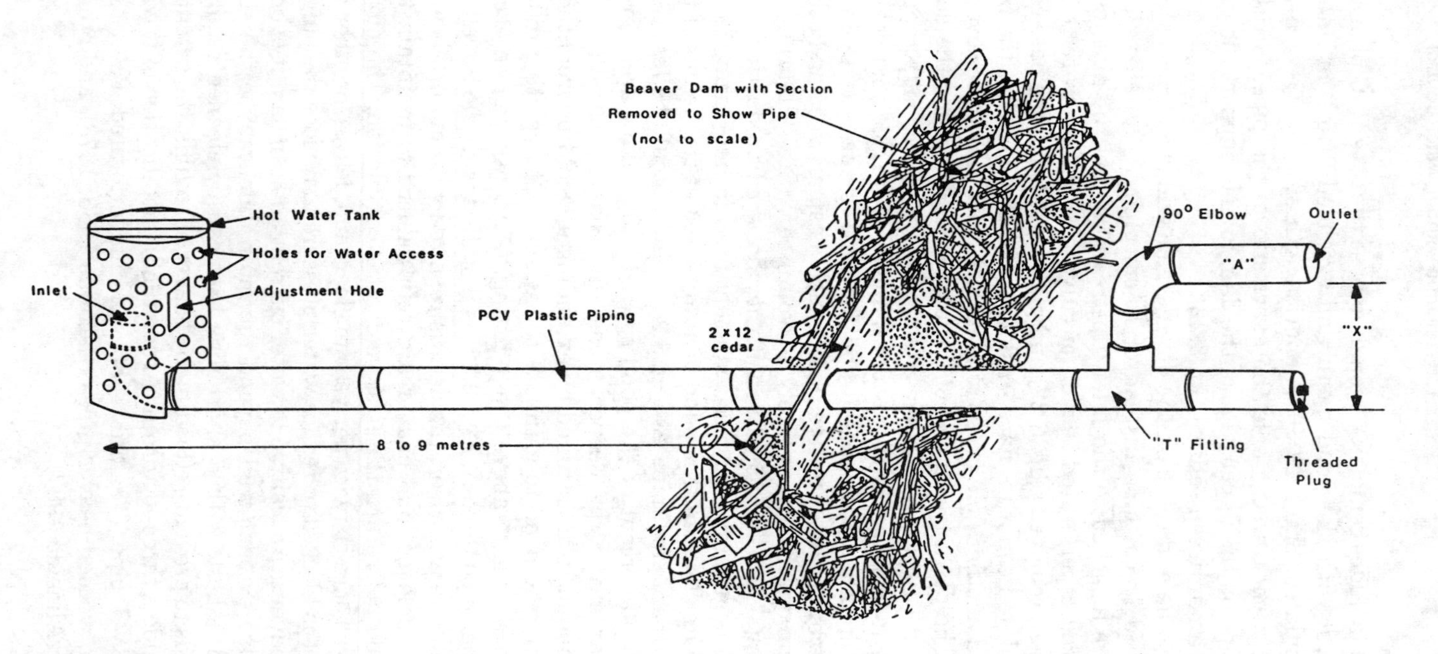

Figure 3.22 Beaver dam with water level control structure (*Kierstead undated*).

Chapter

4

Vegetation Management —Physical and Chemical

Vegetation control in the management of wetlands involves the impoundment or the wetland itself and the surrounding uplands which provide food and cover for wetland wildlife. Some type of controlled disturbance 5 years or less apart generally is beneficial, unless natural disturbance occurs and a no-management scheme is best (Kadlec and Smith 1989). Various categories of control methods exist, but many, if not most, involve a combination of methods.

WATER-LEVEL MANIPULATION

Water-level manipulation is the most important technique to manage wetland plant communities. Often it is integrated with other control methods, depending on various physical and chemical properties associated with specific wetland types and plant species (Fredrickson and Taylor 1982). Water-level manipulation affects wildlife directly as well as indirectly through plant control.

The main objective in water-level control is to maximize the wildlife habitat diversity. Greater management flexibility will be provided for habitat and wildlife diversity through habitat changes, control of plant succession, production of food crops, and control of nuisance wildlife. Any practice that retards or reverses plant succession in impoundments usually benefits waterfowl. Distribution and zonation of wetland vegetation are determined by numerous physical and chemical factors including water depth, salinity, soil texture, pH, and organic content (Prevost 1987). Managers need to experiment with each impoundment to understand soil and water characteristics, patterns of water availability, seed stocks, response of vegetation, and chronology of migration of waterfowl and shorebirds. A computer program can assist in management actions involving early or late spring drawdown, summer disking, summer farming or harvest, and early or late fall flooding (Auble et al. 1988).

Two methods of water management are used where water supply and control structures allow adequate control of water levels (Addy and MacNamara 1948, Atlantic Waterfowl Council 1972). First, the principle of permanent or constant water level is used where water is not acidic and is free of turbidity and stain. Advantages usually include better nesting sites, better brood cover, better muskrat habitat, better water conservation, more irrigation water, slower ecological succession, better fishing, more economical operation, and better mosquito control, especially along the coast. Constant water levels encourage growth of sago pondweed and other submergent plants which attract diving ducks (Weller 1987). Second, the principle of drawdown to control water level is used where water is acidic, saline, turbid, and stained and soils are of low quality.

In northern prairie wetlands, drawdowns are used to promote emergent vegetation used by breeding waterfowl as cover and nesting material when reflooded the following spring (Merendino et al. 1990). In southern areas, drawdowns are used to promote annuals used by migrating and wintering waterfowl after fall flooding.

In general, drawdowns are used when the marsh has opened up excessively, i.e., most aquatic vegetation has disappeared due to high water, overeating by herbivores (mainly muskrats), or plant disease (Kierstead undated). Drawdowns stimulate germination and rapid growth of aquatic plants (Kadlec 1960). Water levels are raised to kill marsh vegetation which has become too dense and extensive. In marshes supporting muskrats, a substantial reduction in the number of muskrat houses indicates the need for a drawdown.

Drawdown simulates the natural flood-dry cycle of many wetlands. As the marsh dries, dead emergent and other plants oxidize and decompose, releasing nutrients into the soil. New emergent plants develop from seed lying dormant for many years on the marsh bottom, because the seed dries out enough to germinate. Plants become established, and the marsh refloods with more fertility and balance for wildlife.

Hydrologic and seasonal variables influence the schedules for water-level manipulation (Gordon et al. 1989). Hydrologic variables include frequency and duration of flooding, water depth, water temperature, dissolved oxygen within and outside the impoundment, turbidity, and salinity. Seasonal factors include tide schedules, lunar phases, and local weather patterns. Wind speed and wind direction influence tidal amplitudes.

Controlled Marsh

Specific objectives for drawdowns include the following (Linde 1969):

1. Improve breeding and brood-rearing habitat for waterfowl by increasing water depths to improve interspersion of cover.
2. Kill sedge-grass monotypes and woody vegetation by flooding to change plant succession from moist-soil and upland types to shallow- and deepwater aquatics.
3. Provide mudflat areas for moist-soil plants produced by natural or artificial seeding of waterfowl food plants after summer drawdown.
4. Till the bottom of the impoundment for planting food crops after spring drawdown.
5. Kill undesirable plants by summer drawdown, drying, and disking of the impoundment's bottom.
6. Allow desirable cover plants to recover via drawdown after being lost from wave action, ice action, and continuous deepwater flooding.
7. Increase production of invertebrates for laying ducks, ducklings and goslings, and shorebirds by flooding to increase aquatic and emergent vegetation.
8. Increase duck and shorebird predation on invertebrates and minnows by a fall drawdown.
9. Kill carp and bullheads by draining in winter after freezing.
10. Reduce carp reproduction by a drawdown right after spawning to strand eggs and fry in shallow temporary pools and expose them to predation.
11. Reduce populations of muskrat and nutria via winter drawdown.
12. Consolidate the bottom, releasing nutrients which accumulate with continuous flooding and age of impoundment, thus improving light penetration and plant growth as colloidal soil particles settle.
13. Reduce growth of unrooted aquatic plants by flushing the pond with fresh water from a nutrient-free source (Applied Biochemists, Inc. 1979).
14. Flush the system of excess salts (Hindman and Stotts 1989).

Water levels

Water manipulation to control (encourage or discourage) vegetation is similar for inland and coastal fresh or saline marshes, except that saline areas also need control of salinity. Marshes vary considerably in response to changes in water level, even within the same state or province. The plant composition and soil fertility in wetlands vary geographically. Thus, broad principles of water-level manipulation will apply, but specific rules must be applied with modification. Water levels on marshes up to 405 ha generally are manipulated with greater ease and efficiency (Beule 1979).

Depths for waterfowl vary from 15 cm to 3.7 m, with 0.3 to 1.2 m being optimum, depending on the flyway and location therein (Atlantic Waterfowl Council 1972). Impoundments over 800 ha and 2 m deep should be managed mainly for fish and perhaps diving ducks. Ponds for diving ducks can be 6 to 9 dm deep at the deep end (Linde 1985), but could range 6 to 24 dm deep, averaging 9 to 12 dm (Lokemoen et al. 1984). Water over 15 dm deep has too few plants to attract breeding ducks. For puddle ducks, water could be 3 to 18 dm deep, averaging 4.5 dm. High mink populations are associated with muskrats, and high muskrat populations with cattails in water 46 to 61 cm deep (Dozier 1950, Errington 1963). For maximum use by puddle ducks, muskrats, and nutria, 50 to 75 percent of the pond should be 2 to 4.5 dm deep (Eng et al. 1979). For diving ducks, coots, beaver, otter, and alligators, water should be deeper. For shorebirds, about 20 percent of the pond should be less than 3 dm deep when the pond is full (Proctor et al. 1983a,b,c). Water 4.5 to 12 dm deep throughout usually produces the 50:50 ratio of open water to emergent cover generally desirable for ducks (Farmes 1985). But canvasbacks prefer wetlands with less than 33 percent emergent vegetation (Stoudt 1982), and Canada geese prefer a ratio of at least 90:10 of open water to emergents (Ringelman 1990). Cover conditions are optimum for muskrats with 50 to 80 percent of the wetlands basin covered by emergent vegetation, and 20 to 50 percent permanent non-fluctuating open water, with a gradient of less than 1 percent in riverine habitat (Allen and Hoffman 1984). For diversity of both plant and animal species, one guideline would be to have 30 percent of the impoundment 1.5 to 4.5 dm deep, 30 percent 4.5 to 9 dm deep, 20 percent 9 to 15 dm deep, 10 percent over 15 dm deep, and 10 percent in islands (New undated).

Management of water levels on controlled marshes basically involves mudflats, or moist-soil areas; shallow marsh, or hemimarsh; and deep marsh, or open-water marsh (Bookhout et al. 1989). The timing and level of water manipulation affect the control of desirable and undesirable vegetation, invertebrates, and nuisance wildlife as well as foraging by waterbirds and certain other wildlife. Essentially, to promote desirable plants, drawdowns are used to encourage a hemimarsh for cover, submerged aquatic vegetation for food and as a medium for invertebrates, and moist-soil plants for food on exposed mudflats. The moist-soil plants can be consumed by waterfowl as greens without reflooding or mainly as seeds after reflooding. Of all natural habitat types, moist-soil plants attract and support the most waterfowl.

The best marsh complex contains all three water levels. Marshes managed mainly for duck hunting probably should have less area in deep marsh. Marshes managed mainly for refuge and resident wildlife should have more deep marsh and shallow marsh (Bookhout et al. 1989).

Mudflats. Mudflats are managed for dense stands of moist-soil plants. These are mostly annual emergents such as barnyardgrass, panic grass, American bulrush, squarestem spikerush, smartweeds, redroot flatsedge, beggarticks, millets, nutsedges or chufa, and rice cutgrass, often in conjunction with shallow-marsh management (Bookhout et al. 1989, Hindman and Stotts 1989). In most marshes of the Great Lakes, the vegetation association of walters millet, chufa, and nodding smartweed is the most common and attractive food source for waterfowl (Bookhout et al. 1989). Zones of similar soil moisture during germination often result in monotypic stands of mudflat plants. The absence of such zonation often results in three or four plant species of different height in the same area, producing diversity from multiple life-forms of vegetation. The mudflat condition can be maintained for more than 10 years. But the spring drawdown used for mudflat management also encourages undesirable plants, such as willow and purple loosestrife, which might need 4 to 5 years of control with high-water and mechanical or chemical control. Sometimes 1 to 2 years of deepwater management, drying, and disking or application of herbicides will suffice.

During the first year of drawdown and on agriculturally tilled basins, mudflat species of plants dominate until eliminated by reflooding (Harris and Marshall 1963, Stewart and Kantrud 1972, Weller and Voights 1983). Prolific seed producers, they generally do not inhibit growth of emergents. Sheet-flooding wetland basins dominated by mudflat plants is a common technique during fall and winter (Fredrickson and Taylor 1982).

Shallow marsh (hemimarsh). Shallow marshes provide the best combination of food and cover for waterfowl (Bookhout et al. 1989). Shallow marshes distribute waterfowl use better than mudflats do, generally throughout migration, compared to maybe 1 to 2 weeks for mudflats.

The management goal is to produce nonpersistent emergent vegetation, preferably seed-producing annual grasses (Stutzenbaker and Weller 1989). This sometimes requires an annual spring drawdown in

April or May so that seed-producing plants can germinate and mature before being reflooded in late summer or early fall. Long-term flooding tends to eliminate annual seed-producing plants and encourage floating and submergent plants. Moist-soil conditions are maintained during summer. Grasses and sedges are encouraged with drier soils; smartweeds and spikerushes with wetter soils (Gordon et al. 1989). If complete drying is needed, helped by extended summer drought, periodic partial summer drawdowns are best afterward for germinating submerged aquatics.

Moist-soil annual plants are uncommon because few germinate under water. Most are emergent perennials like cattail and bulrush. Submergent plants include sago pondweed, curly-leaf pondweed, water milfoil, coontail, and bladderwort, which provide much food from foliage, seeds, tubers, and associated invertebrates (Bookhout et al. 1989).

The main goal of water control, aided by muskrat control, is to produce a hemimarsh condition, i.e., a 1:1 ratio of emergent plants to open water, with interspersed patches 0.1 to 0.2 ha, which suits the greatest variety of wildlife (Weller and Spatcher 1965, Weller 1975, Bookhout et al. 1989, Pederson et al. 1989, Verry 1989), although some patches of at least 4 ha might be useful as flight paths (Ambrose et al. 1983). For mallards in cattail marshes, Ball and Nudds (1989) recommended a 1:1 ratio of cattail to open water so that patches of about 0.15 ha are produced. Hemimarsh management provides broad cover and benefits migrating and staging birds. Molting birds and migrants use submerged plants and associated invertebrates (Fredrickson and Drobney 1979, Kantrud 1986b). Molting and staging waterfowl use extensive stands of emergents, floating root clumps, muskrat houses, and sheltered mud bars as storm shelter and loafing areas (McDonald 1955, Kantrud 1986b).

A partial drawdown in spring (after the spring thaw in northern areas or after northern migration of ducks in southern areas) encourages regrowth of perennials from rootstock and seed, as temperatures increase in shallow water (Beule 1979). Or, water levels, from spring and winter precipitation, can be held high to suppress expansion of plant communities (Bookhout et al. 1989). Water levels are held 10 to 30 cm deep during the growing season and adjusted up or down 15 cm at a time as migrating or wintering waterfowl consume each layer of seeds. When monitored carefully, shallow marshes can be maintained 6 to 8 years before changes are needed.

Deep marsh. Deep marsh is 30 to 120 cm deep across the main basin during the growing season (Bookhout et al. 1989, Pederson et al. 1989). Such areas are used more by diving ducks than by dabbling ducks, although widgeon, gadwalls, and shovelers use them. Dike maintenance is a problem due to muskrat and wave damage, so deep marshes are used more to control undesirable vegetation than to promote waterfowl use. Few undesirable plants can withstand 2 to 3 years of water at least 76 cm deep. With depletion of emergent perennial plants, muskrats emigrate, and the marsh is dewatered for moist-soil management of shallow marshes or mudflats. High water levels eliminate vegetation, but the seed bank can be lost through sedimentation and/or water currents flushing out the seed (Smith and Kadlec 1986).

Maintenance of some deep-marsh habitat enhances wetland diversity. With few carp, growth of preferred submergent vegetation (e.g., wildcelery and water milfoil) will attract diving ducks (e.g., canvasbacks and redheads).

Types of drawdowns

The ratio of vegetation to water in an impoundment should be kept between 30 and 70 percent, with 50 percent optimum (Verry 1985a, 1989). Four categories of dispersion are useful to compare impoundments and guide the management of specific impoundments (Knighton 1985): good interspersion, fair interspersion but shrubs dominate emergents, poor interspersion with too much open water, poor interspersion with not enough open water. An impoundment is kept in the first condition for ducks. If it is in either of the other conditions, it is drawn down to improve the habitat. A general problem in achieving management goals is indicated if most pools are in the last three conditions. Estimates of conditions are made visually from a small airplane after July 14. The impoundment must be at normal pool. The estimate is made only for the zone surrounding the elevation 3 dm below normal pool (Knighton 1985) because emergents in stable water die out below 6 dm (Atlantic Waterfowl Council 1972, Verry 1985a). This elevation is marked permanently, preferably with wood duck houses (Fig. 4.1).

For maximum waterfowl production, the water level must be maintained at normal pool during nesting and brood rearing. But drawdowns are sometimes needed during some of this time, ultimately to improve the habitat for waterfowl and other wildlife. Managing a group of impoundments in rotation is most effective, as is subdividing a large impoundment into units by cross dikes equipped with water control structures to permit independent control of water levels for the development of feeding grounds. Systems should contain at least five impoundments with staggered management (Table 4.1), although at least seven are better to maximize management options (Fredrickson and Reid 1986, Reinecke et al. 1989). Thus, an impound-

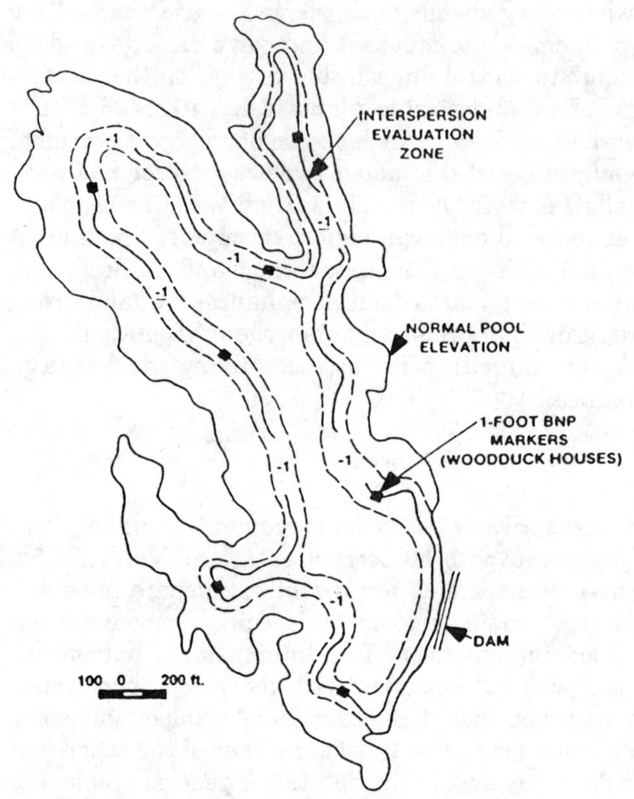

Figure 4.1 Impoundment overview illustrating the location of an interspersion evaluation zone centered at 1 ft (3 dm) BNP (below normal pool) marked with wood duck nest boxes (Knighton 1985).

TABLE 4.1 Suggested Habitat Manipulation for a Hypothetical Complex of Five Moist-Soil Impoundments

	Season		
Impoundment	Spring	Summer	Fall
1	Late partial drawdown	Complete drawdown	Late shallow flood
2	Midseason complete drawdown	Disk and irrigate	Late shallow flood
3	Early partial drawdown	Early complete drawdown	Early shallow flood
4	Early complete drawdown	None	Early shallow flood
5	Late complete drawdown	None	Shallow disk and flood

SOURCE: Fredrickson and Reid (1986), Reinecke et al. (1989).

ment can be drawn down perhaps more than 1 year to achieve the objective, while at least one impoundment in the area is maintained at normal pool. Such an arrangement mitigates loss of breeding tradition developed by waterfowl for a specific pool.

Water levels are regulated for vegetative growth and diversity during the second (first reflooded) season, with limited concern for wildlife in the impoundment at this time (Weller 1987). Emergent vegetation gradually declines with stable or high water levels, due to depredation by muskrats for food and houses.

Drawdowns can be cyclic or noncyclic, complete or partial, fast or slow, early or late. Usually some combination is used. For example, four types of drawdowns are used at Necedah National Wildlife Refuge in Wisconsin (personal communication, D. Nord, Necedah National Wildlife Refuge 1989). (1) An impoundment is drawn down completely the last week in May for volunteer production of moist-soil plants (mainly smartweeds, beggarticks, and millet) and reflooded in steps beginning September 15 to feed migrant ducks. (2) An impoundment is drawn down completely beginning August 1 to raise spikerush for geese which eat it when it is small, tender, and succulent. With an earlier drawdown, the spikerush gets too tough. (3) An impoundment is drawn down in steps the last week in September so that migrant ducks can feed on invertebrates, minnows, and seed. When ducks start to dip to reach food, the water level is lowered another 15 cm, perhaps in October, to expose additional invertebrates and minnows and another layer of seed. (4) An impoundment is drawn down completely in winter after freeze-up to kill carp and bullheads and to solidify the bottom.

Drawdowns at Necedah average once every 3 years, but that varies. Some impoundments are drained every 2 years to discourage perennials and encourage annuals. One pond is drawn down in late summer every year for production of spikerush for geese. Submerged aquatics occur the second year after drawdown, but not the first year. Invertebrates occur both years, because most populations die in winter but grow rapidly in spring (Peterson et al. 1989).

Cyclic and noncyclic drawdowns. Cyclic drawdowns are conducted at regular intervals, maybe every 5 years or more (Table 4.1), without a regular on-site evaluation each time (Knighton 1985, Fredrickson and Reid 1986). Noncyclic drawdowns require a regular on-site evaluation each time to determine need. Impoundments managed for annual seed-bearing plants often need annual drawdown (Table 4.2) (Johnson and Montalbano 1989), with reflooding begun in August or September and completed by September, early October, or even November in the South.

TABLE 4.2 Waterfowl Food Plants Resulting from Drawdown in Freshwater Wetlands in the Southeastern United States

Location	Time of drawdown	Food plants
Tennessee	Late April to early May	Smartweed (*Polygonum* spp.), millet (*Echinochloa* spp.)
South Atlantic and	February to March	Smartweed
Gulf coasts	Late summer	Dwarf spikerush (*Eleocharis parvula*)
Lake Mattamuskeet NWR, North Carolina	April	Dwarf spikerush, smartweed, fall panicum (*Panicum dichotomiflorum*)
South Carolina	February to March	Redroot (*Lachnanthes caroliniana*), smartweed, panic grass, flatsedge (*Cyperus* spp.)
	Spring	(*Panicum* spp.) Smartweed, panic grass, millet, flatsedge
	Summer	Smartweed, millet
Georgia	January, May, and June	Panic grass, spikerush (*Eleocharis* spp.), smartweed
Louisiana	May	Spikerush, paspalum (*Paspalum* spp.)
Florida	February	Watershield (*Brasenia shreberii*)
	February	Spikerush, smartweed, millet
	March	Spikerush

SOURCE: Johnson and Montalbano (1989).

Wildlife management in coastal and freshwater marshes uses similar strategy, but the short-term cycle of drawdown is not as pronounced in coastal marshes (Mitsch and Gosselink 1986). The cycle in sequence is generally as follows (Weller 1978):

1. The cycle begins with a spring drawdown when the marsh is in the open stage with little emergent vegetation. Seedlings will germinate on the exposed mudflat.
2. The water level is increased slowly in late summer or early fall to reduce loss of plants from flotation or shading due to turbidity.
3. The drawdown cycle is repeated the second year to establish a good stand of emergents. Management by the fall is related more to wildlife use than to plant growth.
4. Maintaining low water levels for several more seasons encourages growth of perennial emergents such as cattail.
5. Maintaining stable, moderate water levels for several years pro-

motes growth of rooted submerged aquatic plants and associated invertebrates.

6. Eventually the emergent vegetation dies, again producing the open stage of the marsh, when the drawdown cycle begins again.

Addy and MacNamara (1948) recommended the following general procedures for drawdown:

1. The maximum water level is maintained in new areas for 8 to 9 months from the end of October to mid-June or early July, depending on latitude, to reduce undesirable plants unless the pond is very shallow. With a newly diked salt marsh, the water level might have to be held at maximum for about 2 years to kill the saltmarsh vegetation. Burning and disking the sod before flooding probably will help.
2. The pond is drained abruptly about mid-June or early July, depending on latitude, to expose the bottom. Some water can be left to retain important fish.
3. The wet bottom is sown immediately with seeds of desirable plants such as wild millet and smartweed and perhaps planted with the tubers of chufa, preferably near the upper edge to allow enough time for development of the tubers.
4. If summer rains provide insufficient moisture for plant growth, the pond is irrigated briefly with shallow flooding.
5. When the plants have matured and the seed has ripened, the pond is flooded in increments of 15 to 30 cm to allow fall-migrant ducks to feed. As a layer of seeds is consumed, the pond is flooded to another level of seeds for the ducks. The last rise in water level can be withheld until winter freeze-up or spring to provide food for spring migrants.

Drawdowns every 2 to 4 years are generally best for waterfowl (Linde 1969). A regime used successfully in Wisconsin is to draw down 2 or 3 years in a row, then skip 1 or 2 years. A common practice is to expose about half the bottom for at least 3 months during the growing season every 2 or 3 years (Green et al. 1964). Such drawdowns reduce the production of some species of submerged and floating leaved aquatics during the first year of reflooding, while other species, e.g., water smartweed and northern naiad, grow well. Annual drawdowns ultimately result in production of volunteer vegetation undesirable for waterfowl (Linde 1969). Each succeeding drawdown produces emergents which close in the mudflats that were so productive originally. Changing the water depth in a moist-soil impoundment at ap-

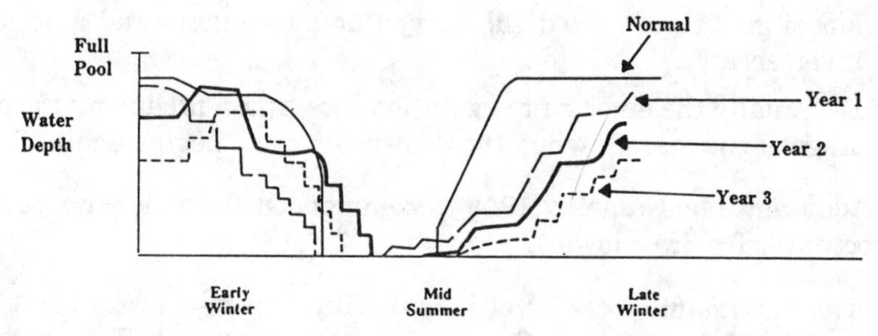

Rationale

Normal - Typical midsummer drawdown to establish moist-soil vegetation. Fall and winter flooding for waterfowl.

Year 1 - Gradual drawdown to optimize use by late spring migrants. Gradual reflooding for rails and waders.

Year 2 - Gradual drawdown lasting into midsummer to optimize use by late spring, migrant waterfowl, shorebirds, and waders. Gradual reflooding in fall to optimize use of seed resources.

Year 3 - Increasing water depths in spring to make food resources available. Gradual drawdown by late spring, followed by gradual reflooding in fall to shallow depths.

Figure 4.2 Flooding regimes suggested for seasonally flooded wetlands in the Midwest (Fredrickson and Reid 1988b).

propriate times ensures that the resources produced are used effectively (e.g., Fig. 4.2) (Fredrickson and Reid 1988b).

In Maine, the cycle generally maintains fairly constant water levels from ice-out until early August when a drawdown of 15 to 30 cm encourages new seeding and plant growth. In mid-September the water level is raised as before or even higher to attract migrating waterfowl and to provide cover for other wildlife during winter. Every 5 years the marsh is drained completely during spring and summer to rejuvenate it (Peppard 1971). In Minnesota, 1- or 2-year drawdowns at 5- to 10-year intervals are used to maintain emergent marshes, depending on water depths and cover types (Harris and Marshall 1963). In southern Ontario, drawdowns are done every 7 to 10 years (Kierstead undated). Tidal impoundments are drawn down every 1 to 3 years, beginning in late March and completed by early April, to flush out excess salts and reestablish emergents, with reflooding begun in mid-May to encourage growth of widgeongrass, muskgrass, and fennelleaf pondweed (Hindman and Stotts 1989).

Complete and partial drawdowns. Complete drawdowns drain the impoundment. They are used when little emergent vegetation occurs in the middle of the impoundment (Weller 1987) and the marsh needs complete rejuvenation, such as tilling and planting, or the dike or control structure needs repair (Linde 1969). Loss of vegetation is due to muskrat eatouts, winter kill, plant disease at times, and high water levels often in combination with excessive carp populations. Complete drawdowns are difficult to achieve because most water control structures are not installed low enough to drain the basin of the impoundment. The degree of drawdown depends on the shape of the basin and the availability of water. A complete drawdown allows naturally occurring seed to germinate and established flood-stressed emergents and even submergents to recover. Cracking of the bottom muds and decomposition of bottom vegetation are ideal for most plants.

Partial drawdowns are used to encourage desirable vegetation or to discourage herbivores (Weller 1987). They are used where the amount of vegetation is reduced substantially, wildlife use has declined, or vegetation is stressed by the water level. Water levels are reduced to meadowlike depths and retained or even reduced in late summer, with gradual reflooding in early fall to a freeze-proof depth, unless muskrats are to be discouraged. Vegetative propagation of emergents and germination and growth of submergents are encouraged in early summer, especially at the marsh perimeter. As vegetation recovers, water levels are regulated to permit nesting or consumption of plants. Partial drawdowns of 30 to 45 cm below normal pool will maintain a willow shrub border that waterfowl broods use (Verry 1989). In the northern Lake States partial drawdowns should begin by July 15 and in the southern Lake States perhaps by July 1. Maintaining normal pool level longer will kill willow shrubs. If alder is the only shrub, partial drawdowns can begin June 15. Partial drawdowns probably are preferable to complete drawdowns for the following reasons (Linde 1969):

1. Some water will be left for broods of waterfowl.
2. Invertebrates will not be reduced to levels so low that few are available the next year as food for broods.
3. Volunteer millet and smartweed will establish readily on higher ground because the groundwater table will not be reduced so drastically. With complete drawdowns, such heavy growths will establish only in the deeper portions of the impoundment where the soil is slow to dry out. But then with reflooding, this food crop is unavailable to the ducks because the water is too deep there.
4. The ratio of plants to open-water area usually is better with a partial drawdown because complete drawdowns conducted annually result in profuse vegetative growth that completely closes in all open-water areas.
5. Muskrats will survive better.

Fast and slow, early and late drawdowns. Both complete and partial drawdowns can vary in starting date, duration, frequency, and rate of drainage and reflooding. With fast or slow drawdowns, early drawdowns usually produce more seed. Late drawdowns produce greater species diversity and higher stem density (Fredrickson and

Taylor 1982). Soils dry quickly then, and germination can occur in soils which remain saturated long enough near the receding water. But reducing water levels slowly during July and early August can cause exposed pondweeds to seed out more (Keith 1961). Drawdowns in May in northern prairie wetlands and perhaps elsewhere benefit habitat for breeding waterfowl the most because they maximize shoot, cover, and seed production of desirable species (hardstem bulrush and alkali bulrush) during the first season, allow deeper flooding (30 to 50 cm) the next year, and minimize establishment of undesirable species (cattail and purple loosestrife) (Merendino et al. 1990, Merendino and Smith 1991). June drawdowns maximize first-season establishment of cattail and purple loosestrife. Because late drawdowns in July and August reduce establishment of first-season shoot and cover and production of first-year seeds, flooding must be shallow the next year to prevent die-off of established vegetation.

Slow drawdowns drain the impoundment gradually or by increments during a 2-week period or more (Fredrickson and Taylor 1982). Fast drawdowns drain the impoundment within a few days, producing extensive excellent stands of similar vegetation simultaneously, dominated by upland and wet-meadow species (Harris and Marshall 1963). But wetland wildlife is forced from the area almost immediately. Fast, late drawdowns might produce less desirable vegetation than fast, early drawdowns, especially if temperatures exceed about 32°C and flooding depends on rainfall (Fredrickson and Taylor 1982). Then saturated soils dry out within a few days, and little germination occurs. Complete drying in Michigan produced abundant growth of sedges and woolgrass but limited growth of more desirable annuals. Irrigation by flooding slightly will maintain a moist condition for plant growth (Green et al. 1964). After spring drawdown to promote germination and growth of desired aquatic plants, mosquitoes lay eggs on the damp soil. In coastal areas after germination of these aquatics, impoundments are reflooded, then quickly drained to remove mosquito larvae and expose them to predatory fish, then quickly reflooded before new eggs are laid, and maintained 10 to 20 cm deep to prevent egg laying (DeVoe and Baughman 1987).

Early drawdowns provide emergent seedlings with a longer growing period, which increases survival and density (Linde 1969). Drawdown in May in Ohio had the best plant associations, did not interfere with duck nesting, probably allowed muskrats to produce two litters, and needed fewer years than other drawdowns in other months for annual weeds to replace semiaquatic plants (Meeks 1969). Drawdowns in late June in Wisconsin encourage annuals and discourage emergents (Linde 1969). Late summer (August 1 in Wisconsin) drawdowns produce moist-soil plants for geese. Early fall drawdowns expose invertebrates and minnows to migrant ducks. Winter drawdowns after freeze-up kill carp and bullheads, solidify and aerate the bottom, and apparently improve germination of seed and recovery of submerged aquatics (Green et al. 1964). Drying of the soil and decomposition of the vegetation to release nutrients might take most of the growing season and extend over winter (Weller 1987). To encourage aquatic growth, fall drawdowns should begin late enough to inhibit seed germination of emergent plants, but early enough to dry the top layer of soil to stimulate germination of aquatic seed the next spring (Green et al. 1964). Sometimes the drawdown in late fall or early winter (posthunting) can be extended to late summer (prehunting) before reflooding, to dry the marsh completely but not adversely affect duck hunting, although muskrat trapping will be adversely affected (Weller 1987), a desirable goal if eatouts are a problem. Movement of muskrats to wetter areas will increase predation and probably inhibit reproduction. Summer drawdowns reduce the breeding population of muskrats, and winter drawdowns reduce the area's habitability, forcing muskrats to move, thus exposing them to increased mortality from trapping, predation, and freezing. Summer drawdowns produce various results due to variation in individual wetlands such as plant species and competition; basin morphology, hydrologic characteristics, and soil types; and water and soil chemistry (Pederson et al. 1989).

Reflooding

Reflooding generally occurs when the exposed marsh is covered with dense aquatic plant growth (Kierstead undated). This usually occurs in late summer to late fall, before the hunting season. In northern areas, deeper parts of the marsh are reflooded at least 1 m deep over winter to provide unfrozen habitat for muskrats. To decrease the muskrat population, reflooding is avoided until early spring, before breeding birds arrive, when spring melt water allows ready reflooding.

Gradual reflooding reduces flotation of emergents, direct scouring of other plants, and mortality of plants due to muddy, turbid water and increases use of seed by ducks if reflooding occurs during fall migration. Except to remove undesirable emergents, reflooding should not occur when ice is on the marsh, as it freezes around plant stems which are torn out with the rising water.

Fine-tuning

To determine the optimum water management level of each pool, it is drawn down in summer (year 1) once every 3 to 5 years depending on when the quality of the vegetation declines, so that annuals will grow on the exposed bottom (personal communication, T. Meier, Wisconsin Department of Natural Resources 1991). Beginning in the end of August in Wisconsin, the pond is reflooded to normal pool over a 1-month period so ducks can eat the different layers of seed. The water level is kept at normal pool over winter. Then in May (year 2) it is lowered to about one-half normal pool and kept that way over the summer. In late August it is raised gradually 15 to 20 cm and held over the winter. The next spring (year 3) it is lowered to 10 to 15 cm above the previous summer's partial-drawdown level. Then the following year (year 4) if the vegetation is too thick, the water is raised 8 to 15 cm to create more open water. At that point the optimum water level is close to being obtained for that particular pond. Annual adjustments thereafter must be based on the previous year's plant response.

Water is added to newly constructed ponds conservatively for efficiency. The scenario is similar to that of old ponds after the first year. About 30 to 38 cm of water (half normal pool) is added to the new pond to see what grows. Water is added gradually over the next 2 years to destroy undesirable vegetation.

Monitoring the impoundment to maintain the hemimarsh condition can be done every 1 or 2 years (Verry 1989). Where more than 15 impoundments are managed, monitoring can be done in groups of 5 to 12 per year on a 3- to 5-year basis. The 30- to 45-cm depth contour at normal pool should be observed for a change in the ratio of vegetation to open water.

Impoundment potential of inland fresh marshes

Knighton and Verry (1983) and Knighton (1985) developed a key to help determine the type of water-level manipulation for impoundments in Minnesota. The key applies generally across the forested region of the western Great Lakes. It probably can be used for other areas with similar habitat characteristics and elsewhere with modification. The key relates erosion, floating mats of vegetation, beaver problems, and dependability of water supply with inland wetland types from Shaw and Fredine (1956). Management classification is as follows:

1. Organic soil, extensive, over 15 cm thick.
   2. Fibric surface soil over 15 cm thick without underlying hemic and/or sapric layer over 3 dm thick; or extensive floating mats...*Floating mats.*
   2. Hemic or sapric soil over 3 dm thick possibly underlying fibric surface layers maybe at least 1.8 m...*Deep peats.*
1. Mineral soil or fibric organic soil 15 cm thick or less.
   3. Soil strongly calcareous* in upper 3 dm.

*Specific conductance cannot be used before construction to identify the source and amount of water if the surface soil is strongly calcareous. Management classification is still useful but more tentative without specific conductance.

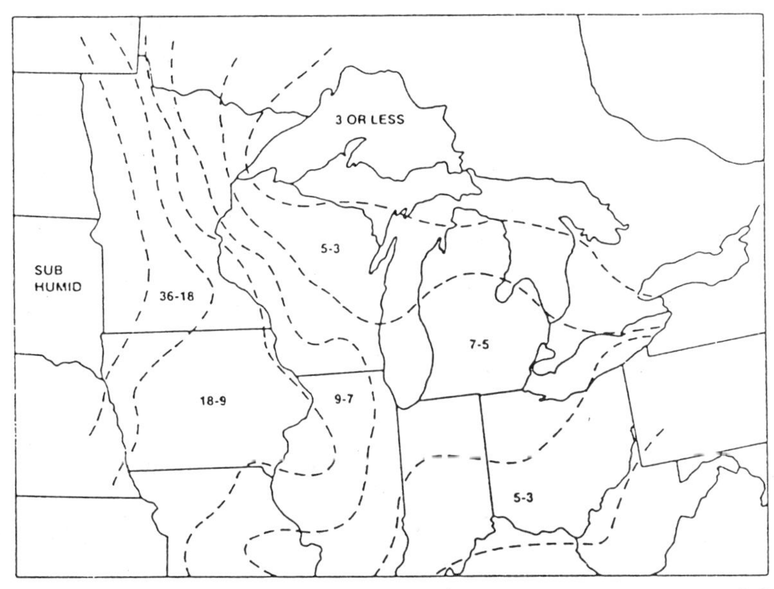

Figure 4.3 Zones with a range of watershed-to-impoundment-area ratios needed to supply enough summer surface runoff to maintain water level at normal pool (Verry 1985a). For example, the 5–3 zone encompasses ratios of 5:1 to 3:1.

4. Ratio of watershed area to impoundment area is 0.75 or less times map value (Fig. 4.3)...*Do not build.*

4. Ratio of watershed area to impoundment area is greater than 0.75 times map value (Fig. 4.3).

5. Ratio of watershed area to impoundment area is 1.5 or less times map value (Fig. 4.3)...*Weather-controlled.*

5. Ratio of watershed area to impoundment area is greater than 1.5 times map value (Fig. 4.3)

6. Ratio of watershed area to impoundment area is 3 or less times map value (Fig. 4.3)...*Manager's option.*

6. Ratio of watershed area to impoundment area is greater than 3 times map value (Fig. 4.3)...*Beaver-controlled.*

3. Soil not strongly calcareous in upper 3 dm.

7. Ratio of watershed area to impoundment area is 0.75 or less times map value (Fig. 4.3), and specific conductance is 25 micromhos/cm or less...*Do not build.*

7. Ratio of watershed area to impoundment area is greater than 0.75 times map value (Fig. 4.3), and specific conductance is greater than 25 μS/cm.

8. Specific conductance is 100 μS/cm or less... *Weather-controlled.*

8. Specific conductance is greater than 100 μS/cm.

9. Ratio of watershed area to impoundment area is 3 or less times map value (Fig. 4.3)...*Manager's option.*
9. Ratio of watershed area to impoundment area is greater than 3 times map value (Fig. 4.3)...*Beaver-controlled.*

Floating mats. A floating mat usually results if water is impounded on a site that has a fibric organic soil layer over 15 cm thick and supports herbaceous plants or bog (ericaceous) shrubs. Such soil does not dry out even with repeated drawdowns. Roots bind the mat together but apparently do not penetrate the mineral soil beneath to anchor the mat, which floats with each flooding unless trees and tall shrubs are present to anchor it. Thus, water-level manipulations are ineffective unless mats are not extensive and can be disassembled by blasting, digging, or cutting with a piece of equipment called a *cookie cutter.* (See "Cutting" in this chapter.) Complete drawdown cycles have virtually no impact on mat vegetation in impoundments less than 101 ha (Verry 1985a). Wetlands dominated by floating mats persist as inland fresh meadows (Type 4).

Although floating mats are unproductive, their interspersion might make the impoundment productive if the mats do not cover more than 70 percent of the surface area when the water level is 3 dm below normal pool. Small scattered floating mats help create diversity and might be used as nesting islands, but they reduce the open-water area and can plug the spillway, causing flooding and structural damage.

Trees and tall shrubs on organic soils are killed when flooded for prolonged periods. Willow survives with regular partial drawdowns made before July 15. Speckled alder survives with partial drawdowns before June 15, but drawdowns before June 15 are critical for waterfowl broods which do best with the water level at normal pool.

Deep peats. Spring runoff of about 28 m^3/s can severely erode peat over 15 cm deep around the edge of large (over 405-ha) impoundments held at normal pool over winter. If ice has frozen into the organic soil, such spring flows will lift 15 cm of soil and roots with the ice, pulverize it with wind-pushed ice, and flush it out with overflow. This action deepens the shallow zone used heavily by wildlife around the edge of the impoundment.

To prevent such erosion, a winter drawdown can start right after the waterfowl season ends. But to protect muskrat populations, the drawdown should start after 7.6 cm of ice has formed. Lowering the water level 6 to 9 dm helps retain spring floods and leaves a collapsed 7.6-cm layer of ice on the shallow areas, which will melt in spring before peak flow of snowmelt.

Do not build. Impoundments are not built on sites with a severely limited water supply because they would be empty most of the time.

Weather-controlled. If the water supply is inadequate to refill the impoundment after a drawdown, it is managed as near to normal pool as possible. Normal weather fluctuations provide natural full and partial drawdowns often enough to eliminate planned drawdowns usually. In normal and dry years during the growing season, such areas become inland fresh meadows (Type 2) or inland shallow marshes (Type 3), or approximate inland deep marshes (Type 4) in wet years (Shaw and Fredine 1956, Knighton 1985). (See Table 2.4.) A drawdown might be desirable after several wet years. But refilling depends on precipitation. Usually, water levels decline progressively through summer.

Meadow emergents are the prevailing plants at water elevations near normal pool (Knighton 1985) (Table 4.3). These are perennials, mainly sedges and various grasses. Peak biomass occurs at about 3 dm below normal pool (Fig. 4.4), shifting downward another 3 dm and increasing during dry years or if the water level is deliberately lowered. If a viable seed supply exists, seed invades on soil continuously or intermittently exposed for 2 years. In impoundments with an unmanageable water supply, meadow emergents outcompete and help exclude shrubs after flooding.

Mudflat emergents (Table 4.3), typically annuals and biennials, need unvegetated moist soils to germinate and mature, prevailing 6 to 9 dm below normal pool during drawdown, where competition with meadow emergents is reduced (Fig. 4.4). Mudflat emergents produce much seed valuable as food for wildlife, but flooding is deleterious to seed viability, plant growth, and survival. Flooded dead plants provide ideal habitat for many invertebrates that feed on the microorganisms active on decaying vegetation. Several of the mudflat emergents, biennials particularly, germinate in late summer, even if water levels remain high, but then grow best if water levels remain low the next year.

Marsh emergents (Table 4.3) typically are perennials that need a stable water level. Even then they represent only a small percentage (2 to 10 percent) of the total herbaceous biomass. They are most abundant at about 3 dm below normal pool or lower during drier years (Fig. 4.4).

Floating-leaved plants (Table 4.4) need relatively stable water levels. *Submergents* (Table 4.4) establish high densities in residual pools 9 dm below normal pool, invading higher elevations as the water level increases. Dramatic reductions occur the first year of reflooding, but consecutive wet years produce reinvasion.

TABLE 4.3 Emergent Plants in Minnesota

Common name	Scientific name
Mudflat	
Water plantain	*Alisma plantago-aquatica*
Foxtail	*Alopecurus aequalis*
beggarticks	*Bidens* spp.
Lambsquarters	*Chenopodium album*
Water hemlock	*Cicuta maculata*
Water hemlock	*C. bulbifera*
Willow-herb	*Epilobium adenocaulon*
Fireweed	*Erechtites hieracifolia*
Joe-pye-weed	*Eupatorium maculatum*
Boneset	*E. perfoliatum*
Bedstraw	*Galium* spp.
Avens	*Geum* spp.
Jewelweed	*Impatiens biflora*
Wood nettle	*Laportea canadensis*
Bugleweed	*Lycopus uniflorus*
Bugleweed	*L. americanus*
Cowwheat	*Melampyrum lineare*
Mint	*Mentha arvensis*
Monkeyflower	*Mimulus ringens*
Tufted loosestrife	*Naumburgia thyrisiflora*
Bindweed	*Polygonum cilinode*
Curltop ladysthumb	*P. lapathifolium*
Smartweed	*P. sagittatum*
Dotted smartweed	*P. punctatum*
Cinquefoil	*Potentilla* spp.
Yellow cress	*Rorippa islandica*
Dock	*Rumex* spp.
Skullcap	*Scutellaria galericulata*
Skullcap	*S. lateriflora*
Goldenrod	*Solidago* spp.
Hedge hettle	*Stachys palustris*
Marsh St. Johnswort	*Triadenum virginicum*
Nettle	*Urtica dioica*
Robust	
Reed	*Phragmites communis*
Broad-leaf cattail	*Typha latifolia*
Tall meadow	
Sweet flag	*Acorus calamus*
Blue flag	*Iris versicolor*
Reed canarygrass	*Phalaris arundinacea*
Wildrice	*Zizania aquatica*
Short meadow	
Sedge	*Carex* spp.
Three-way sedge	*Dulichium arundinaceum*
Grass	Gramineae
Bent grass	*Argrostis* spp.
Bluejoint	*Calamagrostis* spp.
Cutgrass	*Leersia oryzoides*
Rice grass	*Oryzopsis* spp.
Meadow grass	*Poa* spp.
Rush	*Juncus* spp.

TABLE 4.3 Emergent Plants in Minnesota (*Continued*)

Common name	Scientific name
Narrow-leaved marsh	
Spikerush	*Eleocharis* spp.
Horsetail	*Equisetum sylvaticum*
Horsetail	*E. palustre*
Horsetail	*E. fluviatile*
Softstem bulrush	*Scirpus validus*
Bulrush	*S. subterminalis*
Common threesquare	*S. americanus*
Bulrush	*Scirpus* spp.
Woolgrass	*S. cyperinus*
Burreed	*Sparganium fluctuans*
Burreed	*S. chlorocarpum*
Sphagnum	*Sphagnum* spp.
Broad-leaved marsh	
Water arum	*Calla palustris*
Bellflower	*Campanula aparinoides*
Water smartweed	*Polygonum natans*
Marsh cinquefoil	*Potentilla palustris*
Buttercup	*Ranunculus pensylvanicus*
Yellowwater crowfoot	*R. gmelini*
Duck potato	*Sagittaria latifolia*
Water parsnip	*Sium suave*

SOURCE: Knighton (1985).

A perimeter of *shrubs,* mainly willow and alder, occurs at 6 dm above normal pool, but is mostly absent at and below normal pool (Fig. 4.4). Willow is more vigorous than alder and can resprout after appearing dead from a period of flooding. But meadow emergents recover more quickly after flooding, displacing the shrubs.

Manager's option. Impoundments with a moderate water supply are likely to be the most successful for water-level management. Problems with beaver are minimal because the average time needed for a complete water exchange during the growing season is 1 week. A visit once a week is adequate to remove any dam built by beaver.

Water levels can maintain impoundments as inland fresh meadows (Type 2), shallow marshes (Type 3), or deep marshes (Type 4). But the effects of water-level manipulation are influenced by the type of soil and relief of the impoundment's bottom. Plant composition on heavy soils is hard to alter because such soils do not dry out readily. Interspersion of vegetation and open water is influenced by the irregularity in the impoundment's bottom. Uniform basins develop monotonous plant communities. In uniform basins, plant communities tend to segregate into concentric zones relative to the increasing depth of the basin. Drawdowns shift the zones up or down the slope, but any mixing

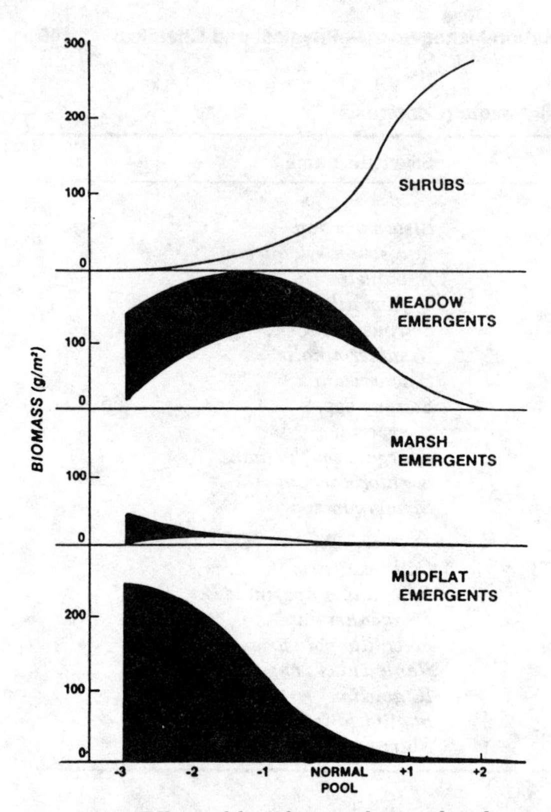

Figure 4.4 Effects of drawdown and normal pool water levels on distribution of shrub and emergent biomass in impoundments with an inadequate water supply for precise water level management. For emergents the upper line approximates distribution of biomass during full drawdown and the lower line approximates biomass when the water level is near normal pool throughout the year (Knighton 1985). [For metric equivalence, see metric conversion table in the front of the book.]

TABLE 4.4 Floating-Leaved and Submergent Plants in Minnesota

Common name	Scientific name
Floating-leaved	
Watershield	*Brasenia schreberi*
Bullhead lily	*Nuphar variegatum*
Waterlily	*Nymphaea tuberosa*
Submergent	
Coontail	*Ceratophyllum demersum*
Moss	
Water milfoil	*Myriophyllum* spp.
Floatingleaf pondweed	*Potamogeton natans*
Pondweed	*Potamogeton* spp.
Bladderwort	*Utricularia vulgaris*

SOURCE: Knighton (1985).

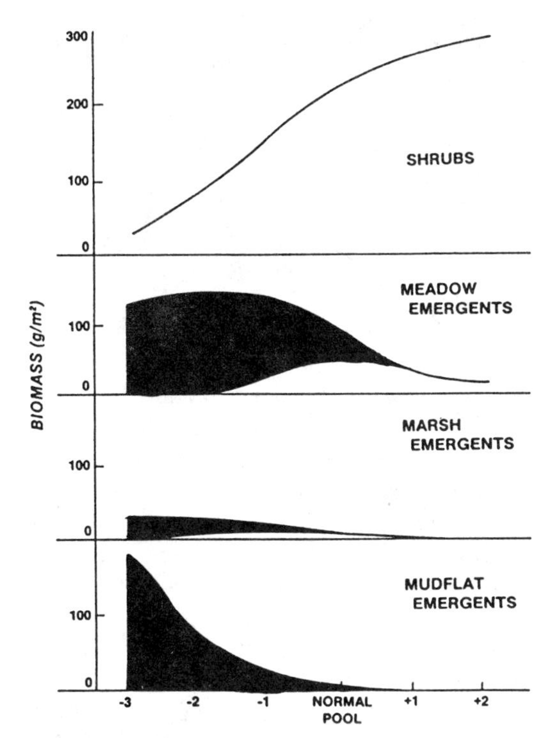

Figure 4.5 Effects of drawdown and normal pool water levels on distribution of shrub and emergent biomass in impoundments where water level management is least difficult. For emergents the upper line approximates distribution of biomass during full drawdown and the lower line approximates biomass when the water level is near normal pool throughout the year (Knighton 1985).

of slopes is short-lived. Interspersion is achieved most effectively through drawdown by extensive scraping and piling of soil into small islands mainly along the contour 3 dm below normal pool.

Drawdowns sharply increase the biomass of emergent plants from normal pool to 9 dm below normal pool, where the increase is most noticeable because no emergents occur 9 dm below normal pool (Fig. 4.5). But these disappear during the first year after drawdown. Drawdowns on 5-year cycles produce three habitat conditions:

1. High biomass and some change in composition of emergents during drawdown
2. High biomass of dead, submerged, and decaying emergents during the first year after drawdown
3. Low biomass of living and dead emergents during all other years

Full drawdowns generally encourage reestablishment of the plant community there before impoundment, mostly grasses and emergent sedges. Partial drawdowns, to about 3 dm below normal pool for part of the growing season, encourage grasses, sedges, and willows. Annual exposure of the pool bottom will discourage emergent marsh plants such as arrowhead and burreed, floating-leaved plants such as waterlily, and submerged plants such as pondweed. These vegetative forms, and shrubs, thrive only at normal pool and with a partial drawdown less than 3 dm below normal pool.

Meadow emergents (see Table 4.3) dominate with drawdowns (Fig. 4.5), but the biomass is less than in weather-controlled impoundments (see Fig. 4.4). Few meadow emergents occur at water levels below normal pool. Drawdowns lasting more than 1 year increase the biomass of meadow emergents. Low biomass at 3 dm below normal pool can be increased substantially with two consecutive years of drawdown.

Marsh emergents (see Table 4.3) increase somewhat during drawdown but disappear with reflooding and generally have low biomass (Fig. 4.5). Stable water levels for several consecutive years favor marsh emergents in the zone between normal pool and 3 dm below normal pool.

Mudflat emergents (see Table 4.3) are absent during years of normal pool and increase during drawdown when mudflats are exposed (Fig. 4.5). Weather-controlled impoundments produce more mudflat emergents than do impoundments with a manager's option. Weather-controlled impoundments periodically dry out from lack of adequate precipitation. Impoundments with adequate precipitation (manager's option) probably have fewer viable seeds, especially at lower elevations where flooding is more persistent. Also seed production is less dependable with less frequent drawdowns or partial drawdowns which occur later in the season.

Floating-leaved plants (see Table 4.4) establish very slowly, are sensitive to drawdown and competition with emergents, and thus have low density in impoundments with cyclic, complete drawdowns at least 5 years apart. *Submergents* (see Table 4.3) comprise 90 percent of the herbaceous biomass at least 9 dm below normal pool, but decrease to 4 percent as water increases to 3 dm below normal pool.

Shrubs, mainly willow, occur at all elevations below normal pool (Fig. 4.5). At 3 dm below normal pool, the biomass of shrubs is about 0.5 that at 6 dm above normal pool and less than 0.1 that at 9 dm below normal pool. This is substantially more than grows in weather-controlled impoundments (see Fig. 4.4), due to reduced competition with meadow emergents that occurs during partial annual drawdowns when the water level is held at normal pool during the first half of the growing season. For willows to remain healthy, partial drawdowns should begin before July 16, and before June 16 if alders are desired.

Beaver-controlled. Impoundments controlled by beaver respond as do impoundments with manager's option. They can be maintained as inland fresh meadows (Type 2), shallow marshes (Type 3), or deep marshes (Type 4). But much time (almost daily) and much energy are spent removing beaver blockages at spillways, which otherwise would raise the water level unduly despite special structures intended to compensate for or thwart beaver dam building. The problem is not as acute on large impoundments as on rivers and large streams. (See "Beaver Ponds" in Chap. 3 and "Beaver" in Chap. 5.)

Impoundment potential of back-flood projects

Foraging and nesting sites are provided for ducks by impounding native hay meadows in shallow marsh and wet meadow areas (Stewart and Kantrud 1972), especially in the northern Great Plains, so that spring runoff can be retained until early summer (Pederson et al. 1989). The area must be next to a large marsh, river, lake, or reservoir; have porous soils for ready draining; and have soil and water salinities below 10 ppt. By means of dikes and control structures, the area is flooded in spring (before April 15 in Manitoba) 30 to 100 cm deep or less, generally for up to 50 days by runoff or supplemental pumping, so that the root zone is saturated 1 m deep, as determined by inserting a steel probe. Water 10 to 30 cm deep (Bookhout et al. 1989, Heitmeyer et al. 1989) is used by dabbling ducks and geese feeding on both vegetation and the dramatic increase in invertebrates. A complete drawdown in early summer furnishes nesting habitat. The hay can be harvested in middle to late summer. Periodic burning in fall and reflooding in early spring before natural wetlands thaw provide habitat for early migrants, and extend use, especially if several such impoundments occur together and are flooded sequentially.

Although control structures connect the backwater area to the main water source, backwaters are separated from the main water source, especially rivers, as far as feasible, to reduce water-level changes caused by seepage and floods (Matter and Mannan 1988). Backwaters should have areas several decimeters deeper than the lowest likely river drawdown. Best sites are on inactive floodplains, terraces, or other areas away from the main river channel, with enough vegetation or levee to protect from a 20-year flood (Woodward-Clyde Consultants 1980).

Impoundment potential of coastal and salt marshes

Impoundment dikes and water control structures can be maintained economically on only a small percentage of coastal marshes due to the prevalence of unstable soil types or foundation and exposure to catastrophic storms (Baldwin 1968). Management in salt marshes (30 to 35 ppt salinity) focuses on reducing salinity to 10 to 20 ppt for optimum biomass production (Neely 1960, Baldwin 1968, Morgan et al. 1975), similar to brackish marshes, or reducing tidal influence (Gordon et al. 1989). For management purposes, the brackish zone is divided into three salinity regimes and associated plant communities (Table 4.5). In the brackish-salt marsh (20 to 30 ppt) and brackish

TABLE 4.5 Water Management and Associated Plants in Southeastern Coastal Impoundments

Marsh zone	Water management	Food plants	Competing cover plants
Fresh marsh (< 1 ppt)	Early spring drawdown with saturated alluvial soils; late fall flooding to $\bar{x}$ depth of 22 cm	*Polygonum punctatum* *P. hydropiperoides* *P. arifolium* *P. sagittatum* *Aneilema keisak* *Eleocharis quadrangulata* *Scirpus validus* *Peltandra virginica*	*Zizaniopsis miliacea* *Typha latifolia* *Scirpus cyperinus* *Juncus effusus* *Alternanthera phloxeroides*
	Early spring drawdown with moist to dry alluvial soils; late fall flooding to $\bar{x}$ depth of 22 cm	*Panicum dichotomiflorum* *Echinochloa crusgalli* *Cyperus* spp.	*Erianthus giganteus* *Eupatorium capillifolium* *Sesbania exaltata* *Erechtites hieracifolia* *Ambrosia artemisiifolia*
	Early spring drawdown with moist to saturated organic soils; late fall flooding to $\bar{x}$ depth of 22 cm	*Lacnanthes caroliniana* *Panicum verrucosum* *Polygonum* spp. *Cyperus* spp. *Panicum* spp.	*Cladium jamaicense* *Centella asiatica* *Liquidambar styraciflua* *Persea borbonia* *Acer rubrum* *Cyrilla racemiflora*
	Semipermanently flooded to depth of 15 to 122 cm	*Eleocharis quadrangulata* *E. equisetoides* *Brasenia schreberi* *Nymphaea odorata* *Ceratophyllum demersum* *Najas* spp. *Chara* spp. *Potamogeton* spp.	*Panicum hemitomon* *Zizaniopsis miliacea* *Typha latifolia* *Alternanthera philoxeroides* *Ludwigia* spp. *Nuphar* spp. *Nelumbo lutea* *Nymphoides* spp. *Cabomba caroliniana* *Utricularia* spp. *Myriophyllum heterophyllum*
Intermediate marsh (1–5 ppt)	Early spring drawdown with saturated alluvial soils; late fall flooding to $\bar{x}$ depth of 22 cm	*Scirpus robustus* *S. validus* *Echinochloa walteri* *Polygonum punctatum*	*Spartina cynosuroides* *Typha angustifolia* *Alternanthera philoxeroides*

TABLE 4.5 Water Management and Associated Plants in Southeastern Coastal Impoundments (*Continued*)

Marsh zone	Water management	Food plants	Competing cover plants
	Early spring drawdown with moist to dry alluvial soils; late fall flooding to $\bar{x}$ depth of 22 cm	*Panicum dichotomiflorum* *Setaria magna* *Cyperus* spp.	*Aster* spp. *Erechtites hieracifolia* *Spartina cynosuroides* *Sesbania exalta* *Baccharis halimifolia*
	Early spring drawdown with moist to saturated organic soils; late fall flooding to $\bar{x}$ depth of 22 cm	*Lacnanthes caroliniana* *Scirpus robustus* *Panicum* spp. *Cyperus* spp. *Echinochloa walteri* *Setaria magna*	*Cladium jamaicense* *Typha angustifolia* *Spartina cynosuroides*
Brackish marsh (5–20 ppt)	Early spring drawdown with saturated alluvial and organic soils; late spring or early summer complete drawdown; early or midsummer reflooding with gradual increase in water depth; fall lowering of water levels to 22 cm	*Scirpus robustus* *Ruppia maritima* *Eleocharis parvula* *Leptochloa fascicularis* *Sesuvium maritimum*	*Spartina cynosuroides* *S. bakeri* *S. patens* *Typha angustifolia* *T. domingensis* *Distichlis spicata* *Juncus roemerianus* *Scirpus olynei*
Brackish salt marsh (20–30 ppt)	Early spring drawdown with saturated alluvial and organic soils; late spring to early summer complete drawdown; early midsummer reflooding with gradual increase in water depth; fall lowering of water levels to 22 cm	*Ruppia maritima* *Eleocharis parvula* *Sesuvium maritimum*	*Spartina alterniflora* *Juncus roemerianus* *Distichlis spicata*
Salt marsh (30–35 ppt)	Semipermanently flooded to depths of 15 to 122 cm	*Nymphaea mexicana* *Potamogeton pectinatus*	*Typha domingensis* *Typha augustifolia*
	Flooded with water salinity <5 ppt	*Najas quadalupensis* *Chara* spp.	*Scirpus californicus* *Nelumbo lutea*
	Early spring complete drawdown with alluvial soils; midspring reflooding to sustained depths of 61 cm; fall lowering of water levels to $\bar{x}$ depth of 22 cm	*Ruppia maritima* *Eleocharis parvula*	*Spartina alterniflora* *Juncus roemerianus* *Distichlis spicata*
	Early spring complete drawdown with muck soils; midspring reflooding to sustained depths of 61 cm; fall lowering of water levels to $\bar{x}$ depth of 22 cm	*Chara hornemannii* *Ruppia maritima*	*Spartina alterniflora*
	Early spring drawdown with moist organic soils; late summer to early fall reflooding to $\bar{x}$ depth of 22 cm	*Sesuvium maritimum* *Ruppia maritima*	*Spartina alterniflora* *Distichlis spicata* *Salicornia* spp. *Pluchea purpurascens* *Aster subulatus* *A. tenuifolius*

SOURCE: Gordon et al. (1989).

marsh (5 to 20 ppt), the focus is on submerged aquatic and certain persistent emergent plants. In the intermediate marsh (1 to 5 ppt), the focus is on moist-soil plants. In the fresh marsh (below 1 ppt), the focus is a combination of management for submerged aquatic, certain persistent emergent, and moist-soil plants.

Regularly flooded salt marshes. On the East and Gulf coasts, the normal tidal state of much of the lowest-elevation salt marsh supports saltmeadow cordgrass (Baldwin 1968). In southeastern marshes of high soil salinity, sea purslane is the only seed-producing plant encouraged by drawdown (Gordon et al. 1989). Widgeongrass, a hardy, excellent food for ducks, is cultured by continuous diking of this lowest zone on protected sites, along with several bulrushes on shallow edges. A good pond of widgeongrass probably provides more waterfowl food with the least management effort than any other type of impoundment. For best growth, 46 to 61 cm of brackish water, one-third sea strength to discourage tropical cattail, should occur over a firm bottom of organic or mineral soil free of silt and detritus (Baldwin 1968, Gordon et al. 1989).

In salinity concentrations of 10 ppt, alkali bulrush competes well with cattail, which is stunted. Alkali bulrush germinates and survives on pickleweed flats if the salt concentration in the soil is reduced by flushing with winter floodwater. In less brackish areas, alkali bulrush is maintained by draining the pond as soon as seed is produced, usually about mid-June in central and southern California. The plants then go dormant for the summer, sprouting again when reflooded in the fall, sometimes producing a second crop of seed. Dry periods during summer discourage cattail (Miller 1962).

Optimum biomass of widgeongrass is produced when water salinity is 10 to 20 ppt (Neely 1960, Baldwin 1968, Morgan et al. 1975). After the annual or biennial spring drawdown, the pond is reflooded to 30 to 60 cm in late March to mid-April, when salinity in tidal creeks generally is lower (Gordon et al. 1989). Control structures located up the estuary, where salinity typically is less, can be used to reflood impoundments of at least 40 ha or impoundment complexes with multiple control structures. Control structures are adjusted in late spring and summer after rainfall to help maintain desired salinity, and during spring, summer, and early fall to circulate tidal water through the marsh to reduce water temperature and algal blooms.

A complete drawdown of 2 to 4 weeks in early spring usually reduces turbidity from unstable soils, often found in saltmarsh impoundments, without excessive acidification, or in late summer or early fall if widgeongrass is killed by filamentous algae or late summer die-off. To control filamentous algae, young mullet in summer migration are let into the impoundment to feed on them (Baldwin 1968).

Drawdowns of proper duration also help achieve the desired interspersion of emergents and open water. Rainfall from tropical storms reduces salinity from fall reflooding. Muskgrass coexists with widgeongrass and can be managed similarly, although it is more tolerant of turbidity, salinity, wave action, and filamentous algae. Sea purslane on highly organic soils is managed by a late May drawdown and early fall flooding.

Food production can be maintained in permanently flooded ponds if salinity is held within the tolerance of desired plants. Holding salinity between 10 and 20 ppt in California causes widgeongrass to become dormant, a reduction in objectionable emergents, an increase in alkali bulrush if water depths are 30 to 38 cm, and a reduction of carp (Miller 1962).

Flooding and drawdown in California vary with location (Fig. 4.6) (Heitmeyer et al. 1989). Permanently flooded ponds should be 2 to 3 m deep, perhaps drained every 5 to 10 years to control cattail or tules (hardstem bulrush) by prolonged drying, grazing, burning, mowing, or disking. To reduce stagnation conducive to botulism, water often is circulated from August to November. An outbreak of botulism requires removal of dead birds, drainage, and flushing. Ponds also need draining and treatment of emergents within 5 years if under summer water management.

Seasonally flooded ponds in California (Fig. 4.6) should be kept 10 to 30 cm deep. Drawdowns in late winter or early spring (January to March) promote dock, slender aster, and smartweeds. Drawdowns in April and May promote pricklegrass, swamp timothy, and watergrass (millet). Drawdowns in May and June promote tules, cattails, and cocklebursor alkali bulrush. Flooding small ponds in late summer several weeks before normal fall flooding might increase the biomass of invertebrates by providing brood stock ponds for them and is generally consistent with mosquito control practices of limited water fluctuations from March to October.

Completely drying soils in some areas under tidal influence (e.g., Suisun Marsh, California) causes drastically reduced pH, soil oxidation, fast decomposition of marsh litter, and subsidence. Flooding with alkaline water causes suspension of iron which precipitates as ferric hydroxide, causing "red water," toxic to some plants and invertebrates. Such conditions are avoided by maintaining high soil moisture year round with water about 10 cm below the soil surface in water delivery perimeter ditches (Gordon et al. 1989, Heitmeyer et al. 1989), by flooding and drying ponds in alternate years, or by flooding ponds year round (Fig. 4.6). Flooding and drying in alternate years might be the best strategy to reduce acidity and control undesired plants during dry periods. Regimes of high soil moisture and permanent water increase growth of cattail, tules, and alkali bulrush, which might be undesirable.

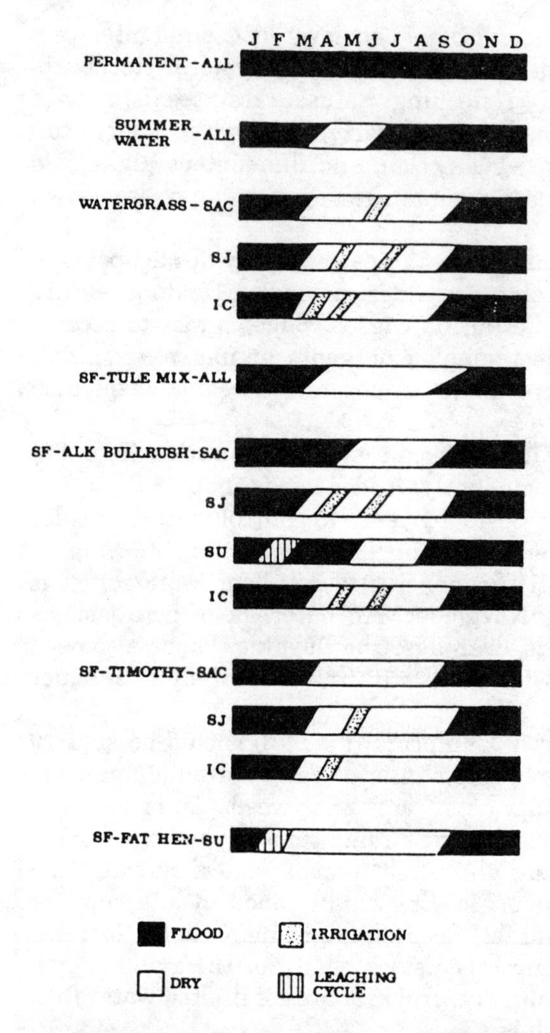

Figure 4.6 Flooding and draining schedules for managing various wetland habitat types in California. Diagonal lines represent the chronological range of flooding or drying. ALL = all valley regions, SAC = Sacramento Valley, SJ = San Joaquin Valley, SU = Suisun Marsh, IC = Imperial-Coachella valleys, SF = seasonally flooded. Irrigations consist of saturating soils for 1–2 wk; leaching consists of flooding and drying ponds repeatedly, with each inundation lasting ≤1 wk (Heitmeyer et al. 1989).

Seasonally flooded ponds in California should be drained in early spring to stimulate germination of pricklegrass and swamp timothy (Fig. 4.6). Seed production is increased with at least one irrigation in midsummer, when swamp timothy is flowering and the leaves show necrosis. Ponds drained in late spring to produce watergrass (millet) and alkali bulrush need one summer irrigation for existing stands and two for newly planted stands.

Moist-soil plants are encouraged in California usually by flooding from September through March. Water circulation through the ponds reduces salinity, which encourages sprangletop, dock, pricklegrass, and swamp timothy.

Soft-bottom ponds should be drained to oxidize and solidify the soil. But prolonged drying should be avoided, as it results in the formation of sulfuric acid by sulfide oxidation in cat clays (Baldwin 1968). If fresh water entering saline impoundments reduces salinity to 10 percent or less sea strength, plant succession is characterized by invasion of the valued sago pondweed and other pondweeds (*Potamogeton* spp., *Najas* spp., *Zannichellia* spp.), muskgrasses, coontail, and eventually the undesired deepwater tropical cattail, white waterlily, and other pest plants. Widgeongrass will recover if the pond returns to stronger salinity. Chamberlain (1960) described water-quality tolerances of southern aquatics in succession from saline and alkaline to fresh and acidic habitats.

Fluctuating the salinity in managing coastal impoundments is valuable for specific food plants (Baldwin 1968, Gordon et al. 1989). If sites freshen to less than 5 ppt and are maintained as deepwater areas, banana waterlily grows (see Table 4.5). Rainfall, runoff, and late winter tidal flooding maintain suitable salinity, especially on barrier islands with diked saltmarsh creek channels and sloughs with soft mud soils and alkaline waters. Desirable submersed aquatics, such as sago pondweed, bushy pondweed, and muskgrass, commonly grow with banana waterlily. With its buried hibernacula, it can survive prolonged salt intrusion from periodic flooding; competing cattails, bullwhip bulrush, and other pest plants cannot survive and might be controlled effectively by periodic flooding with highly saline water (Baldwin 1968, Cely 1979).

Food production can be maintained in permanently flooded ponds if salinity is held within the tolerance of desired plants. Organically stained water should be removed periodically. Excessive runoff from extensive wooded areas should be diverted from nearby impoundments of widgeongrass. Surplus fresh water, especially if stained and acidic, is deleterious to widgeongrass and promotes undesirable freshwater plants such as water hyacinth, alligatorweed, water pennywort, fragrant waterlily, and cattail. Periodic flushing with salt water held long enough to control noxious vegetation will give salt-tolerant plants a competitive advantage (Stutzenbaker and Weller 1989).

Irregularly flooded salt marshes. On the East and Gulf coasts, salt marshes flooded irregularly are generally characterized by a dominant cover of black needlerush (Baldwin 1968). Such sites are at higher elevation than regularly flooded salt marshes and are therefore harder to keep flooded with a water depth best suited for widgeongrass. Pumping might be desirable. Ditch plugs stabilize water levels to convert areas of needlegrass to widgeongrass on unstable peat soils and exposed sites in potholes and waterways.

Water level and salinity fluctuate readily in these shallow impoundments (Baldwin 1968). Reduced salinity encourages invasion by species of plants tolerant of shallow water and exposed soil, such as the desirable bulrushes, wild millets, dwarf spikerush, sprangletop, giant foxtail, annual panic grasses, as well as the undesirable cattails, alligatorweed, coarse perennial grasses, and brush. Use of saline tidal fluctuation will reduce the successional decline of habitat resulting from too much fresh water. Concentrating on the production of saltmarsh bulrush benefits waterfowl because, like widgeongrass, it tolerates salinities that most pest plants cannot. Monthly shallow-water fluctuations, increasing from 0 to 20 cm, then decreasing, provide proper salinities and water levels for saltmarsh bulrush (Neely 1960). Although tolerant of similar salinities and water levels, southern bulrush should not be introduced because saltmarsh bulrush and widgeongrass are better food plants for ducks.

River marshes cause fluctuating salinity in the zone of irregularly flooded salt marshes of an estuary (Baldwin 1968). Impoundment results in periodic freshening of the water, which encourages the transitory dwarf spikerush, softstem bulrush, and annuals mentioned previously. Such impoundments require the wildlife manager to check the salinity often for adjustment.

In brackish-saline wetlands of south coastal areas, management focuses on production of widgeongrass, dwarf spikerush, and sea purslane (Gordon et al. 1989). In intermediate marsh impoundments, water levels are lowered gradually during late February and early spring to maintain moist soil conditions through September (see Table 4.5). Water control structures should maintain water levels 10 cm below the marsh floor in perimeter ditches, unless tidewater salinity is more than 5 ppt.

During spring drawdown, saltmarsh bulrush often dominates before saltmarsh (walters) millet, fall panic grass, dotted smartweed, flat sedges, and giant foxtail develop (Gordon et al. 1989). Drawdowns resulting in soils that are too dry encourage undesirable species such as big cordgrass, fireweed, and asters. Soils that are too wet or shallow flooding might promote undesirable species such as alligatorweed, narrow-leaved cattail, southern (tropical) cattail, and southern wildrice.

In intermediate marsh, impoundments are drawn down after the growing season, in September, to dry the marsh, which is then burned and reflooded 20 to 30 cm deep (Gordon et al. 1989). With incomplete burning, flooding 50 cm deep with gradual reduction to 20 to 30 cm causes standing plants to weaken and fall, thus increasing food availability to ducks.

In brackish marsh, impoundments are drawn down in March or April to maintain a saturated soil condition without excessive drying which causes acidification or formation of cat clays (see Table 4.5). Saltmarsh bulrush is encouraged. Periodic rains in April and occasional flushing with tidal water prevent drying. To stabilize marsh soils and minimize turbidity with reflooding, impoundments are drained in April for 1 to 2 weeks. In May they are flushed twice daily with tidal water 15 to 20 cm deep, 5 to 15 ppt salinity, reflooded in 10- to 15-cm increments during new-moon and full-moon tides each month until mid-October, then drawn down to 22 cm deep. Such water manipulation encourages widgeongrass, dwarf spikerush, saltmarsh bulrush, and, in lower-elevation open areas, sprangletop and discourages filamentous algae which can shade and kill widgeongrass. Rainfall which dilutes salinity too much is removed during ebb tides or retained to dilute high salinity after drought.

Sea purslane is encouraged on suitable soils of brackish impoundments by draining the impoundment to about 20 to 30 cm below bed level by April 1 until September 1, when it is reflooded 20 to 25 cm deep (Swiderek et al. 1988). Flooding for 1 or 2 years during the growing season to encourage widgeongrass might prevent eventual encroachment of undesirable emergents after several years of management for sea purslane. Late summer flooding of dewatered impoundments containing mature sea purslane produces late crops of widgeongrass, which reduces problems of summer die-outs, algae competition, and damage by waves.

The desirable saltmarsh bulrush produces much seed in marshes with 3 to 7 ppt salinity (Neely 1962), but not with more than 20 ppt (Palmisano 1972). To produce saltmarsh bulrush and decrease undesirable perennial emergents such as big cordgrass and tropical cattail along southeastern coasts, management procedures involve fluctuating water levels every 30 to 60 days during the growing season and flooding 2 to 5 dm deep (Landers et al. 1976). When salinity rises from summer drought and saltmarsh bulrush is stressed, managers should produce sea purslane instead (Gordon et al. 1989).

In brackish-salt marshes of 20 to 30 ppt salinity, extensive stands of sea purslane often invade open ponds during extended spring drawdown (Gordon et al. 1989). Then saturated soils are maintained until sea purslane matures (late July to early August), when the pond is reflooded (as late as August or September) to produce good crops of dwarf spikerush and widgeongrass (Neely 1962, Prevost 1987), although the undesirable smooth cordgrass gradually increases.

Typical management schedules for brackish marshes less than 40 ha often result in stands of saltmarsh bulrush that are too extensive and dense to be attractive to waterfowl. If they are flooded 45 to 60 cm deep for one growing season, openings will result, as will growth of widgeongrass and dwarf spikerush, thus improving use by waterfowl (Gordon et al. 1989). Landers et al. (1976) reported results of deeply flooded impoundments (1 to 1.5 m deep) with water levels gradually raised during the growing season to encourage widgeongrass and results of shallowly flooded impoundments until water levels cyclically fluctuated to encourage saltmarsh bulrush (Table 4.6).

TABLE 4.6 Type of Management, Vegetation, and Main Duck Foods in Diked, Well-Managed Impoundments in South Carolina*

Management type†	Vegetational coverage (%)		Major plant foods‡		
	Dominant	Abundant	Plant	%V	%O
		Fresh Water			
I	*Typha* spp. (25)	*Polygonum punctatum* (6)	*Polygonum punctatum*	42	81
	Spartina cynosuroides (15)	*Scirpus robustus* (6)	*Cyperus odoratus*	5	58
	S. bakeri (15)	*Nymphaea odorata* (5)	*Scirpus robustus*	3	17
	Scirpus validus (12)	*Ceratophyllum* spp. (4)			
		Alteranthera spp. (3)			
		Cladium spp. (2)			
II	*Lachnanthes caroliniana* (28)	*Persea* spp. (9)	*Lachnanthes caroliniana*	25	73
	Panicum spp. (23)	*Acer rubrum* (6)	*Panicum dichotomiflorum*	20	32
		Liquidambar styraciflua (6)	*Cyperus erythrorhizos*	14	30
		Sesbania spp. (6)	*Panicum verrucosum*	9	28
		Polygonum spp. (5)			
		Titi (mixed shrubs) (4)			
		Andropogon spp. (4)			
		Taxodium spp. (3)			
III	*Panicum* (mostly *P. dichotomiflorum*) (23)	*Cuphea* spp. (9)	*Panicum dichotomiflorum*	23	74
	Alternanthera spp. (12)	*Polygonum* spp. (7)	*Lachnanthes caroliniana*	18	26
		Juncus spp. (6)	*Polygonum punctatum*	17	31
		Spartina cynosuroides (6)	*Nyssa sylvatica*	9	14
		Sesbania spp. (6)	*Cyperus* (3 spp.)	5	12
		Zizaniopsis miliacea (6)			
		Cyperus spp. (6)			
		Diodia virginiana (3)			
		Erianthus spp. (3)			
		Baccharis spp. (2)			
		Scirpus cyperinus (2)			

Brackish Water					
IV	*Ruppia maritima* (58)	*Spartina alterniflora* (7)	*Ruppia maritima*	18	63
	Typha spp. (20)	*Juncus roemerianus* (6)	*Panicum dichotomiflorum*	10	21
		Scirpus robustus (3)	*Scirpus robustus*	7	23
		Panicum spp. (2)	*Cyperus odoratus*	6	25
		Spartina bakeri (2)	*C. polystachyos*	6	10
			Panicum verrucosum	6	6
V	*Scirpus robustus* (28)	*Echnichloa* spp. (8)	*Polygonum punctatum*	51	37
	Typha spp. (22)	*Alternanthera* spp. (6)	*Scirpus robustus*	19	81
	Spartina cynosuroides (15)	*Eleocharis parvula* (5)	*S. olneyi*	5	54
		Zizaniopsis miliacea (3)	*Panicum dichotomiflorum*	5	31
		Scirpus olneyi (3)	*Echinochloa* spp.	5	31
		Panicum dichotomiflorum (2)	*Cyperus polystachyos*	5	29
		Grasses (misc.) (2)	*Eleocharis parvula*	3	15

*Does not include 59 ducks from unmanaged or poorly managed marshes.

†Management type: I, summer drawdown to saturated bed on typical alluvial soils to encourage *Polygonum* spp. and *Echinochloa* spp.; II, summer drawdown to saturated bed on peat soils to encourage *Lachnanthes caroliniana*; III, thoroughly drained in growing season to encourage *Panicum* spp., grazed; IV, deeply flooded impoundments with water levels gradually raised during the growing season to encourage *Ruppia maritima*; V, shallowly flooded impoundments with water levels cyclically fluctuated to encourage *Scirpus robustus*.

‡*V*, volume; *O*, occurrence.

SOURCE: Landers et al. (1976).

Salt meadows and salt flats. Much of the southern coastal marshes consists of salt meadows of saltmeadow cordgrass and Olney threesquare and flats either barren or covered with glasswort and saltgrass (Baldwin 1968). Undiked, the meadows provide good grazing for geese. When diked, these marshes are hard to flood via direct rainfall or tides due to their high elevation. Average depths on impounded saltmeadow cordgrass are 18 cm in permanent brackish impoundments and 36 cm in permanent freshwater impoundments (Chabreck 1960). A summer drawdown on open fresh meadow is used to grow annual wild millet and sprangletop. With prolonged drawdown or prolonged freshening, such sites succeed to coarse grasses, cattail, and brush. This situation can be prevented by manipulating salinity and water level periodically, plus burning, disking, mowing, and grazing.

Similar to coastal marsh management, marshes in the Great Basin are managed relative to the fresh or saline quality of underlying sediments (Kadlec and Smith 1989). Impoundments are built to spread river water across salt flats, thus reducing salinity in the upper layers (Figs. 4.7 and 4.8). Desirable plants grown in water 45 to 60 cm deep include sago pondweed, curly-leaved pondweed, horned pondweed, muskgrass, and widgeongrass. Shallower sites contain alkali bulrush,

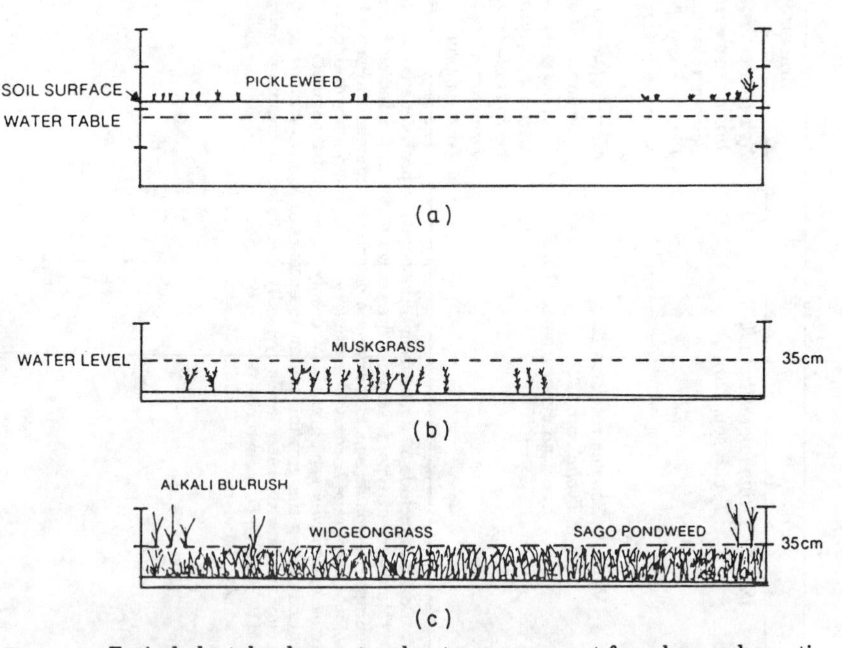

Figure 4.7 Typical plant development and water management for submersed aquatic plants in marshes of the Great Basin: (*a*) salt flat with high soil salinities and little vegetation, (*b*) intermediate stage following flooding, (*c*) final stage with low soil salinities and dense submersed aquatics (Kadlec and Smith 1989).

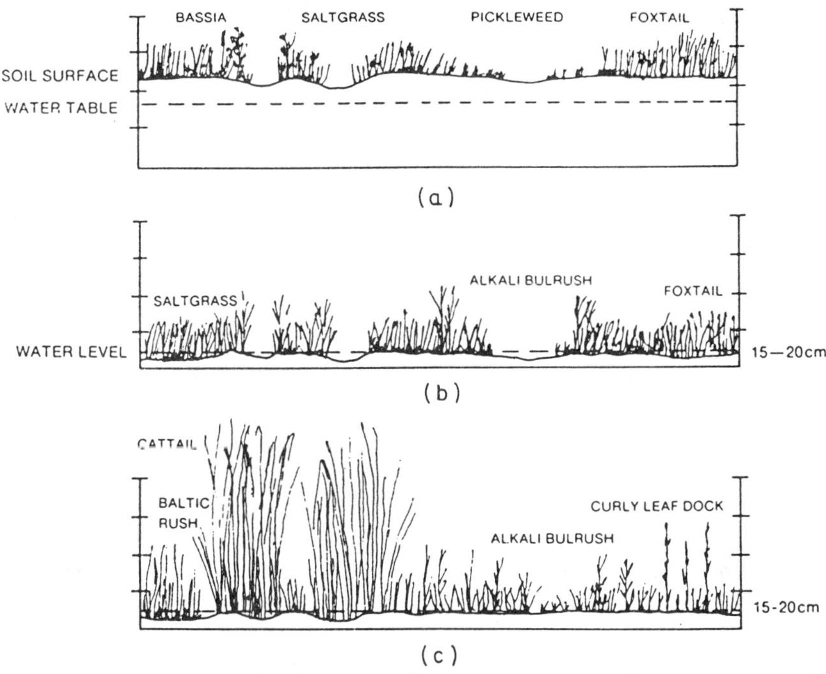

Figure 4.8 Typical plant development and water management for emergent aquatic plants in marshes of the Great Basin: (*a*) upland site with vegetation adapted to high soil salinities, (*b*) intermediate stage with water fluctuating 0–20 cm, (*c*) long-term (> 6 years) effects of flooding 0–20 cm and reduced soil salinities (Kadlec and Smith 1989).

hardstem bulrush, Olney threesquare, and pickleweed. Moist-soil plants such as smartweeds and red goosefoot become prevalent in less saline marshes. Bolen (1964), Duebbert (1969), McCabe (1982), and Smith and Kadlec (1983) listed other marsh plants in the area. The topography of low-gradient salt flats can be altered by contour furrowing to produce conditions suitable for emergents with minimum use of fresh water.

With commercially available equipment, managers can estimate salinity levels easily by measuring conductivity, which relates closely to salinity (Kadlec and Smith 1989). In saline marshes, deep flooding with fresh water (0.5- to 1.0-mS/cm conductivity) pushes salts down into sediments. Salts return to the surface with periodic flooding and drying. Drawdowns that promote moist-soil plants are successful in freshwater marshes (below 8-mS sediment conductivity during drawdown) in the Great Basin; procedures are similar to those elsewhere (Kadlec and Smith 1989). By varying the water levels, a desired salinity level of 18 mS can be achieved, although 0.25 mS is best

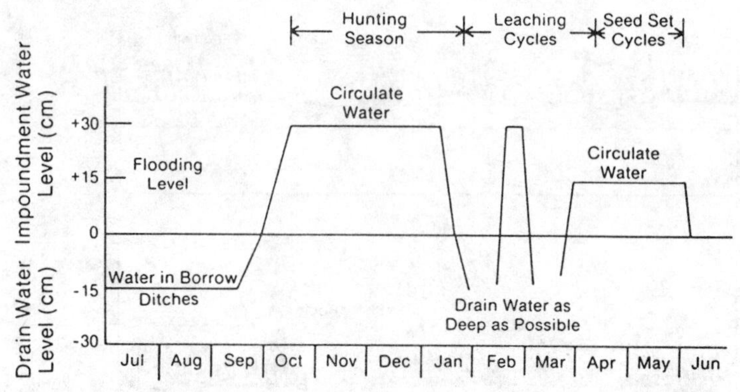

Figure 4.9 Management plan for alkali bulrush (Kadlec and Smith 1989).

for sago pondweed and widgeongrass. Then the water depth is adjusted to about 30 to 45 cm to promote emergents such as hardstem bulrush and cattail or about 45 to 60 cm for submergents such as sago pondweed. Drawdowns that maintain a few centimeters of water will keep salinities low (10 to 25 mS), preventing establishment of salt cedar and allowing establishment of preferred species such as hardstem bulrush and alkali bulrush (Fig. 4.9). Guidelines to calculate water needs of management areas vary with water quality (mainly salinity), precipitation, and evapotranspiration (Christiansen and Low 1970).

Coastal fresh marshes. Coastal fresh marshes contain slightly saline (transitional) to fresh marshes, especially susceptible to saltwater intrusion (Baldwin 1968). The shallow-water tidal marshes of the East and Gulf coasts contain the undesired coarse grasses of common reed, big cordgrass, and maidencane; the desired mixed sedge, bulrush, and spikerush; and sawgrass, arrowarum, smartweed, duck potato, and cattail. The deeper-water fresh marshes contain cattail, giant cutgrass (southern wildrice), wildrice, pickerel weed, spatterdock, watershield, and white waterlily. The undesirable water hyacinth, water lettuce, and alligatorweed are dominant sporadically. Along southern coasts, food habits of waterfowl have shifted to special food plants such as smartweed, Asiatic dayflower, watershield, spikerush, redroot, and banana waterlily in managed impoundments, and rice and weed seeds in ricefield rotations.

After impoundment, the original coarse vegetation must be reduced to encourage the more desired but less aggressive plants. Methods include combinations of water fluctuation, mechanical measures, burning, herbicides, and grazing (Baldwin 1968). Systematic water fluctuations in new impoundments will encourage Asiatic dayflower, smartweed, chufa, softstem bulrush, squarestem spikerush, and dwarf spikerush.

Where freshwater can be manipulated easily, early spring drawdown (February to March) allows smartweeds to germinate before annual grasses and to grow ahead of successively increased water levels. Late spring drawdown causes annual grass seed to germinate. Burning these grasses in fall just before flooding greatly increases use by ducks of the seeds of panic grasses, wild millets, and giant foxtail. Spring drawdown also promotes invasion of cattail and brush, but water levels sustained at 60 cm or lower eliminate all but tropical cattail (Baldwin 1968).

By allowing the soil to dry enough for cultivation or burning before plants turn green in spring, late winter drawdowns encourage smartweeds, wild millet, Asiatic dayflower, tearthumb, spikerushes, panic grasses, redroot, rice cutgrass, and arrowarum (Miglarese and Sandifer 1982). Then a small amount of water is introduced level with the prepared bed, but not ponding over the soil, and is held there during the growing season. Soil kept too dry during the growing season encourages plume grass, beggarticks, tearthumb, woodawn grass, foxtail grass, alder, and willow—all of moderate value to waterfowl. Full flooding during the growing season encourages such undesirables as southern wildrice (giant cutgrass), cattails, pickerelweed, alligatorweed, and lotus. Impoundments can be managed for ducks and crayfish if reflooding is delayed until fall (Chabreck et al. 1989).

Tidal freshwater impoundments of the southern coast should be dewatered in late winter or early spring as most wintering waterfowl leave the area (see Table 4.5), and allowed to dry to prevent establishment of competing persistent emergents such as Virginia arrowarum, golden club, and southern wildrice (Gordon et al. 1989). Drawdowns in late February promote optimum production of desirable smartweed and Asiatic dayflower. Irrigating the marsh bed in the impoundment during late April and maintaining moist soils encourages spikerushes and smartweeds, but reduces water quality, encourages undesirable emergent plants, and can kill plants from high water temperature (which can be corrected with drainage and additional germination and growth, time permitting). Relatively drier soils promote grasses and sedges. Tidal freshwater wetlands are managed best with twice-daily exchange of tidal water, which improves water quality.

Moist-soil impoundments should be drained to about 10 cm below the marsh bed in early September to hasten plant senescence for burning (Gordon et al. 1989). Excessive drying could cause a root burn. After burning to achieve a proper ratio of cover to water, impoundments are reflooded 22 cm deep. The most desirable plant community is produced by alternating years of moist and dry soil conditions. Several years of moist soil produce undesirable stands of plants such as cattail, woolgrass bulrush, and southern wildrice. Several years of dry soil produce undesirable stands of plants such as plume grass, dogfennel, goldenrod, and asters.

Redroot can be managed in organic, acidic soils (see Table 4.5) with a drawdown in late February or early March to allow oxidation for several weeks, followed by rotary cultivation or disking every 3 years to improve rigor, followed by moderate soil moistening throughout the growing season. Soils can be moistened in middle to late April during years without rotary cultivation. The main competitor often is sawgrass, which can be retarded by avoiding standing water and saturated soil (Gordon et al. 1989).

Landers et al. (1976) reported results of three freshwater management types (see Table 4.6): summer drawdown to saturated bed on typical alluvial soils to encourage smartweed and wild millet, summer drawdown to saturated bed on peat soils to encourage redroot, thorough draining in the growing season to encourage panic grasses.

If the water supply allows, freshwater impoundments can be flooded deep enough for submerged aquatics to grow (Baldwin 1968). Initial invaders often are muskgrasses, which are readily eaten by waterfowl but do not survive winter well. Therefore, normal succession to bushy pondweed, or sago pondweed if the water is more alkaline, is preferred. Ponds can be maintained for years in the muskgrass-pondweed stage if fed by rivers. If drainage from acidic uplands decreases the pond's alkalinity, the undesirable bladderwort, fanwort, and coontail become the dominant aquatics, although the desirable watershield tolerates the same water quality. An introduced stand of watershield can be maintained dominant for several decades with water 30 to 91 cm deep and spot treatment of white waterlily with herbicides.

White waterlily speeds plant succession. If alligatorweed, water primrose, or water hyacinth become established, a worthless floating mat and then shrub bog develop. Once the rotten bottom of an impoundment begins surfacing, the only solution is drainage, oxidation, and reversion of the impoundment to an earlier stage of succession (Baldwin 1968).

Moist-soil impoundments can be converted to deepwater (0.6 to 1.0 m) or semipermanently flooded impoundments (see Table 4.5) by raising dikes if necessary, burning the wetland in late winter, flooding with trunks kept open till fall to accumulate all tidewater, and setting flashboard risers to retain all rainfall (Gordon et al. 1989). The impoundment is dried from February to June every 3 to 4 years, converted to moist-soil management for at least one season, and then burned to reduce flotation of organic material. Deepwater management increases habitat diversity; improves production of wood ducks, mottled ducks, and common moorhens; and benefits wildlife dependent on permanent water.

Ricefields. Widespread use of coastal impoundments for economic culture of rice provides opportunities to feed ducks (Baldwin 1968). After the second crop of rice is harvested (October or November), fields should be disked or rolled with a water-filled drum to create openings in the dense stubble. After irrigation dikes are repaired, fields are flooded about 20 cm deep, often by pumping with low-lift diesel pumps in the 30-cm-diameter class that can pump 11,355 L/min (Wesley 1979, Reinecke et al. 1989). Deep-well pumps also are used, although they are not as productive as the low-lift bayou pump. In general, up to 61 ha should be flooded for waterfowl in any one area, with about 20 percent disked or chopped with a heavy spiked chopper called a *water buffalo* pulled behind a tractor. The average amount of waste rice after the second crop is harvested is 140 kg/ha (range 73 to 214) (Hobaugh 1984). Unhunted rest ponds of at least 4 ha should be scattered throughout the rice prairies (Hobaugh et al. 1989).

Newly harvested ricefields should be flooded from about October 15 to 25 to March and fallow fields from about November 25 (before food deterioration) to March (McGinn and Glasgow 1963, U.S. Soil Conservation Service 1979, unpublished report). Flooded fields of at least 4 ha that are away from human disturbance are most attractive to ducks. Rice straw should not be baled, because many seeds occur in and on the straw. When the stubble is too high and continuous, landing strips should be mowed or disked. Planted fields contain at least 190 kg/ha of waste grain; fallow fields contain more than 336 kg/ha of seed from native plants. In addition to waste grain, grass seeds such as paspalums, panicums (panic grass), barnyardgrass, and junglerice become available to ducks if the ricefield is flooded during winter (Neely and Davison 1971).

Some impoundments can be drained and then cultivated for millets or even corn, which also conditions the soil and reduces pest plants, except alligatorweed (Neely and Davison 1971). To produce desired marsh and aquatic plants, such impoundments are returned to their normal wetland cycles.

Plant response

The species composition of moist-soil plants that pioneer on exposed mudflats after a drawdown depends on the composition of seeds in the soil, which is related to the previous plant community there and available seed banks (Fredrickson and Taylor 1982, Smith and Kadlec 1983). Management will vary with the production of desirable or undesirable species, the composition of which is predictable based on the previous plant community.

The response of plants to manipulation of water levels depends on the timing of annual drawdowns, and the stage of succession in the impoundment, i.e., the number of years since the area was flooded continuously or disturbed by disking or plowing. Water-level manipulation controls vegetation by reversing, retarding, or advancing succession, depending on the plant species involved (Table 4.7) (Cooke et al. 1986, Leslie 1988).

TABLE 4.7 Responses of Various Aquatic Plants to Water-Level Drawdown

	Increased			Decreased			No change		
Species	A*	W	S	A	W	S	A	W	S
Alternanthera philoxeroides	x	x	x						
Bidens sp.			x						
Brasenia schreberi					x	x			
Cabomba caroliniana			x		x	x			
Carex spp.			x						
Cephalanthus occidentalis			x						
Ceratophyllum demersum	x	x		x	x	x		x	x
Chara vulgaris		x	x			x			
Cyperus spp.	x								
Eichhornia crassipes		x	x	x	x				
Eleocharis baldwinii			x			x			
Eleocharis acicularis			x		x				
Elodea canadensis		x			x			x	
Elodea densa					x	x			
Elodea sp.					x				
Glyceria borealis		x							
Hydrilla verticillata		x							
Hydrochloa caroliniensis				x					
Hydrotrida caroliniana					x				
Jussiaea diffusa					x				
Leersia oryzoides		x	x						
Lemna minor	x								
Lemna sp.						x			
Limnobium spongia						x			
Myriophyllum brasiliense						x			x
Myriophyllum exalbescens					x			x	
Myriophyllum heterophyllum						x			x
Myriophyllum spicatum		x			x			x	

TABLE 4.7 Responses of Various Aquatic Plants to Water-Level Drawdown (Continued)

	Increased			Decreased			No change		
Species	A	W	S	A	W	S	A	W	S
Myriophyllum sp.					x				
Megalodonta beckii		x							
Najas flexilis	x	x	x					x	x
Najas quadalupensis				x	x				x
Nelumbo lutea						x		x	x
Nuphar advena					x				
Nuphar luteum						x			
Nuphar macrophyllum		x							
Nuphar polysepalum									x
Nuphar variegatum					x				x
Nuphar sp.					x				
Nymphaea odorata			x			x			x
Nymphaea tuberosa					x				
Panicum sp.	x								
Polygonum coccineum		x	x		x				
Polygonum natans								x	
Pontederia cordata							x		
Potamogeton americanus		x							
Potamogeton amplifolius		x			x				
Potamogeton crispus								x	
Potamogeton diversifolius		x					x		
Potamogeton epihydrous		x						x	
Potamogeton foliosus		x						x	
Potamogeton gramineus		x						x	
Potamogeton natans		x				x			
Potamogeton pectinatus		x						x	
Potamogeton richardsonii		x						x	
Potamogeton robbinsii					x				
Potamogeton zosteriformis		x						x	
Potamogeton spp.	x								x
Ranunculus tricophyllus								x	
Sagittaria graminea							x		
Sagittaria latifolia							x		
Sagittaria sp.		x							
Salix interior		x							
Scirpus americanus				x					
Scirpus californicus				x					
Scirpus validus		x	x						
Sium suave		x							
Sparganium chlorocarpum						x		x	x
Spirodela polyrhiza					x				
Typha latifolia		x				x	x	x	
Utricularia purpurea						x			
Utricularia vulgaris					x				
Utricularia sp.					x				x
Valisneria americana		x		x			x		

*A = whole-year drawdown, W = winter drawdown, S = summer drawdown, x = response.
SOURCE: Cooke et al. (1986).

The presence of desirable or undesirable plants is governed by the natural range of the plant species (seed availability), water chemistry, soil type and moisture, light absorption, season and duration of drawdown, and amount of stranded algal debris (Harris and Marshall 1963, Atlantic Waterfowl Council 1972). Best production of emergent aquatics occurs with a combination of early season drawdown, rich soil types, slow rates of mudflat drainage, and small amounts of stranded algae. Plants that interfere with the maximum production of food and cover are considered undesirable. Undesirable species tend to dominate in later successional stages after repeated annual drainage of impoundments (Fredrickson and Taylor 1982). Plants are needed for food and cover, but food plants generally furnish cover also. Thus, management of water levels should focus on food plants. Moist-soil plants vary in successional stage, germination date, seed production, and cover (Table 4.8). Each species of plant must be judged by its assets and liabilities for wildlife management.

Desirable. Moist-soil plants can be encouraged by a complete drawdown requiring tilling and planting or by a partial or complete drawdown relying on volunteer growth or seeding without tillage (Linde 1969). Volunteer growth depends on seed reserves, which usually are adequate except for saline wetlands and impounded bays of lakes (Pederson and Smith 1988). For tilling operations, the drawdown should begin early enough to allow the soil to dry enough for use of equipment, mainly disking, usually with the water table at least 30 cm below ground (Linde 1985). Soil disturbance greatly increases the natural production of plants such as certain smartweeds, millets, and chufa (Green et al. 1964). The drawdown date varies with weather, but occurs in late April in the north central states (Linde 1969). If gravity drainage will not reduce the groundwater table enough, water should be pumped out of peripheral ditches, perhaps in a series of pumpings as seepage refills the ditches. Where no tillage is involved, the drawdown can start later, such as the last week in May for the north central states, so that broadcasting seed by hand or airplane begins about June 1 (Linde 1969). Volunteer growth appears as soon as the water table recedes below the surface, because better aeration and higher temperatures stimulate germination (Green et al. 1964). Adequate soil moisture for best growth usually is provided by maintaining water levels within 10 to 20 cm of the mudflat's elevation (Bookhout et al. 1989) or even below 10 cm in California (Heitmeyer et al. 1989).

A series of water-level reductions spaced over several weeks will expose a relatively narrow range of bottom elevations with each reduc-

TABLE 4.8 Successional Stage, Germination Dates, Food and Cover Value, and Potential Seed Production of Selected Moist-Soil Plants

Plant	Succesional stage		Germination			Value*		Best seed production						
								Drawdown				Moisture		
	Early	Late	Early	Middle	Late	Food	Cover	Early	Middle	Late	None	Dry	Moist	Wet
Pondweeds *Potamogeton* spp.						+	0				x			
Common burhead *Echinodorus cordifolius*		x		x		+	+						x	x
Sprangletop *Leptochloa* spp.	x				x	+	+			x			x	
Rice cutgrass *Leersia oryzoides*		x				+	+		x	x				x
Crabgrass *Digitaria* spp.	x					+	+		x	x		x		
Panicum *Panicum* spp.	x					+	+		x	x		x		
Common barnyardgrass *Echinochloa crusgalli*	x		x	x		+	+	x	x				x	
Barnyardgrass *Echinochloa muricata*	x			x	x	+	+		x	x			x	
Broomsedge bluestem *Andropogon virginicus*		x			x	0	+					x		
Redroot flatsedge *Cyperus erythrorhizos*	x				x	+	+			x				x
Spikerush *Eleocharis smallii*	x	x	x	x	x	+		x	x	x			x	
Beakrush *Rhynchospora corniculata*		x			x	+	+						x	x
Fox sedge *Carex vulpinoidea*		x			x	+	+						x	

TABLE 4.8 Successional Stage, Germination Dates, Food and Cover Value, and Potential Seed Production of Selected Moist-Soil Plants (*Continued*)

	Succesional stage		Germination			Value*		Best seed production						
								Drawdown				Moisture		
Plant	Early	Late	Early	Middle	Late	Food	Cover	Early	Middle	Late	None	Dry	Moist	Wet
Common rush *Juncus effusus*		x			x		+	x					x	x
Poverty rush *Juncus tenuis*		x			x		+	x					x	x
Black willow *Salix nigra*	x		x			0	0	x					x	
Curly dock *Rumex crispa*	x		x		x	+		x				x		
Pennsylvania smartweed *Polygonum pensylvanicum*	x		x		x	+	+	x					x	
Curltop ladysthumb *Polygonum lapathifolium*	x		x		x	+	+	x					x	
Swamp smartweed *Polygonum hydropiperoides*		x				0								x
Toothcup *Ammannia coccinea*	x				x	+	+			x			x	x
Purple loosestrife *Lythrum salicaria*	x		x	x	x	0	0	x	x	x		x	x	
Marshpurslane *Ludwigia* spp.	x	x	x			+			x	x	x		x	x
Morningglory *Ipomoea coccinea*		x		x	x	+			x	x		x	x	
Lippia *Lippia lanceolata*	x			x					x			x	x	

Trumpetcreeper *Campsis radicans*		x			0	0			x	x	
Buttonweed *Diodia virginiana*	x		x		+		x	x		x	
Common buttonbush *Cephalanthus occidentalis*		x			+	+		x			x
Joe-pye-weed *Eupatorium serotinum*		x		x	0	+	x		x		
Aster *Aster* spp.		x		x	0	+			x		
Common ragweed *Ambrosia artemisiifolia*	x		x	x	+		x		x		
Common cocklebur *Xanthium strumarium*	x		x	x	0	+	x	x		x	
Beggarticks *Bidens* spp.	x		x	x	+		x	x	x	x	
Sneezeweed *Helenium flexuosum*		x	x	x	0		x	x	x	x	

*Response is indicated by x; a plus sign indicates substantial value, and a zero little or no value, as food or cover.
SOURCE: Fredrickson and Taylor (1982).

tion. The resulting successive moist-soil conditions will produce maximum seed germination because excessive drying at higher elevations is prevented. For example, three weekly 46-cm water-level reductions beginning June 7 in Wisconsin will produce mudflats that can be seeded aerially to millet 3 days after each exposure (Linde 1969).

Production of wild millet in impoundments in California requires complete spring drawdown (Green et al. 1964). Preparation of the seedbed usually is not needed on new ponds. But the presence of emergent vegetation requires drying of the soil and disking of the sod before seeding. On dry ground the seed is irrigated by shallow flooding several times during summer as needed or as the water supply permits. In fall before migration, the impoundments are flooded, usually by October to November in the South (Hindman and Stotts 1989). Deeper areas of the wetland, usually excluded by dikes from the shallow impoundments, are reserved for the development of aquatics such as sago pondweed and as a possible source of irrigation water.

To encourage alkali bulrush in states on the Pacific Coast, a drawdown exposes mudflats on which the seedlings take root. The mudflats are then irrigated with a cover of 2.5 to 7.6 cm of water until the crop matures, after which the impoundment can be kept dry until flooded just before the hunting season (Green et al. 1964). Dry areas are irrigated by shallow flooding several times during summer, with the water removed within 1 or 2 h after the soil is saturated. In California, alkali bulrush develops if the mudflat is seeded, and the area is flooded 30.5 cm deep for 2 weeks, drained, then reflooded periodically to 7.6 cm deep to keep the mudflat moist until the young plants become established (Miller 1962). In southern coastal marshes, saltmeadow cordgrass might replace Olney threesquare, choice food of snow geese and muskrats, if growing in mixed stands. Burning followed by flooding 10 cm or less deep with water of 5- to 10-ppt salinity encourages Olney threesquare and discourages saltmeadow cordgrass (Chabreck et al. 1989).

Food production of moist-soil plants usually exceeds that of submerged aquatics (Atlantic Waterfowl Council 1972). Best results for submerged aquatics are achieved when the water level is held at 3 to 6 dm during the production season.

Highest seed production of many sedges and annual grasses occurs consistently during early successional stages (Fredrickson and Taylor 1982). Many forbs, e.g., cocklebur, also produce a high volume of seed, but should be controlled because their faster growth displaces many more desirable moist-soil plants. Other forbs, such as beggartick, should be encouraged for high nutritive value, seed production, and cover value. Drawdowns in Wisconsin commonly result in excellent volunteer growth of desirable food plants such as millets, smartweeds, chufa, dock, rice cutgrass, and pigweeds (Linde 1969).

After drawdown and the plants have grown 10 to 15 cm, as much of the impoundment as possible is reflooded to 2 to 5 cm for 2 to 3 days. If many plants are shorter and submerged and do not reach the surface within 2 to 3 days after flooding, water levels must be lowered. Species such as smartweeds, barnyardgrasses, and sedges respond well to shallow flooding. Species such as beggarticks, panic grasses, and crabgrasses are less tolerant. To manage desirable and undesirable species, identification of seedlings is required. Fredrickson and Taylor (1982) illustrated and listed key characteristics of eight common moist-soil seedlings: sprangletop, curltop ladysthumb, fall panicum, sedge, barnyardgrass, hairy crabgrass, cocklebur, beggarticks. As the desired plants grow, water levels can be increased gradually to 15 to 20 cm maximum, or about one-third of the total height of newly established moist-soil plants. Water levels are reduced immediately if plants develop a light green color, an indication that the water probably is too deep. Experience will indicate the water tolerance of plants in an area so that water levels can be adjusted accordingly (Fredrickson and Taylor 1982).

Winter drawdowns apparently improve recovery of submerged aquatics and the germination of seeds (Green et al. 1964). Late fall and overwinter drawdowns improve production of aquatics, due to removal of carp and other bottom-disturbing fish which uproot plants and cause turbidity, reducing light penetration and plant growth. Such drawdowns should begin late enough in the fall to inhibit germination of emergents, but early enough to permit the top layer of soil to dry so that aquatic seeds will germinate the next spring.

Undesirable. Monocultural stands of plants usually have minimal wildlife value and are undesirable. Most are perennials or woody vegetation harder to control than annuals (Reid et al. 1989).

Some of the same strategies of water control to encourage desirable vegetation can be used to discourage undesirable vegetation (Fredrickson and Taylor 1982, Reid et al. 1989). But after a drawdown, some undesirables, such as cocklebur and aster, germinate before the desirables. As soon as desirables such as annual grasses, smartweeds, or sedges germinate, as much of the bottom as possible is covered with about 1 cm of water for 24 to 48 h, which is enough to kill or stunt cocklebur and irrigate the desirables. As the desirables develop, flooding can be increased gradually to control cocklebur and other undesirables on higher ground before they shade out the desirables.

Undesirable perennials such as broomsedge bluestem can be controlled by shallow flooding (10 cm) until midsummer and joe-pye-weed by flooding in late summer and early fall when it blooms. In areas where few desirables and extensive stands of undesirables occur, such as cocklebur, disking and reflooding will set back succession to an earlier and more productive stage of seed-producing plants (Fredrickson and Taylor 1982). Development of reed canarygrass is minimized with high water levels maintained through spring (Ball et al. 1989). Woolgrass is controlled by an early spring drawdown every 3 to 4 years, deep plowing, and fall flooding (Hindman and Stotts 1989). Purple loosestrife seedlings 20 cm tall will be killed if covered by water 30 cm deep for 5 weeks (Thompson 1989). A drawdown before May 15 in northern regions reduces germination of purple loosestrife and promotes species adapted better to cooler temperatures (Merendino et al. 1990).

Large areas can be flooded to shallow depths with little water in flat basins, where immediate control of undesirable plants can be achieved in optimum contour intervals of 15 to 20 cm (Fredrickson and Taylor 1982). But frequent inspections are needed to control undesirable plants without damaging desirable plants due to the extended period needed to flood an area.

The proper interspersion and limited stem density of cattail provide cover for ducks, food for muskrats which eat the rhizomes, and house materials for muskrats. But cattail tends to become too dense for ducks and thus needs control as it fills in the open-water areas.

Poorly drained mineral soil invites establishment of cattail by seed where sedge or grass does not compete (Knighton 1985). Once established, cattail expands vegetatively and endures several years of continuous flooding. Persistent rainfall or heavy-texture soils reduce the effectiveness of drawdowns.

Organic soils have inherently poor drainage and invite cattail. Cattails flourish on wet soils, organic or mineral, and at normal pool less than 6 dm deep. Water depths over 6 dm discourage cattail and southern wildrice (Gordon et al. 1989). Overwinter drawdown and reflooding in late winter help control cattail and waterlily because the bottom freezes, killing some roots. Mainly the roots are torn loose as the frozen bottom rises with reflooding (Beard 1973, Nichols 1974, Beule 1979). Most effective control of cattails is accomplished with other methods used in conjunction with water manipulation, although drawdowns in interior salt marshes control cattail, replacing it with alkali bulrush as salinity levels increase (Nelson and Dietz 1966). Alkali bulrush dominates at sediment salinity levels of 10 to 25 mS (Kadlec and Smith 1989). In saline basins, complete drawdowns control plants effectively by maintaining the water table within several centimeters of the soil surface and concentrating soil salts (Nelson and Dietz 1966). Where cattail cannot be controlled by increasing soil salinity or flooding, partial drawdowns are used, which prevent exposure of the soil to air and alternating hot and cold temperatures needed for best germination (Pederson et al. 1989). In northern regions, drawdowns in May reduce cattail establishment (Merendino et al. 1990).

Stands of big cordgrass are reduced by flooding 30 to 60 cm deep or by salinities over 12 ppt. It is eliminated by burning and flooding, or mowing and compaction with an all-terrain marsh masher, followed by deep flooding (Prevost 1987). Smooth cordgrass is controlled by a complete drawdown for one growing season; burning removes dead stems (Gordon et al. 1989). But soil pH can be reduced, temporarily reducing plant growth.

At southern latitudes, early drawdowns, drying, and disking every 5 to 7 years remove woody seedlings and small saplings adapted to wet sites (Fredrickson and Taylor 1982). But seed-producing plants then might need irrigation by shallow flooding to stimulate germination during dry seasons. In northern regions, woody growth can be prevented or reduced by late drawdowns and shallow flooding. Species such as willow, dogwood, and buttonbush, which spread into water areas remaining during summer drawdowns, can be controlled by raising the water into their branches for 1 month after leaves are out (New undated). Control of woody growth generally needs other methods used with water-level manipulation.

Flooding can stimulate growth of many woody species. After drawdown, impoundments should be examined for woody seedlings before summer flooding begins, in case disking is needed. Excessive accumulation of plant litter in the impoundment can be burned, usually in early spring after drawdown and drying and before new germination. Then the impoundment is reflooded.

Animal response

Diked marshes produce three to five times more muskrats than areas with fluctuating water (Wilson 1968). Densities peak 3 to 4 years after drawdown and then decline (Neal 1968, Weller 1978, Bishop et al. 1979, Kroll and Meeks 1985).

A flowchart developed for Missouri (Fig. 4.10) (Fredrickson and Taylor 1982) lists marsh conditions to attract wetland and upland wildlife; it can be modified for other areas. The chart contains four water conditions: deep (over 15 cm), medium (15 cm), shallow-water mudflat, and dry. These conditions are not necessarily maintained for extended periods because wetland plants and wildlife are adapted to water fluctuations in natural wetlands. To attract shorebirds into habitat for waterfowl, a series of water manipulations is used. First, in early spring the impoundment is drawn down to 5 cm deep. Second, in summer the drawdown is completed, the bottom is disked to eliminate undesirable plants, and the pond is reflooded 5 cm deep. Third, in late fall or winter after the shorebirds have migrated, the pond is reflooded 15 cm deep. Various strategies, including no action, exist for various conditions.

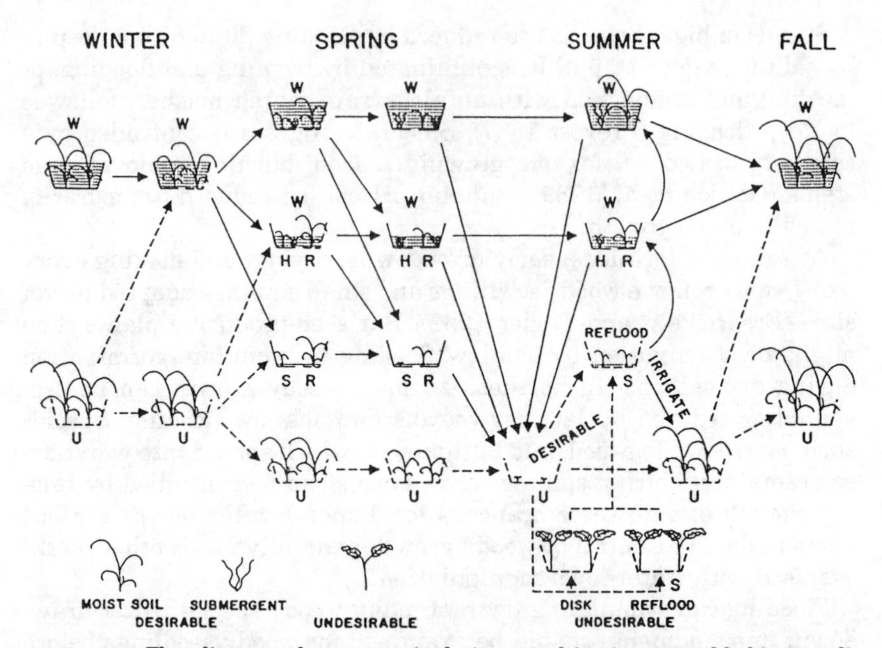

Figure 4.10 Flow diagram of water manipulations resulting in seasonal habitat conditions that attract five wildlife groups (W = waterfowl, H = herons, R = rails, S = shorebirds, and U = upland wildlife, including deer, turkeys, raptors, small mammals, and passerines) (Fredrickson and Taylor 1982). The level of shading in the wetland basins depicts the water depth: deep flooding (> 15 cm), a medium level (15 cm), and a mud and shallow water interface. Dashed arrows represent water manipulations that flood dry basins. Solid arrows and basins represent water levels from continuous flooding. Vegetative conditions within the basins are depicted by three plant groups: desirable moist-soil grasses, sedges, and forbs; (2) desirable submergent species; and (3) undesirable herbs and woody growth. The size of the plants relative to water depth depicts growth stage and robustness. The final stage of spring or summer drawdown depends on the development and composition of the plant community. Disking, reflooding, or both controls undesirable seedlings. When desirable seedlings tolerate higher water levels, they are irrigated to provide a continuous supply of wetland habitat for summer wildlife.

Fall flooding and winter impoundment. Most nonwaterfowl wildlife species are excluded from the deeper (0.5-m or more) water used by most diving ducks, which requires substantial, costly dikes (Fredrickson and Taylor 1982). Thus, shallow-water management for dabbling ducks tends to benefit other wildlife more than deepwater management does (Table 4.9).

Fall flooding of moist-soil areas is timed to the arrival of the earliest

TABLE 4.9 Habitat Conditions Attracting Various Vertebrates to Moist-Soil Impoundments

	Foods				Water depth,* cm	Openings		Vegetative cover			
Vertebrate group	Vertebrates	Invertebrates	Seeds	Browse		Water	Mudflat	Rank	Short	Dense	Sparse
Amphibians		x			0–20	x	x		x		x
Reptiles	x	x			0–50	x		x	x	x	x
Grebes	x				25 +	x			x		x
Geese			x	x	0–10	x	x		x	x	x
Dabbling ducks		x	x		5–25	x	x	x			
Diving ducks		x	x		25 +	x					
Hawks	x				NA				x	x	x
Galliforms		x	x		D-M			x	x	x	x
Herons	x	x			7–12	x			x		x
Rails		x	x		5–30			x	x	x	
Coots			x	x	28–33	x			x		x
Shorebirds		x			0–7	x	x		x		x
Owls	x				D-M				x	x	x
Swallows		x			NA	x			x		x
Sedge wrens		x			NA			x		x	
Nesting passerines		x	x		NA			x	x	x	x
Winter fringillids			x		NA			x	x	x	x
Rabbit				x	0			x		x	
Raccoon	x	x	x		0–10	x	x	x	x	x	x
Deer				x	0			x			

*D-M = dry to moist; NA = not applicable (use of units does not depend on flooding or specific water depths).
SOURCE: Fredrickson and Taylor (1982).

migrant ducks, usually blue-winged teal and pintails, rather than to a certain calendar date. Impoundments are flooded 10 to 25 cm deep, suitable for dabbling ducks and Canada geese. Generally, reflooding in late August accommodates early migrating ducks and shorebirds (Johnson and Dinsmore 1986). Irregular topography in an impoundment provides diverse depths ideal for a variety of wildlife species (Table 4.9), although shallow irrigation of moist-soil plants is difficult after complete drawdown.

Early spring drawdown. Wetland management for breeding and migrating waterfowl in the upper Midwest is compatible with management for rails (Johnson and Dinsmore 1986). Early spring drawdowns are timed to the arrival of shorebirds, which varies with latitude and the phenology of the species that nest on or migrate through the area (Fredrickson and Taylor 1982). In southeastern Missouri, for example, pectoral sandpipers and lesser yellowlegs arrive early to mid-April. Drawdown initiated generally before April 15 avoids disrupting territorial establishment and breeding.

The drawdown reveals mudflats nearly devoid of vegetation, because most of the emergent vegetation has been eaten by waterfowl or often flattened by waterfowl, wind, and waves (Frederickson and Taylor 1982). Shorebirds, raccoons, and other wildlife are attracted to such mudflats interspersed with water 1 to 5 cm deep. Smaller, shorter-legged birds use the shallower areas; larger, longer-legged birds use the deeper water (Fig. 4.11). On some sites within each impoundment drawn down, especially those with very shallow water, rails and late-wintering passerines find cover in new growth of spikerushes and old clumps of soft rushes, bulrushes, and stems and blades of grasses and sedges. Deeper water, especially from late drawdowns, contains submerged, decaying, and regenerating vegetation with scattered emergents, attractive to wading birds, rails, and late-migrating or resident waterfowl. Minnows, amphibians, and invertebrates usually concentrate in or near submerged vegetation such as waterstarwort, regenerating swamp smartweed, or marshpurslane. Emergent cover consists of grasses, rushes, sedges, cattails, and other plants, which when flooded provide ideal habitat for insect populations which attract insectivorous birds and bats.

Slow (gradual) drawdowns are better than fast drawdowns because new mudflats are exposed continually, extending the period of use by shorebirds and other wildlife (Frederickson and Taylor 1982). For most shorebirds, partial drawdowns are better than complete drawdowns to a moist-soil condition (Rundle and Fredrickson 1981, Johnson and Dinsmore 1986). Shorebirds prefer shallow water (1 to 5 cm) interspersed with exposed saturated soil, i.e., water levels held at

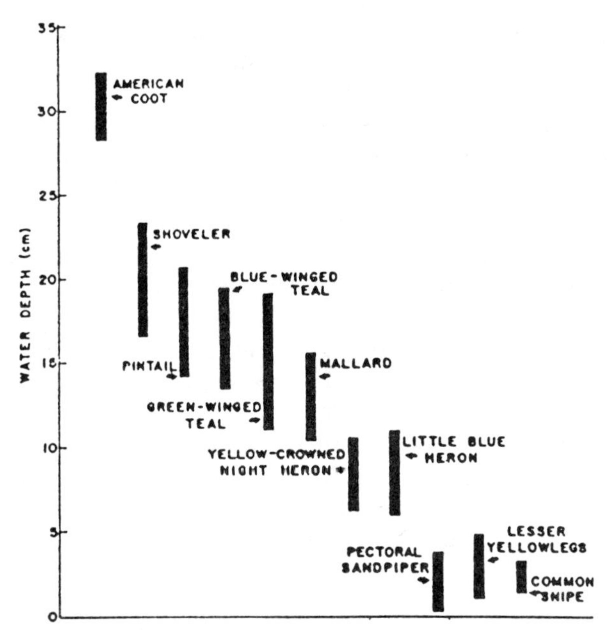

Figure 4.11 Water depths used by common waterbirds in seasonally flooded impoundments (Fredrickson and Taylor 1982).

depths that maximize the edge between shallow marsh and moist-soil sites. Such conditions encourage valuable annual and perennial seed producers. Besides these wetland seed producers, the best habitat for shorebirds includes a mosaic of robust, moderately robust, and fine emergents for both food and cover. Soras might use emergent stands more if a high density of floating and submerged residual vegetation is present. Dense floating residual cover, found mostly in stands of moderately robust or fine emergents, provides habitat for invertebrates which have to be near the surface to serve as food for soras, which have short bills.

Impoundments with spring drawdowns are unavailable as nesting and brood habitat for ducks and coots, but exposed invertebrates, especially crustaceans, and lower vertebrates provide food for waterfowl and a variety of other wildlife. Observation towers can be positioned near the lowest point of the impoundment so they face upslope to take full advantage of the waterfowl and shorebird activity during the drawdown (Fredrickson and Taylor 1982). As habitat conditions deteriorate with declining water levels, times of drawdown can vary in different impoundments to maintain concentrations of shorebirds for longer periods. Rotation of management options within a group of impoundments provides the greatest benefit to wildlife and people.

Late spring drawdown. Drawdowns in middle to late May allow muskrats to raise one litter and minimally disturb nesting birds (Kierstead undated). Late spring drawdowns are most effective if done in two phases (Fredrickson and Taylor 1982). To take advantage of mudflats, the first phase is begun when herons or other birds such as rails or swallows arrive. For example, drawdowns in Missouri begin when yellow-crowned night herons and little blue herons arrive. Water levels are lowered to 5 to 15 cm initially and kept there until plants germinate on the mudflats. After germination the drawdown is completed.

Herons like open water with sparse emergent vegetation and much floating and submerged vegetation. Rails use emergent vegetation in shallow and deep water. Swallows feed on emerging insects. Resident and late-spring-migrating waterfowl feed on insects and other invertebrates. If spring rainfall is impounded for later drawdown, water should be accumulated until it is at least 35 cm deep to inhibit growth of aquatics such as swamp smartweed and creeping marshpurslane (Rundle and Fredrickson 1981).

Summer irrigation and flooding. Maintaining proper shallow-water conditions is made difficult by rainfall and evaporation. Hence, the main goal is interspersion (Rundle and Fredrickson 1981). Areas that are too deep will ensure future habitat if impoundments dry out. Initially dry areas can absorb excess water from heavy rains and floods.

When rainfall is adequate for optimum plant growth, areas intended for upland wildlife are not reflooded until fall (Fredrickson and Taylor 1982). During dry summers, shallow reflooding is needed to saturate the soils at the highest sites. Water can be removed 1 or 2 h after saturation. By irrigating the highest sites first, the lower sites are irrigated with the overflow water. Typical plants in such areas include panic grass, crabgrass, beggarticks, ragweed, and aster. Game birds such as ringneck pheasants, bobwhites, and turkeys use dewatered moist-soil areas for brooding and foraging. Deer use the areas for nurseries and foraging. Cottontails and other small mammals find food and cover, but irrigation can destroy nests. Passerine birds are attracted to the new plant growth in impoundments.

Impoundments used all year mainly for wetland wildlife are reflooded as soon as the desirable vegetation can tolerate it. When plants are tall enough, flooding continues until water is 5 to 15 cm deep. Resident waterfowl, herons, rails, and some passerines feed and often breed on wetter sites. Some passerines breed on drier sites. Harriers and other raptors hunt for prey. Deer, turkeys, and pheasants wade in the impoundment for food and water. Muskrat, mink, raccoon, otter, and other furbearers use the flooded areas.

Late fall and winter drawdowns. Drawdowns during late fall dry out the bottom of the impoundment, causing desiccation and freezing of the roots of some undesirable plants (Beard 1973). The drawdown should occur after migration, with reflooding in spring before migration. Such drawdowns also reduce muskrat, carp, and bullhead populations and aerate the soil.

Record keeping

To facilitate planning and maintain continuity in personnel changes, an annual record of water-level management should be maintained for each impoundment (Fig. 4.12).

Greentree Reservoir

The best greentree reservoirs contain uneven terrain offering gentle, dry ridges that ducks use as loafing sites (Rudolph and Hunter 1964).

Impoundment Number ________ Year ________

Type of Manipulation: (1) Flood (2) Drawdown

Season of Manipulation: (1) Winter (4) Summer
(2) Early Spring (5) Early Fall
(3) Late Spring (6) Late Fall

Notes on Manipulation:

Date	Water level	Stoplog elevation	Notes

Animal Response:

Species	Arrival	Departure	Notes

Figure 4.12 Sample data sheet for moist-soil management (Fredrickson and Taylor 1982).

Average depth is 30 to 46 cm with maximum depth less than 9 dm (Mitchell and Newling 1986).

Annual flooding of dormant timber eventually might increase soil moisture, causing the forest community to convert to vegetation characteristic of wetter habitats (Mitchell and Newling 1986). If that is undesirable and if several greentree reservoirs exist in the system, flooding should be alternated so that each winter one or more impoundments are left unflooded to dry out, which simulates natural conditions and might aid decomposition and nutrient cycling (Reinecke et al. 1989). To encourage seedling establishment, no flooding should occur 2 to 3 years after a year of good acorn production.

Hardwood bottoms being regenerated under even-aged management should be withheld from flooding for 2 years to allow trees to establish (Payne and Copes 1986). Flooding occurs in fall usually when leaves begin to change color, a period that usually needs 6 to 8 weeks to complete; drainage in late winter or early spring is completed before the growing season begins (Mitchell and Newling 1986). South of 36° latitude, median dates are November 1 to flood and March 10 to drain (Wigley and Filer 1989). North of 36°, median dates are October 15 to flood and April 1 to drain. Water retention probably should not be more than 4.5 months in the southeast (Hall 1962). Because only a few centimeters of water or saturated ground during the growing season can cause permanent damage of timber within one or two seasons, thorough draining is essential (Hunter 1978).

Where other wildlife compete strongly with ducks for mast, the reservoir should be filled rapidly, to cover the mast and prevent premature depletion, and then lowered periodically to obtain full use of the acorns by ducks (Rudolph and Hunter 1964). A reservoir with heavy concentrations of ducks during fall should be flooded gradually to prevent ducks from depleting acorns too early (Hunter 1978). Water levels on large, deep reservoirs are increased in stages to avail ducks of mast otherwise too deep. The feather edge of slowly rising water attracts ducks (Hunter 1978). Water should be drained slowly to concentrate invertebrates for migrating waterfowl and to retain nutrients (Reinecke et al. 1989). Changing the water depth in a greentree reservoir at appropriate times ensures that the resources produced are used effectively (e.g., Fig. 4.13) (Fredrickson and Reid 1988b).

PRESCRIBED BURNS

Prescribed burns in wetlands accomplish four basic objectives (Wright and Bailey 1982, Gordon et al. 1989):

1. They make new green shoots, roots, and rhizomes of grasses and sedges available to geese.

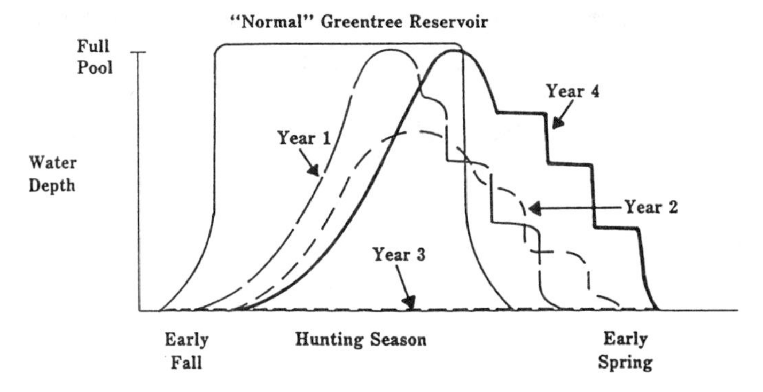

Rationale

Year 1 - Gradual flooding to provide access to food resources on a continuing basis throughout the hunting season. Gradual drawdowns commence before the end of the waterfowl season.

Year 2 - Gradual flooding but water levels never reach full pool. Gradual drawdown extending into spring.

Year 3 - No flooding.

Year 4 - Similar to Year 1, but full pool not reached until end of hunting season. Gradual drawdown extending well into spring.

Figure 4.13 Flooding regimes suggested for southern greentree reservoirs (Fredrickson and Reid 1988b).

2. They expose fallen seed for ducks.
3. They make wetlands more suitable for ducks, muskrats, and nutria by eliminating sour marsh conditions of flooded and decomposed organic matter, and impenetrable growth of climax species of plants such as common reed, bulrush, sawgrass, cordgrass, and cattail, and by promoting growth of good seed producers.
4. They create deep pools and edge for nesting and feeding waterfowl.

Plants vary in their response to burning (Table 4.10). Burns do not seem to affect the seed bank (Smith and Kadlec 1985). Fire on nutrient-rich sites does not change species composition; fire on nutrient-poor sites results in even more oligotrophic bogs low in species diversity (Mallik and Wein 1986). Without water control, burning seasonally flooded wetlands probably is unjustified because of reduced snow-trapping ability of burned vegetation and the crushing effect of snow on finer-stemmed plants in unburned areas which produces the open or semiopen aspect (Kantrud 1986a, Pederson et al. 1989). Bibliographies have been published about fire in wetlands (Kirby et al. 1988) and northern grasslands (Higgins et al. undated).

Types of Prescribed Burns

Prescribed burns in coastal and inland marshes are of three general types: cover or surface or wet burns, root burns, and peat burns (Lynch 1941, Smith 1942, Uhler 1944, Hoffpauer 1968, Miglarese and

TABLE 4.10 Response* of Various Emergent Plants to Grazing, Mowing, Burning, and Tillage

	Grazed					
Plant species	Light	Moderate	Heavy	Mowed	Burned	Tilled
Mudflat species						
Beggarticks *Bidens* spp.						I
Cocklebur *Xanthium* spp.			I			I
Kochia *Kochia scoparia*						I
Goosefoots *Chenopodium* spp.						I
Pigweeds *Amarthus* spp.			D			I
Wild buckwheat *Polygonum convolvulus*						I
Smartweeds *Polygonum* spp.			P			I
Yellow foxtail *Setaria glauca*			I			I
Wet-meadow species						
Water plantain *Alisma* spp.		P	P			P
Bluejoint *Calamagrostis canadensis*			D			D
Northern reedgrass *Calamagrostis inexpansa*	D	D				D
Sedges *Carex* spp.	P, D	D				P
Saltgrass *Distichlis stricta*	P, I	P, I	P	P		D
Wild barley *Hordeum jubatum*	I	I	I			P
Baltic rush *Juncus arcticus balticus*	P, I	I				D
Fowl-meadow grass *Poa palustris*	D	I, D				D
Prairie cordgrass *Spartina pectinata*	I, D	I, D				D
Shallow-marsh species						
Sloughgrass *Beckmannia syzigachne*	P			P	P	
Slough sedge *Carex atherodes*	P	P, D	P, D	D	P	D
Spikerush *Eleocharis palustris*	P	P, I	P, I	P	P	D

TABLE 4.10 Response* of Various Emergent Plants to Grazing, Mowing, Burning, and Tillage (*Continued*)

Plant species	Grazed					
	Light	Moderate	Heavy	Mowed	Burned	Tilled
Shallow-marsh species *(continued)*						
Tall mannagrass *Glyceria grandis*	P	P	P			D
Marsh smartweed *Polygonum coccineum*		D		P	P, I	P
Common threesquare *Scirpus americanus*	P	P				D
Giant burreed *Sparganium eurycarpum*		P, I	P			D
Whitetop *Scolochloa festucacea*	P	D	D	I	I, P	D
Deep-marsh species						
Common reed *Phragmites australis*	P	P	D	P	P, D	D
Hardstem bulrush *Scirpus acutus*		P	I			D
River bulrush *Scirpus fluviatilis*	P, I	P	P			P
Alkali bulrush *Scirpus maritimus*		P	P			P
Softstem bulrush *Scirpus validis*		P	P, I			P
Cattail *Typha* spp.	P	I	D	D	P, D	D

*P (present) indicates the species at least tolerates and might be favored by the disturbance; I (increase) indicates a greater abundance in disturbed than in undisturbed wetlands; D (decrease) indicates a lesser abundance in disturbed than in undisturbed habitats; multiple responses reflect different observations.

SOURCE: Stewart and Kantrud (1972); Millar (1973); Neckles et al. (1985); Fulton et al. (1986) as reported in Pederson et al. (1989).

Sandifer 1982). In general, direct effects of burning on wildlife are negligible (Lyon et al. 1978), although variable (Higgins et al. undated). Root burns and peat burns kill some wildlife but ultimately increase wildlife abundance by improving habitat. Localized use of prescribed burns converts solid stands of reed, cattail, and sedge to productive food, nesting cover, and brood cover.

Cover burns can be clean or spotty and are light burns conducted when water levels are at or above the root horizons. By removing vegetative debris, cover burns temporarily release valuable plants that have an earlier growing season than objectionable plants, thus providing high-quality food to muskrats, nutria, ducks, and grazing geese. Cover burns also render the marsh more accessible to muskrat and nutria trappers, control cattail and cordgrass somewhat, and stimulate seed production of some plants. For example, where saltwater cordgrass and saltmarsh bulrush occur together, a cover burn initially produces the more desirable bulrush in greater abundance because it grows faster than the more dominant cordgrass. But the bulrush could be damaged by geese and ducks grubbing and feeding on rhizomes, producing an eatout requiring years to recover.

Root burns are hotter, designed to control and replace low-value marsh plants such as southern wildrice, common reed, sawgrass, cattail, cordgrass, mints, river bulrush, and other unproductive sedges and generally designed to remove climax vegetation which usually is relatively useless to waterfowl. Root burns are conducted when marsh soil has dried 8 to 15 cm deep. Soil that is too dry will burn too deeply, killing desirable plants. For example, Olney threesquare and saltmarsh bulrush grow deeper than saltmeadow cordgrass. A shallow root burn kills the cordgrass and allows the more desirable bulrush to dominate for 2 to 3 years.

Peat burns are made during drought to convert a marsh into an aquatic habitat or during late summer or early fall when the air temperature is hot, vegetation and ground are dry, and the fire will burn hot (Linde 1969). Peat fires burn holes in the marsh floor that fill with water, creating potholes and ponds. The fire burns down to the claypan, other mineral soils, or subsurface water level, which extinguishes it. Within 6 months aquatic plants develop. Ponds should be 0.2 to 0.8 ha (Yancey 1964). Burns that are too large produce large ponds with wave action which inhibits the growth of fringe vegetation. Wave action also inhibits the growth of aquatics for years due to turbidity caused by colloidal action of clay and water if the burn exposes extensive claypan. Peat burns are more common in freshwater marshes, because saltwater marshes contain soils with too little organic material and too much salt. Cattail, spikerush, and wild millet can vegetate the borders of ponds created by deep peat burns. If there is no flooding or drought, these plants spread rapidly, attracting ducks (Lynch 1941, Hoffpauer 1968). Peat burns are difficult to control, some burning for days or weeks, and should not be set unless the marsh can be flooded to extinguish the fire.

Fires can be low-intensity (cool) or high-intensity (hot) (Linde 1969, 1985). Conditions for a low-intensity fire are high humidity and fuel moisture content, fuels with low stem density and flammability, cold air temperature, a head fire driven fast enough by a relatively strong wind to produce an incomplete burn, and a back fire which beats the flames to the ground as it moves into a relatively strong wind. Conditions for a high-intensity fire are the opposite.

Timing

Preflooding burns are useful in removing ground litter and woody material that will stain impoundments, so long as some woody cover is left for molting waterfowl (Linde 1985). Other burns generally should occur just before the growth of desirable species (Rutherford and Snyder 1983). Burns which remove dead material and expose new growth, but do not increase carrying capacity substantially, should not be used (Hughes and Young 1982). Burning of emergents is most effective if the burned area is flooded immediately. A vapor zone develops about 7.6 cm above a wet marsh, which protects the exposed burned stubble which then sprouts quickly (Hoffpauer 1968). Excessive burning (1) concentrates nesting ducks into unburned cover, thus increasing their vulnerability to predators; (2) reduces overwinter cover and the marsh's ability to catch and retain the drifting snow that is so important in areas of low annual precipitation, especially if conducted in fall or late summer (Ward 1968, Rutherford and Snyder 1983); and (3) reduces the insect prey base (Opler 1981). Marsh and surrounding uplands should be rotated for burning (Hackney and de la Cruz 1981, Opler 1981), with perhaps one-third burned annually (Smith 1960, Kirsch and Higgins 1976), especially during early spring before the nesting season. Early-nesting ducks will use residual vegetation before new growth appears on the rest of the area.

Extensive burning generally should be avoided (1) during and just before the nesting season, normally some time during February to August (especially April, May, and June) depending on latitude, unless thick cover prevents duck nesting, (2) where erosion is a problem, such as in areas where heavy cover protects the marsh from rushing storm tides (Yancey 1964), and (3) during a drought year (Kjellsen and Higgins 1990). Some authorities in North America recommend burning in early spring before the nesting season, especially in coastal areas (Cartwright 1942, O'Neil 1949, Allen 1950, Yancey 1964, Daiber 1974, Prevost 1987). On uplands, spring burning can encourage the more desirable warm-season grasses and legumes and can discourage cool-season grasses (Ohlsson et al. 1982, Owensby 1984), if burns occur after emergence and preliminary growth of cool-season grasses but before major growth of warm-season grasses (Rutherford and Snyder 1983). Ohlsson et al. (1982) recommended that areas dominated with cool-season grasses (e.g., redtop, timothy, fescue, orchard grass) generally be burned in late summer (August).

Shortgrass prairie should be burned infrequently, and only during moist periods (Kjellsen and Higgins 1990). Generally, frequent spring burns (between May 15 and June 15 in South Dakota) in a tallgrass prairie increases height and density of warm-season grasses such as big bluestem, little bluestem, indiangrass, and switchgrass. Earlier burns are better in warm springs. Reclaiming long-rested or abused lands might need two or three burns every 5 years. Maintaining grasslands in good condition might need one burn every 3 years or so, depending on annual precipitation. For mixed prairie cool-season grasses, burning should occur between late March and mid-May or about August 15 to September 15 in South Dakota. Generally, cool-season grasses are more productive after early spring burns. Most tame grasses and legumes are cool-season species, and should be burned soon after snow melt (about February 15 or March 15 to May 1 in South Dakota).

In Texas coastal marshes, most wildlife interests prefer burns during September to November to remove robust vegetation and encourage cool-season sedges (*Scirpus* spp.) (Stutzenbaker and Weller 1989). Spring burns are prescribed along the south coast (Louisiana, Mississippi, Alabama) usually after draining the marsh in late February or early March for several weeks to control competing plants with a hot, clean root burn over a dry marsh bed (Chabreck et al. 1989). Hot, clean burns in fall remove too much emergent cover and create too much open water; they should be avoided then.

In the Chesapeake Bay region, to encourage Olney threesquare, American bulrush, and saltmarsh bulrush and to discourage saltgrass and saltmeadow cordgrass, burns are conducted in late winter but before March 1, and not during droughts when the marsh is flooded with salt tides with over 15-ppt salinity (Hindman and Stotts 1989). Salinity should be 7 to 12 ppt. Burning during December to January facilitates trapping of muskrat and nutria. Burns can be done every 3 years after a hard frost between mid-November and mid-March. Big cordgrass and saltgrass are controlled by draining the marsh during February and March, burning with a hot fire, then flooding 30 to 60 cm deep from March to October; but the growing season is sacrificed (Prevost 1987, Gordon et al. 1989). Flooding about 10 cm deep after burning saltgrass kills most of it, but not alkali bulrush, hardstem bulrush, or cattail (Smith and Kadlec 1985). A marsh of big cordgrass produces enough litter to sustain an annual winter burn; a marsh of black needlerush burns well only every 3 to 4 years (de la Cruz and Hackney 1980).

In coastal freshwater and brackish marshes subject to hurricanes, burning is delayed until about October 15 after the peak of the storm season. Burning in salt marshes should be avoided (Chabreck et al. 1989). Along the Atlantic coast and in the South, burns can be conducted in the winter (Allen and Anderson 1955, Givens 1962, Schlichtemeier 1967) or fall and winter (Smith 1942, Singleton 1951,

1965, Perkins 1968, Prevost 1987). Winter burning is used in Wisconsin (Linde 1969, 1985) and Nebraska when ice is 23 to 30 cm thick and 5 to 10 cm of snow covers the surrounding range, preventing ignition of organic soils (Schlichtemeier 1967). Irregular, patchy burns in winter on semidry sedge or grass areas increase edge and access by waterfowl during the nesting season (Linde 1985).

Winter and early spring burns before ice-out probably conflict least with other interests, but are difficult to do with much snow cover. Late summer or fall burns are best if done in strips to reduce conflict with hunting by preserving some cover. One approach is to divide that portion of the marsh scheduled for burning into four equal parts, with the first burn 5 to 10 days before the waterfowl hunting season and the other three at 15-day intervals (Singleton 1965).

Spring burns might control exotic cool-season grasses (e.g., bluegrass) better than fall burns (Kadlec and Smith In press). Controlling problem cool-season grasses might be accomplished best by rotating habitat units for burning annually; each unit is burned every 5 years. (See "Nesting Meadows" in Chap. 5.)

To control woody vegetation, hot-weather burning in late summer or early fall is better than during the dormant season (Linde 1985). Hot fires also convert sphagnum and leatherleaf bogs to more useful grasses and sedges. Cattail is controlled best by draining and then burning in summer and then reflooding (Mallik and Wein 1986). Saltmeadow cordgrass on the Atlantic and Gulf coasts is removed by burning at low water levels, with immediate reflooding, which favors Olney threesquare (Sipple 1979). Reed and cattail are eliminated by root or peat burns in late summer or are temporarily controlled until early summer by cover burns during late fall (after heavy frost) to early spring, which leave long strips and islands standing for nesting cover (Ward 1942, Beule 1979) and space for waterfowl movements (Schlichtemeier 1967). Control of reed requires a combination of burning in late fall after the growing season, crushing with a bulldozer (or plowing, which is more expensive), and then flooding 30 cm deep within 3 weeks, with water 1 m deep over summer (Clay and Suprenant 1987). If reed germinates during ensuing drawdowns, immediate reflooding 30 cm deep will kill seedlings. Burning alone is reasonably effective if done in winter just before the coldest time of year or in spring after the shoots emerge (Howard et al. 1978).

Burning the marsh attracts snow geese; burning adjacent uplands attracts Canada geese and keeps cattle out of the marshlands. Spring cover burns provide succulent vegetation for muskrats; summer or fall burns reduce muskrat density by destroying house-building material and attracting cattle which destroy muskrat houses (Daiber 1986). Burning a wide area with a steady wind results in a clean cover burn which provides extensive vistas attractive to feeding geese which can produce eatouts that attract ducks (Daiber 1986). Burning during a diminishing wind or in damp weather results in a spotty (wet) cover burn, which provides cover for birds and predators in unburned patches and feeding sites for marsh ducks and shorebirds in open spots.

Wet cover burns rotated every 2 to 3 weeks during late September to January attract and hold wintering snow geese, with reduced marsh damage from eatouts (Yancey 1964, Hoffpauer 1968, Perkins 1968). To provide tender greens, especially Olney threesquare, for snow geese and white-fronted geese wintering in Louisiana, wet burns over standing water are conducted in fall and winter 1 month apart beginning 2 to 3 weeks (late September to early October) before the geese arrive (U.S. Soil Conservation Service 1977). Pintails, gadwalls, and widgeon feed upon small aquatic grubs stirred up by the geese and on the root system debris floating on the surface of the small potholes developed by feeding geese. Cover burns in August in small interspersed patches seem to benefit migratory birds by exposing invertebrates, while also controlling mosquitoes by faster water evaporation from removal of shading grasses (Bradbury 1938).

Pocosins should be burned every 3 to 6 years in areas of 20¼ ha or less for habitat variety in January to February for a cool fire (Ash et al. 1983; personal communication, A. Boynton, Wildlife Forester, North Carolina Wildlife Resources Commission 1983). A shallow water table is a prerequisite to burning to protect the peat, with burning occurring just before a rain.

Preburn Preparation

Experience in prescribed burning is best gained by apprenticing with experienced people, not by experimenting. Advice should be sought from people experienced in prescribed burns rather than from experienced firefighters, who are generally too cautious and waste time by trying to burn when conditions are too moist (Wright and Bailey 1982). Property insurance provides liability protection (McPherson et al. 1986). Procedures presented come from Wade and Lunsford (1989) and Wisconsin Department of Natural Resources (1990), with exceptions indicated. All burns need a prescribed fire plan (Higgins et al. 1989, Wade and Lunsford 1989).

Site preparation

Firebreaks are made several weeks before the burning day to avoid overuse of personnel and equipment during the main fire, preferably after leaf fall to reduce the possibility of a breach from combustion of fallen material in the firebreak (McPherson et al. 1986, Wisconsin Department of Natural Resources 1990). Whenever possible, natural breaks are used, such as roads, ponds, streams, or cultivated fields. Otherwise the firebreak is plowed, disked, rotovated, or dozed down to mineral soil. In sensitive areas where such firebreaks are contrary to land management, other methods are used as firebreaks: very short mowed grass, trails, damp areas of herbaceous plants or grass, wetdown areas, snow cover, northern damp hillsides, previously burned areas, chemical or foam firebreaks.

Chemical retardants, foam retardants, or water should be applied on the fireline just before fire initiation (Martin et al. 1977, Higgins et al. 1989). Foam is best because it expands and extends the amount of water available 3 to 10 times, it smothers and insulates, and it is more persistent and visible than water. Retardants obviate tillage scars in native prairie and tundra.

Width of firebreaks depends on type of fuel. Ideally, firebreaks consist of two parallel strips of exposed mineral soil 30 to 90 m apart, the distance depending on fuel load (Fig. 4.14). For herbaceous vegetation 3 to 9 dm high, the exposed soil should be 1.8 to 2.4 m wide, with the 30- to 90-m fireline burned preferably about 1 month before the main burn (Wright and Bailey 1982). In areas of relatively light fuel loads, such as short grasslands, the firebreak can be a 3-m-wide rotovated strip beside a 6-m-wide mowed strip to remove excess fuel for starting the burn. The fireline (area between dozed lines of firebreak) is installed in the fall after the vegetation goes dormant and is burned in late March every 5 years (personal communication, J. Kier, Wisconsin Department of Natural Resources 1989).

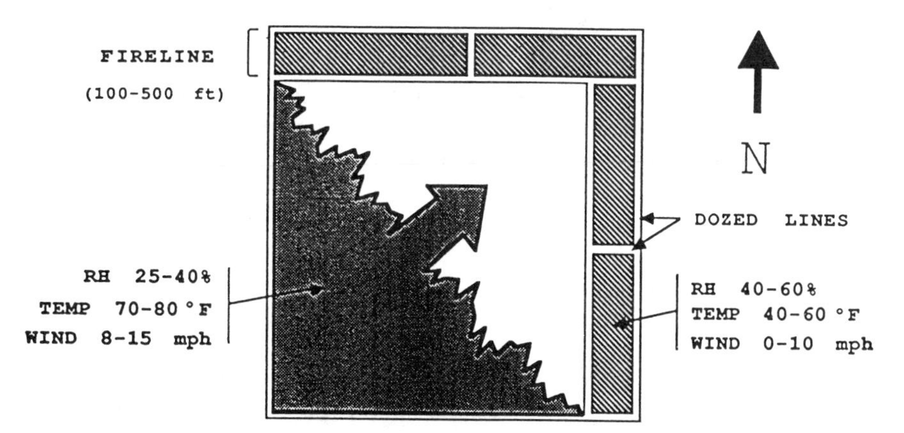

Figure 4.14 Generalized fire plan for prescribed burns. Width of fireline depends on fuel type and amount (McPherson et al. 1986).

Plowlines are as shallow as possible on contour in hilly areas and as straight as possible, winding gently around excluded objects and corners. The area is subdivided into logical burning jobs. Parallel interior plowlines 183 to 244 m apart, 90 degrees to the prevailing wind, provide good control in dry marsh for an 8-h burn (Linde 1969). Where firebreaks can be plowed or natural breaks occur, eight of these divided areas can be burned in 1 day if fired in succession after the baseline in the preceding area is established as safe. Brush, heavy fuel pockets, and standing trees are removed near the fireline. For an 810-ha burn on grassy uplands, three strips usually are needed to burn out all firelines (Wright and Bailey 1982). Woody material is mixed with soil along the fireline because holdover fires could break out of control days or weeks later. Special care is used to expose mineral soils in peat areas. Wider breaks are preferred around project boundaries.

A detailed map of the area to be burned and the surrounding area includes cover types, control lines, water access points, topography, location of possible hot spots or trouble areas, location of personnel and equipment, firing pattern, and location of high-value items such as water control structures, fences, buildings, blinds, and utility poles. All people involved get copies of the map.

Fuel

The most important factor affecting fire behavior is the fuel load. The decision to burn is made at least 1 year before, to build up enough fire fuel. Areas to be burned are not grazed for one growing season before the burn. Preburn treatments, such as mechanical or herbicidal, might be desirable.

Accumulations of 2.2 mt/ha of fine fuel are recommended, but less than 1.1 mt/ha will suffice if the fuel is dominated by continuous sod-forming grasses rather than intermittent bunch grasses (Rasmussen et al. 1986). Those guidelines probably apply to most herbaceous vegetation, including marsh. With low fuel loads, relative humidity must be lower and temperature and wind speed higher to reduce labor and to ensure a satisfactory burn. Reduction of woody vegetation such as live shrubs by fire requires substantial fine fuel from herbaceous vegetation.

Pile burning is accomplished under proper conditions if the pile of debris is dry and virtually free of soil, which produces incomplete combustion and lingering smoke. Windrow burning is not recommended because of the soil mixed in, the severe ecological consequences, and the danger of trapping personnel between intensively burning windrows.

Variables influencing fire spread and intensity include species of

TABLE 4.11 Time Needed for Various Size Fuels to Dry Out or Gain Moisture

Moisture class, H TL FM*	Size class, cm	Time needed to approach equilibrium, h	Fuel moisture, %†
1	0.00–0.61	1 (0–2)	7–8
10	0.62–2.53	10 (2–20)	10–12
100	2.54–7.59	100 (20–200)	17
1000	≥ 7.60	1000 (200–2000)	—

*H TL FM = hour time lag fuel moisture.
†Ideal for burning in slash.
SOURCE: Wisconsin Department of Natural Resources 1990.

fuel, size, volume, weight, arrangement, and moisture content (Wisconsin Department of Natural Resources 1990). Fuel size is classified relative to time to gain moisture or dry out. Fine fuels, such as grasses and leaves, need 1 to 2 h, twigs fluctuate daily, large logs might need a summer of drought (Table 4.11). Fuel that is too moist is hard to ignite and burns too slowly. Fuel that is too dry burns too fast and too hot, and it might be damaging. The maximum fine-fuel moisture for burning at 1-h time lag fuel moisture is 30 to 40 percent, but 7 to 20 percent is preferable (Countryman 1964, 1971, Mobley et al. 1973). Spot fires are certain at fine-fuel moisture of less than 5 percent, rare at 11 percent. As a general rule, 1-h time lag fuel moisture of 7 to 8 percent corresponds to a relative humidity of 40 percent, the minimum at which firebrands usually cause a problem in dry grass.

The 10-h time lag fuel moisture is a good indicator of the small fuels which will carry the fire. When the 10-h time lag fuel moisture is 6 to 15 percent, prescribed fires will carry well, burning quickly at 6 percent but almost stopping at 15 percent. Ten-hour time lag fuel moisture sticks (Wisconsin Department of Natural Resources Equipment and Development Center, Tomahawk) can be used to estimate the moisture condition of dead fuels of this size class if placed at the site 10 to 14 days before burning and weighed periodically. Fuel moisture of live fuels also should be considered, due to their higher moisture content which can dampen the fire. Burning usually starts about 10 a.m. after the sun has dried the dew off the vegetation (Linde 1969).

Weather

Air temperatures should be below 15.5°C. Danger from firebrands increases exponentially above that (Bunting and Wright 1974). Relative humidity should be 20 to 40 percent (Britton and Wright 1971, Lindenmuth and Davis 1973), unless winds are less than 10 km/h and temperatures below 4.4°C. Winds should be 10 to 24 km/h and steady. Burning is risky on calm days because light and variable winds, which are unpredictable, often are forecast then (Wisconsin Department of Natural Resources 1990). In grassy uplands of west Texas, weather conditions for burning firelines should be 4.4 to 15.6°C, 40 to 60 percent relative humidity, and 0- to 16-km/h wind; for head fires, weather conditions should be 21.1 to 26.7°C, 25 to 40 percent relative humidity, and 13- to 24-km/h wind (McPherson et al. 1986). Fires started in the afternoon become progressively cooler and more manageable toward evening. Thus, ignition in the morning usually produces a better burn if the fuel moisture content is high (Rutherford and Snyder 1983). Fires should be deferred for at least 5 days after a rain of at least 0.25 cm.

Smoke management

Prescribed burning can contribute to air pollution. Guidelines to reduce air pollution include the following (Wisconsin Department of Natural Resources 1990):

1. Weather forecast information is used to predict smoke and fire behavior.
2. Burns are avoided during pollution alerts or during temperature inversion, which reduces dispersal of smoke.
3. Air pollution regulations must be followed.
4. The local fire control office, adjacent landowners, and nearby residents are notified to inform them and to learn of public reaction and special problems (e.g., respiratory ailments).
5. Burns are avoided when wind will carry smoke toward highways, airports, populated areas, or other sensitive areas such as schools, hospitals, and nursing homes.
6. Burns should occur when the atmosphere is slightly unstable so smoke will rise and dissipate; fire control problems will occur if conditions are too unstable.
7. Burns along roads must be mopped up as soon as possible.
8. Test fires are used to test smoke behavior.
9. Back fires are used when possible to consume fuel better and to produce less smoke.
10. Burns during the growing season should be avoided, because green vegetation produces more smoke.
11. Night burns are avoided because visibility is reduced; smoke drift is less predictable; downslope winds prevail at night, causing smoke to concentrate in low areas; and humidity greater than 80 percent causes moisture and smoke to form fog.

12. Burns on windless days are avoided because smoke tends to stay near the ground and because variable winds often occur then, carrying smoke in unpredictable directions.
13. Emergency plans must be made, such as controlling traffic on nearby roads if the wind changes direction.

Personnel and equipment

The number of people needed depends on the size of the area to be burned, fuel characteristics, and experience. Usually 6 to 12 people will suffice (Wright and Bailey 1982); they are divided into a fire boss, ignition crew, and pumper crew (McPherson et al. 1986).

Equipment includes the following (Wright and Bailey 1982, McPherson et al. 1986):

1. Four two-way radios
2. Six shovels, rakes, pulaskis
3. Six swatters or backpack spray pumps
4. Two pickups (one with a 379-L slip-on pumper and 15.24 m of hose with handheld adjustable spray nozzle)
5. Five drip torches (or fusees, which stay lighted better in high winds)
6. 114 L of fuel (70 percent to 30 percent diesel-gas mixture)
7. A standby D-7 dozer or tractor with rotovator, which is faster than the dozer
8. Two belt weather kits
   a. Psychrometer (temperature and relative humidity)
   b. Bottle of water
   c. Anemometer (wind speed)
   d. Notepad and pencil
   e. Compass
9. Fence cutters

Other items could include (1) extra torch fuel, (2) extra engine fuel and oil, (3) first aid kits and air splints, (4) fire shelters, (5) drinking water, (6) smoke masks and respirators, (7) goggles (for chemicals), (8) binoculars (1 pair per vehicle), (9) foam units, (10) chemical retardants (wet or dry), (11) wetting agent, (12) tool kit, (13) road grader, (14) tow chains, cables, or ropes, (15) handyman jacks, (16) chainsaw (Higgins et al. 1989). Clothing which will melt or flame easily, such as nylon, polyester, and plastic, should not be worn. Safety clothing should include the following (Higgins et al. 1989):

1. Hightop leather boots 20 cm or higher, or work shoes with nonslip soles and leather laces, but not steel-toed

2. Cotton or wool socks
3. Cotton underwear
4. Nomex slacks, loose fitting, with hems lower than shoe tops
5. Leather belt or natural fiber suspenders
6. Nomex long-sleeved loose fitting shirt
7. Leather gloves
8. Hard hat
9. Goggles
10. Belt pack case with aluminum emergency fire shelter

Other items could include canteen, food, lip balm, handkerchief, ear plugs, ear muffs, face shield, map, knife, matches, flashlight, extra batteries, first aid kit, AM radio, and flares.

The fire boss places the pumper crew in the area of greatest potential for fire escape, i.e., where terrain and fuel conditions alter wind patterns and fire behavior. The suppression (pumper) crew of one to four people, usually large enough on a burn of less than 405 ha, is responsible for patrolling downwind of the fire to extinguish spot fires. All ignition stops until a spot fire outside the treated area is controlled. The ignition crew serves as a backup suppression crew. The lead torch is followed by two to three people with suppression equipment, with a pumper following them.

Ignition can be done by ground or aerial technique. The most commonly used ground ignition is the drip torch, consisting of a reservoir tank, spout, and wick. The fuel can be a mixture of gasoline and diesel oil (3:1) or kerosene and diesel oil (2:3) (Scifres 1980). For best control, ignition crews consist of one person setting the fire followed by one person with a portable radio (Wisconsin Department of Natural Resources 1990). (Dense smoke can interfere with optimum radio functioning.)

Aerial ignition is used mainly on extensive uplands rather than on wetlands. Aerial ignition can be accomplished by use of (1) a delayed-action ignition device (DAID) consisting of an 18-cm-long 30-s safety fuse (match) dropped by hand, (2) a back-fire starter cartridge fired from a 12-gauge shotgun (Leege and Fultz 1972), (3) an aerial ignition device (AID) consisting of a plastic sphere (similar to a Ping–Pong ball) charged with potassium permanganate activated by ethylene glycol when dropped (Gnann 1985), or (4) a helitorch consisting of a 208-L barrel containing a mixture of gasoline and alumagel and a positive displacement pump and igniter to provide a spark that ignites the mixture as it drops from the barrel (Florence 1983, Stevens 1985).

All methods are used from a helicopter flying 61 to 91 m above ground at about 48 km/h. Florence (1983) and Stevens (1985) discussed use of the helitorch in detail, including coordination with the ground crew. Aerial ignition does not cause significant direct mortality to wildlife (Folk and Bales 1982).

Firing Patterns

Escape routes for personnel and wildlife must always be provided. Converging fires that will trap them must be avoided.

Fireline

If used, an extensive fireline (area between dozed lines) generally is burned on the north and east sides of the site, e.g., if the head fire will be burned with a southwest wind (McPherson et al. 1986). The strip head fire (see Fig. 4.17) is the most common firing pattern for firelines.

The first strip of fire is ignited within a few decimeters of the dozed strip, the second strip of fire 1.5 to 6.1 m upwind of the first. A narrow strip of fire is used in heavy fire fuels, a wider strip in light fuels. The third and subsequent fire strips are wider, because the chance of spot fires decreases as the width of the burned-out area increases. Torch carriers are staggered, and they maintain visual or verbal contact with torch carriers on each side to prevent the fire from surrounding a slow-moving torch carrier.

Firelines in low-volatile fuels can be burned safely with five people. A fire boss, two-person pumper crew, and five-person ignition crew can burn 4.8 km of 30-m-wide fireline in 4 h.

Main fire

Firing patterns vary with conditions (Mobley et al. 1973, Wisconsin Department of Natural Resources 1990). If firelines are established on the north and east sides of the site, wind direction should be west, southwest, or south for the head fire (McPherson et al. 1986). Fires move as (1) a head fire, in the same direction as the wind; (2) a back fire, in the opposite direction to the wind; and (3) a flank fire, at a right angle to the wind (Wisconsin Department of Natural Resources 1990). Often the different firing patterns are combined. For example, the head fire may be ignited after the back fire has burned into the burn area about 15 to 18 m in nonvolatile (mostly herbaceous) fuel or 30 m in more volatile fuel (Fig. 4.15) (Scifres 1980). Fire whorls can develop when a head fire runs into a back fire, especially with fast winds, producing firebrands that cause spot fires outside the burn

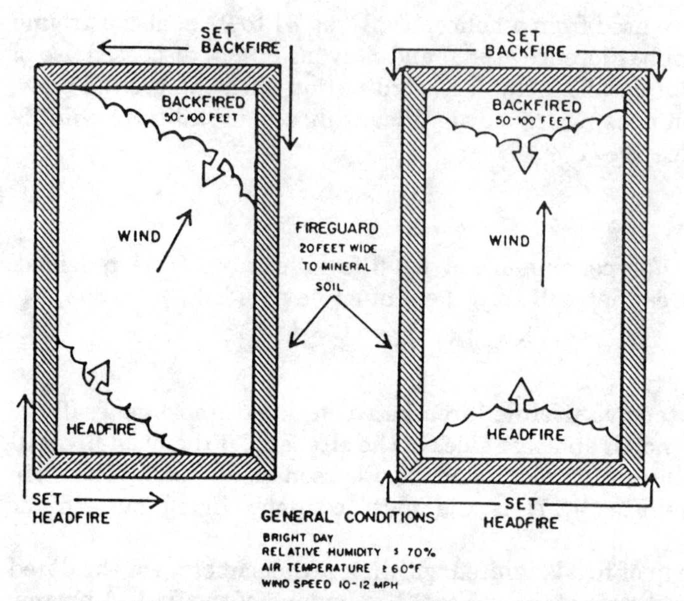

Figure 4.15 Ignition of back fires and head fires in oblique winds (left) and direct winds (right) (Scifres 1980).

area. Then back firing is completed up to a plowed strip before the head fire is ignited.

When the wind is blowing in line with the intended direction of the fire, two ignition people begin at the center of the windward side of the area and walk toward the corners, igniting along the way (Fig. 4.15). With the wind blowing diagonally across the burn area, the igniters begin back firing at the leeward corner, ignite down each of the two adjacent sides, and then similarly along the windward sides (Scifres 1980).

Back fire. The back fire (or backing fire) burns into the wind, starting along a prepared firebreak such as a plowed line, road, stream, or impoundment. To burn a large area, interior lines are plowed every 201 to 402 m and ignited about the same time (Fig. 4.16). A back fire is the easiest and safest type of prescribed burn, if a steady wind speed and direction prevail. Fire burns into the wind at 20 to 60 m/h, with little variation regardless of wind speed. (Fires burning downslope behave similarly to back fires.)

Back fires can be used in heavy fuels to ignite large areas in little time, and normally they result in minimum scorch, but they can result in root damage or severe duff reduction because they remain on

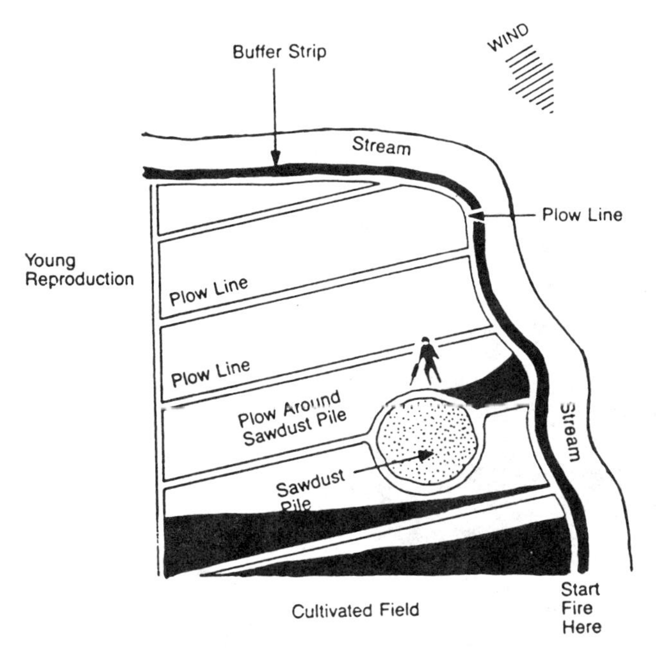

Figure 4.16 Back fire with interior plow lines (Wisconsin Department of Natural Resources 1990).

site longer. They need continuous fuels but produce less smoke due to more complete combustion. They do not burn well with fire fuel moisture greater than 20 percent. They are expensive, especially if interior plowlines are needed, which also reduces flexibility of direction, because they take longer to burn. A back fire usually is used with other firing patterns to establish a base control line.

Head fire. The head fire is set at right angles to the wind, burning with the wind, usually as a strip head fire (Fig. 4.17) which permits quick firing, similar to burning the fireline, and good smoke dispersal. A series of lines of fire are set progressively upwind of a firebreak, so that no individual line of fire will intensify before reaching a firebreak or another line of fire. Strips usually are 20 to 100 m apart, often with the base control line treated first with a back fire. Altering the angle of strip fire with the baseline will compensate for minor changes (up to 45°) in wind direction and can be adjusted for fuel density and composition. Fire intensity is reduced when a series of spots or strips 3 to 6 dm long is used instead of a solid line of fire. Major changes in fuel type are treated separately. Head fires can be used in

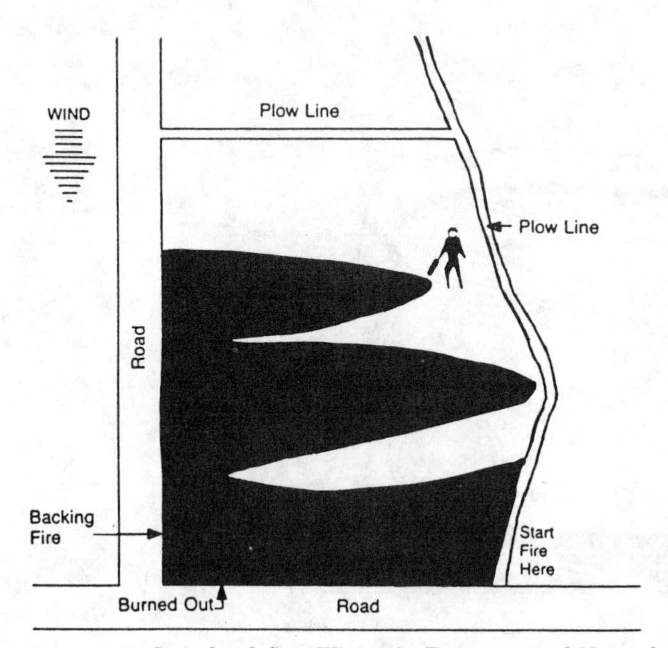

Figure 4.17 Strip head fire (Wisconsin Department of Natural Resources 1990).

flat fuels such as hardwood leaves and in most other fuels including light or discontinuous fuels, but not in very heavy fuels. Head fires can be used with low wind speed, i.e., just enough to give direction, and with high relative humidity and fuel moisture. They generally are safer than perimeter fires because there is less chance of spotting. A single person can ignite strips progressively; otherwise close timing is needed to ensure safety.

A head fire can be used to move over the entire area without stripping, if the area is small with light, even fuel distribution. Nonstrip head fires reduce fire intensity that occurs at junctures of strip fires. A back fire wide enough to control the head fire must be used downwind.

Perimeter fire. The perimeter fire is set around the entire burn area and allowed to sweep over the area (Fig. 4.18). As with other techniques, first a base control line on the downwind side of the burn area is secured with flank and/or back fires. Area ignition is a variation used during light winds. The perimeter fire is used when a relatively hot fire is needed to kill stems or reduce logging slash. But care is needed to control firebrands because a perimeter fire can develop strong and often violent convection columns, providing good smoke

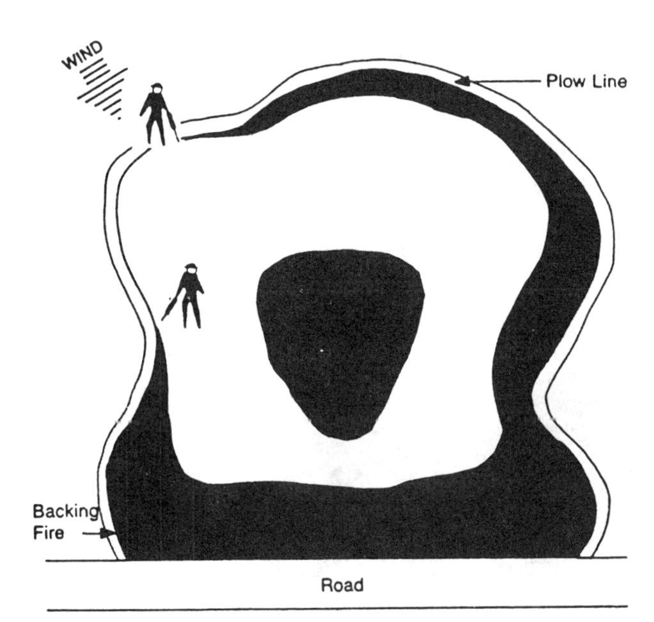

Figure 4.18 Center perimeter fire (Wisconsin Department of Natural Resources 1990).

dispersal but causing serious spotting problems. Perimeter fires can be used with variable wind directions, low wind speed, high humidity, and light, discontinuous fuels.

The center perimeter fire (Fig. 4.18) is a variation that provides a hot burn with light and variable winds. It consists of first setting at least one spot fire near the center of the burn area to help pull the perimeter fire, set next, toward the center. It provides good smoke dispersal with a strong convection column which rises from the center of the burn unit and thus has less chance of spotting than a perimeter fire. Close timing and caution are needed to prevent endangering the center torch person. Mechanical methods, i.e., fusee or aerial ignition, are safer for firing the center.

Flank fire. The flank fire consists of lines of fire set into the wind which burn outward at right angles to it (Fig. 4.19). It is used (1) on small areas or to burn a large area fairly fast when a strip head fire would be too intense, (2) on rolling or hilly terrain, (3) to supplement a back fire in light fuel, or (4) to secure the flank of a strip head fire, back fire, or perimeter fire as it progresses. Fuels should be light to medium. Wind direction must be constant. Fast ignition occurs, but

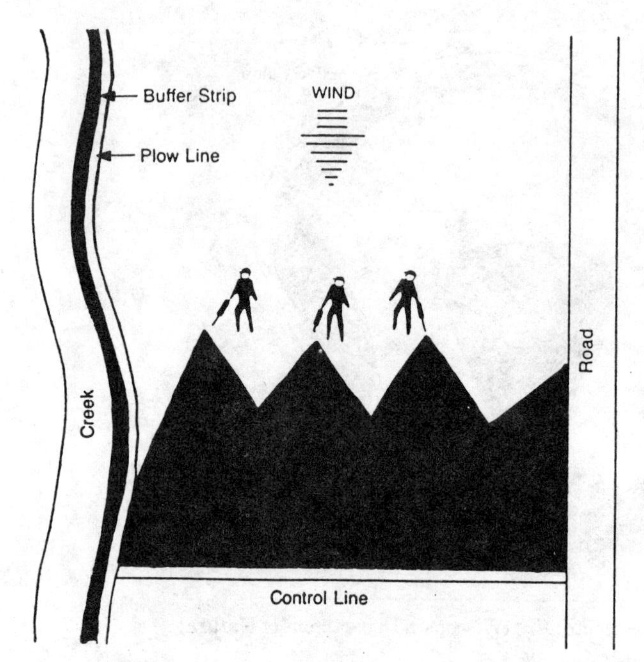

Figure 4.19 Flank fire (Wisconsin Department of Natural Resources 1990).

good knowledge of fire behavior and expert coordination of crews are needed.

Spot fire. The spot fire consists of a series of fires set at predetermined spacing throughout the area (Fig. 4.20). Experience determines the spacing, usually 40 by 40 m, which is far enough apart to prevent too many junction zones of high-intensity fire, but close enough to prevent individual spots from developing into hot fire heads. Area ignition is fast because the spot fires burn in all directions. Spot fires are less intense than perimeter or strip head fires. Spot fires can be used with light, variable winds. Fuels should be uniform and light to medium. Spot fires can be used with low temperatures and high fuel moisture. An experienced crew is needed.

Chevron fire. The chevron fire is used in hilly areas, with lines of fire started simultaneously from the apex of a ridge point or ridge end, progressing downhill (Fig. 4.21). Essentially it is a flank fire concept, except the lines of fire are not parallel and thus the fire is not as in-

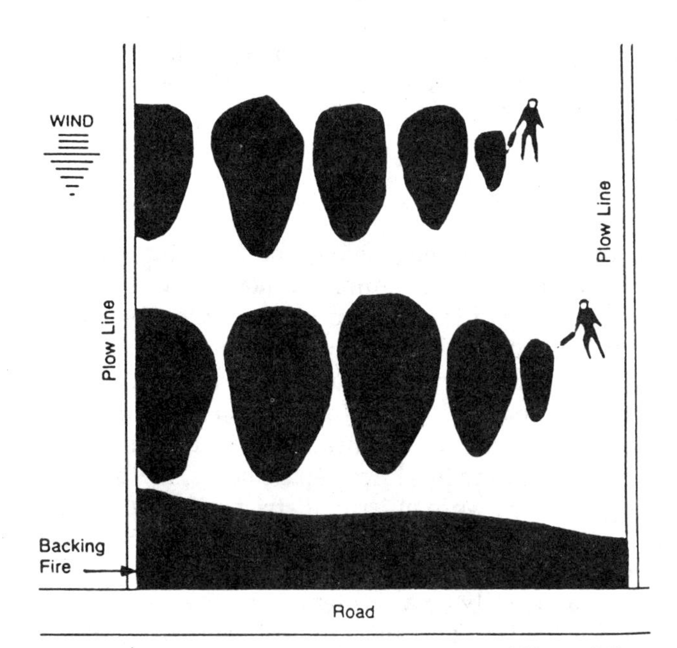

Figure 4.20 Spot fire (Wisconsin Department of Natural Resources 1990).

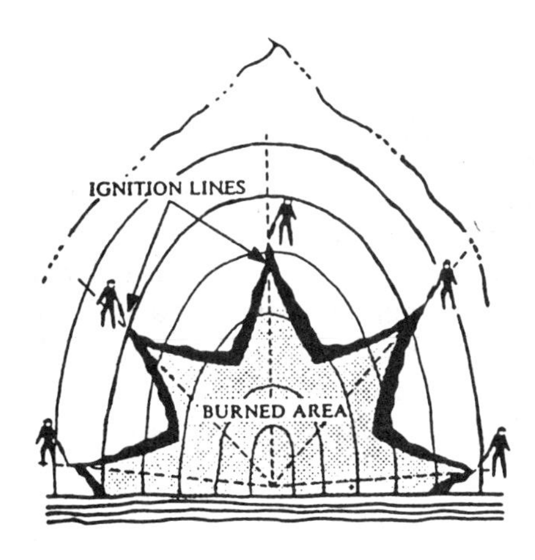

Figure 4.21 Chevron fire (Wisconsin Department of Natural Resources 1990).

tense. Junction zones of intense fire are small because any two lines of fire converge at a point rather than along a line. The chevron fire is unsuited to most topography and should be combined with other techniques. Topography usually restricts use of vehicles. Other firing patterns are used to secure baseline and flanks.

Burn

The decision to burn is made the day before the burn after a spot weather forecast is obtained. All personnel are notified to stand by with equipment. The morning of the burn, a spot weather forecast is again obtained, and if it is favorable, all personnel are notified to arrive at the site for briefing, familiarization tour, and equipment check. Additional equipment is assembled in case of equipment failure. The briefing includes discussion of action to contain any wildfires. To prevent undue concern, in most cases local fire departments, law enforcement agencies, emergency fire wardens, the district office, and adjacent landowners are notified. If needed, a burning permit is obtained.

A test burn is discussed during the briefing and then made just before each prescribed burn, usually by allowing a head fire to burn into a plowed strip to see if it holds. Such burns indicate weather and fuel moisture suitability, fire intensity and behavior, mop-up required, and smoke dispersal patterns. The final decision to burn is made after the test burn is completed.

Report

A final report on the prescribed burn includes pictures (preferably colored slides) in sets of three from fixed and identified points, taken before and after the burn each year for 4 years. Subjective and, if possible, quantitative measurements are made of various parameters for wildlife, soil, ground layer, shrub layer, and trees. Itemized lists of people, equipment, and expenses are kept to determine the overall cost of the burn. Preburn and postburn maps are included. To evaluate the burn, the report includes information from the burning proposal submitted before the burn as well as information during and after the burn:

Location	Preferred firing pattern(s)
Size	Objectives
Cover type	Date of burn (month/season)
Soil type	Temperature range
Aspect	Humidity range

Slope	Wind direction
Topography	Wind speed
Year of last burn	Flame height
Fuel load	Time of ignition
Fuel moisture	Time completed
Firebreaks	Percentage of area burned (grass, brush, slash)
Personnel needs	Percentage of crown scorch
Equipment needs	Bark scorch height, bark char height

MECHANICAL TREATMENT

Water zones for managing aquatic vegetation can be classified as deep (over 76 cm), intermediate (30 to 76 cm), and shallow (2.5 to 30 cm). Management guidelines for control of cattail and other emergents include herbicides and cutting for deep water, cutting on ice and crushing for intermediate water, and crushing, herbicides, creating openings, and bottom contouring for shallow water (Beule 1979).

Openings can be created from shoreline to intermediate zone by removing plants or soil in irregular strips 20 m wide at the base, interspersed with strips of cattail or other emergents 75 to 100 m wide, with gently sloping sides less than 12 dm deep (Fig. 4.22). Openings in emergent vegetation can be randomly spaced circles at least 0.1 ha to reduce aggregations of breeding ducks and to allow diving ducks to take flight or can be shaped in sinuous strips to increase edge and reduce visual encounters between conspecific pairs of ducks (Kaminski and Prince 1981). Frozen soil from lowered water levels can be excavated and deposited on uplands, spread evenly, and seeded in spring to recommended grasses. Other permanent openings can be dug along the marsh edge near blocks of good upland nesting cover, but not near uplands used for agriculture.

Bottom contouring in the shallow zone at the 6-dm level before flooding includes digging channels 10 to 20 m wide at the base and 9

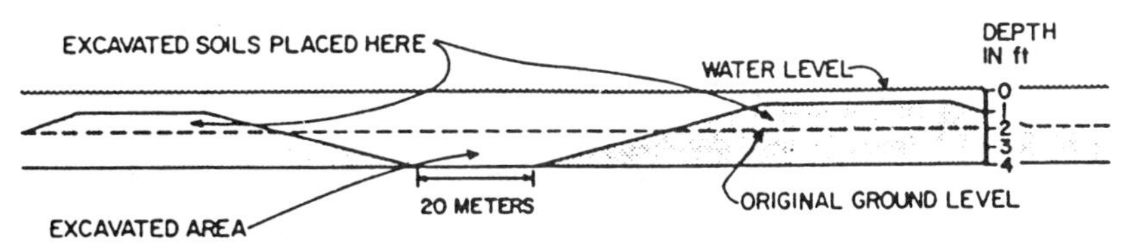

Figure 4.22 Bottom contoured to create openings in vegetation of the intermediate zone (Beule 1979).

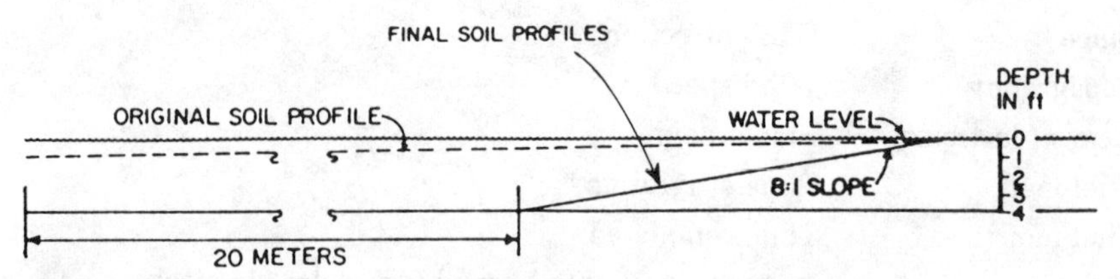

Figure 4.23 Bottom contoured to create openings in vegetation of the shallow zone (Beule 1979).

to 12 dm deep when flooded. Soils can be pushed to either side to form parallel ridges of shallow water 15 to 30 cm deep, connected by gently sloping sides (Fig. 4.23). So that the deepened channels can drain and fill freely when water levels are manipulated, they should connect with the river system flowing through the marsh or should end at the outlet structure. When flooded, the channels remain as open water, and emergents dominate the ridges and shallow slopes, to be controlled somewhat by muskrats and nutria, which are controlled by predators, trappers, and winter drawdowns. Efforts to control vegetation in the shallow zone often are temporary because seeds are ready to germinate at the shoreline or on other exposed soil, water depths are optimum for cattail growth and shoot production, and muskrats usually are excluded when the water freezes solid (Beule 1979). During construction of the dam or after drawdown, microtopography can be created by excavating a washboard pattern 30 cm deep (Kadlec and Smith 1984, Verry 1989). (See "Landscaping" in Chap. 2.)

Cutting

Wetlands

Cutting or removing aquatic plants can be done with hand tools and with mowers attached to floating machines, with wheeled and tracked machines operating on the bottom of a drawn-down impoundment, and with machines operating from the bank. But harvesting can have lethal and sublethal effects on plants and animals through killing and through modifying the ecosystem (Engel 1990).

Hand tools. For removal of select aquatic plants from small areas, a length of barbed wire or chain dragged by two people through small ponds, marshes, ditches, or potholes will break off or uproot weeds (Payne and Copes 1986). Long-handled scythes, sickles, and grass hooks also can be used (Robson 1974). Cut plants and filamentous algae are removed with rakes and forks. A hedge trimmer, with a 6-dm reciprocating blade attached to one end of a 2-m shaft and driven by a small gasoline engine at the other end, can be used for cutting stems 1 cm or less in diameter. Slung from the shoulder, the trimmer can cut plants on the bank and in water up to 6 dm deep.

Floating machine. The most common system for cutting aquatic plants is a version of the reciprocating mower bar similar to the type used on agricultural machinery (Nichols 1974). Such machines cannot cut emergent plants except in relatively deep water (at least 30 to 60 cm). The cutting bar can be mounted on a boat or be part of a large harvesting system that includes transport barges and shoreline unloading apparatus. If not loaded into barges by conveyor belts, the cut plants are left in the lake, raked, towed to shore in a floating enclosure, or floated to shore via current or wind (Koegel et al. 1978). The wide range in equipment requires care in selecting a suitable machine (Cooke et al. 1986).

Width of cutting blades ranges 1 to 3 m, which can cut up to 3 m deep (Nichols 1974). Typically, the harvester is a highly maneuverable, low-draft barge with one horizontal and two vertical cutter bars, a conveyor to convey cut plants to a hold on the craft, and another conveyor for rapid unloading. Cutting rates of 0.2 to 0.4 ha/h and storage capacity of about 6 to 23 m^3 vary with the size of the harvester (Koegel et al. 1974, Cooke 1988). Small, boat-mounted harvesters are the most maneuverable for operating in shallow water (Nichols 1974). Cooke et al. (1986) compared types of aquatic plant cutters and harvesters.

In the southern United States, the sawboat and the Kenny crusher are used to destroy water hyacinth and alligatorweed (Robson 1974). The sawboat shreds the floating mats of vegetation in the water. The Kenny crusher lifts the floating mats out of the water onto a wide conveyor belt on a barge, which drops the plants between two large rollers that crush and deposit the plants into the water again. If deoxygenation and other results of organic pollution are a problem, dead material is removed from the water.

A cookie cutter can remove vegetation to create open water and can cut new channels or clear existing channels of mud or silt (Schnick et al. 1982). It consists of a 7.9-m aluminum boat powered by a 174-hp diesel engine and two five-cylinder radial hydraulic motors to turn two 12-dm-diameter rotary cutting and propelling blades. Top speed is 9.7 km/h. The cookie cutter can remove all species of floating and submerged aquatic plants and brush up to 5 cm in diameter and can operate in water up to 46 cm deep. It can cut ditches through overgrown vegetation such as cattail. It has greater production than a dragline and can work in areas where a dragline cannot (Ducks Unlimited Canada 1977).

Wheeled and tracked machines. Plants vary in their response to mowing (see Table 4.10). Emergent plants, mainly cattail, but also common reed, whitetop, and sedge, are mowed as close as possible to the basin substrate to facilitate inundation of the stubble during the growing season to reduce shoot density 50 to 93 percent and flowering shoot density 64 to 97 percent (Kaminski et al. 1985). A greater than 50-hp standard tractor or crawler tractor with 5- by 10- by 60-cm oak cleats bolted to the tracks, pulling a rotary mower, is most efficient for cutting dead and live material (Nelson and Dietz 1966, Kaminski et al. 1985). A rotary cutter can mow up to 6.5 ha/day of heavy cattail. By removing the steel chain curtain attached to the rear of most rotary mowers, the cutting efficiency and dispersal of clipped plants are increased. Other equipment improvements include a radiator screen to prevent engine overheating and half-tracks or front and rear dual wheels to improve support over soft substrates (Nelson and Dietz 1966).

Dense stands of cattail and perhaps other emergents can be controlled by burning in winter, then shredding remaining plant stalks with a rotary mower, then flooding the cut stalks for at least 2 weeks in spring (Weller 1975, Murkin and Ward 1980). Extensive monotypic stands of cattail should be controlled or fragmented when carbohydrate reserves are low (Linde et al. 1976), generally in June when production of fruiting heads is heavy (Fig. 4.24). Three cuttings below the

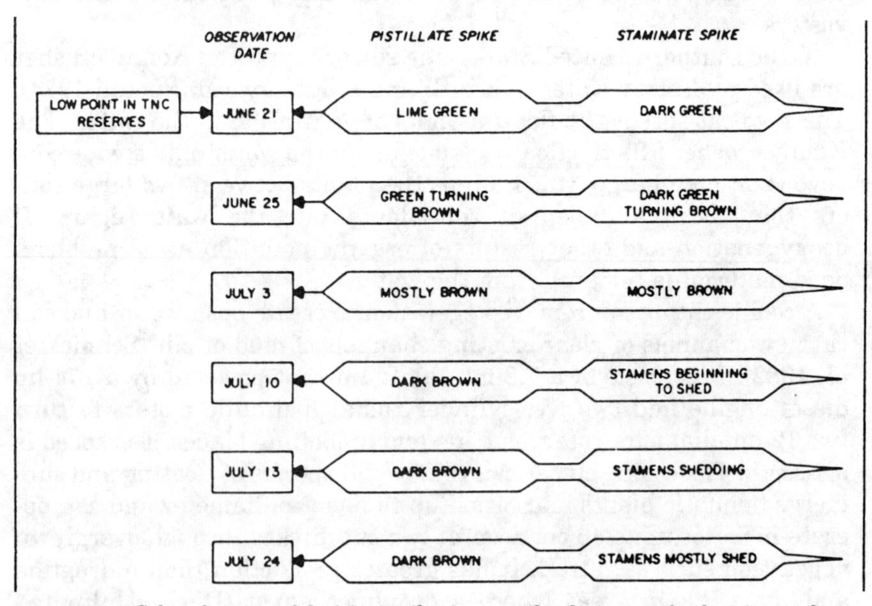

Figure 4.24 Color changes of fruiting spike in cattail relative to the low in total nonstructural carbohydrate (TNC) reserves (Linde et al. 1976).

water then kill nearly all the underwater biomass (Sale and Wetzel 1983), although two cuttings at ground level 6 weeks apart, the first in July, will suffice if then covered with at least 7.6 cm of water (Nelson and Dietz 1966). About 75 percent of the cattail is killed if mowed after the heads are well formed but not mature and then mowed about 1 month later when the new growth is 6 to 9 dm high (Lopinot 1986). One cutting will suffice if the cattail is cut when all leaf growth is completed (late July), all stems are cut as low to the ground as possible, and the area is flooded with at least 7.6 cm of water (Beule 1979). Cutting cattails on ice is successful if the stubble is covered with about 18 cm of water after the thaw (Weller 1975), or if the water level is lowered in the fall before cutting on ice and then raised over the stubble in spring (Buele 1979). Cattail also is controlled by a March drawdown, then burning, then plowing or disking (which also reduces woody growth), then reflooding from October to November over winter until an April drawdown (Hindman and Stotts 1989).

Cutting reeds for two or three consecutive years in August or September or twice yearly for 2 years seems to reduce the density for about 8 to 10 years (Larsson 1982, Cross and Fleming 1989). On wet sites a cookie cutter or rotary ditch digger will create openings in reeds with rhizome-packed substrates. On drier sites bulldozers, brush cutters, disks, mowers, rototillers, plows, crushers, and dredges can create openings and control reeds. The most effective treatment of reed is to combine mowing, burning, and then disking at least twice with a heavy disk (81-cm blade) pulled with a 400-hp tractor (Cross and Fleming 1989).

To destroy dense stands of cutgrass, the marsh is drained, mowed in late July or early August, burned 2 to 3 weeks later when the hay is dry, and flooded immediately and as deep as the control structure allows, until the following spring (Neely and Davison 1971). Various types of wheeled and tracked low-ground-pressure equipment have been developed for use in marshy areas (Green and Rula 1977, Willoughby 1978). Roby and Green (1976), Brown (1977), Willoughby (1978), Larson (1980), and Long et al. (1984) compared specifications for rubber-tired and tracked tractors, with illustrations of revegetation equipment and descriptions of techniques, capabilities, and limitations.

Plants such as willow, buttonbush, cocklebur, and cattail are favored by management of reservoirs for hydroelectric production, irrigation, or flood control (Griffith 1948). These plants are undesirable for waterfowl and can be eliminated by periodic mowing with a D-6 dozer pulling an Athens fire plow, which will cut woody growth 1.8 to 3.7 m tall, so that food plants can establish.

Machines operating from the bank. Weed-cutting buckets come in sizes up to 4 m long, with smaller buckets attached to backhoe booms of wheeled tractors and large buckets attached to backhoe booms of large tracked tractors (Robson 1974). The bucket comprises a series of curved, vertically arranged bars connected to a horizontal bar on top and a reciprocating cutter bar at the bottom. The bucket is lowered next to the soil and moved toward the excavator, cutting the weeds which collect in the bucket and then depositing them on the bank.

Rotary mowers can be attached to the hydraulic arm of a tractor if slopes are steep or towed (e.g., a batwing) if slopes are gentle. Such mowers are used on dikes in late summer after the growing season to reduce the amount of residual cover in spring so that ducks will not nest on dikes used so efficiently by predators as travel lanes.

Uplands

Standard rotary mowers, including batwings and brush hogs, or sickle-bar mowers are towed by standard farm tractors to cut brush and herbaceous plants. Rotary mowers vary in size and pattern, having two to four blade designs. Mowing should occur after July 15 by which time most ground-nesting birds have fledged, but before September 1 so that grasses and forbs are rejuvenated and will grow enough to produce residual vegetation for nesting in spring. The cutter bar is set to leave 10 to 15 cm of stubble, except where closer cropping is preferred for feeding and nesting of species such as robins, killdeers, and horned larks (Ohlsson et al. 1982). Except for upright annuals, mowing is relatively ineffective in controlling weedy growth. Repeated mowing provides temporary control of tall perennial weeds and small-stemmed brush by depleting carbohydrate stores (Vallentine 1989). The heavy brush hogs can cut off and chop up any tree the tractor pulling it can run over. Mowing should occur every 3 years, with one-third of the area mowed each year and only one side of a road mowed in any given year (Voorhees and Cassel 1980). Along road rights-of-way, vegetation should be removed after cutting, and ditch bottoms, back slopes, and interchange areas should be unmowed.

Mowing or burning cheat grass attracts Canada geese (Kadlec and Smith 1989). Cutting and fertilizing grasslands attracts white-fronted geese (Owen 1975).

Crushing

Annual crushing will maintain openings in emergent vegetation, eventually killing cattail and other emergents if timed to the low period of carbohydrate storage (see Fig. 4.24) (Beule 1979, Linde et al.

1976, Linde 1985). Openings from crushing last 4 years if the area is dewatered to accommodate crushing and then flooded later with at least 15 cm of water to cover the crushed stems (Nelson and Dietz 1966, Beule 1979). The crusher consists of a 208-L drum with eight angle-iron cleats welded at equal intervals along the drum. Metal blades 10 cm wide with sharpened edges are bolted to the cleats (Beule 1979). The crusher weighs 227 kg when filled with water (Nelson and Dietz 1966). A 303-L water heater, constructed in the same way, is longer and works better (Beule 1979). In drained areas, 8 to 12 ha of cattail per day can be crushed (Nelson and Dietz 1966). All-terrain vehicles (ATVs) can pull the crusher through water 15 cm deep (Beule 1979).

Disking

To break up stands of sod-forming grass too dense for use by ground-dwelling wildlife, disking can be done without mowing or after mowing to help prevent regrowth of woody vegetation. Disking might be needed every 3 to 4 years (Ohlsson et al. 1982). Plants vary in their response to tillage (see Table 4.10).

Disking cattail is ineffective because of resprouting, and flooding in spring causes disked cattail tops to float, completely covering the water surface (Beule 1979). Cattail and purple loosestrife were controlled in New York by draining the impoundment in May or early June, disking in August with a Rome disk pulled by a D-8 dozer, and reflooding in fall (personal communication, G. F. Mattfield, New York Department of Environment and Conservation 1984). Disking will break up, aerate, and expose drained soils to sunlight, renewing fertility which stimulates plant and animal growth upon reflooding, and retarding filling in of wetlands by breaking down organic deposits (Burger 1973).

Marshes in the southeastern United States with small stands of Asiatic dayflower will produce extensive stands of this valuable duck food through annual disking (Neely 1968). Asiatic dayflower needs partial shade. It does well around wooded edges of marshes and grows in the open marsh in the shade of the taller rushes and other emergents, if not crowded out. Disking it in February and March reduces competition, leaves enough for shade, and incorporates the seed into the soil. Water is retained until ducks have migrated in fall. It is removed to allow disking in February and March and returned to keep a wet-soil condition throughout summer.

After water control is established in a newly developed marsh, disking prepares the marsh for planting by cutting up the thick mat of peat, which is then burned, and the marsh is disked and burned again, perhaps four times in one season (Neely 1968). Woody growth is retarded with 30.5- to 45.7-cm farm disks (Griffith 1948).

Snipe are encouraged in the southeastern United States by early fall disking of freshwater or brackish water marshes allowed to grow native vegetation during spring and summer (Neely 1959). The field is flooded shallowly (puddled) during late fall and winter to produce an undulating pattern of rills and troughs ideal for feeding snipe.

Dozing and Draglining

Before a newly constructed marsh is flooded, or after an impoundment is dewatered and dry enough, a dozer or dragline can be used to scrape the soil in areas of dense aquatics to develop openings in the water when flooded. If it is operated from the shore or dike, or if amphibious, the dragline tolerates wetter soil conditions than the dozer. Bottoms can be contoured at this time to vary the depth as well as the location of aquatics.

Dredging

Dredging generally is done to deepen areas silting in, but can be used, incidentally and directly, to reduce populations of aquatic plants. Dredging in shallow water 1 m or less must be done maybe every 2 years because macrophytes or chara returns the season after treatment (Born et al. 1973). Depth of dredging to control regrowth of macrophytes is described roughly by the equation $y = 0.83 + 1.22x$, where y = maximum depth (in meters) of plant growth and x = average summer water clarity (in meters) with a Secchi disk. The dredging depth should exceed y (Dunst 1980).

Dredging can be mechanical or hydraulic (see Table 2.14). Mechanical dredges can be land-based or water-based for operation. Hydraulic dredges use a pump to lift sediment from the bottom of the impoundment as a slurry transported by boat or pumped through a pipeline to the disposal site (Pierce 1970). Excavation can be done either on the dry bottom if the impoundment can be dewatered or under water with mechanical dredges operating from shore or with mechanical or hydraulic dredges floating on the water. Mechanical dredges cause more turbidity, but incorporate less water in the dredged material than hydraulic dredges do. With watertight buckets, mechanical bucket dredges can reduce turbidity 30 to 70 percent (Barnard 1978).

A turbidity barrier called a *silt curtain* can be placed downstream from, downcurrent from, or around a dredging operation to control the dispersion of near-surface turbid water (Barnard 1978). Silt curtains are impervious floating barriers of flexible, nylon-reinforced polyvinyl chloride (PVC) fabric maintained vertically by flotation segments at the top and a ballast chain along the bottom. Joined 30-m sections are held by anchor lines in a U-shaped or circular configuration, but are ineffective in currents over 1 kn (50 cm/s). Typical operations have curtains 150 to 450 m long in a U shape or semicircle, 300 to 900 m long in a circle or an ellipse.

Considerations which can rule out some methods of dredging as impractical or too costly are (1) access to the pond and shoreline area and characteristics of the shoreline, (2) location and distance to disposal sites, (3) location and area to be dredged in the pond, (4) original water depth and volume of water present, (5) final water depth required, (6) type and volume of material to be removed, (7) inflow to and outflow from the pond, and (8) capability of draining and refilling the pond. Before dredging proceeds, ecological impacts must be considered as well as guidelines for procedures to dredge, transport, and dispose of the dredged material (Morton 1977, Allen and Hardy 1980). The U.S. Army Corps of Engineers has acquired the most expertise (Murphy and Zeigler 1974, Johnson and McGuinness 1975, Environmental Effects Laboratory 1976, Bartos 1977, Hammer and Blackburn 1977, Barnard 1978, Barnard and Hand 1978, Brannon 1978, Coastal Zone Resources Division 1978, Eckert et al. 1978, Haliburton 1978, Hirsch et al. 1978, Hunt et al. 1978, Lunz et al. 1978, Montgomery et al. 1978, Palermo et al. 1978, Soots and Landin 1978, Smith 1978, Spaine et al. 1978, U.S. Army Engineer Waterways Experiment Station 1978, Walsh and Malkasian 1978, Walski and Schroeder 1978, Willoughby 1978, Wright 1978, and others). (See "Dredge Fill Marshes" in Chap. 2.)

Land-based dredging

Land-based dredging is accomplished mechanically from the shore or on the dewatered pond bottom. Mobile or crawler (tracked) cranes can be used. Although slower and requiring a low-bed tractor-trailer for highway transportation, the crawler crane is more common because it can negotiate rougher terrain. The five basic attachments for a crane are the dragline, clamshell, shovel front, backhoe, and pile driver. All are useful in wetland work. All but the pile driver can be used in dredging.

The dragline consists of a crane with a perforated bucket, suspended by cable from a long boom. The bucket is cast from the shoreline into the impoundment and dragged by another cable toward the crane, filling up with sediment en route (see Fig. 2.42). Large draglines can cast the bucket 30 to 38 m, but require wide unobstructed areas for operation and stable, level ground on shore. Trucks might be needed to transport disposed material elsewhere.

The Sauerman bucket is a combination dragline and dozer because the bucket has no bottom and thus cannot be used for loading, although a standard dragline bucket can be used. A truck winch or crane hoist on one side of the pond is connected to an anchor (e.g., deadman) on the other side of the pond, and the bucket is dragged back and forth across the pond (Fig. 4.25, also see Fig. 2.44). Maximum practical reach is about 305 m, permitting the bucket to travel 152 m from both shores. The hoist or track cable and anchor must be moved often to cover the entire pond bottom. Loading disposed material is difficult because of its viscosity.

Water-based (floating) dredging

Typical floating mechanical dredges are the clamshell, dipper, and bucket ladder operating from a barge (Fig. 4.26) (Stakhiv undated).

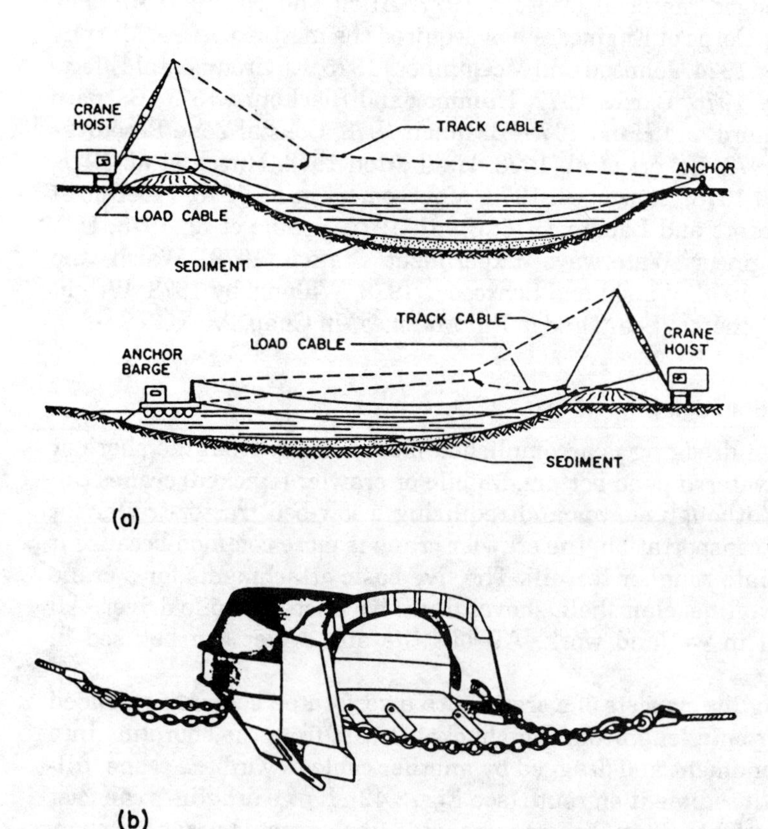

Figure 4.25 Installations for a Sauerman bucket (*a*) and Sauerman lightweight bucket (*b*) (Pierce 1970).

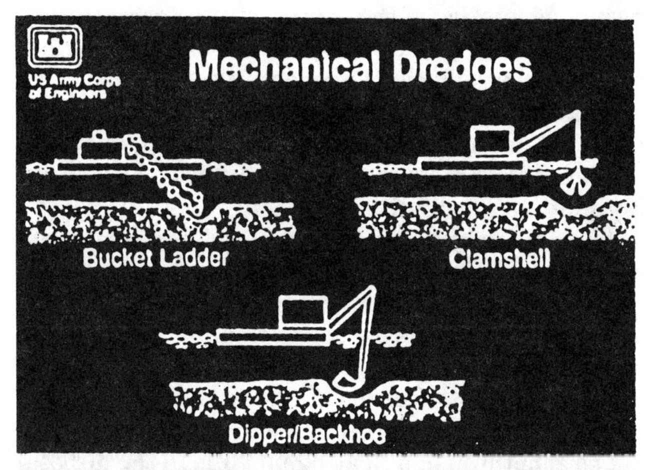

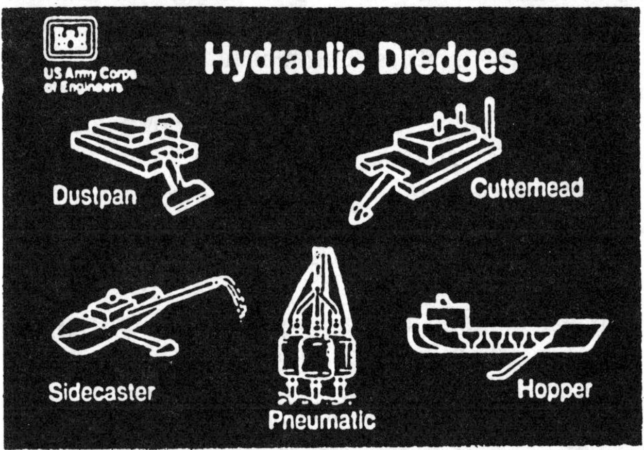

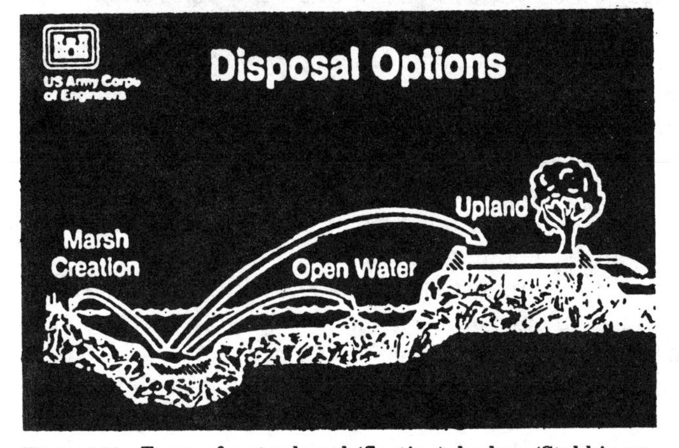

Figure 4.26 Types of water-based (floating) dredges (Stakhiv undated).

The dipper is a floating adaptation of the shovel front attachment to a crane, used mainly in excavating consolidated materials such as clays and rock, for which the backhoe attachment also is used. The clamshell (grab) is a jawed bucket suspended by cable from the boom of a crane, used to dig soft materials or remove stumps, logs, and boulders. The bucket-ladder is an endless chain on tracks, with buckets attached, used mostly in levee construction.

Water-based mechanical dredges cannot transport dredged material far and must use adjacent barges for depositing dredged material, which must then be moved to the disposal area. Thus, most floating mechanical dredges are not adaptable for use in small or medium-size inland impoundments and lakes.

Hydraulic dredges consist of a device to loosen the bottom materials and a pump to suck the loosened material in a slurry of 80 to 90 percent water from the bottom through a floating pipeline to the disposal site or unloaded into open water as a dredge fill marsh or island. Typical hydraulic dredges are the dustpan, cutterhead, sidecaster, hopper, and pneumatic (Fig. 4.26). Unconventional dredging systems might be useful in wildlife work, but are not commonly used or available: mud cat, waterless dredge, delta dredge, bucket wheel dredge, cleanup system (Barnard 1978), or amphibious dredge (e.g., Quality Industries, Inc., Thibodaux, LA). The hydraulic dredge most useful to wildlife work is the cutterhead (Pierce 1970, Cooke et al. 1986).

Hydraulic dredging is unsuitable for shallow water. The draft of the hull and the size of the cutterhead determine the minimum depth of water for operation. A 30-cm dredge needs water up to 1.1 to 1.2 m deep (Pierce 1970). Cutterhead dredges are described by the size of their discharge pipe, which varies from 15 to 91 cm; pipes from 15 to 36 cm are common for inland lakes.

A cutterhead dredge operates by advancing forward in a pivoting swinging motion from vertically mounted 3- to 12-dm pipes, called *spuds,* long enough to penetrate the solid bottom, located at both rear corners of the hull, and raised or lowered by cable or hydraulics. The spuds are alternately lowered, and the hull swings to the limit of its arc. Rate of progress is a function of type of material being dug and depth of cut by the cutter mounted off the bow.

The discharge pipeline connected to the cutterhead consists of a floating pipeline assembled from sections 6.1 to 18.3 m long perhaps leading to a shore pipeline assembled from 4.3- to 18.3-m lengths, which leads to the disposal site (see Fig. 2.43). Depending on length of pipeline, booster pumps might be needed to pump the slurry along.

Rototiller

A bargelike rototiller with a hydraulically operated tillage device can be lowered in water 3 to 4 m deep for tearing out roots (Cooke 1988). In a dewatered impoundment, a tractor or marsh buggy pulling a cultivator or rotary tiller can remove 90 percent of the roots of undesirable plants.

EXPLOSIVES

To create open water in solid stands of emergent vegetation on marshes, explosives can be used. (See "Blasted Potholes" in Chap. 3.)

BOTTOM BLANKETING

Bottom blanketing is accomplished by spreading a 15- to 20-cm layer of sand or gravel on areas of the pond bottom, sometimes over sheets of perforated black plastic, to reduce the growth of macrophytes (Nichols 1974). Work is done during construction of the impoundment, or the sand or gravel is spread on the ice to sink. Bottom blanketing might reduce production of benthic invertebrates and be inadequate for long-term control of macrophytes (Born et al. 1973). Gravel blanketing provides desirable spawning substrate for some species of fish (Payne and Copes 1986).

SHADING

Macrophytes can be shaded out by staking various types of covers over the bottom or on the surface or by applying dyes or increasing turbidity to reduce light penetration and plant growth.

Sediment Covers

Sediment covers have limited application because of their expense and relatively small coverage. Polyethylene plastic sheeting is too light to remain in place long and too difficult to apply and secure to the sediment (Cooke et al. 1986). Hypalon, a synthetic rubber, and polyvinyl chloride (PVC) are effective and easier to install, but too expensive. Aquascreen, a fiberglass screen with polyvinyl chloride, is the best, but too expensive. Dartek, a black nylon, and polypropylene (e.g., Typar) are cheaper and relatively effective, but buoyant. All materials can provide seasonal control of macrophytes if properly installed. The most effective screens are gas-permeable fiberglass mesh screening (62 apertures per square centimeter), polypropylene, nylon, and to a lesser extent burlap, which lasts one season (Perkins et al. 1980, Perkins 1984, Cooke et al. 1986, Cooke 1988).

Sediment covers are placed flush with the sediment with no ballooning or pockets after installation and are staked or securely anchored during winter or spring drawdown before plant growth. Flocculent sediments, rock bottoms, gravel, or hard clay requires cement blocks or link chain sewn into the edges of the fabric. A scuba diver is used in deep water. Stakes can be made by bending one end of a 0.6-cm-diameter steel reinforcing bar into an L-shaped handle and sharpening the other end, to facilitate penetrating a double layer of screen and sediment until the L end is flush with the sediment. The roll of screening can be supported by a reel built into the stern of a boat, unrolled by two people, and staked on each side every 1 to 2 m. A roll of Dartek is first soaked in water 12 to 24 h to make it more flexible. Depending on siltation rates, covers might need cleaning to prevent plant fragments from rooting.

Surface Covers

Surface covers have limited application because of their expense and relatively small coverage. Surface shading with opaque polyethylene sheeting will reduce or eliminate the light needed for photosynthesis in small water areas. Plastic sheeting 8 mil thick and 279 m^2 or less can be suspended on floats, with each corner and edge weighted or anchored against wind, or it can be spread over the bottom (Mayhew and Runkel 1962). Sheeting controls pondweeds and coontail after 18 to 26 days but not chara or emergents after 30 days, and filamentous algae can revegetate up to 30 days after removal of the sheeting.

Dyes

Whereas sheeting treats individual, separate areas, a dye treats the whole pond (Mayhew and Runkel 1962, Dawson and Kern-Hansen 1979). Dyes are ineffective in water less than 1 m deep where light penetration is adequate for plant growth (Lopinot 1986). Water gained or lost will need additional dye. Nontoxic dyes containing inert coloring matter are used. The number of treatments depends upon water inflow and outflow, length of growing season, water fertility, water clarity, etc. Dyes are not widely used.

Turbidity

Turbidity functions as a dye in reducing light penetration. Inorganic turbidity occurs where large populations of carp disturb the water during feeding and spawning. But encouraging such turbidity is impractical, with major ecological consequences. Organic turbidity results from application of fertilizer to the water, causing an increase in phytoplankton which shade submerged plants.

FERTILIZING

In the southeastern United States, inorganic fertilizer applied to ponds will control submerged vegetation by producing a bloom of microscopic (planktonic) algae that shades out submerged plants. In northern states, fertilizing might not be recommended because the decay of the planktonic algae depletes oxygen. This compounds situations of severe winters with heavy ice and snow cover which cause fish to suffocate, resulting in winter kill (Payne and Copes 1986), which might be desirable because fish are undesirable in most wetlands. Potential negative effects from using fertilizers include stunted fish, summer kill, increase in filamentous algae in the North, algae bloom, and upset food chain.

Triple superphosphate should be applied at about 20 kg/ha of surface water before the growing season, with additional applications every 10 days or as needed to maintain the bloom that will obscure a white disk 30.5 to 45.7 cm below the water surface (Lopinot 1986). For general fertilization projects where adequate soil and water analysis data are unavailable, complete fertilizers (N-P-K) should be used of 8-8-2 breakdown at 112 kg/ha of surface water, or 20-20-5 at 45 kg/ha, at 2- to 5-week intervals 8 to 12 times per year or as needed (Payne and Copes 1986).

Fertilizer can be poured into the water in a thin stream from a boat, broadcast by hand from the bank of small ponds, or placed on stationary or floating platforms 15 to 30 cm below the water surface (one platform per 6 ha). On small ponds, a single line of fertilizer poured from the boat on each side of the pond will suffice (Davison et al. 1962). Fertilizing can begin in late winter or early spring, although winter fertilizing works, too. (See "Fertilizing and Mulching" in Chap. 5.)

LIMING

In acid ponds, low populations of plants and invertebrates occur. These populations can be increased by adding lime, i.e., calcium carbonate or calcium oxide every 3 to 6 years (Payne and Copes 1986). A self-feeding rotary drum will hold a week's supply (8.2 to 9.1 mt) of limestone, to release up to 10.9 mt/h (Zurbuch 1984). About 4.5 mt/ha of agricultural lime will increase the pH by 1 point (e.g., from 5.5 to 6.5) (Payne and Copes 1986). In the southeastern United States, 1121 to 4474 kg/ha of agricultural lime produces good results in water with a total hardness of less than 15 ppm, as does 50.4 kg/ha of hydrated lime which is more water-soluble. The effects of acid precipitation can be mitigated in some waters with applications of a lime slurry (Opler et al. 1989).

Methods of application include spreading the limestone by truck on ice; blowing it from a truck, boat, or airplane; or using a calibrated silo activated automatically by pH or streamflow. Limestone can be applied to shorelines, either in accessible locations or in a 30-m band around the entire pond at the rate of 5 to 10 mt/ha (Cooke et al. 1986). Fraser and Britt (1982) described methods of liming acidified waters and the effects on aquatic ecosystems.

Limestone also can be used to reduce landslides on slopes of levees constructed of highly plastic clays (Townsend 1979) and to reduce staining from tannic acid from decaying woody vegetation (Burger 1973). (See "Liming" in Chap. 5.)

DILUTION AND INACTIVATION

To reduce algal bloom and growth of macrophytes, nutrients can be reduced by dilution or inactivation (Nichols 1974). Nutrient-rich water can be diluted by releasing it and flushing with nutrient-poor water (Born et al. 1973, Cooke et al. 1986). A flushing rate of 10 to 15 percent of the impoundment volume per day is considered adequate. Flushing will increase the pH of acidic water (Neely 1962, 1968) (see "Liming" in Chap. 5). To reduce salt and nitrate concentrations in coastal areas, maximum tidal flushing and circulation occurs from February to April, after which water levels should be partially stabilized for nesting and brood-rearing habitat (Whitman and Cole 1987). Half the impoundment is flooded, and the rest continues to be flushed by tides until November. (See "Impoundment Potential of Coastal and Salt Marshes" in this chapter.)

Flushing can work for freshwater marshes, too, but desirable nutrients can be flushed out by spring floodwaters and the water supply often lacks the carbonates needed to overcome acidity. Anaerobic conditions in winter cause solution of iron and manganese, which are toxic to growth of some plants, and dissolved phosphorus. Minimal outflow until the water is reoxygenated will immobilize the iron, manganese, and phosphorus to be retained in the bottom muds (Kadlec and Wentz 1974). Water should be withheld from the marsh until the dissolved iron is less than 20 ppm (Cook 1964).

Eutrophication results from too much phosphorus in the water, which causes nuisance accumulations of planktonic and filamentous algae and other plants (Peterson et al. 1973, Dunst 1980). Alum (aluminum sulfate) will precipitate phosphorus if applied in a concentration of 200 mg/L of pond water, and plant growth will decrease. Alum slurry produces good results when injected through a manifold at about 3 dm below the water surface, behind a propeller-driven boat. Liquid alum also can be used (Cooke et al. 1986). Equipment consists basically of slurry tanks, a freshwater supply for filling the tanks, a mixer, an application pump, and a distribution manifold. Alum can be added to small ponds from shore with a pump and hose or in blocks of ferric alum suspended in cloth bags at middepth from anchored floats in the pond, dissolved, and replaced as needed (Cooke et al. 1986).

Turbid water results from suspended particles, reducing light penetration and impeding plant growth. Clay soils have the smallest soil particles, which help seal the impoundment and dike from leakage but also can remain in suspension, causing turbidity. Suspended clay (colloidal) particles can be precipitated with carbonic acid, created by decaying submerged vegetation, thus clearing the cloudy water (Green et al. 1964). If one-half to two-thirds of the bed of a turbid impoundment is exposed, enough moist-soil vegetation usually can be produced to supply the amount of carbon dioxide needed to precipitate the suspended clay soil particles when reflooded. Turbidity caused by suspended clay can be precipitated out by applying gypsum evenly at 195 kg/1000 m^3 of water (Atlantic Waterfowl Conference 1972).

HERBICIDE

Herbicides are used as a last resort due to general lack of specificity and potential environmental problems. Herbicides can have lethal and sublethal effects on plants and animals through killing and through modifying the ecosystem (Engel 1990). They usually are used to accomplish the following objectives in marsh management (Rollings and Warden 1964):

1. Create open-water areas in dense, emergent vegetation for general use by all waterfowl and nesting by coots and diving ducks
2. Create open loafing areas in shoreline vegetation and on nesting islands and inaccessible sandbars

3. Destroy emergent vegetation used as predator travel lanes to reach nesting islands
4. Reduce dense nesting cover on dikes to reduce duck nesting there and consequent nest losses to predators using the dikes as travel lanes
5. Facilitate maintenance of dikes, ditches, canals, and water control structures by destroying woody vegetation
6. Control algae to reduce potential algae poisoning and to improve light penetration and growth of food plants which also provide cover for invertebrates

Plant control with herbicides is relatively simple and inexpensive, but can have undesirable effects such as toxicity to nontarget species of plants and animals as well as persistence in the environment. Organochlorine pesticides persist longer in the environment than organophosphorous and carbamate pesticides and are thus more hazardous to wildlife. Little evidence exists that organophosphates and carbamates cause significant population changes to any species of wildlife (Smith 1987). Johnson and Finley (1980) and Mayer and Ellersieck (1986) reported levels of acute toxicity of various chemicals to various freshwater animals.

The Weed Science Society of America (1989) listed 148 major herbicides available in the United States and Canada. The U.S. Environmental Protection Agency (1974 plus updates) listed herbicides approved for use in the United States, but some of these might not be approved by Canada or the individual states and provinces. Schnick et al. (1982) compiled other information about herbicides from Applied Biochemists, Inc. (1979), U.S. Fish and Wildlife Service (1979), and the Weed Science Society of America (1979). Steenis et al. (1968) and Hansen et al. (1984) described herbicides used in wetland management. The list of herbicides approved for use by federal, state, and provincial governments varies annually or even more often as new herbicides are discovered and approved or as new information indicates approved herbicides are dangerous to public health and are then disapproved. Federal, state, and provincial laws and regulations must be followed (Hansen et al. 1984) as well as the directions on the label of the product. Restricted-use pesticides may be applied in the United States only by applicators certified with the U.S. Environmental Protection Agency. Laws and regulations governing the use of herbicides are listed in U.S. Environmental Protection Agency (1974 plus updates) and Hansen et al. (1984). The U.S. Federal Aid Advisory Review Board must approve use of herbicides to be purchased with, or applied on lands purchased with, U.S. Federal Aid funds (Rutherford and Snyder 1983). Advanced planning is needed because approval takes several months and probably will require an environmental impact assessment or an environmental impact statement.

Types of Herbicides

Herbicides can be either selective for specific plants or types of plants or nonselective. Herbicides used for wildlife management fall into two general categories: foliar herbicides and root-absorbed herbicides (Linde 1985). Foliar herbicides are absorbed through the leaves and must be applied to actively growing plants. Burning or cutting immediately before or after spraying produces poor results. Treatments in late summer produce poor results. To ensure translocation, water-soluble foliar sprays should remain on the plant at least 6 h before a rain.

Root-absorbed herbicides are absorbed from the soil. Mowing and then burning the foliage and ground debris before treatment expedite absorption, as do moderate rains following treatment. But heavy rains after treatment produce undesirable runoff.

Foliar and root-absorbed herbicides can be either translocated or systemic herbicides or contact herbicides (Hansen et al. 1984). Translocated herbicides move to other parts of the plant from the point of application. Systemic herbicides alter normal plant (or animal) biological functions such as growth, respiration, or photosynthesis. The two terms (*translocation* and *systemic*) often are associated. Contact herbicides affect only the part of plant contacted.

Plant growth inhibitors are applied as foliar sprays to prevent additional growth or flowering and seed production. These compounds are used to reduce the vigor of perennial weed grasses while maintaining soil stability, e.g., on shorelines.

Concentrations of aquatic herbicides often are expressed in parts per million (ppm or p/m), for example, 1 L of active ingredient (ai) or acid equivalent (ae) to be mixed with 1 million L of water (Hansen et al. 1984). The herbicide and dosage vary with the species of plant to be controlled. A foliar herbicide might need a surfactant added to it to accelerate the effect (activators), retard settling of solid particles in suspension (deflocculators or dispersants), suppress surface foam and entrained air (antifoaming agents), increase area of contact on a leaf or stem (spreaders), increase retention of sprays on plants (stickers), increase spreading and penetration by reducing surface tension on plants (wetting agents), or stabilize suspension of one liquid in another liquid (emulsifiers).

Herbicides are available in liquid or dry form (Hansen et al. 1984). Liquid forms can be high concentrates of active ingredient (at least 958 g/L) or low concentrates (240 g/L or less) and in either emulsifiable concentrate form or solution form. An emulsifiable concentrate consists of a water-insoluble herbicide dissolved in an organic solvent and an emulsifier, which stabilize suspensions of droplets of one liquid in another liquid that would not mix otherwise. The emulsifier usually is fuel oil or diesel oil. Emulsifiers reduce drift from ground or aerial applications. Solutions are mixtures of herbicide and water of low alkalinity usually.

Dry forms of herbicide can be wettable powder, soluble powder, or granular. Dry forms are expressed as a percentage of the active ingredient. Wettable powders range from 15 to 95 percent active ingredient, but usually at least 50 percent, as with soluble powders. When added to water, wettable powders form a suspension rather than a true solution as soluble powders do. For thorough mixing, wettable powders need continuous agitation in the spray tank.

Granular herbicides are made by applying a liquid form of active ingredient to coarse particles such as clay, ground corncobs, vermiculite, or ground walnut shells which absorb the herbicide. They are applied either directly to the soil or over the plant.

Controlled-release (slow) formulations are relatively recent and, because they gradually dissolve to release the herbicide in small amounts over extended time, provide longer control.

Herbicides can be divided into three categories based on registration restrictions regarding their use near aquatic areas (Schnick et al. 1982): (1) herbicides approved for aquatic use in water supporting food fish, (2) herbicides approved for aquatic use in water without food fish and having other possible restrictions, and (3) herbicides approved for nonaquatic use only. Schnick et al. (1982) listed copper chelate, copper sulfate, 2,4-D, diquat, endothall, glyphosate, and simazine in the first category; amitrole, dalapon, diuron, and picloram in the second category; and atrazine and bromacil in the third category. Plants that can be controlled by various herbicides are listed in DeVaney (1967), Applied Biochemists, Inc. (1979), Schnick et al. (1982), Hansen et al. (1984), Gangstad (1986), Lopinot (1986), Weed Science Society of America (1989), and others. Many states and provinces have publications describing control of vegetation by herbicides (e.g., Burkhalter et al. 1974, Lopinot 1986, Woehler 1987). Cooke (1988) listed some common aquatic weed species and their response to some common herbicides (Table 4.12). Gangstad (1986) described herbicidal, environmental, and health effects of selected aquatic herbicides. Planktonic algae (water bloom), filamentous algae (pond scum), and microscopic algae can be controlled with copper sulfate, copper chelate (Applied Biochemists, Inc. 1979), and other chemicals (Lopinot 1986).

Application

Best results from herbicides accrue from well-maintained, carefully adjusted equipment and well-trained operators who have an interest

TABLE 4.12 Response of Common Aquatic Weeds to Herbicides

	Diquat	Endothal	2,4-D	Glyphosate (Rodeo)	Fluridone (Sonar)
Emergent Species					
Alligatorweed *Alternantherca philoxeroides*	—	—	Yes*	Yes	Yes
Water willow *Dianthera americana*	—	—	Yes	—	—
Mannagrass *Glyceria borealis*	Yes	No	No	—	—
Purple loosestrife§ *Lythrum salicaria*	—	—	Yes	Yes	—
Reed *Phragmites* spp.	—	—	—	Yes	—
Buttercup *Ranunculus* spp.	—	—	Yes	—	—
Arrowhead *Sagittaria* sp.	No	No	Yes	—	Yes
Bulrush *Scirpus* spp.	No	No	Yes	—	Yes
Cattail *Typha* spp.	Yes	No	?	Yes	?
Floating-leaved Species					
Watershield *Brasenia schreberi*	No	No	Yes	—	No
Water hyacinth *Eichhorniae crassipes*	Yes†	—	Yes	—	No
Duckweed *Lemna minor*	Yes	No	No	—	Yes
American lotus *Nelumbo lutea*	No	No	?	No	—
Cowlily *Nuphar* spp.	No	No	Yes	Yes	Yes
Waterlily *Nymphaea* spp.	No	No	?	Yes	Yes
Submerged Species					
Coontail *Ceratophyllum demersum*	Yes	Yes	Yes	—	Yes
Stonewort *Chara* spp.	No‡	No‡	No‡	No‡	—
Elodea *Elodea* spp.	Yes	?	No	—	Yes
Hydrilla *Hydrilla verticillata*	—	—	—	—	Yes
Water milfoil *Myriophyllum spicatum*	Yes	?	Yes	No	Yes
Northern naiad *Najas flexilis*	Yes	?	No	No	Yes
Bushy naiad *Najas guadalupensis*	Yes	?	No	—	Yes
Large-leaf pondweed *Potamogeton amplifolius*	?	Yes	No	—	—

TABLE 4.12 Response of Common Aquatic Weeds to Herbicides (*Continued*)

	Diquat	Endothal	2,4-D	Glyphosate (Rodeo)	Fluridone (Sonar)
Submerged Species (continued)					
Curly-leaf pondweed *Potamogeton crispus*	Yes	Yes	No	—	—
Waterthread *Potamogeton diversifolius*	No	Yes	No	—	—
Floatingleaf pondweed *Potamogeton natans*	Yes	Yes	Yes	—	Yes
Sago pondweed *Potamogeton pectinatus*	Yes	Yes	No	—	Yes
Illinois pondweed *Potamogeton illinoiensis*	—	—	—	—	Yes

*Yes = controlled, No = Not controlled, — = information unavailable, ? = questionable control.

† = Plus chelated copper sulfate.

‡ = Controlled by copper sulfate. §Thompson (1989).

SOURCE: Cooke (1988).

in and aptitude for the work (Hansen et al. 1984). Operators should know weed identification; growth stages; herbicides, surfactants, and application rates; calibration and adjustment to change in conditions; sprayer maintenance; spray timing, map reading, and land descriptions; record keeping; and safety.

Precautions

The following precautions should be followed when herbicides are used for wildlife management (Rollings and Warden 1964, Applied Biochemists, Inc. 1979):

1. Use herbicides only where needed and where benefits outweigh potential hazards.
2. Use the safest herbicide registered, and buy only from reputable dealers.
3. Treat the smallest area possible.
4. Apply when potential hazards to wildlife are lowest.
5. Apply uniformly only at recommended rates, and mix only as much chemical as needed.
6. Use water-based sprays rather than oil solutions or water-oil emulsions.
7. Follow directions on the label.
8. Avoid prolonged breathing of vapors or contact with skin or clothing.

9. Protect herbicides from freezing or excessive heat by storage in the original container in a locked, cool, dry area.
10. Discard—do *not* reuse—empty containers and unused herbicide, according to safe, prescribed methods.
11. Hold treated waters the time specified on the label before releasing.
12. Repair leaky water control structures before applying herbicide.
13. Patrol treated canals and connecting ditches often enough to prevent deliberate or accidental opening of diversion structures.
14. Label bulk tanks, and nurse tanks in the field, with the proper registered label.
15. Use the proper application equipment, clothing, and respirator with trained personnel, certified and licensed, if required.
16. Identify the problem weeds.
17. Choose the proper application equipment.
18. Avoid drift by avoiding windy days and setting sprayer for coarse spray.
19. Treat from shoreline outward to avoid trapping fish and other organisms.
20. Treat 33 to 50 percent of the area at a time in heavily infested smaller ponds, with 1 to 2 weeks between successive treatments.
21. Clean application equipment; wash and change clothes.
22. Post signs in treated areas and notify residents.
23. Use biodegradable, short-lived herbicide, if possible.

The physical texture of soil (sand, silt, clay, organic matter) influences movement of herbicide through it (Hansen et al. 1984). The large soil particles of sand allow ready leaching, which might cause contamination of groundwater, streams, and lakes. Silt is finer and less porous than sand and thus retains herbicide better than sand. Clay has the finest soil particles and is the least porous of the soils, thus retaining herbicide the longest—maybe too long in some cases. Organic matter in soil decreases porosity and leaching. (See "Soils" in Chap. 2.)

Dosage and size of treatment area

Most aquatic herbicide labels give dosage relative to surface area expressed in hectares (acres) or volume expressed in cubic meters (acre-feet). For example, a 1-ha (2.5-acre) pond averaging 1 m (3.3 ft) deep contains 10,000 m^3 (3.25 acre · ft) of water. For most small lakes and ponds and those with uniform bottom slope, average depth is approximated by dividing the maximum depth by 2 (Applied Biochemists, Inc. 1979, Lopinot 1986). For other water areas, soundings are taken with an electronic echo sounder, long pole, or weighted rope marked in decimeters and averaged. Surface area is determined for linear ponds by multiplying the average length by the width in meters (feet) and dividing by 10,000 m^2/ha (43,560 ft^2/acre), for circular ponds by multiplying 3.14 (pi) by the radius in meters (feet) squared and dividing by 10,000 m^2/ha (43,560 ft^2/acre), and for triangular ponds by multiplying width by length in meters (feet), dividing by 2, and dividing by 10,000 m^2/ha (43,560 ft^2/acre).

Shoreline treatment requires determination of the length of the shoreline by using a map measurer on a pond map and multiplying that length by the width of the treatment area. Without a map, the treatment area can be divided into rectangles or triangles to determine the area of each before adding them.

Let's say that the pond is 150 ft by 400 ft by 2.2 ft deep containing 3 acre · ft, that the herbicide label recommends a dosage of 3 ppm active ingredient (ai) herbicide, that 10 gal of water be mixed with each pound of active ingredient, and that the herbicide formulation has 2 lb/gal active ingredient. The amount of spray solution for the 1.5 acre · ft of pond to be treated is 2.72 × area × dosage × mix + (2.72 × area × dosage)/formulation = 2.72 × 1.5 × 3 × 10 + (2.72 × 1.5 × 3)/2 = 122.4 + 6.12 = 128.52 gal. [The number of pounds of active ingredient per acre-foot of water equal to 1 ppm of water is 2.72, that is, 2,720,000 lb of water in 1 acre · ft (Hansen et al. 1984).]

Let's say the nozzle chart recommends a pressure of 20 lb/in^2 for this application and that each of 20 nozzles on the boom releases 2 gal/min of spray, or 20 gal/min total. The time required to release 128.52 gal is 128.52/20 = 6.42 min. If one-half of the pond is to be covered, or (150 × 400)/2 = 30,000 ft^2, then 4673 ft^2/min (which is 30,000/6.42) must be covered to deliver the 128.52 gal of herbicide in 6.42 min. The distance that must be traveled to deliver the 128.52 gal of herbicide in 6.42 min at a rate of 20 gal/min is 4673/20 = 234 ft/min, which is 2.66 mi/h (234/88 ft/min is 1 mi/h).

Copper chelate (Cutrine Plus) with 0.2 ppm of active ingredient applied at the rate of 11.23 L/ha of surface will control planktonic algae; 11.3 L/10,000 m^3 [0.6 gal/(acre · ft)] will control filamentous algae. Only the upper 6 dm of water is treated to control planktonic algae. Macroscopic algae usually need five or more successive treatments of 0.25-ppm copper sulfate, with 1 to 2 days between treatments (Ohlsson et al. 1982). Filamentous algae need two to three treatments of 0.5 ppm copper sulfate, 2 weeks apart. Copper sulfate is sprayed from a boat over the surface (Table 4.13) or onto the shoreline (Table 4.14) (Ohlsson et al. 1982, Lipinot 1986).

TABLE 4.13 Application Rates for Copper Sulfate

Pond volume*		Desired concentration of copper sulfate: 0.25 ppm		0.5 ppm		1.0 ppm†	
		Weight* of copper sulfate to apply					
acre · ft	gal	lb	oz	lb	oz	lb	oz
0.1	30,000	0	1	0	2	0	5
0.2	65,000	0	2	0	4	0	8
0.3	98,000	0	3	0	7	0	13
0.4	130,000	0	4	0	9	1	1
0.5	163,000	0	6	0	11	1	6
0.6	196,000	0	7	0	13	1	10
0.7	228,000	0	8	0	15	1	13
0.8	261,000	0	9	1	1	2	3
0.9	293,000	0	10	1	4	2	6
1.0	326,000	0	11	1	6	2	10
2.0	652,000	1	6	2	12	5	5
3.0	978,000	2	1	4	1	8	1

*1 acre · ft = 43,560 ft^3 = 1233 m^3 (1 ft^3 = 0.0283 m^3); 1 gal = 3.79 L; 1 lb = 454 g.
†Do not apply this rate on ponds with fish.
SOURCE: Ohlsson et al. (1982).

TABLE 4.14 Dosage (kg and lb) of Copper Sulfate for Shoreline Treatment of Filamentous Algae

Width of treatment area, ft	Length of shoreline, ft: 500	1000	2000	3000	4000	5000	5280
50	3	6	12	18	25	32	34
100	6	12	25	38	50	62	66
150	9	18	37	56	74	93	98
200	12	24	49	74	99	124	131
250	15	30	61	93	124	155	164
300	18	36	74	112	149	187	185
Width of treatment area, m	**Length of shoreline, m: 125**	**250**	**500**	**750**	**1000**	**1250**	**1500**
15	1.1	2.2	4.4	6.6	8.8	11.0	13.2
30	2.2	4.4	8.8	13.2	17.6	22.0	26.3
45	3.3	6.6	13.2	20.0	26.3	32.9	39.5
60	4.4	8.8	17.6	26.3	35.1	43.9	52.7
75	5.5	11.0	22.0	32.9	43.9	54.9	65.9
90	6.6	13.2	26.3	39.5	52.7	65.9	79.0

SOURCE: Lopinot (1986).

Hansen et al. (1984) presented calculations and calibrations for liquid and dry applications in ponds, lakes, reservoirs, irrigation and drainage channels and banks, rights-of-way, and other noncrop areas:

1. Sprayer precalibration of operating parts and nozzle tips and screen
2. Sprayer calibration for swath width, tank capacity, pressure, boom capacity, time, speed, distance traveled, and test run
3. Spray distribution pattern
4. Calibration for spot spraying
5. Calibration for pour or drip liquid aquatic herbicides
6. Calibration of equipment for dry herbicides
7. Calibration of granular applicators

Timing

Weeds, i.e., plants too competitive for preferred plants, go through four stages of growth which influence timing of herbicide application: seedling, vegetative, flowering, maturity (Hansen et al. 1984). For herbicide application, plants are divided into two main categories—grasses and broadleaf plants—and into three groups relative to life cycle—annuals, biennials, and perennials (woody and herbaceous). Although aquatic plants can be vascular (referred to as *macrophytes*) or nonvascular (referred to as *algae*) and some are not true perennials, most common aquatic plants behave as perennials. Aquatic plants are categorized as emersed, submersed, and floating-leaved.

The seedling stage of annuals, biennials, and perennials is readily controlled by herbicide because seedlings are small and succulent so less energy is needed for control. The best time to control summer annuals is in spring or summer when they germinate. The best time to control winter annuals is in fall when they germinate. Biennials usually germinate in spring, develop to a rosette stage in fall, overwinter, and develop a seed stalk the following spring. While the seedling stage of biennials is controlled the most easily, the rosette (vegetative) stage can be controlled in fall with herbicide, helped by the effects of winter weather and heavy demand for nutrients the following spring.

The vegetative stage of annuals can be controlled with difficulty; control is mediocre for perennials. Control of the flowering stage is usually impractical for annuals and biennials. A good translocated herbicide is effective during the flowering stage of perennials. Control during maturity (seed set) is ineffective for annuals, biennials, and perennials (Hansen et al. 1984).

Aquatic application should occur early on sunny days when the water temperature is above 15.5°C (Lopinot 1986), or midday for emergents, so the foliage above water is free from dew or rain (Gangstad 1986). Several successive monthly or annual treatments might be needed. Fish kills can result if the entire pond is treated at once, due to oxygen depletion from decaying plants. To reduce fish kills, one-half is treated and the rest is treated 1 to 2 weeks later. Reduction of plants reduces the invertebrate population used as food by fish. Control of one weed can promote growth of another weed immune to the chemical used.

The canopy of woody perennials must be uniformly developed of similarly aged fully formed leaves not so mature that a heavy cuticle limits herbicide absorption (Scifres 1980). To expedite absorption and translocation, the plant system must be physiologically active. Condition of the plant will vary annually with environmental conditions and is therefore more indicative of the plant's response to herbicides than the calendar is.

Original growth of perennials, both herbaceous and woody, is generally easier to control with translocated herbicides than is regrowth (Scifres 1980). Because the spray cannot penetrate dense, tall brush, cutting it and then treating the sprout growth with herbicide are more efficient (Linde 1969).

The optimum stage of plant growth for chemical treatment is influenced by the application method and the chemical formulation (Scifres 1980). Spraying of stumps, or frill cuts, can be done any time of year. Basal spraying of tree trunks is done between full foliage development and the first fall rains. Sprays are not applied in climates where growth is renewed in early fall following late rainfall, but spraying can be resumed in late fall and completed throughout winter dormancy.

Herbicides injected into the soil usually are most effective during spring. Herbicides in dry form might be most effective if applied in late winter just before the spring rains, which stimulate plant growth and dissolve the herbicide, moving it into the soil for root uptake with soil water, or if applied in early fall before fall rains stimulate additional growth in such areas.

Aquatic application

In ponds over 20.25 ha, 61- to 91-m-long strips of submersed aquatic plants should be left for food and cover (Lopinot 1986). Emersed plants should be treated in patches, with scattered clumps left intact. About 50 percent of the shoreline should contain clumps of emergent vegetation. Because ponds vary in chemical composition, a specific herbicide will not always produce the same results in every pond.

Hand application of herbicide is effective in ponds of 0.4 ha or smaller (Applied Biochemists, Inc. 1979). Pellets can be sprinkled along shorelines or from a boat by hand or spread with a hand or motorized seeder. Diluted liquid formulations can be applied with a sprinkling can in infested areas or poured directly into the water in the wake of a boat. A backpack hand sprayer is effective in small ponds (Lopinot 1986).

Many low-horsepower siphon and centrifugal pumps with a Y connection on the intake side are suited for using undiluted chemical which mixes with pond water inside the pump (Applied Biochemists, Inc. 1979). These pumps usually involve a spray hose with nozzle and a chemical intake and/or water intake. A simpler arrangement with one intake is used with diluted chemical. These arrangements deliver spray from the boat over the water surface. Orchard and tractor sprayers and truck-mounted sprayers or tank trucks can be used from the shoreline.

For extensive application of liquid herbicide, a boat bailer can be attached to an outboard motor so that the herbicide is drawn into the water in the wake of the motor (Lopinot 1986). Another method is to mount a boom on a boat and fit it with deep hoses through which herbicide is pumped from a tank (Hansen et al. 1984).

For extensive application of granular herbicide, a battery-operated granular fertilizer spreader and hopper can be mounted onto the boat (Applied Biochemists, Inc. 1979). The Gandy spreader contains feed hoppers, delivery hoses, and delivery spouts mounted along a tool bar (Hansen et al. 1984). A small battery-powered electric motor operates in each hopper at a constant speed to maintain constant, uniform output.

Although time-consuming, a good distribution pattern by boat for uniform coverage in a small pond is to apply one-half the herbicide at right angles to the other half (Hansen et al. 1984). On larger areas, a zigzag pattern is most efficient (Fig. 4.27), with a buffer lane between each treated lane. On large areas, both buffer lanes and treated lanes should be 15 to 30 m wide. On smaller areas such as ponds, buffer and treated lanes are narrower to conform to the treatment area. Guideposts or marker buoys are set out to guide the operator.

In some areas commercial applicators can be hired to furnish and apply the necessary herbicide. On large marshes or ponds, aerial application with helicopter or fixed-wing airplane can be used, with appropriate guideposts or marker buoys set out for guidance.

When herbicides such as glyphosate (Rodeo over water, Roundup over land) are used on weeds such as purple loosestrife, spraying is best in late summer, but neighboring plants must be avoided; they are needed to close in the space occupied by the dying loosestrife clump (Thompson 1989). Broadleaf herbicides, e.g., 2,4-D, also control loosestrife and do not harm the monocots usually associated with loosestrife. But 2,4-D must be applied in early growth stages (late May

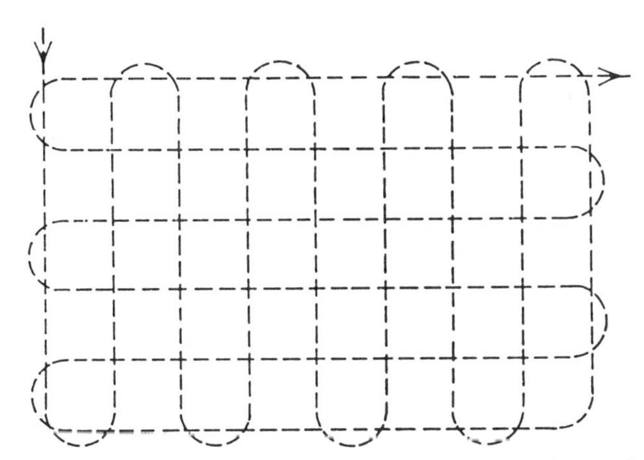

Figure 4.27 Pattern recommended for spraying herbicide from boat in ponds (Hansen et al. 1984).

to early June); plants can be overlooked then due to the absence of flower spikes. Common reed, e.g., is so widespread that glyphosate (Rodeo) should be applied by helicopter twice in late summer at 4.7 L/ha first, then at 2.4 L/ha some 15 to 30 days later, with dead stems then removed by burning (Jones and Lehman 1987, Cross and Fleming 1989). Spraying the same area in two successive years and then spot-treating work best. Missed spots can be located from infrared photographs. Aerial spraying of glyphosate (Rodeo) at 8.5 L/ha kills cattail (Pederson et al. 1989). Dalapon also kills cattail if sprayed in spring at the rate of 3.4 kg of dalapon with 227 g of liquid detergent and 467 L of water per surface hectare, perhaps with additional treatment (Hill 1990).

Terrestrial application

In general, herbicide can be applied in broadcast treatment and individual plant treatment (Scifres 1980). Broadcast treatment is accomplished through aerial application with fixed-wing and rotary-wing (helicopter) aircraft and through ground application. Care must be used to avoid spraying the weed's nearest neighbors, needed to close in the space occupied by the dying weeds, especially with individual plant treatment.

Individual plant treatment is accomplished with the following methods (Hansen et al. 1984):

1. Foliar spraying (also for broadcast treatment).
2. Soil treatment.

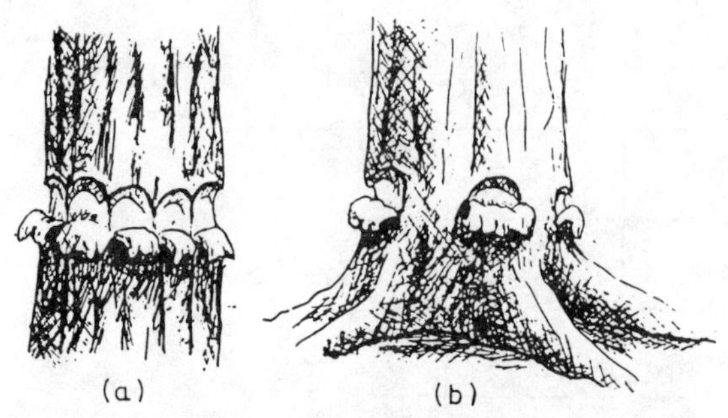

Figure 4.28 Frill-cut method (*a*) and notch or cup method (*b*) for treating trees with herbicide (Hansen et al. 1984).

3. Basal spraying to the lower parts of trunks and crowns.
4. Frill-cut spraying into continuous axe cuts deep into the sapwood around a tree (Fig. 4.28).
5. Notch (cup) spraying into notches at the base of tree trunks made by prying out the chips from two downward axe cuts, one above the other (Fig. 4.28). Trees 7.6 to 15.2 cm dbh should have two notches; larger trees should have notches every 25.4 to 40.6 cm around the circumference.
6. Injection 2.5 to 10.2 cm apart, depending on species and chemical used, into the base of a tree via a pipe with a chisellike bit on the lower end (Fig. 4.29) or at a convenient height of the tree trunk via a hatchet with a built-in calibrated pump (Fig. 4.30). Herbicides should be water-soluble salts often used at full strength.

Figure 4.29 Basal tree injection of herbicide in continuous frills or space cuts at the tree base or at waist height (Hansen et al. 1984).

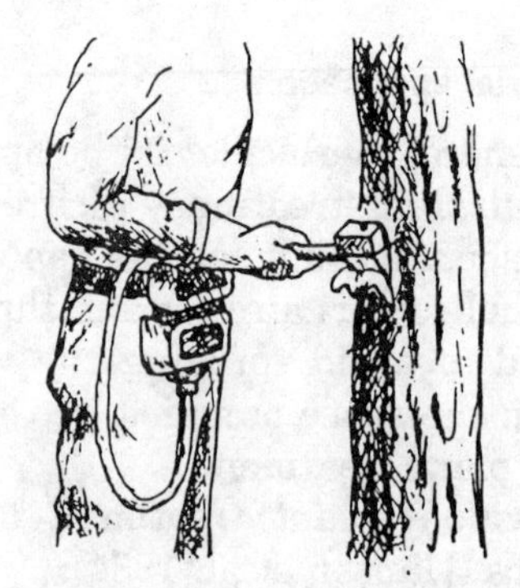

Figure 4.30 Injection of herbicide by hypo-hatchet (Hansen et al. 1984).

Figure 4.31 Stumps treated with liquid, granules, or pelleted herbicide (Hansen et al. 1984).

7. Stump treatment of close-cut stumps and root collars (Fig. 4.31), most effective immediately after cutting.

Vallentine (1983, 1989) outlined the methods available to apply herbicides:

I. Foliage application
  A. Spray
    1 Broadcast
      a Aerial (fixed-wing or helicopter)
      b Ground
        (1) Nondirectional (boom sprayers and mist sprayers)
        (2) Directional
          (a) In row (rowed plants physically protected from spray)
          (b) Strip (chemical seeded preparation for interseeding)
    2 Spot
  B. Wipe-on (rope wicks, rollers, sponge bars as on end of drip torch)
  C. Dust
II. Stem application (individual plant)
  A. Trunk base spray (can be enhanced by use of frills or notches)
  B. Trunk injection
  C. Cut-stump treatment
III. Soil application
  A. Broadcast (spray, pellets)
  B. Grid ball (spaced placement of pellets)

C. Individual plant
   1 Soil injection (liquid)
   2 Soil surface placement (around stem base or spread under canopy)

Herbicides (e.g., tebuthiuron) can be used to control excessive woody or forb vegetation that inhibits grasses and duck nesting (Linde 1969, Kadlec and Smith In press). Herbicides also can be used to control grass where dense grass inhibits forbs. Woehler (1987) listed guidelines for herbicide application in treating woody plants in Wisconsin (Table 4.15). Vallentine (1989) listed methods and herbicides to control woody and herbaceous rangeland plants.

For broadcast spraying, the optimum application rate is determined by formulation, climactic, environmental, plant, and other factors interacting together (Vallentine 1989). Local recommendations are useful. The application rate must be within the range specified on the EPA-approved herbicide label.

Primary factors affecting herbicide persistence in soil are photodecomposition, microbial decomposition, chemical decomposition, adsorption on soil colloids, leaching, volatility, and surface movement or runoff (Scifres 1980). Herbicides persist longer in cool, dry, poorly aerated soils and in those with less clay and organic content. Thus, more herbicide is needed on clays than on sandy soils. Soil texture, herbicide solubility, and the relative affinity of the herbicide for the soil complex influence the rate of leaching, which influences herbicide persistence and groundwater contamination. The influence of runoff on herbicide persistence depends on herbicide rate, the time between herbicide application and the first rainfall, intensity of rainfall, slope of land, soil texture, and vegetative cover. Surface movement of pellet formulation might be somewhat higher than for spray.

On large areas (over 40 ha) aerial application of herbicide is fast and economical. Most aircraft (fixed-wing and helicopter) cover a swath at least 12.2 m wide at 145 to 193 km/h (Scifres 1980). Spray width is about the width of the wingspan plus 10 percent, generally. Herbicide can be applied in dry form such as pellets or as spray, with spray volume usually 2.6 to 7.8 L/ha of water or oil/water emulsion but as low as 0.7 to 1.3 L/ha in some situations (Vallentine 1983). To reduce drift, aerial sprays are applied when wind speed is under 8 km/h. The coarser the droplets sprayed, the less drift that occurs. Markers are installed to guide the pilot for uniform application. Stewart and Gratkowski (1976) discussed equipment used to reduce herbicidal drift with aerial application.

Ground sprays should be applied when the air temperature is below 32°C, wind speed is under 16 km/h, and relative humidty is less than

TABLE 4.15 Guidelines for Herbicide Applications to Treat Nuisance Woody Plants in Wisconsin

Application method, period, and appropriate conditions	Procedure*
Low-volume foliar spray June–September Aerial application to wooded tracts	Apply a diluted concentration (1½ to 3%) to foliage with aircraft, mist blowers, or sprayers. This method is nonselective and affects all plants. A drift control agent is necessary to protect nontarget areas.
High-volume foliar spray June–September Woody plants not forming a closed canopy, areas accessible to equipment	Apply a diluted concentration (1½ to 3%). Equipment varies from small, handheld or backpack sprayers to agricultural crop sprayers.
Foliage roller wiper June–September Dense, uniform shrub growth ≥1.8 m	Apply to carpet-covered wipers that wet foliage on contact. Equipment varies from handheld to tractor-mounted sprayers.
Basal concentrate Any time except during snow cover; drift limited, selective for stems ≤12.7-cm diameter, brush clumps	Apply with a spot gun that delivers 2 to 4 mL of concentration per load. Apply 2 mL/2.5 cm of basal tree diameter. A 6- to 9-dm sterile zone is created around the treated trees for two growing seasons, and nontarget trees might be killed if their roots extend into treated zones.
Basal bark dilute Any time except during snow cover; drift limited, selective for stems ≤12.7-cm diameter, brush clumps	Apply by saturating the lower 38 to 51 cm of treated stems with a diluted mixture. The method results in a slow response (symptoms might not develop for 1 to 2 months) and requires a fuel oil carrier, and sterile zones might remain ≤2 years around the base of treated stems.
Frill girdle April–October Trees ≥12.7-cm diameter	Apply 1 mL of undiluted herbicide into axe cuts spaced every 5 cm, about waist height around the tree. The method is selective but labor-intensive.
Frill with notches Any time except during snow cover and heavy sap flow in spring	Space cuts every 10.2 to 15.2 cm or one cut per 5 cm of stem diameter. Place undiluted herbicide into notches. The method is selective but labor-intensive.
Cut stump April–December Saplings and trees ≥5-cm diameter	Apply herbicide ≤1 week after cutting, avoiding heavy spring sap flow. Use low-volume sprayer or paint stumps with a brush, and add a coloring agent to identify treated stems. The method is selective and will not affect nontarget species.
Tree injector May–September Trees ≥12.7-cm diameter	Apply with a special tree injector that delivers 1 mL per cut, and inject every 5 cm of circumference at base of tree. The method is selective and will not affect ground cover.
Hypo-hatchet May–November Trees ≥12.7-cm diameter	Apply at one cut per 2.5 cm of stem diameter. The method is selective and labor-intensive and requires special hypo-hatchet equipment.
Granular pellets Any time except during snow cover; any tree or shrub	Apply by hand or cyclone spreader near trunks; quantity of herbicide used depends on formulation and tree density. Herbicide will affect ground-layer nontarget species wherever pellets are dropped.

*Proper storage and use of herbicides and an assessment of their risk to the environment are the responsibility of the user.

SOURCE: Woehler (1987).

30 percent (Scifres 1980). Spray volume usually varies from 13 to 104 L/ha depending on need, but 26 L/ha is common (Vallentine 1983).

The simplest type of sprayer is the backpack sprayer, comprising a pressurized 19-L spray tank with a hand pump and an adjustable handheld gun sprayer. Backpack sprayers are designed mainly for spot treatment or small, complete coverage of undesirable vegetation.

Power-operated sprayers are designed with or without booms (Scifres 1980). Swath width for boomless sprayers should be 9.1 m or less, with nozzles high enough for the spray to cover the vegetation canopy completely. Lower spray pressure generally reduces drift. A spray pressure of 1.4 kg/cm^2 produces satisfactory results generally, delivering at least 93.6 L/ha with boomless sprayers, depending on size of nozzle orifice, which is the rate generally needed for adequate foliage coverage. Hansen et al. (1984) described the design and operation of various aquatic and terrestrial weed sprayers.

Power-operated sprayers have a large supply tank, a pressure regulator to control spray pressure and flow, and a pump powered by a gasoline engine or power takeoff (Linde 1969). Power-operated sprayers are of several types.

1. Boom sprayers have horizontally mounted tubes with a series of spray nozzles placed at intervals.
2. Brodjet or T sprayers are versatile, with a single cluster of nozzles which emit a fan-shaped spray on each side of the sprayer head, covering at least 11 m, depending on spray pressure.
3. Gun sprayers have a handheld sprayer gun with a trigger control, allowing easy change of direction, which is useful in dense brush where a constant spray pattern cannot penetrate efficiently.
4. Mist sprayers emit a fine mist spray in a 12.2-m swath to control brush 0.9 to 1.8 m tall most effectively on large areas where drift is relatively unimportant.

To operate at the desired application rate, all power-operated sprayers should be calibrated if they operate across the terrain at a predetermined speed. With the tank filled with water, a test run is made at 3.2 to 8.0 km/h over a specific area, with the spray pressure set usually at 2.5 kg/cm^2 for low-pressure spraying, to determine the liters per hectare used.

Chapter

5

Vegetation Management —Biological

Biological methods to manage vegetation for improving wildlife habitat include (1) controlled grazing, (2) managing species of wildlife that alter the habitat or are attracted to it and provide food for or prey excessively on wildlife targeted for increase, and (3) plant propagation.

CONTROLLED GRAZING

Wetlands

Marsh

The purpose of livestock grazing as an aquatic and semiaquatic habitat modification technique is to open up dense patches of cover to improve composition so that diving ducks and other waterbirds can penetrate it for nesting (Rutherford and Snyder 1983). The effect of grazing on wetland vegetation varies (see Table 4.10). Grazing tends to reduce certain undesirable perennial plants and favors annuals (Chabreck et al. 1989). Trampling and smashing of vegetation and moderate grazing are recommended for marsh edges with solid stands of tall, rank vegetation such as cattail, phragmites, river bulrush, hardstem bulrush, cordgrasses, and willows (Bossenmaier 1964, Kantrud 1986a). Ideally, cattle should be used for 2 to 3 months in late winter and early spring (Rutherford and Snyder 1983). Cattle should be excluded from freshwater and brackish marshes during July, August, and September (Chabreck et al. 1989). Extensive removal of rank growth on shores around large lakes and ponds must be avoided, because ducks will not readily use such windswept, wave-washed, unsheltered areas and erosion accelerates. Cattail and river bulrush are reduced 25 percent by the grazing and trampling of 1.5 to 2 AUM—animal unit month (Pederson et al. 1989). Grazing works best after a burn when cattle eat new shoots. Stands of cordgrass are opened for feeding waterfowl, and woody growth is controlled. (See also p. 295.)

Sheep can be controlled in a close band by a herder more easily than cattle and can eliminate undesirable plants more readily by grazing more closely (Ermacoff 1968). Sheep can be forced into dense stands of herbaceous vegetation to open potholes. They also firm the dike, fill expansion cracks, seal muskrat dens, add fertilizer, and reduce predator cover. Sheep must not trample bare shoreline excessively. Horses control woody vegetation better than cattle or sheep do (Pederson et al. 1989). Intensive sheep grazing, mowing, or both maintain a uniform grass-dominated turf in a young succulent stage attractive to wintering waterfowl, but reduces the diversity of nesting birds (Daiber 1986). To include nesting birds, a habitat mosaic is best produced by combinations of varying intensity and frequency of grazing and trampling, mowing, and even burning.

Riparian

Riparian habitat is probably the single most important, least abundant, and most abused plant community for wildlife (Rutherford and Snyder 1983). In dense shrub-willow riparian areas of high altitude, cattle can improve structural diversity by creating tunnels through the brush (Krueger and Anderson 1985). But mostly, riparian habitat should be fenced and ungrazed. Movements of livestock through riparian areas should be minimized, with stock driveways and trailing areas located away from streamside zones. If stock must be moved through, small groups should be moved slowly, perhaps over stock bridges or revetment at crossings where local problems occur (May and Davis 1982). Where grazing is permitted, the four-pasture rest-rotation system (Table 5.1) seems best in the Southwest (Davis 1982), requiring modification for use elsewhere. But most riparian areas are too small to provide an economical unit within a rest-rotation grazing

TABLE 5.1 Four-Pasture Rest-Rotation Grazing System

	Month grazed (G)			
	Year 1	Year 2	Year 3	Year 4
Pasture	MAMJJASONDJF	MAMJJASONDJF	MAMJJASONDJF	MAMJJASONDJF
1	----GGGGG---	---------GGG	G-----------	GGGGG-------
2	---------GGG	------------	-GGG--------	-----GGG----
3	------------	GGGG--------	----GGGG----	--------GGGG
4	GGGG--------	----GGGG----	--------GGGG	------------

SOURCE: Davis (1982).

system (Melton et al. 1987). Presently, no single grazing strategy for riparian areas functions well under all situations, but some can produce good results under the right conditions (Platts 1989). Table 5.2 was designed for compatibility with fishery needs, and it would apply generally to riparian and other aquatic wildlife. To maintain plant vigor, bank protection, and sediment entrapment, the following guidelines are recommended if riparian areas are grazed (Clary and Webster 1989, Myers 1989, Platts 1989):

1. Grazing in spring should leave 65 percent of the current herbaceous growth and cease by July 15 or earlier at low elevations.
2. Grazing in summer should leave 50 to 60 percent of the current herbaceous growth.
3. Grazing in fall should leave 70 percent of the current herbaceous growth which should be 10 to 15 cm high or even higher on easily eroded stream banks.
4. Small riparian meadows within extensive upland range cannot tolerate grazing.
5. Include a riparian pasture within a grazing allotment or operation to allow the riverline riparian ecosystem to be managed separately from the uplands.
6. Fence streamside corridors to rehabilitate riparian habitat.
7. Change from cattle to sheep on certain areas to improve compatibility with riparian types.
8. Reduce intensity, i.e., stocking rates.
9. Increase distribution of livestock; seasonal dispersal behavior of livestock usually is best from early May through early July and poorest during the hot season between early July and mid-September.
10. Add more rest to the grazing cycle; at high elevations, graze about 28 days/yr and 13 days/yr during the hot season.
11. Remove stock by early August at high elevations (above 1830 m).
12. Where deciduous woody species are important components, limit fall grazing to about 1 year in 4, for about 21 days.
13. Insist upon strict grazing system compliance because a few stray cattle tend to spend much time in stream bottoms.

Uplands

The goal of grassland habitat management for upland-nesting ducks and other birds is to provide the tallest and densest residual vegeta-

TABLE 5.2 Evaluation of Grazing Strategies in Riparian Habitats Along Streams

Strategy	Level to which riparian vegetation is commonly used	Control of animal distribution (allotment)	Stream-bank stability	Brushy species condition	Seasonal plant regrowth	Stream riparian rehabilitative potential	Rating*
Continuous season-long (cattle)	Heavy	Poor	Poor	Poor	Poor	Poor	1
Holding (sheep or cattle)	Heavy	Excellent	Poor	Poor	Fair	Poor	1
Short duation, high intensity (cattle)	Heavy	Excellent	Poor	Poor	Poor	Poor	1
Three herd—four pasture (cattle)	Heavy to moderate	Good	Poor	Poor	Poor	Poor	2
Holistic (cattle or sheep)	Heavy to light	Good	Poor to good	Poor	Good	Poor to excellent	2-9
Deferred (cattle)	Moderate to heavy	Fair	Poor	Poor	Fair	Fair	3
Seasonal suitability (cattle)	Heavy	Good	Poor	Poor	Fair	Fair	3
Deferred-rotation (cattle)	Heavy to moderate	Good	Fair	Fair	Fair	Fair	4
Stuttered deferred-rotation (cattle)	Heavy to moderate	Good	Fair	Fair	Fair	Fair	4
Winter (sheep or cattle)	Moderate to heavy	Fair	Good	Fair	Fair to good	Good	5

Rest-rotation (cattle)	Heavy to moderate	Good	Fair to good	Fair	Fair to good	Fair	5
Double rest-rotation (cattle)	Moderate	Good	Good	Fair	Good	Good	6
Seasonal riparian preference (cattle or sheep)	Moderate to light	Good	Good	Good	Fair	Fair	6
Riparian pasture (cattle or sheep)	As prescribed	Good	Good	Good	Good	Good	8
Corridor fencing (cattle or sheep)	None	Excellent	Good to excellent	Excellent	Good to excellent	Excellent	9
Rest-rotation with seasonal preference (sheep)	Light	Good	Good to excellent	Good to excellent	Good	Excellent	9
Rest or closure (cattle or sheep)	None	Excellent	Excellent	Excellent	Excellent	Excellent	10

*Rating scale based on poorly compatible (1) to highly compatible (10) with fishery needs.
SOURCE: Platts (1989).

tion possible. To accomplish this in some areas, periodic manipulation is needed to disrupt vegetative succession and to restore rigor. Such treatment is spaced as many years apart and in as short a time as possible in the treatment year to minimize disruption to nesting ducks and other wildlife (Kirsch et al. 1978). Mixed grass prairie produces denser and taller residual vegetation if managed by prescribed burning rather than by annual grazing, mowing, or no treatment.

Treatment should occur about every 2 to 4 years for tall-grass prairie on fertile soils where precipitation is 51 to 76 cm/yr, longer in a more xeric mixed grass prairie where precipitation is 30 to 38 cm/yr, and never in xeric desert or semidesert grasslands (Kirsch et al. 1978). In general, the less grazing, the better the height density of residual vegetation is for nesting cover. Small differences in residual vegetation can influence nest site selection by upland-nesting ducks. Patches of shrub cover, such as snowberry, Wood's rose, and sweet fern, will enhance grasslands for nesting ducks.

Grazing generally impacts nesting waterfowl negatively (Molini 1977). Light grazing can improve nesting success by reducing cover for predators, fuel loads that support undesirable grass fires, and brush invasion (Bossenmaier 1964, Rees 1982). Grazing pressure for good management of dabbling ducks and grassland birds in general varies by geographic location, plant cover, past use of the range, soil type, topography, and moisture conditions (Bossenmaier 1964). As a general rule, use by cattle and ducks is compatible when grazing removes only one-half the average amount of the primary forage plants produced annually.

In areas used for waterfowl migration and wintering, grazing is used to maintain pastures in a heavily grazed condition to favor use by geese, widgeon, and coots which eat weeds and closely clipped grass shoots. Cattle grazing in corn stubble break the ears and scatter the kernels, and after a snowstorm they expose the ground by walking and feeding, thus improving feeding opportunities for waterfowl (Jorde et al. 1983, Ringelman et al. 1989). On wintering grounds, grazing increases the availability of seeds of annual food plants (Griffith 1948, Ermacoff 1968, Neely 1968, Sanderson and Bellrose 1969). In waterfowl production areas, light grazing can be used to enhance nesting cover. Light grazing improves dense stands of cattail and other rank emergents for breeding waterfowl by opening them and increasing planktonic algae which are the main food of invertebrates upon which breeding hens and broods feed (Munro 1963, Kantrud 1986a).

In general, vegetation should be left ungrazed in blocks over 32 ha for 3 to 5 years, until height density declines. Then the pasture is grazed,

TABLE 5.3 **Example of 2-Year Rest-Rotation Grazing System to Benefit Waterfowl**

	Year 1			Year 2			Years 3, 4, 5†...		
	Pasture			Pasture			Pasture		
Season	1	2	3	1	2*	3	1	2	3
Spring	Graze	Defer	Graze	Defer	Graze	Defer	Rest	Rest	Rest
Summer	Defer	Graze	Defer	Graze	Defer	Graze	Rest	Rest	Rest
Fall	Graze	Defer	Graze	Defer	Graze	Defer	Rest	Rest	Rest

*In this example, pasture 2 must be large enough to accommodate cattle from pastures 1 and 3 at the same time.
†Length of rest period depends largely on habitat quality.
SOURCE: Rees (1982).

and another pasture is rested in a rest-rotation system (Mazzoni et al. 1983). Treating large blocks reduces predation. Grazing pressure is restricted by fencing and can be more evenly distributed by strategic location of watering areas and salt licks.

Where moisture patterns and growing seasons encourage fall regrowth of vegetation, rest rotation is effective with maximum periods of rest and minimum periods of grazing (Mundinger 1976, Eng et al. 1979, Rees 1982), especially around water sources where cattle tend to congregate (Eng et al. 1979). (See "Stock Ponds" in Chap. 3). On some areas, pastures are grazed moderately in spring and summer, to avoid damage to some grasses then, and moderately to heavily in late summer and fall (Table 5.3), with no burning or grazing afterward for at least 2 years. Cattle are removed early if habitat damage is visible the last few weeks of the season, especially during years of drought or high muskrat populations (Rees 1982).

To eliminate undesirable wetland plants in the southeastern United States, the U.S. Soil Conservation Service recommends controlled grazing on wetlands of at least 20.25 ha, especially where alligatorweed dominates. The wetland should be drained, burned, and then grazed when dry enough. Cattle should be removed in October before they nip the smartweed which is unpalatable to them. Then the area is reflooded 15 to 30 cm deep.

Summer stocking rates are 1 cow or less per 0.8 to 1.6 ha in the northern Midwest and adjacent Canada; 1 cow per 3.2 to 4.9 ha in eastern North Dakota and South Dakota and adjacent Canada; 1 cow per 4.9 to 6.1 ha in central Nebraska, western North Dakota and South Dakota, and adjacent Canada (Bossenmaier 1964); and 1 cow per 2 ha in northwest Iowa (Bennett 1937). Stocking rates of perhaps 1 cow per 2 ha can be used to control tall, rank, wet-meadow vegetation, especially in spring when such plants are most palatable. A stocking rate of 1 cow per 11 ha/yr or 37 cattle-days/ha/yr increases the duck population on stock ponds; overgrazed areas are fenced (Bue et al. 1952). A stocking rate of 37 cattle-days/ha/yr also improves stands of whitetop (Neckles et al. 1985). But even moderate grazing for a few weeks in summer can devastate plants such as alkali bulrush, which is preferred cover for redheads nesting in Utah (Kadlec and Smith In press).

To reduce brush invasion of aspen and willow into tall-grass prairie used by greater prairie chickens and nesting waterfowl, a 4-year sequence is used (Tester and Marshall 1962): first year, spring burning; second year, no treatment; third year, moderate grazing; fourth year, no treatment. In central Wisconsin, brush invasion of aspen, willow, steeplebush, and meadowsweet into grasslands requires control efforts to disturb 15 to 20 percent of managed areas annually through plow, disk, and idle; burning; and mowing and grazing (personal communication, J. Keir, Wisconsin Department of Natural Resources 1989). Farmers are contracted to plow and plant 8 ha of corn for 2 years and grass the third year, leaving 20 percent (1.6 ha) of the corn for wildlife, and to rotate the scheme across a series of 8 ha. The brush in other areas of a 48.6-ha wildlife management area is mowed and the pasture then grazed heavily every 6 years from mid-May when grass is 7.8 to 15.2 cm tall (late May to early June in southern Canada) to late October. Agency personnel burn the areas of 32.4-ha blocks every 5 years, and plow, disk, and leave idle other 8-ha blocks during the growing season (late July in Wisconsin) to convert areas dominated by goldenrod to a more significant grass component.

HABITAT-ALTERING AND OTHER WILDLIFE

Some species of wildlife can be beneficial to waterfowl management in some situations and detrimental in others. Where detrimental, control of such species can improve habitat conditions for others.

Geese

Geese can help control cattail by grubbing and eating sprouts and rhizomes after the stand is drawn down in summer, burned, then flooded just before arrival of the geese (Pederson et al. 1989). Snow geese and Canada geese feed on rhizomes of smooth cordgrass, big cordgrass, Olney threesquare, saltmarsh bulrush, American bulrush, southern bulrush, and seashore saltgrass, often resulting in eatouts that make bulrush seeds more available to ducks (Daiber 1986). Such sites can fill with water used by ducks, but reduce nesting cover. Ponds from small eatouts produce mosquitoes. Ponds from large eatouts breed fewer mosquitoes because the ponds support mosquito-eating minnows.

Beaver

Beaver impoundments generally enhance wetland management, but they can interfere with control of artificial impoundments and warmwater sloughs. In greentree reservoirs, some beaver ponds might need draining because less desirable mast trees such as overcup oak, water hickory, and red ash will gradually replace more desirable but less water-tolerant trees such as pin oak, Nuttall oak, willow oak, and cherrybark oak (Reinecke et al. 1989). In such cases, water levels behind the beaver dams can be controlled with beaver pipes through the dam (see "Beaver Ponds" in Chap. 3) or explosives to blow out the dam (Dickerson 1989), especially in winter when beaver might starve from inability to procure enough food from the browse pile after the ice level drops. A dog can be trained to enter the lodge thus exposed during ice-free periods to force the beaver out the plunge hole to be shot. Other control measures include trapping, snaring, shooting, and use of fumigants or poison bait in the lodge, where allowed. Use of artificial scent mounds to establish an artificial territory shows promise in preventing beaver from inhabiting an area (Welsh and Müller-Schwarze 1989). Best management of beaver involves a fur trapping program. Kierstaed (undated), Arner (1963), Laramie (1963, 1978), U.S. Forest Service (1978), Horstman and Gunson (1983), Miller (1983a), Buech (1985), Teaford (1986), Almeida (1987), and Wiley (1988) described other control methods.

Muskrats and Nutria

Too many muskrats cause extensive eatouts and dike damage from burrows. Too few result in extensive stands of emergent vegetation, especially cattail, that exceed the 1:1 ratio of open water to emergent vegetation desirable for waterfowl. Muskrats use cattail for food and houses. If cattails and bulrushes are not abundant, pieces of root with attached buds can be transplanted into the muck at edges of ponds (Rutherford and Snyder 1983), with cattle and wild ungulates fenced out if necessary. Muskrats will open up dense stands of cattail if an area 4.6 m in diameter is first cleared with a scythe or herbicide and a pyramid is built of three bales of hay or straw placed in water 6 to 9 dm deep for muskrats to use (Hamor et al. 1968). Geese will nest on the old muskrat houses, and ducks use houses and feeding platforms for brooding and loafing. Muskrats and nutria also open up dense stands of reed and southern bulrush, but subsidence and salinity in coastal areas can create permanent openings in such areas (Stutzenbaker and Weller 1989).

Dikes of mineral soil, especially clay, are less susceptible to muskrat, nutria, and even woodchuck damage than dikes of organic soil, especially peat (Linde 1969, Dillard 1982). Dikes of heavy clay are essentially undamaged with high muskrat populations (Cook 1957). Collapsing dens and burrows in dikes need filling with sand, rock, peat, or gravel, perhaps after blasting by placing a 6.8-kg charge of ANFO into a 9-dm hole dug with a posthole digger directly above the burrow (Linde 1969). To prevent muskrat and perhaps nutria burrowing, barriers of Wakefield sheeting or heavy-gauge galvanized chicken wire can be buried upright at least 6 dm deep along problem areas, or the wire can be set flat and fastened down every meter or so. It lasts about 12 years. Animals must be removed first so that they do not tunnel beneath to reach their dens on the other side. Repeated filling of collapsed areas might be cheaper. Winter drawdowns are effective, especially if rock riprap at least 15 cm thick is then placed from about 9 dm below the normal low-water level to at least 3 dm above the high-water level (Linde 1969, Caslick and Decker 1981).

Holes into the burrow can be made with a bar or pipe, carbide pellets inserted through the hole or pipe, and soil used to plug the hole so that the acetylene gas produced will drive out or kill the animals (Linde 1969). Zinc phosphide is the only federally approved toxicant for the control of muskrats and nutria. At 63 percent concentrate for muskrats and 75 percent for nutria, it is applied to 5-cm lengths of fresh carrot immersed in corn oil as an adhesive for the zinc phosphide (Evans 1983, Miller 1983b). The bait is then placed in burrow entrances, on feeding platforms, or on floating platforms spaced 0.4 to 0.8 km apart for nutria, closer for muskrats. Precautions must be used in handling zinc phosphide. Best management of muskrats and nutria involves a fur trapping program. Trapping of muskrats should occur with a density of 2.5 houses per hectare; a density of 6 houses per hectare needs immediate attention to prevent an eatout (Dozier 1953).

Crayfish

Too many crayfish can muddy the water and burrow into the dike, causing leaks. They are controlled with bass or catfish (Dillard 1982) or insecticide placed in the burrow and sealed with sod or soil (Illinois Department of Conservation 1986).

Rough Fish

Control of carp, bullheads, buffalo, and sheepshead often is a combination of using fish barriers, drawdowns, and chemicals (Moyle and

Kuehn 1964). Maintaining habitat for predators such as otter, osprey, kingfisher, great blue heron, cormorant, merganser, etc. is helpful. The most effective strategy is to eradicate the entire fish population chemically and install barriers to prevent reinfestation (Poff 1985). Draining the pond in February and reflooding in March were effective in California (Heitmeyer et al. 1989). Winter drawdowns in northern regions reduce the carp population by creating anoxic conditions and winterkill. Rapid summer drawdowns can be effective in exposing carp eggs to air (Moyle and Kuehn 1964), especially if outlet gates are then shut and small remaining pools are treated with rotenone (Atlantic Waterfowl Council 1972). Rotenone is relatively ineffective against bullheads (Payne and Copes 1986). Reentry of carp can be controlled by mechanical, electrical, or behavioral barriers alone if impoundments are drawn down regularly (Schnick et al. 1982, Poff 1985).

If a drawdown is impractical, emulsified rotenone or a slurry of powdered rotenone and water can be applied throughout, including backwater and pothole areas where carp could escape treatment (Atlantic Waterfowl Council 1972, Payne and Copes 1986). The rotenone can be applied by hand, boat, canoe, truck, and airplane (mainly helicopter). It can be sprayed by power pump and hose or a boat bailer device (Venturi pump) attached to an outboard motor. Two boats and four people per 40 ha are needed (Poff 1985).

Late fall application of rotenone is recommended when the water temperature is 4.4°C (Poff 1985). Partial drawdowns concentrate carp and reduce the water volume treated. Waterfowl nesting apparently is not disturbed by a brief drawdown in early spring for treatment, followed by rapid reflooding (Kadlec and Smith In press). In general, satisfactory concentration is 0.5 ppm for alkalinity of 1 to 35 ppm, obtained by applying 5 percent emulsified rotenone at 0.5 kg/1000 m^3 of water (Atlantic Waterfowl Council 1972, Poff 1985). A toxic concentration of rotenone of 0.75 ppm is needed for 35- to 70-ppm alkalinity, 1.00 ppm for 71- to 100-ppm alkalinity, and 1.00 to 2.00 ppm for more than 100-ppm alkalinity. A complete kill requires uniform application in area and depth and maybe direct application to the lower water levels by pump and submerged hose. Nearby tributaries capable of holding carp should be treated to prevent reentry to the main area. Rotenone persists usually less than 2 weeks except in soft water. Faster detoxification will occur if potassium permanganate is applied at about 1 ppm after treatment with rotenone (Payne and Copes 1986). Most fish toxicants are toxic to aquatic invertebrates, but the effect is temporary (Payne and Copes 1986).

Carp will jump into the impoundment from water on the downstream side of the control gates if that water is near the level of the spillways. Dillon et al. (1971) recommended an overfall of 61- to 91-cm vertical drop from the outlet of mechanical spillways and culverts as a barrier for unwanted fish. Swinging screens, balanced to touch the concrete spillway, allow debris to pass and prevent reentry of carp unless floodwaters from downstream inundate the control structure (Atlantic Waterfowl Council 1972). Fish screens and barrier dams usually are too expensive for use in waterfowl impoundments (Nelson et al. 1978). To be effective against rough fish, barrier dams must have a head differential of at least 1 m (Korschgen 1989).

Predators and Competitors

Habitat improvement of wetlands generally produces an island of habitat highly attractive to and producing unusually high densities of predators such as raccoons, red foxes, striped skunks, opossums, Franklin's ground squirrels, crows, gulls, snapping turtles, and competitors such as blackbirds. Under such circumstances, population management (control) measures of such species might be needed to complement the habitat management program for key species such as waterfowl or to maintain a better balance in biodiversity (Payne et al. 1987). Horstman and Gunson (1983), Timm (1983), Novak et al. (1987), and other sources described various control methods. Dietz (1967) recommended shooting California gulls on inaccessible areas, and placing discarded Christmas trees on accessible areas.

Aquatic Invertebrates

Spring and summer drawdowns increase moist-soil plants which, when flooded in fall, decompose more readily than the more robust emergents such as cattail and bulrush (Godshalk and Wetzel 1978). Fall flooding stimulates hatching of many species of invertebrates feeding on the detritus, especially after disking, with larval forms continuing to develop over winter (Reid 1985). The most practical means of increasing production of invertebrates might be to manage for a diversity of plants with high seed production and finely dissected leaves. For maximum populations of invertebrates, intervals between drawdowns should be 5 years or less (Harris and Marshall 1963, Whitman 1974). Abundance of invertebrates is highest in some greentree reservoirs 6 weeks after reflooding (Hubert and Krull 1973) and in beaver ponds 3 to 5 years old (Reinecke 1977). Mosquito or agricultural pest control measures can affect survival and growth of wetland invertebrates adversely. (See "Open Marsh Water Management" in Chap. 3 and "Water-Level Manipulation" in Chap. 4.)

To expedite colonization of invertebrate prey in newly created isolated impoundments, invertebrates can be introduced by collecting small quantities of submerged vegetation and bottom sediments from nearby natural or well-established similar wetlands without introducing fish or fish spawn (Britton 1982). To encourage diving ducks, mussels, e.g., could be introduced from local stock.

Reducing water levels by 50 percent in late March in impoundments of Great Lakes marshes increases invertebrate biomass substantially, which is exploited by spring-migrating ducks (Bookhout et al. 1989). In seasonally flooded marshes, brood stock ponds of invertebrates (i.e., ponds flooded 1 to 2 months before other ponds) should occur in each habitat so that water and invertebrates from them can be reintroduced into drawn-down ponds to accelerate recolonization (Euliss and Grodhaus 1987). In permanently flooded marshes, midge-eating fish (e.g., threespine stickleback, carp, goldfish) should be discouraged.

Shallow disking of undesirable plants in late summer followed by immediate flooding increases invertebrate populations (Fredrickson and Reid 1988a). To encourage invertebrates, drawdown should not occur in fall during the early stages of leaf litter decomposition exploited by invertebrates. Robust, emergent vegetation cut in winter above ice increases invertebrates after spring thaw.

Creating openings in emergent vegetation accelerates fragmentation of detritus that is attractive to invertebrates. Openings should be of random or uniform distribution rather than clumped, which might cause breeding birds to aggregate too much (Kaminski and Prince 1981). Openings should be at least 0.1-ha circles, to allow diving ducks to take flight, or shaped in sinuous strips, to increase edge and reduce visual encounters between conspecific pairs of ducks. Streams can be improved for production of invertebrates by planting at least 75 percent of their length with woody vegetation, with periodic gaps left unplanted along riffles (McCluskey et al. 1983).

Artificial wetlands, especially in gravel pits, usually are closed systems with no inflow streams to provide fresh energy and nutrients, few trees to provide leaf litter, and a substratum usually of detritus-poor clay, silt, sand, or gravel often with poor conditions for aquatic vascular plants except along shallow margins. Applying straw at the rate of about 10 Mt/ha will provide detritus to support aquatic invertebrates and reduce turbidity by preventing movement of silt (Street 1982). The straw is applied loosely from the windward shore to drift across the impoundment preferably in spring. In large impoundments, barriers of floating poles must be arranged 90° to the wind and each section treated separately to prevent all straw from collecting on the downwind shore. The straw sinks in 10 to 14 days, and decay and colonization by invertebrates occur.

Waterfowl management practices, such as fluctuating water levels, controlled burning, etc., which favor cover-type interspersion and diversity, generally produce diverse invertebrate populations (Schroeder 1973). To reduce siltation, reflooding of impoundments should be slow, with water uncontaminated by pesticides. Extensively grazed, burned, chemically controlled, or flooded areas that reduce plant abundance and diversity also reduce invertebrate abundance and diversity, although invertebrate populations can build rapidly when conditions become suitable.

Other Biota

Biotas with potential for plant control, such as insects, mites, disease organisms, snails, and fishes, need intensive study and rigorous testing to ensure that economic crops are not attacked or that the ecosystem is not otherwise adversely affected (Bennett 1974, Cooke et al. 1986). All have limited application. The water hyacinth weevil has been used to reduce populations of water hyacinth, as has the alligatorweed flea beetle on alligatorweed (Cooke et al. 1986, Cooke 1988). The manatee also has been tried with limited success due to its scarcity, low reproductive rate, high cost of management, and selective grazing.

The reproductively sterile triploid grass carp is authorized for introduction in some areas to control aquatic plants, although permits are needed (Lopinot 1986). Wiley et al. (1987) described stocking rates. But grass carp should not be introduced into waterfowl habitat because they eat certain waterfowl food plants, can increase turbidity, and can change hypereutrophic lakes into phytoplankton-based lakes (Johnson and Montalbano 1989).

PLANT PROPAGATION

Selection

Plantings are expensive and are avoided if natural germination and succession are possible. Generally, only lands with fewer than 15 percent desirable perennial species are considered for seeding (Payne and Copes 1986). Plantings on newly created wetlands are considered if undesirable plants are likely to invade. Former marshlands rarely need seeding with aquatic plants because seed occurs in bottom soils (Ermacoff 1968). Exposure to air by low tide or drawdown stimulates germination (Lynch et al. 1947). In old established wetlands, species that do not occur are absent usually because the area does not meet their needs, so planting perennial aquatics usually is hopeless (Burger

1973). Berger (1990) gave information on restoring wetlands and various uplands.

Before planting, an accurate base map is sketched of the upland or wetland site, including depth contours, bottom types (muck, peat, clay, loam, sand, rock), water clarity, and species of plants already present (Erickson 1964). This map helps determine if plantings are needed, the species and quantity of planting stock needed, and where to plant. To prevent planting in sites with viable seeds, stems, roots, or tubers, the site is visited several times during the growing season to see which plants occur where. Undesirable plants might need eradication before new stock is planted.

Information on wetland plants can be obtained from the Plant Information Network (PIN), which is a computerized service that stores, organizes, and rapidly retrieves information on native and exotic vascular plants of several western states in the United States. Access is obtained from the U.S. Fish and Wildlife Service Western Energy and Land Use Team in Fort Collins, Colorado. References on wetlands are described in Federal Interagency Committee for Wetland Delineation (1989) and Schneller-McDonald et al. (1990).

Selecting a species or mixture of species to plant requires consideration of several factors (Schnick et al. 1982):

1. *Topography, shape, size, and location of proposed site:* Topography influences water depth in wetlands and flooding potential of uplands. Large areas offer more wildlife habitat. Wildlife use, human disturbance, and predation are influenced by location. Shape affects amount of edge habitat and influence of currents and waves.

2. *Soil characteristics:* Survival, growth, and reproduction of plants depend on soil texture, moisture-holding capacity, fertility, pH, salinity, and contaminants (U.S. Army Engineer Waterways Experiment Station 1978). These are checked to determine appropriate plant species that are supportable.

3. *Water quality:* Wetland plants have tolerance limits for physical and chemical characteristics of water: fertility, pH, salinity, turbidity, contaminants, and mineral concentrations. Kadlec and Wentz (1974) compiled the literature on physical and chemical conditions tolerated by various species of wetland plants (Table A.8) in various regions (Fig. A.1), including tolerance for turbidity and pollution (Tables A.9 and A.10).

4. *Water depth and fluctuation:* Wetland plants (emergents, submergents, and floating-leaved) vary in tolerance of water depth and fluctuation (Table A.8), competitive ability, and seedling estab-

lishment due to several factors including light penetration (Kadlec and Wentz 1974).

5. *Current, waves, ice:* Strong currents and waves can tear out existing plants or remove fine sediment, leaving unstable coarse material unsuitable for anchoring plants (Kadlec and Wentz 1974, U.S. Army Engineer Waterways Experiment Station 1978). Ice can tear out plants, especially emergents. Sites open to fetches over 1 km are not seeded; those with fetches over 4 to 5 km are not planted with vegetative propagules (Woodhouse 1979).

6. *Climate and microclimate:* Precipitation, wind patterns, and temperature including frost-free days control distribution of vegetation. Topography, slope, and other factors influencing microclimate control local distribution patterns. Plantings are restricted to species with universal range, such as agricultural crops, or varieties obtained from local sources (Hunt et al. 1978).

7. *Wildlife value:* The value of the plants to the target wildlife species or wildlife in general is of prime importance (Tables A.11 to A.14). In addition to serving wildlife directly as food and cover, marsh plants serve indirectly as food and cover for invertebrates so important to laying ducks, broods, shorebirds, and other wildlife.

8. *Plant location:* Native, adapted species, or mixtures of species when possible, give best results (Coastal Zone Resources Division 1978). Moreover, native species are less likely to spread out of control, a major consideration. To maximize adaptability and reduce genetic contamination, plantings are made from the nearest available source (York 1985). The donor marsh can benefit if transplants are removed so as to create desirable openings. Excessive removal can damage the donor marsh.

9. *Plant species' characteristics:* Plants must be adapted to the soil, water, and climate conditions. Other factors to consider in selecting plants include growth form; growth rate (rapid establishment); life span; reproductive form and flexibility; hardiness; resistance to insects, disease, or other biological agents; competitive ability including inhibition of other species; and need for maintenance, management, or control (Hunt et al. 1978).

10. *Availability:* Second-choice plants must be used if first-choice plants are not obtainable due to time, money, or workforce considerations, especially if they are unavailable or too costly commercially, must be harvested by hand labor, cannot be stored readily, or the seeds cannot be used as propagules (U.S. Army Engineer Waterways Experiment Station 1978).

Stock and Collection

Plants for wildlife can be classified as annuals, perennial grasses and forbs, and woody perennials (Burger 1973). All three groups can furnish both food and cover, but wildlife uses annuals mainly for food and perennial grasses and forbs for nesting, summer cover, and food. Most woody plants are used as thermal and hiding cover, except for some parts of some species that are heavily browsed or used as mast.

Plants use various vegetative structures for reproduction (Kadlec and Wentz 1974). These consist of aboveground structures, underground structures, and seeds. Aboveground structures comprise whole plants and fragments, hibernacula (mostly winter buds), runners (extension of stem above ground), and stolons (extension of stem in ground surface). Underground structures comprise rootstocks, rhizomes (extension of root), tubers (starch-swollen structures of roots, runners, and stolons), and winter buds (for some species).

Foods of ducks and other wildlife and the collection, storage, and handling of planting stock of wetland- and upland-associated wildlife plants obtained from natural or commercial sites have been described by More Game Birds in America (1936), Cottam (1939), Martin and Uhler (1939), McAtee (1939), Moyle and Hotchkiss (1945), Addy and MacNamara (1948), Martin et al. (1951), Erickson (1964), Kadlec and Wentz (1974), Hartman and Kester (1975), Hunt et al. (1978), U.S. Army Engineer Waterways Experiment Station (1978), Bellrose (1980), Chapman and Feldhamer (1982), Long et al. (1984), Wenger (1984), Novak et al. (1987), Platts et al. (1987), and others. U.S. Army Engineer Waterways Experiment Station (1978) provided species-specific information for collecting upland and wetland plants (Table A.15). Hunt et al. (1978) listed commercial sources of plant propagules. Transplanting from natural sites might require a permit. Several forms of planting stock (propagules) are available, with advantages and disadvantages (Table 5.4):

1. *Seed:* Seeding is least expensive for planting large areas. Seed is easy to collect, dry, store, transport, and plant.
2. *Rootstock:* The propagule comprises the root system and small portion of stem usually below ground, perhaps divided into sections or chunks for planting.
3. *Rhizome:* The rhizome is dug and sectioned so that each section contains one or more viable growth points (meristematic tissue).
4. *Tuber:* Tubers are dug near the end of the growing season.
5. *Cutting:* Cuttings are top sections with nodes. Top shoots can be cut from herbaceous marsh plants and planted shallowly.

TABLE 5.4 Advantages and Disadvantages of Propagules

Basis of comparison	Advantages	Disadvantages
Source		
Nursery	Uniform quality Little or no disturbance of natural stands Reduces labor effort	Might increase costs in certain situations Requires planning and ordering in advance Might not be adapted to local condition
Natural stand	Decreases costs Adapted to local conditions	Disrupts natural stands Can be difficult to locate sufficient supply Increases labor effort
Type*		
Seeds	Reduce labor and costs Suitable for large sites Can be stored for several months	Unsuitable for fine-texture materials Wide range of viability, reliability, and success Storage requirements known for relatively few species Restrict harvest time Restrict planting time Cultivation generally required Require advance planning to harvest and store supply
Tubers	Can be harvested mechanically if small Planting effort smaller (can be broadcast)	Large tubers difficult to extract from soil Susceptible to washout
Rootstocks, rhizomes	Maximize use of plant materials	Susceptible to washout Cultivation generally required Limited success
Cuttings	Reduce labor and costs Rapid collection Maximize use of plant materials	Susceptible to washout Must be planted promptly or potted and stored Some disruption of natural stand Lower survival than rooted propagules
Sprigs	Less costly than transplants	Must be planted promptly or potted and stored
Seedlings	Can be planted over longer periods Can be stored in greenhouse or nursery Permit flexibility in coordinating project engineering and planting	Require planning and preparing in advance Increase costs
Transplants	Rapid establishment Increase probability of success, especially on salt marsh sites Stabilize soil rapidly	Highest costs and labor requirements Might be difficult to dig, transport, and plant

*Costs and ability to withstand physical stresses increase from seeds to transplants.
SOURCE: U.S. Army Engineer Waterways Experiment Station (1978).

6. *Seedling:* Seedlings are young plants with weakly developed root systems, obtained from natural stands or by germinating seeds in a greenhouse.
7. *Transplant (sprig):* Transplants are the most successful type of propagule. Transplants are entire plants dug from a natural site and transplanted with some attached soil to another site, preferably when dormant. (Sprigs are small, young transplants.) A convenient size is a root clump 10 to 15 cm or less in diameter with top shoots of compatible size.
8. *Winter bud (turion):* Winter buds are dormant during winter, producing shoots in spring. Winter buds can be above or below ground, depending on species.

Seed

Aquatic and marsh plants vary in seed production, germination, and storage (Table A.16). Seeds of native plants unavailable commercially must be collected (Larson 1980), which requires knowledge of time of ripeness and proper handling, storing, and treatment before planting (Mirov and Kraebel 1939, U.S. Forest Service 1974a). Seed generally is harvested between ripening and shattering (Stanton 1957). Viability can be determined by cracking with a hammer, cutting with a knife, or biting to feel the kernel inside or see the mature embryo (Coastal Zone Resources Division 1978). Seed can be gathered by hand or with simple equipment such as a stick by beating ripened seed heads until they fall on a groundsheet or into a canoe. In shallow marshes seeds can be shaken into a cradlelike canvas bag worn over the shoulders (Martin and Uhler 1939) or collected with a hand-held, motorized Grin Reaper, a boxlike attachment for string line trimmers (Mahler and Walther 1990). A rakelike device can be used to harvest the seeds of bulrush (Sypulski 1943). A rake or net is used to harvest fruiting structures containing seed of wildcelery (Moyle and Hotchkiss 1945).

Seed heads of plants such as bulrush, sedge, and various upland species have firmly attached seeds which must be picked by hand or cut off with a pocket knife, hedge trimmer, or scythe and left to sun-dry so that the chaff can be removed by flailing and winnowing. Ground or ice conditions might permit use of tractor-drawn mowers, modified combines, or the same equipment used to collect seed from cultivated grasses (Erickson 1964, Woodhouse and Knutson 1982). A commercial harvester with tracks (Rogalsky et al. 1971), a similar harvester on a small boat (Dore 1969), or innovative marsh equipment (Willoughby

1978) can harvest wildrice seed and probably seed from millet, smartweed, bulrush, cordgrass, and most grasses and sedges (Kadlec and Wentz 1974).

Seeds of submerged plants are collected along margins of ponds, lakes, and rivers, if undesirable seed can be recognized and discarded by sifting through hardware cloth of suitable mesh. Seeds should be planted immediately or dried before sifting. Ease of collection compensates for possible low germination rate from prolonged wetting and drying. Some species need special handling. Seed from wildrice, e.g., must be kept immersed continuously in cool fresh water until planting (Erickson 1964).

Seeds of most aquatic plants are stored in cool water (Kadlec and Wentz 1974). But seeds from plants such as millet, smartweed, bulrush, and upland species are stored dry and cool (5 to 10°C). Before storage, seeds are tested for viability (U.S. Department of Agriculture 1961, Maguire and Heuterman 1978) and perhaps treated with insecticide and fungicide (Hunt et al. 1978). In general, soft or thin-walled seeds such as wildrice, wildcelery, and pickerelweed are stored wet. Hard-coated seeds such as bulrush, pondweed, and burreed are stored dry (Addy and MacNamara 1948).

To enhance germination, hard, moisture-resistant seed coats can be scarified with a seed scarification machine; by hand with sandpaper; by heating with steam; or by soaking in sulfuric acid for 15 to 60 min, flushed with water, and planted while wet or stored dried (Coastal Zone Resources Division 1978). Scarification can occur for some seeds by heating them with steam or by soaking them for 3 min in sulfuric acid or 0.05 percent solution of sodium hypochlorite or even water in some cases (Kadlec and Wentz 1974). Seeds of many woody perennials must be scarified to break dormancy and induce germination, unless they are planted in fall. Stratification involves mixing 1 part cleaned seeds with 3 to 5 parts sphagnum, sand, peat, or sifted sawdust, keeping the mixture moist by placing it into plastic boxes or bags, and cooling it at 4°C for 30 to 60 days before spring planting.

Vegetative structures

For collecting flexuous plants, drag hooks, hand lines, or modified garden tools such as long-handled rakes and curved pitchforks can be used to dislodge entire plants or parts thereof any time during the growing season (Erickson 1964). These are usually free-floating and nonrooted or sparsely rooted aquatics of the families Ceratophyllaceae, Hydrocharitaceae, and Lemnaceae (Kadlec and Wentz 1974). Free-floating plants such as duckweed can be skimmed or screened from the water with nets or hardware cloth. Large mechanical harvesters (modified barges and dredges with cutting equipment) probably can be used to collect aquatic plants, as can backhoes and perhaps innovative equipment (Willoughby 1978). Many of these plants are perennials which can regenerate from any part with a bud, and they are viable and can be collected throughout the growing season.

Other wetland or upland propagules are collected preferably when the plant is dormant, otherwise at the start or end of the growing season (Erickson 1964, Hunt et al. 1978). Many can be pulled by hand or dug with a spade or sharpshooter. With suitable soil conditions, some plants without taproots can be removed as small clumps or sods with a moldboard plow or homemade device set to shear the soil just below the main root level. The top is cut off to within 10 cm of the rootstock, rhizome, or tuber, and the rootstock or rhizome is divided. Tubers uprooted with a long-handled rake or fork usually float for collection (Fellows 1951). Suckers of adventitious root bud are removed in the dormant season by digging down and cutting the shoots from the parent plant and then transplanting (Yoakum et al. 1980).

Winter buds of species such as wildcelery can be collected from an existing bed by shovel or hydraulic dredge (Korschgen and Green 1988). If winter buds float to the surface, a high-pressure boat-mounted water pump can be used to disturb the substrate enough to dislodge the winter buds for collection by two or more people wading and using dip nets.

Because trees and shrubs are not easily and reliably seeded, transplants are used most often to ensure revegetation (Plummer et al. 1968). Cuttings of upland plants up to 4 years old, mainly trees and shrubs, should be taken from a fast-growing seasoned stem (no new growth evident) just under a leaf node about 15 to 20 cm from the growth point of the stem (Hunt et al. 1978, Swenson 1988). Generally, evergreen cuttings are taken and transplanted in late summer when spring and summer growth is essentially over (Coastal Zone Resources Division 1978) but time allows development of new roots before winter. Softwood cuttings of deciduous plants are taken and transplanted in early summer to use the new spring growth. All but three to five leaves are removed from evergreens or actively growing deciduous cuttings. Hardwood cuttings of deciduous plants are taken and transplanted in late fall after leaves have fallen or are stored over winter.

Cuttings are taken from wood 2 to 4 years old and at least 1 cm in diameter (Platts et al. 1987). Cuttings 1.8 to 2.4 m long are best (Swenson 1988), but shorter cuttings at least 38 cm long will work if at least 20 cm is planted below ground and at least 18 cm is above ground (Platts et al. 1987). Cuttings should be soaked 10 to 14 days after cutting (Swenson 1988), and 50 to 100 cuttings should be bundled and treated with a fungicide (Doran 1957) by dipping the entire stem into a prepared solution or powder, drying, and storing for future plantings (Platts et al. 1987). Cuttings can be stored frozen or just above freezing in plastic bags until 2 to 3 weeks before planting, when frozen cuttings are chilled at 5°C to break dormancy. Cuttings also can be tied in bundles and buried outside upside down in sandy soil or in a refrigerated box of moist peat, sand, or sawdust, to callus the butt ends by spring when they are transplanted to a greenhouse or to containers covered with glass or transparent plastic until roots develop. Rooting hormones can be used to induce root development. Indolebutyric acid is best (Doran 1957, Platts et al. 1987).

All propagules must be kept moist. Aquatic forms are kept in cool water. Others are stored in all-purpose potting soil (1 part soil, 1 part sand, 1 part vermiculite, 1 part bark chips) in a well-drained container (Hunt et al. 1978). Tubers are handled as seeds are, except they are moistened. Dormant deciduous cuttings are stored wrapped in moist peat moss or are buried in sand in a dark room at 5 to 10°C. Evergreen or actively growing deciduous cuttings can be heeled in for rooting. Plants are fertilized with all-purpose fertilizer once each month with 30 mL per 3.8-L pot and are watered.

Site Preparation

Vegetation is established by natural plant invasion (volunteer growth), planting, or some combination thereof (Hunt et al. 1978). Invasion decreases as the distance from the source increases and as the size of the site and the ease with which the propagule can be transported by wind, water, or animals decrease. An efficient method to vegetate a new site might be to transplant just enough stock to develop a stand near or on part of the new site so that seed will disperse onto it.

Site factors (soil, slope, shade, and drainage) and climate determine which plants will thrive. The most important elements of successful plantings are site selection, site preparation, time of planting, planting depth, and soil moisture (Yoakum et al. 1980).

Soft bottoms of marshes are suitable for planting some species, but hand methods generally are used (Kadlec and Wentz 1974). Innovative techniques and equipment might prove useful (Green and Rula 1977, Kruczynski et al. 1978, Willoughby 1978). Excessively soft bottoms sometimes can be consolidated if dried through water control.

Excessively hard bottoms, mostly clay or gravel, need additions of finer material to support plants. Sandy bottoms also are excessively firm, especially from waves, and require adapted plants. Mechanical treatment of sandy bottoms helps.

Competing plants are removed chemically or mechanically (Platts et al. 1987). Interplanting shrubs with herbaceous plants in riparian areas works if the understory competition is reduced. Scalping 51 to 76 cm around a transplant reduces competition during establishment. Applying a thin layer of topsoil throughout or in select areas of upland and riparian sites enhances seedling establishment. Interseeding is accomplished with various drills, disks, scalpers, and spray units by preparing and planting the site without complete removal of existing vegetation.

As with upland sites, riparian and wetland sites generally are improved with tillage by conventional agricultural techniques for planting. Tillage generally involves plowing and disking. Plowing works best on deep soils, disking on shallow, rocky soils with litter (Platts et al. 1987). Disking is used where some residual vegetation is to remain. Disks can be adjusted to dig deeply, leave some vegetation, plow it into the soil, or accept seeding devices. A rotovator might substitute for plow and disk (Linde 1969). Sites developed from dredged material might need leveling with a grader before tillage. Before chemical additions (gypsum, lime, fertilizer) or other substrate modifications are made, a complete soil analysis should be done. The U.S. Forest Service (1969), Rutherford and Snyder (1983), and Long et al. (1984) described land treatment measures for uplands.

Salinity

Most salt-marsh plants will not tolerate salinity over 70 ppt, which is twice that of seawater (Kadlec and Wentz 1974). Many plants do better in soils less saline than those available. Plants of regularly flooded low salt marshes, such as Pacific cordgrass, smooth cordgrass, and mangroves, will grow in salinities up to 35 ppt (sea strength) (Woodhouse 1979).

Propagules should be acclimated to the salinity of the new site (U.S. Army Engineer Waterways Experiment Station 1978). For example, if propagules of saltmeadow cordgrass are dug from a donor marsh with 5-ppt salinity but the new marsh has higher salinity, then the propagules are maintained at 5-ppt salinity until 2 weeks before transplanting, when they are moved to a solution of the same salinity as the new marsh. Large differences in salinity between donor and new sites require gradual acclimation of the propagules.

Highly saline soils can be treated with gypsum (calcium sulfate), planted with salt-tolerant plants, or left to leach over time (perhaps 1 year or more) (Coastal Zone Resources Division 1978, Hunt et al. 1978). Gypsum is useful only when high levels of exchangeable sodium (over 15 percent of the cation-exchange capacity) are present with the salinity. But good internal drainage is needed to leach out the sodium sulfate formed, along with other salts present, which requires time.

Liming

If the pH of the soil is less than 5.5, lime can be applied if practical, although it is not always successful (Kadlec and Wentz 1974, U.S. Army Engineer Waterways Experiment Station 1978). When the soil is dried and then flooded, sulfuric acid is formed on coastal cat clay soils of silty clay high in reduced sulfur. The pH can drop from 7 or 8 to 2.5, which is too acidic for plants. To raise the pH to 5.5 to 6.0, nine to twelve flushings are needed (Neely 1962, 1968). Such soil can be covered with other soil. Otherwise enormous quantities of lime are needed to neutralize the acidity. A pH of 4.5 to 5.0 is the lower limit for most duck food plants.

The amount of lime needed to increase the soil pH is influenced by the initial soil pH, cation-exchange capacity of the soil, level of sulfides, and species of plants to be grown. In general, legumes grow best at a soil pH near 7.0, although as low as 5.5 will produce good growth of legumes such as lespedezas. Grasses are reasonably tolerant of soil acidity.

Plant growth reaction depends on the solubility of the lime (Cooke et al. 1986). Liming materials such as calcium oxide and calcium hydroxide are fast-acting; calcium carbonate (agricultural limestone) is only slightly soluble, so its particle size governs the reaction time. Soil pH can be raised in 3 to 6 weeks if particles of calcium carbonate, calcium oxide, or calcium hydroxide small enough to pass a 100-mesh screen are used with enough soil moisture (Kadlec and Wentz 1974). Conventional agricultural limestone takes several months to raise the pH, but the effect lasts longer. In coastal areas, oyster shells as a source of calcium can be ground or crushed and spread. (See "Liming" and "Dilution and Inactivation" in Chap. 4.)

Fertilizing and mulching

If needed, fertilizer is best applied before propagules are set out, or it can be placed beside each propagule after planting (U.S. Army Engineer Waterways Experiment Station 1978), with care not to burn the roots with direct contact especially of granulated fertilizer. Fertilizer should not stimulate herbaceous vegetation which competes with woody transplants. Slow-release fertilizer tablets can be placed in or near the planting hole (Platts et al. 1987). For hand-planted, bare-rooted nursery stock, about 14 g of 8-8-8 (nitrogen-phosphorus-potassium) or 10-10-10 fertilizer is placed well below the root zone or in slits 15 cm from the stem (Coastal Zone Resources Division 1978). For even distribution of nutrients available to the plants, a good general all-purpose fertilizer such as 13-13-13 is applied initially and again at midseason, or a slow-release fertilizer is applied initially. On lowland cropfields seeded and flooded for ducks, 560 kg/ha of 5-10-10 fertilizer is recommended (Burger 1973).

One application of fertilizer generally is effective for 1 to 3 months and is applied during or just before the growing season and in later years for maintenance. It is added in early spring and summer for warm-season species and at the time of seeding in fall and again in midwinter for cool-season species (Carpenter and Williams 1972). Care must be taken to prevent nitrate contamination of surface and groundwater by overfertilizing, especially on porous soil with high leaching potential. For coastal erosion control, establishment of plantings is helped by ammonium sulfate, diammonium phosphate, Osmocote, or mag-amp fertilizer, at a rate of about 112-336 kg of nitrogen and 112 kg of phosphate per hectare applied at low tide through the second year (Knutson and Woodhouse 1983). Garbisch (1986) described the use of Osmocote to fertilize wetland plants (Table 5.5). Tisdal and Nelson (1975) and Vallentine (1989) described types of fertilizer, uses, and rates of application. Fertilizers can be applied over large areas by fixed-wing aircraft and helicopter spreaders (Larson 1980). (See "Fertilizing" in Chap. 4.)

Mulches often are spread over seeded areas to shelter new seedlings, retain moisture, and reduce erosion. The Estes blower-spreader is a blower-impactor attachment for the hopper of a large truck-mounted rotary spreader, which blows lime, fertilizer, shredded bark, wood chips, or seed up to 38 m horizontally or 23 m up a 60° slope. Fertil-

TABLE 5.5 Use of Osmocote* to Fertilize Wetland Plants

Propagule	Time	Rate
Peat-pot, plug, tuber, bulb	Winter, fall	30 g (1 fluid oz) 18-5-11 (12- to 14-month release)
	Spring	30 g (1 fluid oz) 18-6-12 (8- to 9-month release)
	Summer	30 g (1 fluid oz) 19-6-12 (3- to 4-month release)
Sprig, rhizome	Same	Same, except 15 g (½ fluid oz)

*Osmocote is a controlled-release fertilizer that performs well in saline water and in saturated soils. Use under water involves placing Osmocote in burlap bags beneath transplants.
SOURCE: Garbisch (1986).

izer, mulch, and seed also can be applied in one operation with a hydroseeder (Doerr 1986). Ohlsson et al. (1982) listed various mulches and application rates. A wood chip mulch is applied at the rate of 85 m^3/ha. (See "Mulching" in Chap. 2.)

Planting

Although large, dense plantings are more tolerant of damage by herbivores than are small, sparse plantings, planting programs generally benefit from test or pilot studies that determine feasibility and success. Plants grown for waterfowl ideally should provide both food and cover. Food crops grown for waterfowl consist of those that produce seed and those that produce green forage. Some crops provide both. Seed crops are summer annuals, used mostly by ducks. Forage crops are annual and perennial grasses and legumes, used mostly by geese and to some extent ducks (Givens et al. 1964). Generally, annuals must be propagated by seed, perennials by seed or vegetative propagule. Row crops are more expensive to produce than broadcast crops, but propagating through broadcast seeding is less reliable and requires 50 to 55 percent more seed (Crawford and Bjugstad 1967). Generally, sites planted with a variety of species are preferred, except where physical stress is severe, stabilization is critical (e.g., dike slopes), only one species can tolerate the conditions, or quick cover by a vigorous single species is desired (U.S. Army Engineer Waterways Experiment Station 1978).

Planting vegetative propagules includes selection of vigorous stock, careful removal to reduce root damage, placement to avoid sunscald, proper pruning to reduce transpiration, and support to prevent windthrow. Hartman and Kester (1975) and Williams and Hanks (1976) described techniques in detail. Accepted nursery practices should be followed.

Propagule

Seed. Seeding is done by broadcasting, drilling, hydroseeding (Doerr 1986), aerial seeding (Jensen and Platts 1989), and topsoiling (Garbisch 1986, Clewell and Lea 1989). Seed is broadcast by hand with a hand-operated cyclone seeder on foot (about 35 L/0.5 ha, Kierstead undated) or aboard a small rubber-tired or crawler tractor (Yoakum et al. 1980). Wooden cleats 5 by 10 by 76 cm are bolted to the tracks for better traction in wet areas (Linde 1985). Commercial drill seeders such as grassland or Truax seeders are used to drill fine fluffy seeds into firm soil to a depth 4 times the seed diameter (Herricks et al. 1982).

Wet seeding involves soaking seeds which absorb water, sink, and tolerate water-saturated soils and flooding (Linde 1985). Japanese millet, e.g., is soaked 24 to 36 h before planting and then broadcast by hand from a boat over the partially dewatered area, sinking and lodging in the mud as the water is drawn off.

Conventional seed drills can be used to plant and cover seeds in variously spaced rows. A corn planter is used specifically for corn. Red oak acorns collected during a year of good production last 2 to 3 years and can be planted 5 to 10 cm deep at least 1 m apart in rows 3 to 4 m apart, manually without site preparation or with a modified soybean planter following disking or other treatment, especially in greentree reservoirs (Reinecke et al. 1989). The Range Seeding Equipment Committee (1970), Brown (1977), the U.S. Forest Service (1979), Larson (1980), and Long et al. (1984) described equipment used for seeding.

Seeding rates of small-seed species, such as some grasses, are 1.5 to 7.5 kg/ha planted 2 cm or less deep; rates for large-seed species, such as corn or soybeans, are 60 to 90 kg/ha planted 2 to 5 cm deep (Hunt et al. 1978). Seasonal timing varies geographically.

Extensive mudflats, such as along inundation zones of reservoirs drawn down annually, can be seeded with a helicopter, an aquaseeder which uses hydroseeding equipment, or an air cushion vehicle with cyclone seeder (Fowler and Hammer 1976). The aquaseeder sprays a mix of seed, fertilizer, mulch, and water in one step for distances over 30 m. The aquaseeder and air cushion vehicle can seed 36.5 ha/day, the helicopter 405 ha/day.

Aerial seeding is best for areas over 10 ha (George 1963). A commercial crop-dusting fixed-wing plane or helicopter is used at an altitude of 11 m (Linde 1969). It should have a positive, power-driven, seed-metering spreader to accommodate mixtures of various sizes and weights of seeds (U.S. Forest Service 1979, Larson 1980). If a commercial crop-dusting plane is unavailable, a conventional light plane can be used with a piece of stovepipe extended through an opening in the plane. Seed is poured through a funnel into the pipe as the pilot flies 46 m above the ground (Linde 1985). If a crew is unavailable to guide the pilot in overlapping passes, flags are installed at regular intervals in the area to be seeded, as with crop dusting. Strong winds prevent even seed distribution. As drawdown progresses over several days, narrow strips can be seeded. The U.S. Forest Service (1969, 1979) provided additional details about seeding by hand and aircraft.

Topsoiling involves removal of topsoil from a nearby donor site with a suitable seed bank and depositing it on the project site. This can be done with dragline or backhoe and dump truck; with dozer, front-end loader, and dump truck; or with scraper.

Vegetative. Hand labor is used often for vegetative propagules, especially for trees and shrubs over 30 cm tall. A mechanical planter is used commonly when combined with hand labor to feed propagules into a hopper (Woodhouse et al. 1972, Hunt et al. 1978). Marsh planting is possible from boats, floating walkways, small rafts at high tide (U.S. Army Engineer Waterways Experiment Station 1978), or innovative equipment for soft substrates (Willoughby 1978). Cabbage, tomato, and tobacco planters are efficient for extensive planting of sprigs (Knutson and Woodhouse 1982). Power augers can be used to plant seedlings and plugs in difficult soil.

Plant spacing depends on the quality of the substrate, species and life-form of plant, type of propagule, length of growing season, desired speed of plant cover, and budget (U.S. Army Engineer Waterways Experiment Station 1978). Dense plantings are expensive but desirable if the site is very unstable, if it is subject to heavy physical stress or wildlife pressure, or if aesthetics are of immediate concern. In extreme cases, animal depredation should be controlled by trapping or fencing. A good compromise between full cover and high costs for transplants is a plant spacing of 1 m. For rootstocks and rhizomes, 0.2 to 0.5 m is recommended, and 0.5 m for small sprigs (U.S. Army Engineer Waterways Experiment Station 1978). For establishment of ground cover (grasses and forbs) in 2 years, one plant or clump per square meter is planted, and one plant per 0.5 m^2 for cover establishment in 1 year (Soots and Landin 1978).

For good cover in 2 to 3 years, propagules are spaced from their centers as follows (Hunt et al. 1978): vines, 1.0 m; grasses, 0.2 m for clumps and 0.5 m for stolons; forbs, 0.5 m; small shrubs (under 1.8 m tall), 0.7 m; large shrubs (1.8 to 6.0 m tall), 2.0 m; small trees (under 9 m tall), 2.0 m; and large trees (9 to 24 m tall), 7.0 m. Plants are spaced more closely if soil stabilization is needed within 1 year and are farther apart if a delay in complete cover is acceptable. Plantings in spring can be spaced farther apart than those planted later. Spacing of uprooted cuttings of herbaceous plants should be adjusted for 50 percent mortality. Tubers that are sowed as seeds should be spaced 7 to 10 cm apart. If the goal is nesting habitat for colonial waterbirds, large shrubs and small trees are planted on 4-m centers and large trees on 10- to 15-m centers to allow branches and foliage to spread. After planting, propagules must be watered profusely.

Small tubers are sowed as seeds are. Rhizomes and large tubers are planted in rows 3 to 7 cm deep. Winter buds (e.g., wildcelery) can be placed in a reinforced paper envelope with holes punched in or in a nylon, plastic, or cotton polyester mesh bag (0.64-cm holes) filled with a gravel mixture for weight; closed with a twist tie; kept wet; and planted at 1000 per 0.4 ha 5 to 10 cm in the sediment (Korschgen and Green 1988).

To reduce damage to the donor site so that remaining trees or shrubs can fill the gap, every fifth or tenth tree or shrub should be taken for transplanting (Carothers et al. 1989). If a cutting is planted immediately, the end should be dipped in a rooting hormone and pushed into loose soil or augered holes one-half to two-thirds the length without recurving or J-rooting and with the holes backfilled to avoid air pockets. If the cutting was stored so no roots formed, it should be soaked in warm water for 24 h before being dipped into a rooting hormone and then planted. Plastic tree shelters (Tuley tubes) for seedlings prevent rodent and rabbit damage, reduce deer and elk damage, and enhance growth (Potter 1988).

For rootstocks, transplants, seedlings, and rooted cuttings, a hole should be dug twice as deep and wide as the propagule, then backfilled with loose soil and mulch so that the top shoots are exposed above the soil level. The larger the propagule, the greater the need to plant in a biodegradable pot, or in a burlapped ball, or with a ball of soil around the root system to minimize shock. Transplants, including riparian ones, should be placed in moist but unsaturated soils (Platts et al. 1987). Large numbers of seedlings can be planted readily with a tractor-drawn seedling planter that opens furrows for bare-root tree or shrub seedlings and packs soil around them (Larson 1980). An operator sits in the planter to space the seedlings in the furrows at selected intervals, at a rate of 1000 to 1500 seedlings per hour. Saplings and large shrubs are transplanted with a combination tree spade and tree transport trailer.

Wetlands

Impoundments. Controlled marshes, greentree reservoirs, dredge fill marshes, beaver ponds, and shorelines can be seeded with Japanese millet. Wildrice is a preferred food of ducks, but difficult to grow, with high loss of successful crops to blackbirds and humans. Smartweed, too, is a preferred food of ducks and is tolerant of wet soil, flooding, and frosts. It volunteers in many areas, but seeding might be justified where the season is too short for Japanese millet. Japanese millet is perhaps the best duck food to develop on wetlands (Linde 1985). It occurs across North America; it is a heavy seed bearer, wind-resistant, tolerant of acidic conditions (pH 4.6 to 7.4), more tolerant of salty soil (up to 3 ppt) than any other cereal, and tolerant of most wetland conditions including flooding to prevent frost damage (Coastal Zone Resources Division 1978, Linde 1985, Mitchell 1989). It can germinate and grow on moist loam or sandy soils and in saturated or shallow-flooded soils, continuing to grow when flooded if the leaves are above the surface, which allows flooding before maturity for frost prevention. Early flooding can cause undesirable early use though. A large enough area of millet will produce enough food for ducks and blackbirds. Seeds knocked into the water by blackbirds are available to ducks and can produce a second-year crop without additional seeding. Loss to blackbirds is reduced when the entire stand of millet is flooded after maturity and the water is lowered in 15-cm stages as ducks eat the seeds (Linde 1969). Plantings as late as possible without loss to frost will suffer reduced loss of the crop to migrating blackbirds (Linde 1969).

Sites must be moist, but dry enough for planting Japanese millet in midsummer and wet enough for shallow flooding in fall (Mitchell 1989). If soaked overnight and drained 1 to 2 h to ensure sinking, seeds can be sown on flooded areas. Three plantings of Japanese millet at timed intervals provide food over a longer time if flooding also is staggered. Maturity generally takes 6 to 7 weeks after seeding, although northern states need 10 weeks. Seeding rate is 22.4 to 33.6 kg/ha (Teaford 1986, Mitchell 1989). Soil amendments generally are not needed (Mitchell 1989), but because Japanese millet removes most of the nitrogen and phosphorus from the soil, leguminous crops grown with the millet will help, unless fertilizer is applied at the recommended rate of 336 to 448 kg/ha in a 1-2-1 ratio of standard agriculture fertilizer (Coastal Zone Resources Division 1978). Large areas (up to 162 ha) should be planted in strips 15 to 30 m wide, with areas of open water in between for approaching ducks (Hopkins 1962). Dates for regional planting are as follows (Mitchell 1989):

Southeast	Mid-July to mid-August
Northeast	Mid-June
North-central	Mid-June to late June
Northwest	May to June
California	May to July

Seed can be sown by drilling or broadcasting. In arid climates where soils dry quickly, drilling is best, with seeds planted 0.6 cm or less deep. Millet is planted 2.5 to 5.0 cm deep in the Midwest, 1.2 to 2.5 cm deep in the Southeast, and 0.6 cm deep in ricefields. Seeds can be broadcast onto disked dry land, moist exposed mudflats, or flooded ground, by hand, cyclone seeder, crawler tractor, or airplane. Seed broadcast onto a prepared seedbed or onto land that is dry enough to work is covered with up to 2.5 cm of soil by harrow or drag. Seed should remain uncovered if broadcast onto mudflats or partly drained ponds and lakes. Aerial seeding works well on flooded fields, mudflats, or extensive drained areas that might dry out before completion of land-operated seeding (Mitchell 1989).

For beaver ponds, ground preparation usually is not needed. Fertilizer is not needed, at least for the first 2 years. Two recommended varieties of Japanese millet are *chiwapa,* which matures in 80 to 90 days with a yield of about 2074 kg/ha of seed, and *frumentacea,* which is preferred because it matures in 45 to 65 days with a yield of about 2354 kg/ha (Teaford 1986). Planting *frumentacea* will allow late drainage of beaver ponds, which keeps it in brood habitat longer for late-hatching ducks and reduces weeds. Water is kept off the pond until the millet is 3 dm tall to maturity, during which period it grows well in water one-fourth to one-half its height. If army worms (moth family Noctuidae) are a problem, incremental flooding might create a barrier for them and reduce damage to the growing millet. Incremental flooding at this time also provides brood cover for ducks and helps prevent frost damage (Linde 1969). As the millet matures, the PVC cap is placed on the pipe outlet so that beaver can reflood the pond (Kierstead undated). The pond is drained annually for reseeding. Reseeding after draining might not be needed the next year if surviving seed from the previous year's millet germinates well enough. (See "Beaver Ponds" in Chap. 3.)

Browntop millet and proso millet also receive high wildlife use (Mitchell and Tomlinson 1989) but are more susceptible to water and frost damage and intolerant to flooding (Linde 1985). Buckwheat, field corn, grain sorghums, rice, peas, and beans have been used in shallow fields, but all have disadvantages in yield, maturity time, or tolerance to flooding (Burger 1973). The Coastal Zone Resources Division (1978) listed 250 species of plants useful to wildlife in North America, with synopses for 100 of those.

Emergent plants provide cover and food to ducks and other wildlife. Submergent plants are primary food for diving ducks and are important to dabblers, too. Pondweeds are among the best, with sago pondweed preferred.

Emergents and submergents are planted by pulling or digging in spring, balling the roots in mud to keep them moist, wading in the water or boating, and dropping or embedding them by hand, foot, or pole in small clusters weighted in clay balls or with a stone, nail, or other nonlead sinker, perhaps wrapped in cheesecloth. Long (1-m) stems can be laid on the bottom and covered at intervals with mud so that shoots develop on the aboveground parts and roots on the underground parts (Addy and MacNamara 1948). Planting is easier if the water level is lowered. Planting begins in the deeper parts and progresses into the shallows as the water level is restored gradually (Erickson 1964). Vegetative propagules can be planted in groups to create clumps of plants 1.8 to 3.7 m wide which help reduce wind and wave action, serve as nesting materials and sites for certain ducks and other waterbirds, and serve as photography and hunting blinds.

In wetland restoration, cattails and other plants can be established by spreading live topsoil, i.e., topsoil which has been removed and reused immediately without storage that might kill the seed banks, microflora, and microfauna (Buckner 1988). The donor marsh must have suitable seed banks preferably without undesirable vegetation. Live topsoil should be placed in standing water to kill undesirable plants. Removal of some existing soil might be needed to accommodate the live topsoil. The substrate should be at least 3 dm deep, of clean inorganic or organic material of which 80 to 90 percent by weight will pass through a no. 10 sieve (Garbisch 1986). Removal should be done in a checkerboard pattern to reduce disruption of the donor site and minimize attraction to Canada geese, which might create an eatout.

Cattails also can be established by collecting spikes when seeds are ripe and shattering is imminent (November to December) (Buckner 1988). In March the spikes are poked into the ground where desired, or seed is released into the wind upwind of the project area, or the seed is mixed with wet masonary sand and broadcast by hand.

The bottom should be soft, muddy, and the same depth as the original site (Burger 1973). Submergents such as pondweeds, wildcelery, naiads, and widgeongrass are most productive in clear water less than 1.8 m deep. Where the bottom is suitable in water 0.6 to 1.8 m deep, a handful of medium-sized seed, such as sago pondweed, can be broadcast about every third step to seed 1 ha with about 87 L of seed (Erickson 1964), or 44 kg/ha during late summer and midfall (Rutherford and Snyder 1983). Tubers of sago harvested in spring can be planted in early spring to early summer at 2960 per hectare. Tubers of chufa are broadcast at 56 kg/ha between March 1 and June 15 on mudflats of impoundments, beaver ponds, and other shallow-water areas with seasonal flooding, with water kept off for at least 3 months during the growing season (Mitchell and Martin 1986).

Seeds and vegetative propagules should be soaked in water at least 1 day before being set out, to reduce buoyancy and expedite sprouting. In shallow expansive waters, large areas can become choked with introduced species of tall, coarse emergents such as cattail, reed, or the tall bulrushes like hardstem and river bulrushes. Margins of dikes, islands, and other shorelines become less attractive to ducks for loafing as the density of such emergents increases. Only short plants like the spikerushes should be encouraged along shores. Existing stands of undesirable plants must be eliminated or weakened before new plantings are attempted (Erickson 1964).

Brackish water ponds can be managed for widgeongrass if salinity is at least one-third that of ocean or Gulf waters, or about 10 ppt (1 percent) (Neely and Davison 1971). Brackish water with 3 to 10 ppt salt is too fresh for freshwater plants and too salty for widgeongrass, and can be managed for saltmarsh bulrush. Widgeongrass prefers water 6 dm deep, saltmarsh bulrush a few centimeters deep. If planting is needed, several liters of plants raked out of a nearby pond or purchased from an aquatic nursery and scattered over the water will get them started, especially in spring.

Shoreline vegetation is transplanted in clumps for small areas, or seeded to provide food and cover or erosion control. Grading one shoreline of a fish pond to a 5:1 or 6:1 slope will encourage emergent plants at shoreline and permit seeding attractive grasses (Burger 1973). Wild millet, smartweed, sedges, rice cutgrass, and some panic grasses are among the best shoreline plants. Natural stands of the arctic grass pendont can be used as a source for revegetating shallow ponds and slow-moving streams in the arctic (McKendrick 1987). Pendont accelerates spring thawing of ponds, thus providing open water for birds in early spring. Allen and Klimas (1986) provided additional details for revegetating shorelines.

Cropfields. Planting for species of wildlife using wetlands comprises management of aquatic and semiaquatic habitats as well as adjacent uplands. Semiaquatic habitat contains standing shallow water part of the year, with or without water control. It can be a flat, lowland cropfield with heavy soils capable of cultivation and growing a good grain crop, yet able to hold water. A low permanent dike should be constructed around the perimeter, with a control structure at upper and lower ends of the field and a dependable source of water for fall flooding (Burger 1973). Grading might be needed so that the entire field drains readily. Such a lowland field is planted with corn, commercial rice, or other high-seed-yielding crop, such as millet or smartweed, and flooded shallowly after seeds mature or after the commercial harvest (Davison and Neely 1959). Where water levels cannot be controlled, millet stands sometimes can be created by seeding bottoms or shorelines exposed during natural dry periods.

Riparian habitats. Four main types of woody riparian habitats are recognized in the United States and Canada (Swift 1984): northern floodplain, southern floodplain, elm-ash, and mesquite-bosque. Management of most riparian areas mostly involves protection from human disturbance, specifically road construction, recreational development, improper grazing, reservoir development, lumbering (Thomas et al. 1979), urbanization, agriculture, and stream channel modification (Swift 1984). Streamside zones of mature trees should be at least

30 m wide, and much wider if some harvesting occurs (up to 25 percent of the basal area) (Erman et al. 1977, Hunter 1990, Rudolph and Dickson 1990). Howard and Allen (1989) recommended that protected zones at least 60 m wide should be left on each side of streams wider than about 10 m. On smaller streams the 60-m width could be divided between the two sides.

Riparian ecosystems can be enhanced with natural and artificial channel structures, which affect streamflow hydraulics and sedimentation (DeBano and Heede 1987). Natural structures are cienagas, log steps, and beaver dams. Artificial structures include large dams, check dams, and bank protection structures (bank armor, deflection structures, flow separation structures).

Platts et al. (1987) listed characteristics of 72 herbaceous and 49 woody species that can be planted in riparian areas (Tables A.17 to A.21). Baldcypress planted along shorelines in the Southeast benefits wildlife and reduces mosquitoes (Bates et al. 1978). Desirable riparian plants such as blue paloverde, willow, Fremont cottonwood, velvet mesquite that contains mistletoe, quailbush, fourwing saltbush, and desert blite can be planted in unvegetated areas or after removal of the exotic salt cedar in the Southwest (Davis 1977, Anderson et al. 1978, Swenson and Mullins 1985). Trees and shrubs should be planted in configurations that provide horizontal and vertical diversity, by augering holes to the water table for each plant. Highest survival is ensured by temporarily installing an irrigation system until plantings are 6 to 12 months old. The best system comprises a gasoline-diesel pump, PVC pipe, and free-flowing water (Disano et al. 1984).

To develop riparian habitat quickly in arid regions or to protect stream banks, rooted cottonwood cuttings can be transplanted (Miller and Pope 1984). Narrowleaf cottonwood, plains cottonwood, and coyote willow 10 to 20 cm dbh are cut into 1.8- to 2.4-m lengths; or saplings 1.8 to 2.4 m high are cut down, the branches are trimmed, and the saplings are placed into a barrel of water with a rooting hormone. Each tree should be notched with true north for proper orientation when transplanted. Holes 6 to 12 dm deep are drilled close to the water so that the trees can be set directly in the water table. The bark of the portion below ground is scratched through to allow water to penetrate readily. Soil is tamped firmly around the base, and tar is painted on exposed cut areas to retard disease and insect infestation. Sprouting occurs in 2 to 4 weeks with new trees becoming 1.8 to 2.7 m tall the next summer. Willow slips are planted 3 m apart in wet sites, but farther apart in dry sites to reduce competition (Rutherford and Snyder 1983).

In arid regions, water can be diverted to propagules by destroying competing vegetation and shaping the surface into broad V-shaped ditches to collect water (Swenson 1988). The soil is compacted with heavy equipment, sprayed with a sealant emulsion, or covered with an ultraviolet-protected plastic film which collects water and discourages weeds. Then rhizomes or container transplants of grasses, forbs, and shrubs are machine-planted 25 to 46 cm deep in the bottom of the ditch. Natural establishment of trees, shrubs, and vines in river-bottom floodplains is most practical if the soil is removed close to groundwater by pothole blasting, bulldozing, or similar mechanical methods (Snyder 1988). On gravel bars of intermittent streams in California, willow establishes on fine surface sediment of less than 0.2 cm, Fremont cottonwood does better on coarser sediments of 0.2 to 1.0 cm, and mule fat does well on larger sediments (McBride and Strahan 1984). All species sustain high mortality on streams that dry up completely before September 1 or if they are not on areas protected from high winter flows.

If stream banks are bare and cut back because riparian vegetation has been removed, especially where streams are swift along outcurves, 5 to 10-m-high conifers such as subalpine fir can be cut and placed horizontally on the face of the base stream banks, with tips pointed downstream (Burton et al. 1989). The first tree is placed along the bank at the point furthest upstream, with the tree butt overlapping stable bank for a distance of at least 3 m. Successive trees are placed on the bare bank to overlap the adjacent trees by at least 20 percent. A backhoe can be used for placement during late summer and fall when streamflow is low. Trees should be secured to the bank by a cable connected to a deadman driven 0.6 m into the ground. These trees will catch soil sloughing off the bank. After a year, willow cuttings are planted into the sediments accumulated within the branches of the tree revetment.

For riparian species, the water table should be less than 6 dm below the ground and the salinity less than 6 ppt (less than 3 ppt for cottonwood) (Swenson 1988). Species such as Gooding's willow and Fremont cottonwood are transplanted by cutting poles 7.6 to 15.2 cm wide and 12 to 60 dm long, scoring the bottom 30 to 36 cm with a hand axe, soaking in a rooting hormone (e.g., Routone F, 28 g per 133 L of water) until transplanted, planting into the water table before birds appear, and sealing exposed cut surfaces with tree paint (York 1985, Carothers et al. 1989). Nondormant poles also work if all branches and leaves are stripped. Species such as willow, aspen, and cottonwood are planted in clumps 6 dm apart, with 6 to 15 cuttings per clump and 25 to 50 clumps per hectare (Herricks et al. 1982); or individually 1.8 to 3 m apart; or more closely on eroded land, in two to three rows, by pushing the cutting into the soil by hand, if possible (McCluskey et al. 1983), with 3 to 6 cm left aboveground to prevent moisture loss (Allen and Klimas 1986). Native bushy types of willow (e.g., sandbar willow)

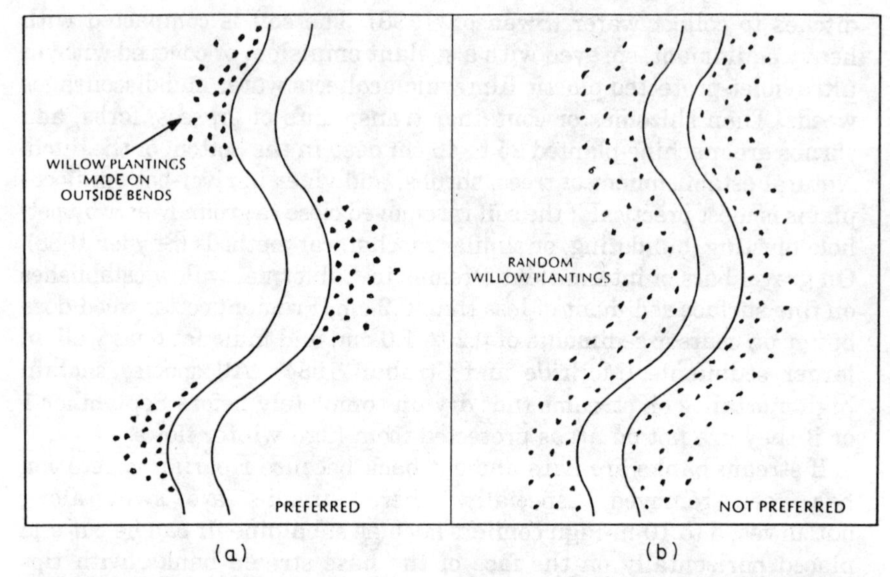

Figure 5.1 Willow cuttings arranged to provide stream bank protection. Selective arrangement in (*a*) is preferred and more cost-effective than random arrangement in (*b*) (Melton et al. 1987).

can be cut about 2 m long and 5 to 8 cm wide and driven in the soil 1.2 m, spaced in uneven rows 0.6 to 1 m apart, with smaller cuttings planted in between (Poston and Schmidt 1981). Melton et al. (1987) recommended taking willow cuttings 30 to 38 cm long, preferably from the previous year's dormant growth; storing them in a moist, cool environment; and planting into a 25-cm slit dug at a 45 to 60° angle, spaced 9 dm apart next to the bank and 1.9 to 2.7 m apart farther back. Willows should be planted only on the outside banks of stream meanders (Fig. 5.1) (Melton et al. 1987). Coniferous seedlings planted in riparian areas usually need some site preparation to reduce competition from sod-forming grasses. Willows are planted between the expected mean and mean high-water levels; alders, poplars, ashes, maples, and elms are planted at and above the mean high-water level (Allen 1978). A 5- by 10-cm welded wire mesh fence 12 dm high around each tree and shrub might be needed as protection from beaver, cattle, and wild ungulates (Carothers et al. 1989). Chicken wire 46 cm high is used against some rodents and rabbits. (See "Greentree Reservoir" in "Timber Sales and Clearing" in Chap. 2.)

For successful planting of Fremont cottonwood on dredge spoil, the water table should be no more than 4.6 m from the surface, with less than 1.2 ppt salt concentration in the soil and ground water (Anderson et al. 1984). Holes should be at least 20 cm wide and dug 3 m deep or to the water table, and preferably leached for 48 consecutive hours before planting the saplings with 85 g of time-release fertilizer. Sites must be weeded. Trees with leaves, girdled by mammals, should be cut down immediately so that they will resprout. In southern California, planting should occur in January, February, and March, and irrigation should begin no later than March for at least 150 consecutive days at about 114 L/da.

Seeding of mixtures in riparian areas is done with drills, cultipack seeders, interseeders, hydro-row dry seeders, or hand planting (Platts et al. 1987). Use of tractor-drawn equipment often is impractical in riparian areas, where hand seeding or helicopters are used. But broadcast seeding of riparian areas should be accompanied by some means of seed coverage.

Coastlines. Coastal shorelines are especially susceptible to erosion due to waves, currents, and tides. Usually plantings are made primarily for stabilization of shorelines and secondarily as food and cover for wildlife. Broader marshes result from greater tidal ranges and gently sloping shorelines (Knutson and Woodhouse 1983). Slopes of less than 1 to 10 percent can be planted, but slopes of 1 to 3 percent are preferable (Knutson 1978). Steep slopes can be leveled with heavy equipment before planting. Lewis (1982a) listed suppliers of coastal plant propagules. Coastal sites can be evaluated for potential stabilization (Fig. 5.2) (Knutson and Woodhouse 1983).

Washout of bare-root seedlings on coastal areas is reduced when plant rolls are used (Allen and Klimas 1986, Meeker and Nielsen 1986). A trench is dug 20 cm wide, 20 cm deep, and maybe 4 m long; lined with burlap 90 cm wide; filled with soil; and planted with seedlings 40 cm apart and about 28 g of 18-6-12 slow-release fertilizer applied to each plant. The sides of the burlap are then stitched together or fastened with hog rings so that stem and leaf blades protrude. This burlap sausage is then covered with soil to the existing grade.

The main erosion control species are smooth cordgrass along the Atlantic and Gulf coasts, Pacific cordgrass along the southern Pacific coast, and Lyngbye's sedge and tufted hairgrass along the northern Pacific coast (Knutson and Woodhouse 1983). Minimum planting width for erosion control is generally about 6 m (Knutson and Woodhouse 1983), although two rows 61 cm apart of young single-stem smooth cordgrass planted between mean high tide and mean low tide, with plants 61 cm apart in the row, will grow into effective shoreline control in one season if fertilized with a timed-release fertilizer tablet (Cutshall 1985). Overstory shading of any woody vegetation should be cleared above the planting area to a distance of 3 to 5 m landward. A barrier of hay bales or coquina boulders or other energy barrier along the lower limits of the tidal range can aid in colonizing the entire zone (Beeman 1983). Sands are easiest to plant, and peats

1. SHORE CHARACTERISTICS	2. DESCRIPTIVE CATEGORIES (SCORE WEIGHTED BY PERCENT SUCCESSFUL)				3. WEIGHTED SCORE
a. FETCH-AVERAGE AVERAGE DISTANCE IN KILOMETERS (MILES) OF OPEN WATER MEASURED PERPENDICULAR TO THE SHORE AND 45° EITHER SIDE OF PERPENDICULAR SHORE SITE	LESS THAN 1.0 (0.6) (87)	1.1 (0.7) to 3.0 (1.9) (66)	3.1 (1.9) to 9.0 (5.6) (44)	GREATER THAN 9.0 (5.6) (37)	
b. FETCH-LONGEST LONGEST DISTANCE IN KILOMETERS (MILES) OF OPEN WATER MEASURED PERPENDICULAR TO THE SHORE OR 45° EITHER SIDE OF PERPENDICULAR SHORE SITE	LESS THAN 2.0 (1.2) (89)	2.1 (1.3) to 6.0 (3.7) (67)	6.1 (3.8) to 18.0 (11.2) (41)	GREATER THAN 18.0 (11.2) (17)	
c. SHORELINE GEOMETRY GENERAL SHAPE OF THE SHORELINE AT THE POINT OF INTEREST PLUS 200 METERS (660 FT) ON EITHER SIDE	COVE SHORE SITE (85)	MEANDER OR STRAIGHT SHORE SITE (62)	HEADLAND SHORES SITE (50)		
d. SEDIMENT* GRAIN SIZE OF SEDIMENTS IN SWASH ZONE (mm)	less tnan 0.4 (84)	0.4 – 0.8 (41)	greater than 0.8 (18)		
4. CUMULATIVE SCORE					

5. SCORE INTERPRETATION			
a. CUMULATIVE SCORE	122 – 200	201 – 300	300 – 345
b. POTENTIAL SUCCESS RATE	0 to 30%	30 to 80%	80 to 100%

*Grain-size scale for the Unified Soils Classification

Clay, silt, and find sand - 0.0024 to 0.42 millimeter
Medium sand - 0.42 to 2.0 millimeters
Coarse sand - 2.0 to 4.76 millimeters.

Figure 5.2 Form to evaluate potential planting success for stabilizing coastal shorelines (Knutson and Woodhouse 1983).

are the most difficult; best growth usually occurs on silts and clays (Woodhouse 1979). Plant species and planting characteristics vary by region (Tables A.22 and A.23). Regional coastal characteristics and species-specific planting characteristics can be found in Kadlec and Wentz (1974), Falco and Cali (1977), U.S. Army Engineer Waterways Experiment Station (1978), Woodhouse (1979), Lewis and Bunce

(1980), Carey et al. (1981), Lewis (1982a), Knutson and Woodhouse (1983), Zedler (1984), and Broome (1989).

Mangroves. Mangroves are coastal trees and shrubs that control shoreline erosion and grow in a complex ecosystem of few plant species and many animal species at the edge of warm seas of the world (Teas 1980). In the United States, mangrove forests consist of buttonwood and three genera and species of mangrove—red mangrove, black mangrove, white mangrove—which grow only along the coast of Florida. White mangroves and buttonwoods tolerate the driest soil, red mangroves the deepest submersion, and black mangroves the highest salinity (Knutson and Woodhouse 1983). White mangroves are the least cold-tolerant and grow the fastest; black mangroves are the opposite.

Seeds and propagules can be harvested directly from the tree, litter under trees, and rack on beaches (Thorhaug 1990). Naturally planted propagules also can be transplanted. Seeds planted after developing in a nursery for up to 1 year have higher success rates. But seedlings can be grown readily in pots (Woodhouse 1979). During winter and early spring, numerous germinated seedlings can be gathered from drift lines for potting. Growth requirements are similar to those for other marsh plants. Saplings 0.5 to 1.5 m tall and 4 to 5 years old survive transplanting best if root-balled and pruned when dug from the field. The root ball diameter should be at least one-half the height of the tree before pruning and about 20 to 25 cm deep. Similar-size plants grown in pots should fare equally well.

Stabilization is accomplished quickly by planting smooth cordgrass so that mangroves will eventually dominate the site through natural invasion if a seed source exists. Otherwise mangrove seed, seedlings, or larger plants can be planted in the cordgrass stand.

Mangroves should be set 0.5 m apart for seedlings and 2 to 3 m apart for saplings, at low tide, in holes large enough to accommodate the root ball, at about the same level in the ground as they were growing (Lewis 1982b, Knutson and Woodhouse 1983). The roots should be covered with soil so that pneumatophores are exposed. The top and side branches of white mangroves and black mangroves should be pruned to about two-thirds of their original length. Lateral buds might not grow on branches of red mangroves pruned to a diameter of more than 2.5 cm. Because of their broad moisture tolerance, white mangroves should be planted with red mangroves or black mangroves to develop a better root mat. Red mangroves grow in soil continuously flooded up to 0.5 m deep to occasional flooding a few centimeters deep. Black mangroves grow at slightly higher elevations. Larger transplants might succeed at lower levels where young plants and propagules cannot tolerate continuous flooding. Wherever warranted, slow-release fertilizer such as Osmocote or mag-amp is applied in the planting hole (see Table 5.5).

Periodic removal of debris might be needed to prevent smothering. After 1 to 2 years, brush and leaves should be added as a source of reduced carbon to help develop the ecosystem (Teas 1981). Plantings remote from well-developed mangrove stands should have biota established from a well-established mangrove forest by bringing in branches, litter, mud, and snails.

Sand dunes. Vegetative zones of sand dunes fall into three general categories (Woodhouse 1978, 1982):

1. *Pioneer zone:* An area of continual sand movement facing the coast, vegetation consists of a few species of grasses, sedges, and forbs that can withstand salt spray, sandblast, sand burial, flooding, drought, wide temperature fluctuations, and low nutrient supply.
2. *Scrub or intermediate zone:* Immediately behind the pioneer zone, plants include the pioneer species, shrubs, and stunted trees at times.
3. *Forest zone:* Immediately behind the intermediate zone, vegetation varies from the dense thickets of trees, shrubs, and vines of the maritime forests of the south Atlantic and Gulf coasts to the coniferous, hardwood, and mixed open forests of the Great Lakes dunes.

Creating new barrier dunes or rebuilding damaged or incomplete dunes can be done mechanically, by moving sand into place with truck, bulldozer, hydraulic pipeline dredge, etc., and grading it to suitable form; or by trapping blowing sand with a sand fence (e.g., snow fence) or vegetation or a combination thereof. (See "Sand Dikes" in Chap. 2.) In all cases, the sand must be stabilized to be a protective device. Erosion control focuses on the pioneer zone of barrier dunes or foredunes, where mostly perennial grasses are used to initiate dune development because of rapid establishment and growth and superior sand-trapping capacity (Seneca 1980). Vegetation is established mainly by vegetative propagule (see Table A.24). Except for the Gulf Coast, American beachgrass and European beachgrass seem the best species to plant (Seneca 1980, Ternyik 1980). All areas need protection from disturbance caused especially by humans (e.g., dune buggies) and livestock (grazing and trampling).

Seagrasses. Seagrasses are marine vascular plants which occur on muddy sand substrates in shallow protected coastal waters (Phillips

1980a). In addition to binding soil and providing wildlife cover, seagrasses are eaten in the United States and Canada by at least 67 species of wildlife: 1 mammal, 35 birds, 1 reptile, 9 fish, 7 crustaceans, 4 mollusks, 6 echinoderms, and 4 annelids (McRoy and Helfferich 1980). Of 47 species of seagrasses occurring in the world (Fonseca et al. 1988), five occur in North America (Phillips 1980b): eelgrass, shoalgrass, turtle grass, manatee grass, and widgeongrass.

Side-scan sonar can be used to locate and map donor beds of seagrass in shallow, turbid water, by searching areas up to 300 m wide at speeds of 3.7 km/h (2 kn) (Gaby 1986). A microwave-based navigation device, called a *Trisponder,* can locate any transect ±1 m to help identify recipient sites rapidly. A Water Witch vessel, normally used for cleaning debris from harbors, can be fitted with a hydraulic lifting arm and face shovel digging apparatus to dig seagrass sod for transplanting, with relatively little permanent damage to the beds. Each shovelful produces about twenty 15-cm plugs.

Other methods of obtaining transplants include removing plugs with a posthole digger (Thorhaug 1980) or a 20-cm-diameter, 90-cm-long PVC sewer pipe with opposing holes through which iron bars are placed for handles (Phillips 1980b). Aquarium cement seals the holes. A wooden cap with a hole is sealed to the top. The pipe is placed over the seagrass and pressed into the sediment about 20 cm, a cork is placed in the cap, and the plug is extracted by withdrawing the pipe. The plug is planted by removing a core of substrate with the same device and inserting the plug.

Seeds can be gathered by scuba, dehisced (separated from the fruit pod), and planted in low-turbulence areas in barren sediment, in benthic algae, or in successional stages of colonizing seagrasses (Thorhaug 1990). They are cost-effective, with full restoration within 4 years if no severe storms occur.

Although regionally variable (Table A.25), spring is generally the best time to plant (Phillips 1980b). The decision of what and where to plant varies regionally also (Fig. A.2). High currents will result in patchy seagrass beds, requiring more transplants than where low currents occur (Fonseca et al. 1988). Carangelo (1988) presented other considerations for creating seagrass beds.

Sod as well as plugs can be transplanted, but sod is time-consuming and therefore expensive to obtain and plant, with substantial damage done to the donor site as well as erosion of sod at the recipient site (Thorhaug 1980, Fredette et al. 1985). Use of turions (single-blade groups with stem and rhizome attached) and seeding is unreliable. Transplants of aboveground adventitious runners work when pressed to the bottom with large metal staples (Derrenbacker and Lewis 1982), as do bare-root transplants of 5 to 15 shoots attached to L-shaped anchors pressed into the bottom (Fredette et al. 1985). The most cost-effective, generally applicable method seems to be planting seagrass in sediment-free bundles of individual short, leafy shoots connected by their rhizome, perhaps including an adventitious meristem on the rhizome anchored in or on the sediment (Fonseca et al. 1988). Osmocote fertilizer (see Table 5.5) improves survival and growth (Orth and Moore 1982, Garbisch 1986).

Where shoalgrass, manatee grass, and turtle grass occur together, transplants of shoalgrass and manatee grass should be made first, followed later perhaps with turtle grass. Shoalgrass grows most rapidly, turtle grass most slowly (Fredette et al. 1985).

Uplands

Uplands associated with wetlands generally are managed for wetland species of wildlife, which benefits upland wildlife, too. Such uplands are managed as nesting cover mainly for ducks (waterfowl production areas) and as pasture for migrating and wintering geese and ducks. Grassland birds, small mammals, grazers, associated predators, and other grassland wildlife also benefit. Uplands on islands are managed mainly as nesting cover for various species of ground- and tree-nesting waterbirds, although other wildlife species also benefit.

Nesting meadows. Although sedge meadows and other wetland emergents on higher and drier wetlands provide substantial nesting cover for dabbling ducks and other waterbirds, upland nesting cover, if next to or near water, is preferred by dabbling ducks, grassland birds, and other upland wildlife. Ducks tend to nest within 400 m of water (Moyle 1964), but most can be accommodated within 100 m if suitable cover exists (Atlantic Waterfowl Council 1972). The ratio of water to managed nesting cover should be at least 1:4, with nesting cover in 32- to 40-ha blocks, when possible, or at least 16-ha ones, to reduce predation and hunter disturbance, rather than in long, narrow strips (Nelson and Duebbert 1974, Petersen et al. 1982). Large contiguous blocks of native prairie provide space and reduce predation (Greenwood et al. 1987). To justify establishing such blocks of cover, a ratio of 2.5 pairs of ducks per hectare within 0.8 km of the water area is needed (Ball et al. 1975). Minimum goals are 2.5 nests per hectare of cover and 70 percent nesting success. In general, grass should be at least 20 cm tall for duck nesting (Duebbert et al. 1981); small ducks such as blue-winged teal need residual vegetation at least 5 cm tall (Duebbert et al. 1986).

Development of uplands for dense nesting cover (DNC) on a waterfowl production area (WPA) includes clearing shrubs, trees, and stones; site preparation; and seeding to suitable species. To reduce predation, the rock piles or woodpiles, denning trees, raptor hunting perches, old foundations, fence lines, gullies, tree seedling seed sources, and clumps of trees and shrubs between nesting cover and brood habitat must be removed (Petersen et al. 1982). Stand quality and longevity are the main goals for establishing DNC (Higgins and Barker 1982). Fields of at least 16 ha that do not contain wetlands or other habitat divisions usually have higher duck nest densities and hatch rates, although no field is too small (Duebbert et al. 1981). Waterfowl nesting habitat more than 700 m from farmland receives less predation from crows (unpublished report, B. D. Sullivan, Department of Wildlife Ecology, University of Wisconsin-Madison, 1990).

Warm-season grasses and legumes germinate at higher temperatures than cool-season grasses do (Meyer 1987). Warm-season grasses provide beneficial nesting and winter cover because they are deep-rooted, more heat-resistant than most exotic grasses, and mainly lodge-free bunchgrasses with a tall, overhead canopy and open travel space at ground level (Burger 1973). But native warm-season grasses are poor competitors with annual weeds and therefore harder to establish than introduced species, although once established, native grasses remain in vigorous condition longer (Duebbert et al. 1981). Introduced grasses are considered as semipermanent cover, native grasses as permanent cover.

DNC can be established on previously tilled or other land with suitable site characteristics. These site characteristics are determined from a standard government soil survey (U.S. Soil Conservation Service 1976, 1978a, Clayton et al. 1977, Canada Soil Survey Committee 1978) and technical guide (e.g., U.S. Soil Conservation Service 1978b). The seed mixture to plant is based on the relative frequency and occurrence of each species listed in the technical guide for a specific range site and vegetation zone. If published guidelines are unavailable, a similar site in the same vegetation zone is examined in climax or near-climax condition to determine the frequency of each major species. Seeding rates (Table 5.6) are then determined for each species (Duebbert et al. 1981, Meyer 1987).

Guidelines for the prairie pothole region—the primary WPA in North America—have general application to other areas with similar soil and moisture conditions (Meyer 1987). Ecotypes can be moved up to 480 km north or 320 km south of their origin with few problems of winter hardiness, longevity, and disease. Movement east or west is governed by annual precipitation and elevation, with 305 m high equaling 282 km north (Duebbert et al. 1981). Where annual precipitation is at least 50 cm, stands of tall warm-season grasses can be established with big bluestem, indiangrass, and switchgrass (Klett et al.

TABLE 5.6 Grassland Seed Mixtures Recommended for the Prairie Pothole Region and Other Areas of Similar Soil and Moisture Conditions

	Seeding rates, kg/ha				
Grassland	Pure stand	Mixed stand	Seeding date	Site	Soil
Introduced cool-season grasses and legumes			< May 15 or August 10–September 20	All sites suited to farming	Moist, well drained
Tall wheatgrass *Agropyron elongatum*	12.3	5.0			
Intermediate wheatgrass *Agropyron intermedium*	11.2	4.5			
Alfalfa *Medicago sativa*	4.5	1.1			
Yellow sweetclover *Melilotus officianalis*	3.4	0.6			
Tall, native warm-season grasses			June 1 to 15	Lowlands, bottomlands, nearly level plains	Deep, fine, well drained to moderately drained
Big bluestem *Andropogon gerardi*	12.3	5.6			
Indiangrass *Sorgastrum nutans*	11.2	3.4			
Switchgrass *Panicum virgatum*	5.6	1.1			
Midheight native grasses			< May 15	Uplands, rolling plains with moderate to steep slopes	Moderately deep, medium-textured, well drained
Green needlegrass *Stipa viridula*	11.2	4.5			
Western wheatgrass *Agropyron smithii*	13.5	4.5			
Sideoats grama *Bouteloua curtipendula*	10.1	3.4			
Little bluestem *Andropogon scoparius*	7.5	1.1			

SOURCE: Meyer (1987).

1984). Where precipitation is less than 50 cm, native grasses of western wheatgrass and green needlegrass produce good nesting cover. Such areas also produce good nesting cover if introduced grasses such as intermediate wheatgrass or tall wheatgrass are planted with alfalfa.

Recommended grassland seed mixtures for the prairie pothole region and areas of similar soil and moisture conditions are of three types (Table 5.6): introduced cool-season grasses and legumes, tall native warm-season grasses, and midheight native grasses (Duebbert et al. 1981, Meyer 1987). Fine soils (clay-silt) to moderately textured soils should be tilled for seeding. Coarse (sandy) soils should be planted with an annual grain crop the first year and the next spring seeded with grass directly into clean, standing stubble. Fields with undesirable vegetation might need at least 1 year of intensive cultivation, herbicide treatment, or both before seeding to grass. Rhizomes of undesirable perennial grasses and weeds can be killed with a final tillage just before fall freeze-up. For successful germination, the seedbed should be compacted with a heavy roller before seeding. The seedbed is ready when a person's foot sinks 1 cm or less into the soil.

Soils exposed to wind or water erosion should be planted to a nurse crop of a fast-growing annual before seeding to the desired perennials. A nurse crop provides shade, protection from wind and water erosion, and reduced competition from weeds during the critical first year until the young perennials can make it on their own (Burger 1973). Oats, barley, or flax should be planted the year before the desired grass mixture, usually at about 1.7 to 2.2 kg/ha, and harvested to leave a stand of stubble. The nurse crop can be planted after site preparation that includes clean-tilled summer fallow. The nurse crop is planted in spring after at least one crop of annual weeds has germinated during year 1 and is killed by shallow tillage. The area can be mowed after mid-July when most nesting is over, leaving a 30- to 38-cm stubble. Herbicide treatments generally are needed to control undesirable plants before the desirable grasses are planted. These usually are seeded directly into standing stubble without disturbing the soil again. Nurse crops provide a seedbed for fall or spring plantings. (Warm-season grasses are not planted in fall.)

Ground or aerial broadcasting of grass seed is not recommended (Meyer 1987). Commercially available drills can be used to plant the various grass and legume seeds evenly, at the proper rate (see Table 5.6) and depth, and to pack soil around the seed. Seeds should be planted 1.5 to 2.0 cm deep in medium to fine soils and about 1.5 to 2.5 cm deep in sandy soils. Grasses and legumes are planted when the moisture content is proper, as determined when the topsoil forms into a ball when hand-squeezed and the ball retains its form. Ducks prefer nesting in zero-tillage (seeding into stubble) areas, especially if seeding is done in fall (Cowan 1982, Duebbert and Kantrud 1987).

Broadleaf annual weeds need spraying with a suitable herbicide if prevalent when grass seedlings are 25 cm tall (Meyer 1987). Introduced cool-season grasses or native warm-season grasses need periodic rejuvenation every 5 to 10 years to maintain optimum vigor (Duebbert et al. 1981). Introduced grasses and legumes need reseeding after tillage every 1 to 2 years. Native grasses do best with prescribed burning or grazing or some combination thereof. Introduced cool-season grasses and legumes should be burned from March 15 to June 15, native warm-season grasses from May 15 to June 15, native cool-season grasses from late March to mid-May or from August 15 to September 15 (see "Prescribed Burns" in Chap. 4). Grazing should not be done annually. Herbicides and fertilizers also can be used. In general, phosphorus improves legumes, and nitrogen improves grasses.

Mechanical rejuvenation includes mowing, disking, spiking, chisel-plowing, and shallow moldboard plowing (Duebbert et al. 1981). Tillage is best for introduced cool-season grasses and legumes. Mechanical treatments causing the greatest soil disturbance, 10 to 15 cm deep, are the most effective. Before mechanical treatment, residual vegetation should be removed by burning, haying, or grazing to facilitate tillage. Then spiking, chiseling, or shallow (10-cm) plowing is followed by light disking or harrowing. Haying should not be done annually. Any rejuvenation treatment should be avoided in the years of and following abnormal conditions such as drought, excessive moisture, heavy snow or ice pack, or excessive harvest by rodents. Such conditions constitute forms of rejuvenation.

In areas planted specifically for wildlife, hay should be cut once per year from July 20 to August 20 to allow nesting ducks and other nesters to succeed and to allow vegetation to recover in late summer for winter cover and residual nesting cover for early spring nesters (Rutherford and Snyder 1983, Alberta Fish and Wildlife undated). Mowing should leave stubble at least 20 cm high as residual nesting cover (Atlantic Waterfowl Council 1972, Duebbert et al. 1981). If the desired species do not appear adequate in the second growing season, procedures should begin over. Where conditions vary substantially from the prairie pothole region, provincial and state wildlife and soil agencies can provide suggestions for seeding mixes and planting conditions.

To attract grassland birds, five strips of grassland 1.5 to 9.1 m wide should be designed so that each year one strip is plowed in sequence, such as 1-3-5-2-4 or 2-4-1-3-5 (Fig. 5.3). Shrub invasion is impeded, and the first year favorite bird food and cover plants such as poverty grass, panic grasses, lambsquarter, ragweed, and smartweed will

Figure 5.3 Plowing pattern to develop natural grassland strips for grassland birds (Ohlsson et al. 1982).

grow. By the third year, a mixture of grasses and forbs such as daisies, asters, goldenrods, milkweeds, etc., will occur (Ohlsson et al. 1982).

Goose pasture. When both hard seed and green forage are available, waterfowl eat green forage mainly during mild weather and late winter, and seeds throughout the colder weather of fall and winter (Hagan 1980). Thus, some combination of crops should be established, in conjunction with dense nesting cover in some areas, to feed fall-migrating and wintering ducks and geese. Spring migrants are not restricted in feeding range by hunting pressure.

Farming operations are expensive because they require time and equipment. Equipment is expensive to purchase, operate, and maintain. Where budgets are limiting, sharecropping is preferable. Cooperating farmers contract with the wildlife agency to plant the crops and harvest about 80 percent, or harvest all of some crops and none of others, or all the hay (mowed from July 20 to August 20) and half the corn, or some other combination. The number of farmers involved varies. For example, Wisconsin's Mead Wildlife Area contracts with about 24 local farmers to farm its 650 ha of cropland which supports 19 impoundments of 2670 ha and 3930 ha of woodland. Standard farming operations are used, with crop rotation and zero tillage where feasible.

Farm crops useful in a wide range of soil and climatic conditions include corn, the grain sorghums (milo maize, kafir corn, hegari, feterita, and other varieties), rice, millets, wheat, oats, barley, rye, ryegrass, alfalfa, certain clovers, buckwheat, soybeans, peanuts, and peas (Givens et al. 1964, Johnson and Montalbano 1989). But proper strains and varieties must be selected. The preferred forage in all localities is wheat, although legumes, especially ladino clover and alfalfa, are more nutritious than grain forages. Corn, rice, and grain sorghums are the preferred seeds, with millet, barley, and wheat almost as good.

Though low in protein, corn produces highly nutritious, good yields readily eaten by waterfowl and is therefore the most satisfactory of all agricultural feeds, especially if planted early enough to allow drying and harvesting before or during the hunting season. Winter wheat can be seeded aerially into standing corn (Hindman and Stotts 1989). Leveled or flooded corn improves the availability of the ears to ducks and reduces landing impediments for geese. But except for flooding, manipulation of standing crops constitutes baiting under federal regulation; hunting of birds attracted to such feed is illegal (Givens et al. 1964).

The grain sorghums are more drought-resistant than corn is and produce well in broadcast stands. (Corn is drilled.) But they are more susceptible to damage by blackbirds, insects, and disease, and in wet weather the grains rot or sprout quickly. Rice, too, is one of the best, but it is expensive to grow, needs special soil and much water, and sustains heavy damage by blackbirds. Wild millet, Japanese millet, and browntop millet have traits similar to those of the grain sorghums, especially rice.

Grain provides green forage in western Canada and United States; wheat and barley provide it wherever planted. Spring planting is required, because fall plantings mature in early summer and rot before migrant waterfowl arrive. Because wheat is preferred, oats and rye are sometimes planted with it to extend the period of use. Oats less than 10 cm high are almost as palatable as young wheat.

Ryegrass is desirable for several reasons. It resists silting and flooding better than any of the small grains. It does not become too coarse when planted early. It tolerates a wide range of soil, moisture, and climate. It can be overseeded in fields that have grown corn and other summer crops because the seed needs little soil coverage for germination. Such fields produce both hard seed and green forage. Ryegrass can tolerate sandy soils, high wind, and salt spray along the Atlantic Coast.

Buckwheat can be grown from Canada to southern Florida, although moist soil and a cool growing season are best. It produces fair yields when broadcast and has a short growing season. It should be planted about 65 days before the first expected frost.

Alfalfa is good, but expensive and sensitive to soil conditions, insect damage, and freezing. Of the clovers, ladino, white dutch, crimson, red, and alsike are used most often in waterfowl management. Ladino and white dutch clovers do not need annual planting when mixed with perennial grasses. To maintain succulence, stands should be grazed or mowed low before waterfowl arrive. If heavy use is anticipated, stands should not be treated, because rank stands sustain less damage. More nutritious than the grasses, legumes can substitute to some extent for hard seed.

By planting corn in perpendicular blocks, the tops of some rows are exposed by prevailing winds during a heavy snowfall. If wide swaths of harvested corn are separated by several rows of unharvested corn, a snow-fence effect results, which reduces snow cover in harvested areas, thus increasing availability of waste corn and providing corn on stalks above deep snow (Ringelman et al. 1989). After a heavy snowfall, cattle grazing in the cornfield will help open up the snow cover and increase availability of corn to waterfowl.

After second-crop ricefields in Texas are harvested, about 140 kg/ha (range 73 to 214) of rice are left (Hobaugh 1984). Waste corn in harvested cornfields averages 203 to 447 kg/ha (Reinecke et al. 1989). Harvesting corn when the moisture content is less than 21 percent increases the amount of waste corn available for waterfowl (Baldassarre et al. 1983, Baldassarre and Bolen 1987). This equation is used to predict the amount of corn lost during normal harvest operations: kg/ha of corn wasted = 1544 − (54 × percentage of moisture content). Burning cornfields after harvest improves the availability of waste corn for ducks, pheasants, and other wildlife, but some unburned patches should remain as cover for pheasants and other wildlife. After use by waterfowl, the burned fields should be disked to reduce soil erosion. Disking between harvest and waterfowl use removes about 77 percent of the waste corn, deep plowing removes about 98 percent, grazing removes about 84 percent, and hand salvaging removes about 58 percent.

Ducks eat about 85 to 114 g of rice or corn per day (Givens et al. 1964). A 10-ha cornfield producing 15,900 kg of corn will support 10,000 geese for 3 weeks. (See Chap. 1.)

In some areas, grit might be a limiting factor (Chabreck et al. 1989). Spreading of sand and shell has been used to attract snow geese in Louisiana's Sabine National Wildlife Refuge and Rockefeller Wildlife Refuge (personal communication, C. Boydstun, U.S. Fish and Wildlife Service 1989). First an area is cleared by burning so that geese will be attracted to the area to grub for roots and thus find the grit. Then about 11 to 14 m^3 of sand is dumped there and spread with a dozer. Depending on the size of the wetland area, about two to three grit sites are used and replenished about every 2 years as needed.

Chapter

6

Artificial Nesting and Loafing Sites

Installing artificial nesting and loafing sites is expensive and sometimes useless. Before they are installed, availability of natural sites relative to space requirements of the target wildlife must be determined or estimated. Development of islands tends to increase carrying capacity, but that is not necessarily so with other artificial structures if enough natural sites exist, unless they provide better protection from predators than the natural sites do.

ISLANDS

Islands provide security from predators and increase suitable nesting area in open-water locations. They should be incorporated into the design and construction of the impoundment.

Earthen Islands

Permanent earthen islands are expensive to build, especially after the impoundment is functional. During construction of the pond or after drawdown, earthen islands can be built with a dragline, dozer, scraper, or dredging (Fig. 6.1) (Jones 1975). Islands in water over 6 dm deep need more material and are more subject to wave erosion. (See "Islands" in Chap. 2.)

Small islands with low profiles are less attractive to predators (Johnson et al. 1978, Hoffman 1988). Because of the richness and higher predator populations in freshwater systems, predators are more likely to swim to an island in a freshwater ecosystem than in an

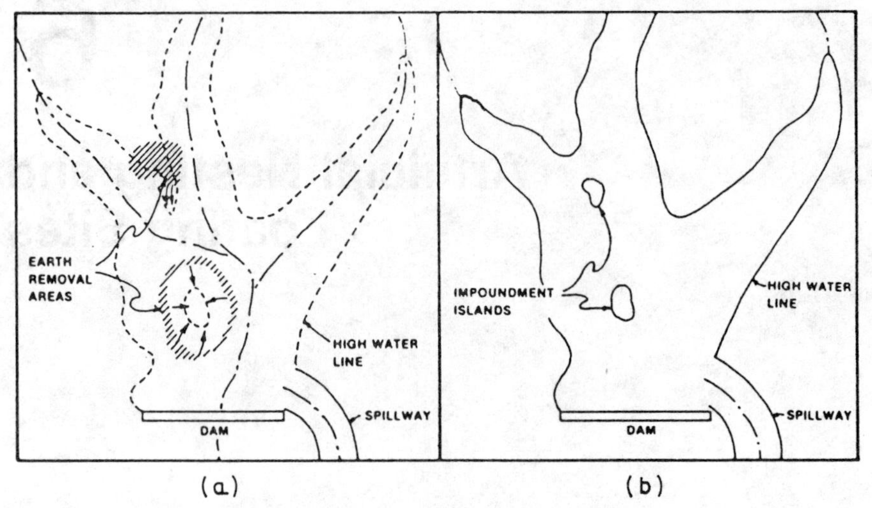

Figure 6.1 Possible areas of earth removal (*a*) and areas of placement to create islands after impoundment (*b*) (Jones 1975 *in* Ambrose et al. 1983).

alkaline basin (Hoffman 1988). Thus fewer nesters are lost by creating smaller (0.04 ha) and more numerous islands in an alkaline basin.

Islands of 0.04 to 0.8 ha can be built after freeze-up by removing ice from the island's site (see Fig. 6.6) and dumping soil into the opening (see Table 6.2) from a borrow pit on nearby uplands (Hoffman 1988). Otherwise they can be pushed up after drawdown, surrounded with a moat 9.1 m wide and 0.9 m deep, with a 3- to 6-m berm between moat and island.

Push-up islands can be constructed from natural high points of ground or the more gentle slopes in the upper end of the pond and/or side drainages (Fig. 6.1). Materials are pushed up and compacted with a bulldozer or moved, dumped, and sloped with a scraper. Scrapers compact and slope islands better, reducing wave erosion. Slopes should be 5:1 from the breaching area of a 3- to 6-m berm (Fig. 6.2) (Eng et al. 1979) or less if in a sheltered area. A freeboard of at least

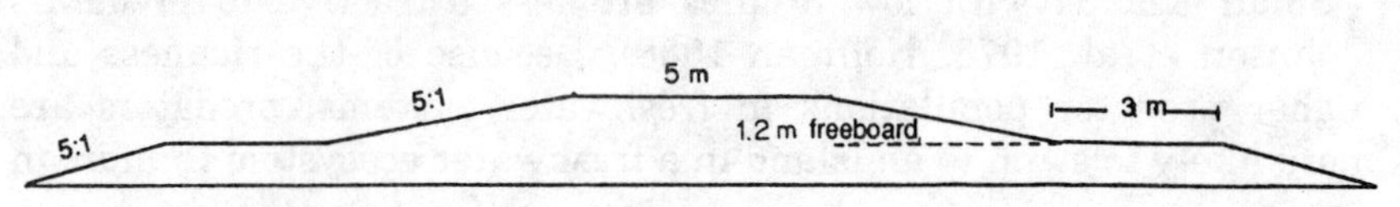

Figure 6.2 Cross section of a push-up island, with recommended size and slopes for waterfowl (Eng et al. 1979).

9 dm reduces nest destruction from settling and periodic flooding. In tidal areas, freeboard also should be at least 9 dm above maximum predicted tide (Harvey et al. 1983).

For maximum use by waterfowl, islands on ponds over 10 ha should be closer to the leeward than the windward side of the mainland, at least 9 m from the mainland (Eng et al. 1979, Ohlsson et al. 1982), in water 0.5 to 0.75 m deep (Hammond and Mann 1956, Jones 1975). Better yet is a distance of at least 30 to 52 m from and parallel to the mainland to avoid direct exposure to prevailing winds and wave action (Jones 1975). Double rows of trees and shrubs can be planted on 1.8-m centers as a windbreak on the shore of the mainland upwind of the island (Ohlsson et al. 1982, Green and Salter 1987), but they might provide cover for raptors and crows that prey on birds. Islands should be 0.5 to 5.0 ha (Duebbert 1982, Higgins 1986) and rectangular for greater shoreline area and ease of construction (Giroux 1981). Smaller islands might not be large enough to support many nests; larger islands might support too many predators. If not parallel to the prevailing wind, the windward side can be protected with emergents (Green and Salter 1987) or riprap (Duebbert 1982), with additional riprap needed if wave action or ice scouring exposes bare soil. The leeward side should have a flat, open shoreline for loafing and access for nesting. Islands should be 60 to 150 m apart (Jones 1975, Giroux 1981, Duebbert 1982, Hoffman 1988). If larger islands are impractical, small, rectangular (25- by 40-m), push-up islands with flat tops are useful (Giroux 1981), at a density of one per 20 ha of wetland (Alberta Fish and Wildlife Division undated), preferably with at least 170 m between island and mainland for maximum nesting success (Giroux 1981). Smaller nesting mounds also can be made, with a 3- to 6-m circular top stabilized with sod strips (Fig. 6.3).

Cutoff islands at least 4.6 to 6.1 m wide can be established when tips of peninsulas, projecting 4.6 to 6.0 dm above the anticipated high-water level, are isolated with minimal earth moving (see Fig. 6.1) by excavating a 91-m-wide, 0.9-m-deep channel across the peninsula (Fig. 6.4) (Eng et al. 1979, Hoffman 1988), especially when the ground is frozen 6 dm deep to support equipment. The material removed could

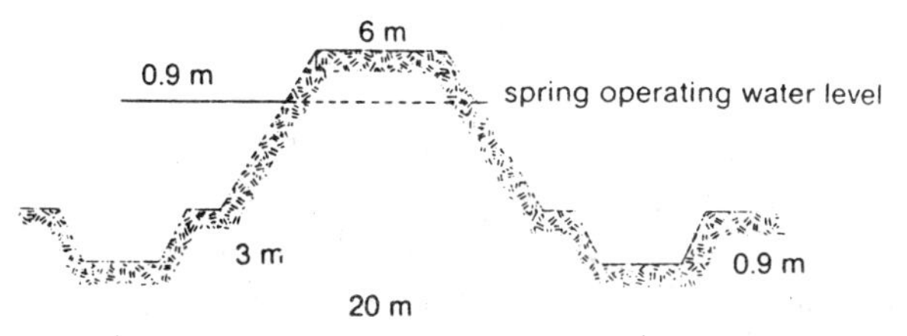

Figure 6.3 Small nesting mound (side view) (Alberta Fish and Wildlife Division undated).

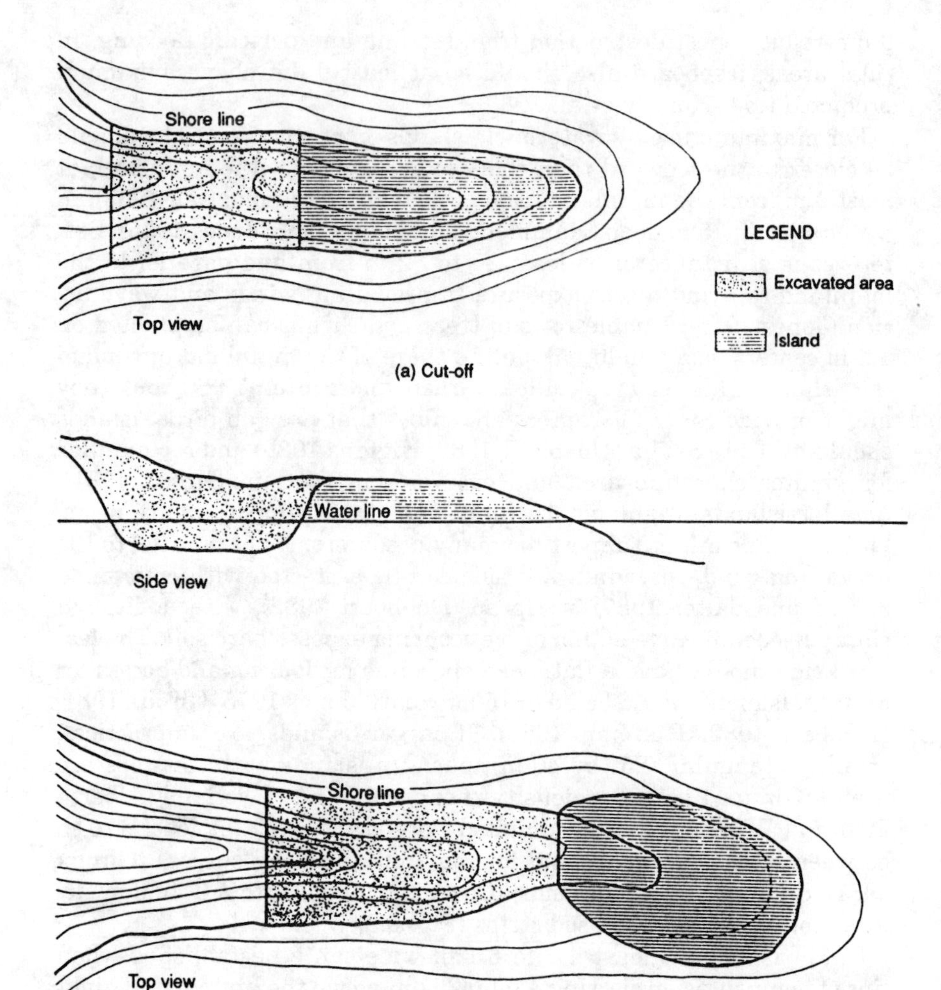

Figure 6.4 Two methods to develop cutoff islands (Eng et al. 1979).

be shaped into a push-up island (Fig. 6.5) (Stoecker 1982) or used in the dam. Advantages of cutoff islands over push-up islands include a more natural slope, largely undisturbed soil that is less susceptible to wave erosion, and established vegetation for immediate nesting cover

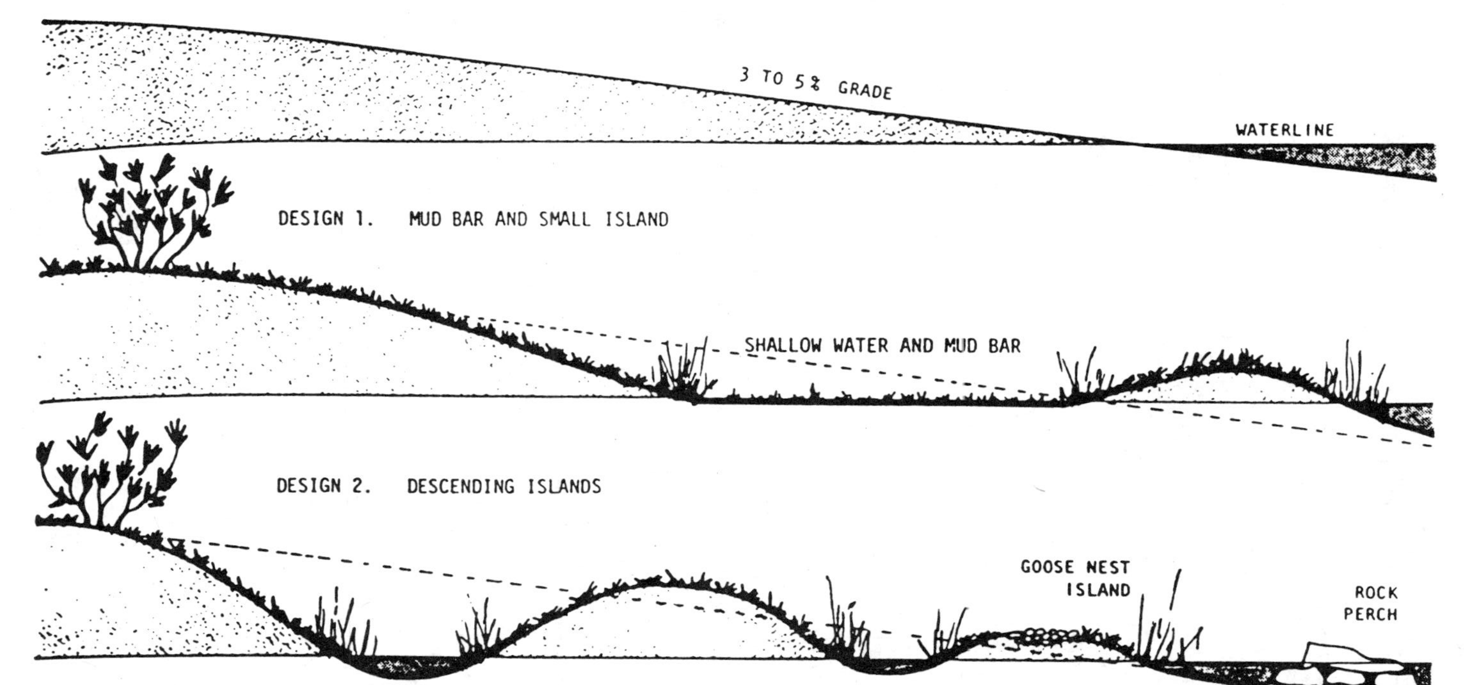

Figure 6.5 Wildlife habitat designs for shoreline peninsulas and arms of cove islands (Stoecker 1982).

and reduced wave erosion. But they usually are closer to the mainland and thus are more susceptible to predators.

If managed for ground nesters, islands should not have trees, which serve as perches for raptors and crows. A grass-legume mixture provides nesting cover (Duebbert and Lokemoen 1976), especially if production of forbs is encouraged (Duncan 1986), with a canopy cover of 50 to 100 percent and a height of 30 to 70 cm for geese (Green and Salter 1987). Plant species most often reported used by ducks in island nesting include Canada thistle, nettle, Wood's rose, snowberry, and shrubs with similar structure (*Rubus* spp., *Ribes* spp.) (Duebbert 1982). Islands should be vegetated with alternating patches of native grass and legumes, with dense shrubs such as willow, dogwood, or gooseberry planted on the windward side (Alberta Fish and Wildlife Division undated).

Islands developed for ground-nesting and arboreal nesting waterbirds from dredged material should be 2 to 20 ha, although 2 to 10 ha might be better, and 1 to 3 m above high water so that the island is high enough to prevent flooding and low enough to prevent wind erosion (Smith 1978, Soots and Landin 1978). Several small islands are better than one big one. Slopes should be 1:30 or less, with no dikes built on the island, although containment dikes usually are needed initially. Construction is best done in fall or winter, which allows time for sorting of fine materials by wind and water without disturbing nesters. Coarser material, e.g., a mixture of sand and shell material, is more stable and generally makes better nesting material for species nesting in bare substrate or sparse herb habitats. Indications are that the formation of a bay or pond increases attractiveness to birds. Generally, the larger the island, the greater the habitat diversity that must be maintained.

The species of birds attracted to an artificial island depends largely on the cover. For example, shingle-topped islands attract terns. Ducks use such islands for loafing, but need dense herbaceous cover for nesting. Coots and grebes prefer vegetation extending into the water margin (Swift 1982).

For natural establishment of vegetation, islands should have the following characteristics (Soots and Landin 1978):

1. Less than 5 km from another vegetated area
2. At least 10 to 15 ha
3. Less than 3 to 5 m above flood or tide stage but accessible to the water table
4. Even or gentle relief in elevation or topography

5. Protected from wind and water erosion
6. Located in fresh water
7. Soil of silt, sand, or sorted pebbles with adequate nutrients and no growth inhibitors
8. No intensive animal use (including human)

Other recommendations include decreasing the slope of the dome for better nesting sites, cutting down or bulldozing trees or shrubs beneath the deposition site on older islands to attract seabirds more readily, and planting vegetation to stabilize the island and attract specific wildlife by managing for a particular successional stage, e.g., sparse grass cover for terns, dense grass cover for gulls, and woody vegetation for herons (Soots and Parnell 1975). Desired results can be obtained in 5 to 17 months for ground nesting and 3 to 10 years for arboreal nesting (Soots and Landin 1978).

Soots and Landin (1978) listed colonial and noncolonial birds nesting on dredged islands (Tables A-26 and A-27). Animals nesting on *bare ground* require no plantings. In fact, excess plants must be removed. Animals nesting in *sparse herbs* (less than 25 percent coverage of low herbaceous plants) benefit from widely spaced plants less than 1 m high. Animals nesting in *medium herbs* (25 to 75 percent coverage of low herbaceous plants) benefit by increasing planting density, by selecting sites that readily grow dense plant cover, or by planting taller plants. Animals nesting in *dense herbs* (over 75 percent coverage of low herbaceous plants) can include some arboreal nesters.

The *herb-shrub* community is a mixture of low herbaceous plants and relatively low, multiple-stemmed woody plants, supporting both ground and arboreal nesters. The *shrub thicket* has a canopy and understory of dense woody plants requiring several years to establish. The *shrub forest* contains densely packed crowns of woody plants requiring years to establish. The *forest* is a climax stage of trees requiring years to establish.

Rock Islands

Rock islands are cheaper than large earthen islands, lending themselves as projects to wildlife associations and volunteer groups (Giroux et al. 1983). They consist of about 21 m^3 of rock and/or gravel deposited on the ice (see Table 6.2) in at least 40 cm of water at least 45 m from shore, to yield an island of about 18.4 m^2 at waterline. Rocks of 15-cm diameter are preferred. Islands should be 100 m apart and about 80 cm above water, with topsoil or sod on top to promote establishment of vegetation and reduce maintenance. To reduce damage from winds, wave action, and shifting ice, islands are placed along the lee shore or in emergent vegetation open enough to permit passage of goslings and ducklings. Before construction, stakes are used to mark potential sites relative to distance from shore, water depth, and bottom consolidation. Wire enclosures containing small rock islands can be used, but are not recommended because the wire eventually deteriorates to become a hazard or nuisance in the water. Single rocks 0.5 to 1.5 m in diameter provide resting and nesting sites for various species of shorebirds and waterfowl (O'Leary et al. 1984).

Culverts

Culverts set on end and filled with soil and sod function as small islands and might be one of the best structures for ducks and geese, because the requisite nest material and cover can be grown, thus minimizing maintenance while resisting ice damage (Higgins et al. 1986, Ball et al. 1988). Concrete culverts probably are more ice-resistant than smooth steel culverts; smooth steel culverts probably exceed rough concrete in resistance to climbing predators, but will rust through in time. Corrugated steel concrete is less desirable aesthetically.

To limit access by climbing predators, culverts should be 1.8 m long in order to extend 0.9 to 1.2 m above the water surface. Culverts 1.2 m long can be used in shallow areas with few climbing predators. The outside culvert diameter should exceed water depth. Culverts with over 122-cm inside diameter can be used concurrently by ducks and geese, are extremely resistant to ice damage, but are difficult to set up due to excessive weight of the 15- to 20-cm walls. Half the culvert can be screened off for ducks, with 1.8-m by 15-cm wire mesh over a wood frame, with flax straw attached by hog rings to the wire for cover the first year. Culverts with less than 76-cm outside diameter are not recommended. A 91-cm culvert usually is the best compromise (Table 6.1) (Ball et al. 1988). Surplus culverts can be obtained free and painted to improve aesthetic quality. Culverts made to order can be built to color and to resist alkali.

Culverts can be installed through the ice (Table 6.2) (Messmer et al. 1986) or in dry wetland basins with a front-end loader or a special boom truck or trailer equipped with a standard culvert harness. Culverts are seated firmly into the bottom with the front-end loader or by prying soil from around the base with a shovel. Culverts tip over if left to melt through the ice. The culvert is filled to the waterline with bottom sediments if they are not saline or alkaline. Otherwise, a sand-gravel mixture will suffice. Topsoil is added to the culvert level, sat-

TABLE 6.1 Weights of Concrete Culverts, kg/linear dm

Inside diameter,* cm	Wall thickness, cm				
	5.1	7.6	10.2	15.2	20.3
61†	29	46	66	NA	NA
76	35	56	78	130	NA
91	41	65	91	148	214
122	53	83	115	185	264

*Construction people refer to inside diameter.
†Outside diameters < 76 cm can be used only if seated > 3 dm into the bottom of sheltered areas.
SOURCE: Ball et al. (1988).

TABLE 6.2 Amount of Ice Needed to Support Various Loads

Thickness of ice,* cm	Permissible load
< 5.1	Stay off
5.1	One person on foot
7.6	Group of people in single file
19.1	1.8-mt truck gross or car or snowmobile
20.3	2.3-mt truck gross
25.4	3.2-mt truck gross
30.5	7.3-mt truck gross

*Ice must be twice as thick if it is soft and slushy. Water flowing beneath ice impedes freezing in rivers (especially the main channel), in lakes at inlets and outlets, and around beaver lodges.
SOURCE: Messmer et al. (1986).

urated for settling, and leveled again. Handfuls of straw dug into the topsoil reduce wind erosion and provide nest material the first year. Then the topsoil is seeded with a suitable grass-legume mixture.

Bales

Tight, cylindrical (round) 1.5- by 1.5-m (681-kg) flax bales double-tied with polypropylene twine and banded with plastic strapping at the baling site function as small islands when set on end in the impoundment. They will last 3 to 5 years for ducks and geese, are cheap, require low maintenance, and look generally natural, but are subject to climbing predators and damage by muskrats and cattle (Ball et al. 1988). Such bales can be loaded from a ramp into a pickup truck and unloaded by hand. Wrapping bales with wire mesh might prolong life, but the wire might trap goslings and is difficult to remove from the marsh later. To increase program efficiency, bales probably should be replaced with more durable structures.

Bales should be placed in water 4.5 to 6 dm deep in a dry marsh after drawdown or through the ice. Water over 9 dm deep reduces the bale's life, and water under 3 dm deep increases the risk of mammalian predation. A hole must be cut in the ice to accommodate the bale, or it will tip over as the ice melts. When the snow is not yet too deep, vehicles can be driven on the ice (Table 6.2). Ice grippers and appropriate safety equipment can be used to cut the ice with a sharp chain saw with 46- to 56-cm bar, which is safer than ultralight models. The chain of the chain saw should be oiled often and inspected regularly for chain and bar wear. A 1.8-m circle or square is cut into segments. If the ice thickness is 20 to 40 percent of the total depth, cut segments of the ice can be slid beneath the ice with an ice spud, which can save over 50 percent in labor. Ice more than 40 percent of the total depth can be removed with a large pair of ice tongs, with a toggle chain and a lever or a front-end loader, or the cut segments can be shoved beneath the ice. If a lever and fulcrum with toggle chain are used, a light power auger is used to drill a hole in the center of each segment. If a front-end loader is used with a toggle chain, the circle or square is not cut into segments and only one hole is drilled in the center with the auger (Fig. 6.6) (Ball et al. 1988). The bale should be positioned quickly in the hole through the ice to avoid wedging, because bales float briefly when first tipped into the hole. Bales which ultimately tip on their sides can be chopped on the upper side to form a depression for nesting.

Brush Islands

Brush islands of limbs 15 to 25 cm in diameter can be crisscrossed on ice with smaller limbs on top in piles 3 to 4.5 m across and high enough to be about 1.2 m above water (Linde 1969). Hay or straw is added after spring breakup. Such islands in quiet sloughs or potholes will last 3 or 4 years.

Floating Islands

Floating islands called *Schwimmkampen* (Hoeger 1988) are commercially available. They consist of aggregates of equilateral triangles 2.3 m on a side, made of 3-dm-diameter welded plastic pipes and reportedly environmentally safe, corrosion-proof, age-resistant (ultraviolet light) plastics such as polyethylene, polyurethane, and neoprene that are resistant to pests and microorganisms. They reportedly withstand high winds, waves, current, collisions, and freezing. Two types are available: vegetated and gravel-topped. The vegetated type contains rot-resistant plant fibers in a multilayered substrate with punched-out planting holes, and it weighs about 30 kg. The gravel-topped type

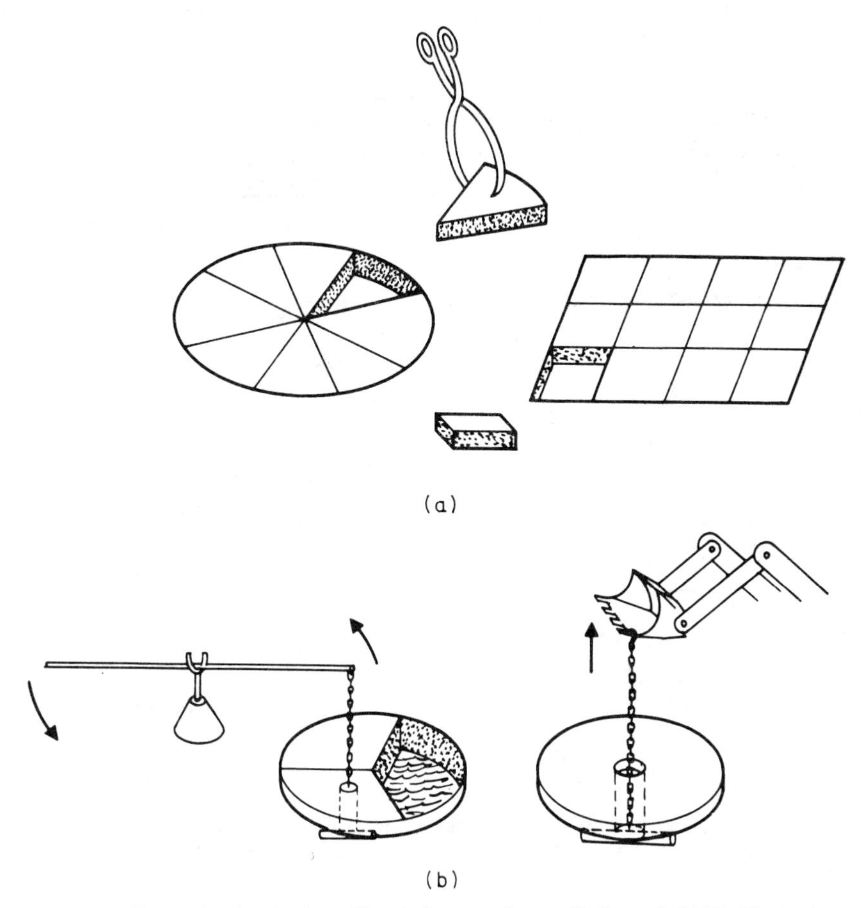

Figure 6.6 Removing ice for installing bales or culverts (Ball et al. 1988). Best solution is to push ice segments under surrounding ice with an ice spud. If ice thickness exceeds 40 percent of total depth, segments must be removed by using (*a*) a large pair of ice tongs or (*b*) a toggle chain and lever or loader bucket.

contains a foam surface for up to 200 kg of gravel (personal communication, S. Hoeger 1990). They are attached in various configurations and moored with poles, buoys, anchors, ropes, chains, weights, concrete blocks, etc. (Fig. 6.7).

BEACHES

Artificial beaches can improve nesting habitat for some species of birds (Swickhard 1974). Least terns will nest on a beach created with sand dumped 61 cm deep over an area at least 21 by 52 m. Common terns need 10 to 30 percent vegetated cover (unpublished report, S. W.

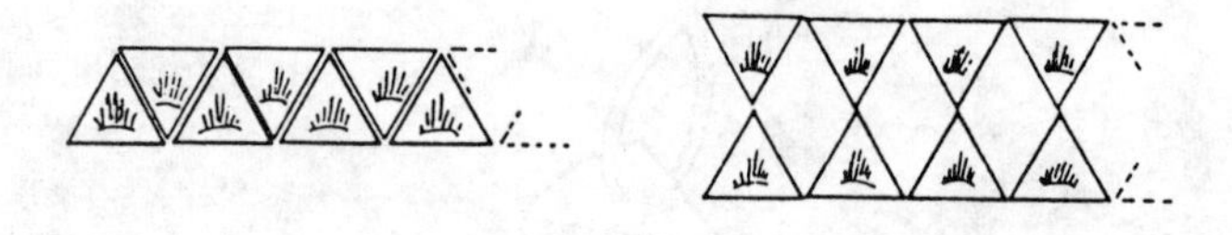

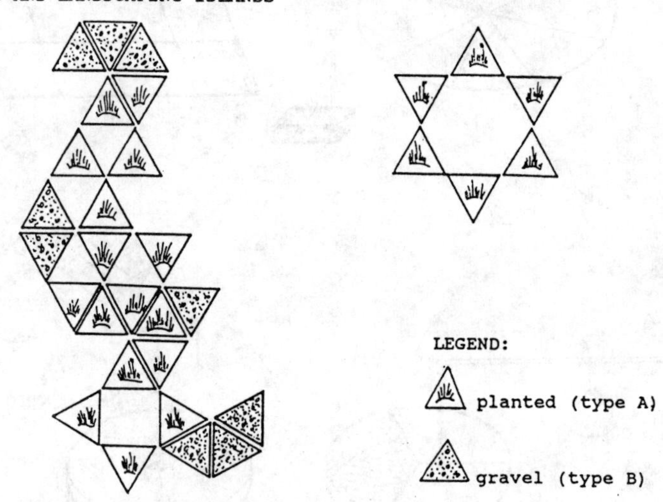

Figure 6.7 Suggestions for assembling *Schwimmkampen* floating islands used as barriers or habitat and landscaping islands as described in Hoeger (1988).

Matteson, Wisconsin Department of Natural Resources 1990). Some species of birds, e.g., arctic terns, common terns, and black-headed gulls, will continue to incubate when their nests are artificially raised with a pad of vegetation inserted under the eggs to protect them from rising water (Merila and Vikberg 1980).

NEST STRUCTURES

The decision is made to install artificial nest structures only after habitat evaluation indicates the need. Artificial nest structures are used to increase nest success because either the habitat is inadequate or predation rates are high. Management of wetlands often results in isolated islands of habitat that attract higher-than-normal densities of predators. Nest success of less than 25 to 30 percent for ducks and less than 60 percent for geese might justify the use of artificial nest structures (Ball et al. 1988).

The focus for waterfowl nest structures is on Canada geese, mallards, and wood ducks, although black ducks, gadwalls, pintails, blue-winged teal, canvasbacks, redheads, common goldeneyes, Barrow's goldeneyes, hooded mergansers, red-breasted mergansers, and black-bellied whistling ducks sometimes use these structures, too. Conscientious record keeping of decisions, work plans, budgets, and annual reports is essential.

Waterfowl

Nest material

The number of artificial nest structures to erect is dictated to a large extent by the number which can be maintained annually, for most require annual maintenance. A program of artificial nest structures often is thwarted by changes in priorities or personnel, because of the 5- to 10-year commitment involved. That much time is needed for success because acceptance by ducks and geese of artificial nest structures seems related to imprinting or learning (Burger 1973). Nesting success influences nest fidelity; previously successful nesters and their offspring home back to the structures (Doty and Lee 1974). Tradition and site tenacity influence use of artificial sites (Giroux et al. 1983). Unless nest material is at peak condition each year, particularly in structures unused the previous year, only previously successful hens are likely to use them, but not naive hens (Ball et al. 1988). Hunting should be disallowed near artificial nest structures or at least timed so that the resident hens have migrated, because hens are most vulnerable to hunting mortality near the nest structure and because the elimination of the pioneering cohort of artificial nest users and their progeny of experienced nesters will jeopardize the success of the program (Doty and Lee 1974).

Maintenance of artificial nest structures involves repair and adjustments, but mainly the assurance that plenty of suitable nesting material is available. To reduce human disturbance, which is a major cause of nest abandonment, maintenance should be done before the nesting season.

Except for sod-filled culverts, the best material for nesting is flax straw because it resists rot and wind loss the best (Ball et al. 1988). It might last two to five nesting seasons (Doty et al. 1975, Ball et al. 1988). The best flax straw is mechanically crimped to soften it. If it is unavailable, flax can be crimped by running over it with a truck or by twisting it manually into bundles 5 to 8 cm in diameter and bending them on a tight radius.

Sod is the only other nest material which alleviates annual maintenance (Ball et al. 1988). Solid mats and strips 25 by 15 to 25 cm are used, but they might not provide adequate cover for ducks in open nests.

Nest material should be 7.6 to 15.2 cm deep (McGilvrey 1968). Lightweight wood material such as cedar mulch lasts well if used with large chips to reduce wind loss, but might not attract ducks to open structures (Ball et al. 1988). A mixture of decorative pine bark chips and expanded shale landscaping rock about 1 cm in diameter lasts over 10 years and is used by geese. Loose vegetation seldom lasts over 1 year, and it can be blown out with a single windstorm. Small bales of grain or flax straw (not alfalfa) last 3 years for geese, but ducks seldom use them. At times, geese will nest in structures with down as the only nesting material, but goslings might be unable to escape such structures. Loss of nest material to wind can be prevented by use of wire (Fig. 6.8) (Messmer et al. 1986), crossbars, or sod collars (Fig. 6.9) (Ball et al. 1988). Vertical cone baskets should be stuffed nearly full of flax straw, with a wreath of it then wired to the basket rim in four or five places (Fig. 6.9*a*). Flax straw also can be wired in four or five places to the bottom of tub-type structures, with additional straw placed on top (Fig. 6.9*b*). Square crossbars of 0.6-cm hot-rolled steel that is wired or welded at joints or 1.3- to 1.9-cm willow sticks that are wired or notched at joints, can be placed over the nesting material (Fig. 6.9*c*), as can a collar of sod (Fig. 6.9*d*).

If the height from the nest material to the rim of a solid-sided structure is over 45° and over 7.6 cm, ducklings and goslings will need help escaping (Ball et al. 1988). Methods include escape holes, 5-cm wooden stairs, wood chips attached with fiberglass to the inside surface of fiberglass structures, and a 45° ramp of 0.6-cm or 1.3-cm mesh hardware cloth, sod, shale, or gravel (Fig. 6.10). Wire mesh must be 1.3 cm or less if ducklings or goslings might walk on it, especially as nesting material disappears. Wire mesh through which ducklings or goslings might try to squeeze must be at least 5 cm at the smallest dimension. Chicken wire (5-cm hexagonal mesh) and 2.5- by 5-cm welded wire are unsuitable, although commonly available.

Types of structures

Geese and mallards. Some 14 types of structures for geese and mallards are used, with numerous variations (Ball et al. 1988). Structures should be as aesthetically pleasing as possible. Marcy (1986) and Ball

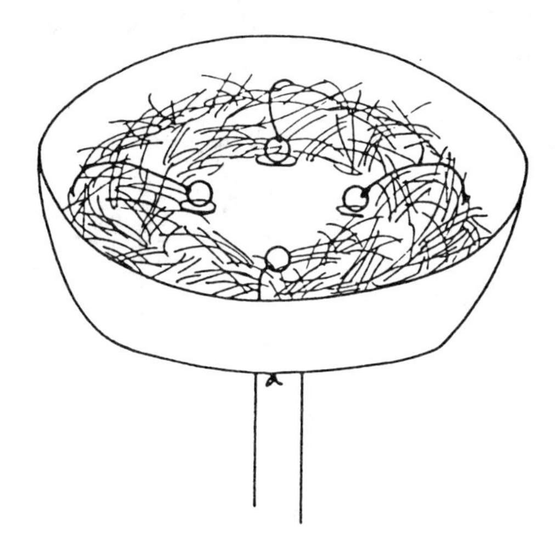

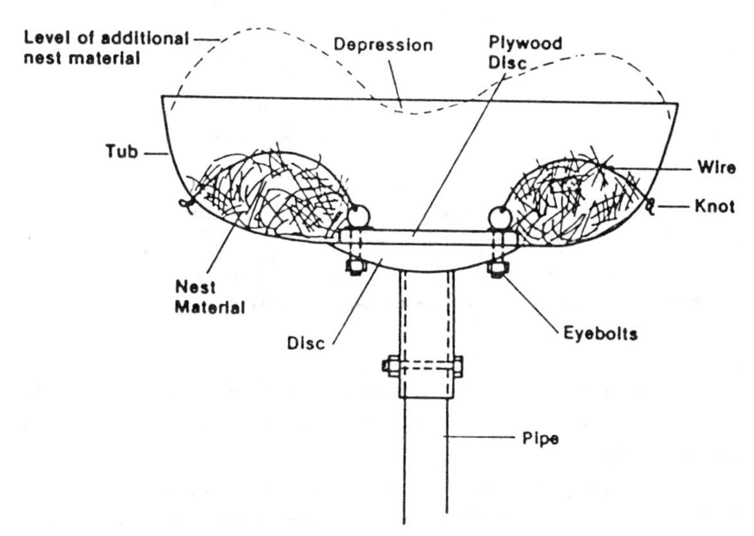

Figure 6.8 For Canada geese, secure a substantial bundle of nest material in the bottom of the tub, then fill it to the top, making a depression in the middle (Messmer et al. 1986).

et al. (1988) described those which are most common or which might solve chronic problems.

The single-post structure for geese probably is the simplest of structures and usually is the easiest to install (Ball et al. 1988). It lasts 12 years on average, compared to 10 years for multiple-post structures.

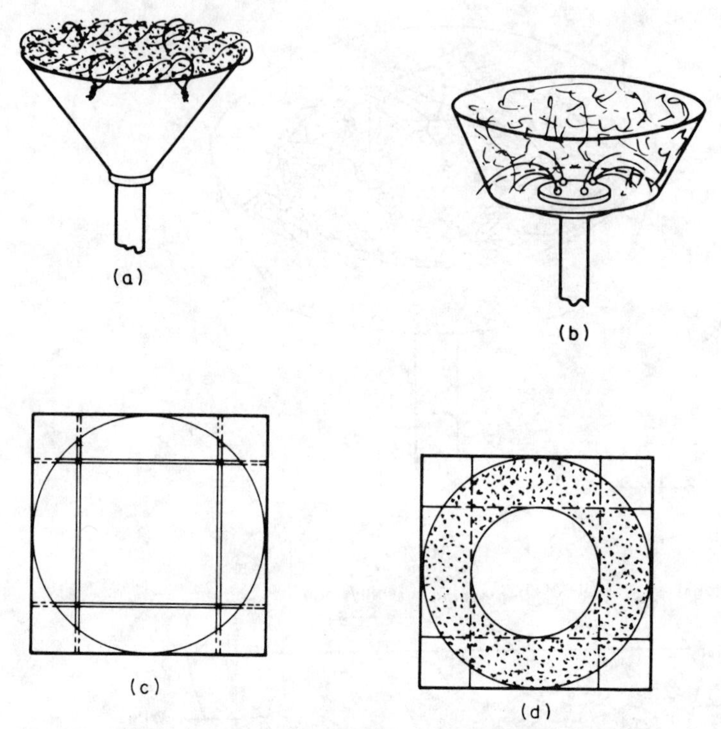

Figure 6.9 Method to minimize wind loss of nest materials (Ball et al. 1988). (*a*) Vertical cone basket stuffed nearly full of flax straw with a wreath of flax straw wired to the basket rim in four or five places. (*b*) Base of flax straw wired into tub-type structure, with additional flax straw added. (*c*) Crossbars of 0.6-cm hot-rolled steel (welded or wired at joints) help to keep nest material from blowing out. (*d*) Collar of sod keeps chips or vegetation from blowing out.

Structures with two or three legs are unstable. Those with four legs are more expensive, require more installation time, and are less appealing aesthetically.

The best nest compartments supported by the single post are a wooden box, commercial fiberglass tub, rejected fiberglass tank end, end of plastic drum, inverted tire, and vertical cone basket (Fig. 6.11) (Ball 1990). Snow fence also has been used (Grieb and Crawford 1967, Will and Crawford 1970). Mallards and hawks will use these structures if geese do not. Loss of goslings to hawks is minimal (Schmutz et al. 1988).

Posts 2.7 to 3.0 m long generally are adequate to drive at least 9 to 18 dm deep to prevent damage from ice. In rocky areas, a probe of 1-cm cold-rolled steel rod is used to locate suitable sites. Additional stability is provided by heavy-walled pipe or standard 5- to 10-cm-diameter pipe beveled at one end for driving and then filled with

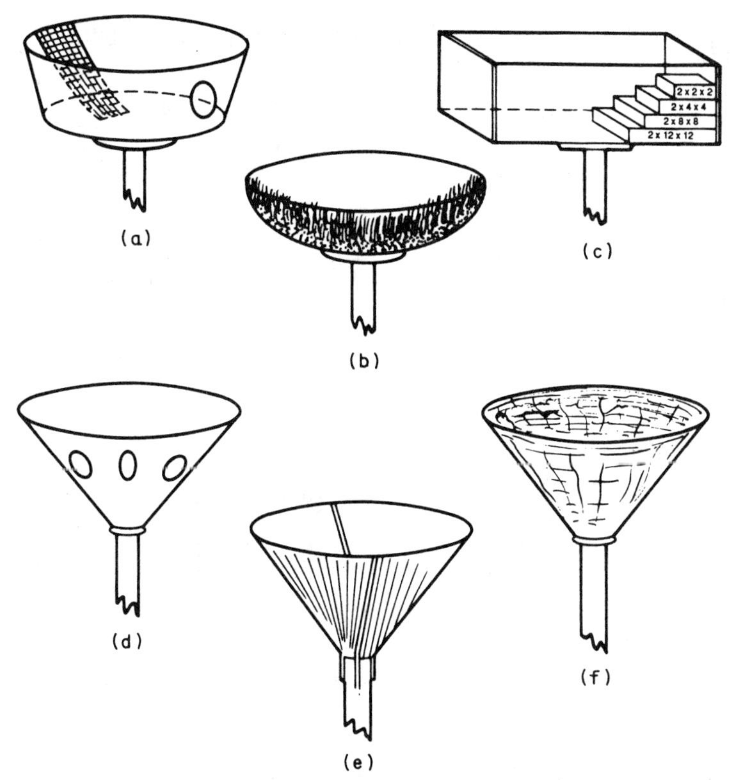

Figure 6.10 Methods to prevent entraping young waterfowl (Ball et al. 1988). (*a*) Escape hole or 45° ramp made of 0.6- or 1.3-cm mesh hardware cloth. (*b*) Ramp of heavy material (sod, shale, gravel). (*c*) Wooden "stair steps." (*d*) Escape holes in fiberglass cone basket. (*e*) Wire mesh (0.6- or 1.3-cm) 45° sides on cone basket. Chicken wire (2.5-cm and 2.5- by 5-cm mesh) causes entanglement and entrapment. (*f*) Wood chips secured by fiberglass to inside surface of fiberglass cone basket.

concrete. Fins of 0.3- to 0.6-cm steel can be welded to the post's bottom end to improve stability, as with steel fence posts. In some difficult situations, 25- to 36-cm-diameter wooden poles can be driven 4.6 m deep with a pile driver on large lakes, marshes, and river flowages.

Trees for geese obviate support posts and provide cover for nesting compartments (Ball 1990). Live conifers are preferred because raccoons seldom climb them. Nest compartments include a steel box supported with reinforcing rod, galvanized washtub, and inverted tire (Fig. 6.12). Trees should be mature conifers within 15 m of the high-water mark, somewhat isolated from other trees. Nest compartments should face the water and be 6 to 13 m above ground (Mackey et al. 1988).

Mallards and to a lesser extent other species of ducks readily accept

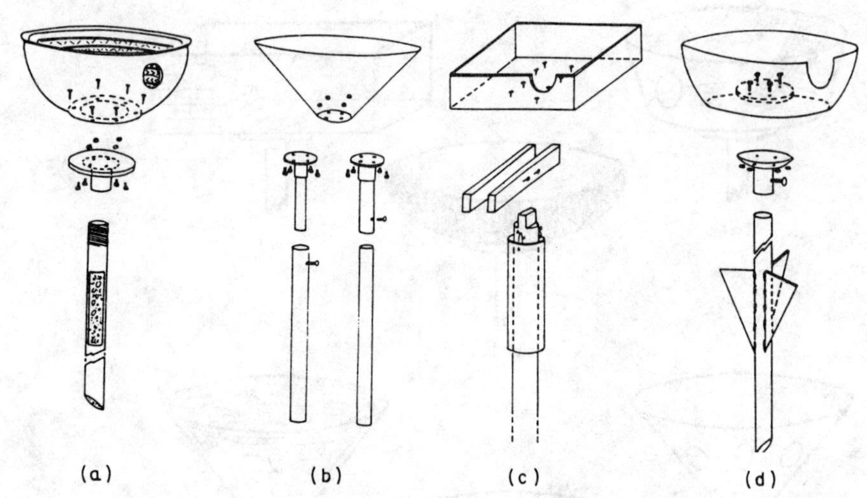

Figure 6.11 Single-post nest structures for Canada geese (Ball 1990). (*a*) Inverted, painted tire attached to threads on the support pipe with a treated plywood disk and a plumbing floor flange. A driving cap is needed to prevent thread damage during installation. The support pipe can be filled with concrete to prevent bending. (*b*) Fiberglass cone basket with welded mounting plate and adjustable ferrule mounts. (*c*) Wooden box with predator guard made of PVC pipe. The box can be built 30 to 46 cm deep with slatted sides to maintain nest material but allow goslings to exit through the 5-cm gaps exposed between slats as the fill level drops. (*d*) Fiberglass tub with a mounting plate made from a farm implement disk. The pole is finned to prevent tipping.

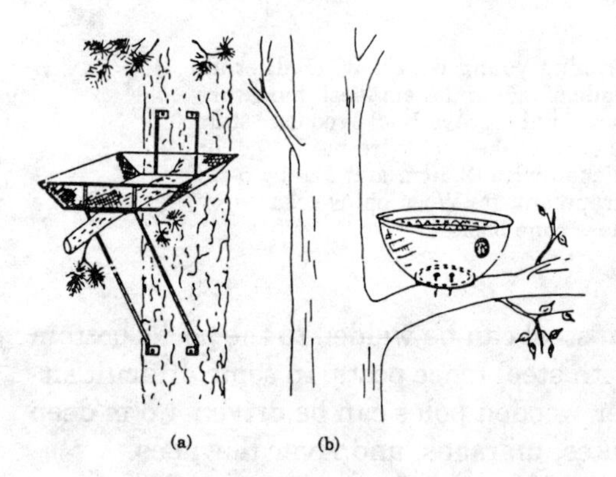

Figure 6.12 Tree nest structures for Canada geese (Ball 1990). (*a*) Expanded steel structure attaches to tree with lag screws and bends to accommodate tree growth. (*b*) Inverted, painted tire with treated plywood disk bottom attached to tree with ring nails. Attachment can be on a large horizontal limb, on a sawed-off vertical limb, or in a crotch. Aluminum nails should be used if logging could occur.

open baskets (Ball et al. 1988). But annual maintenance is essential, and nest abandonment caused by avian predators can be a problem in some areas. Open baskets (Fig. 6.13) (Marcy 1986) must be protected from terrestrial predators by use of a tight-fitting predator guard on the support pipe.

The double-wire cylinder for mallards reduces avian predation and

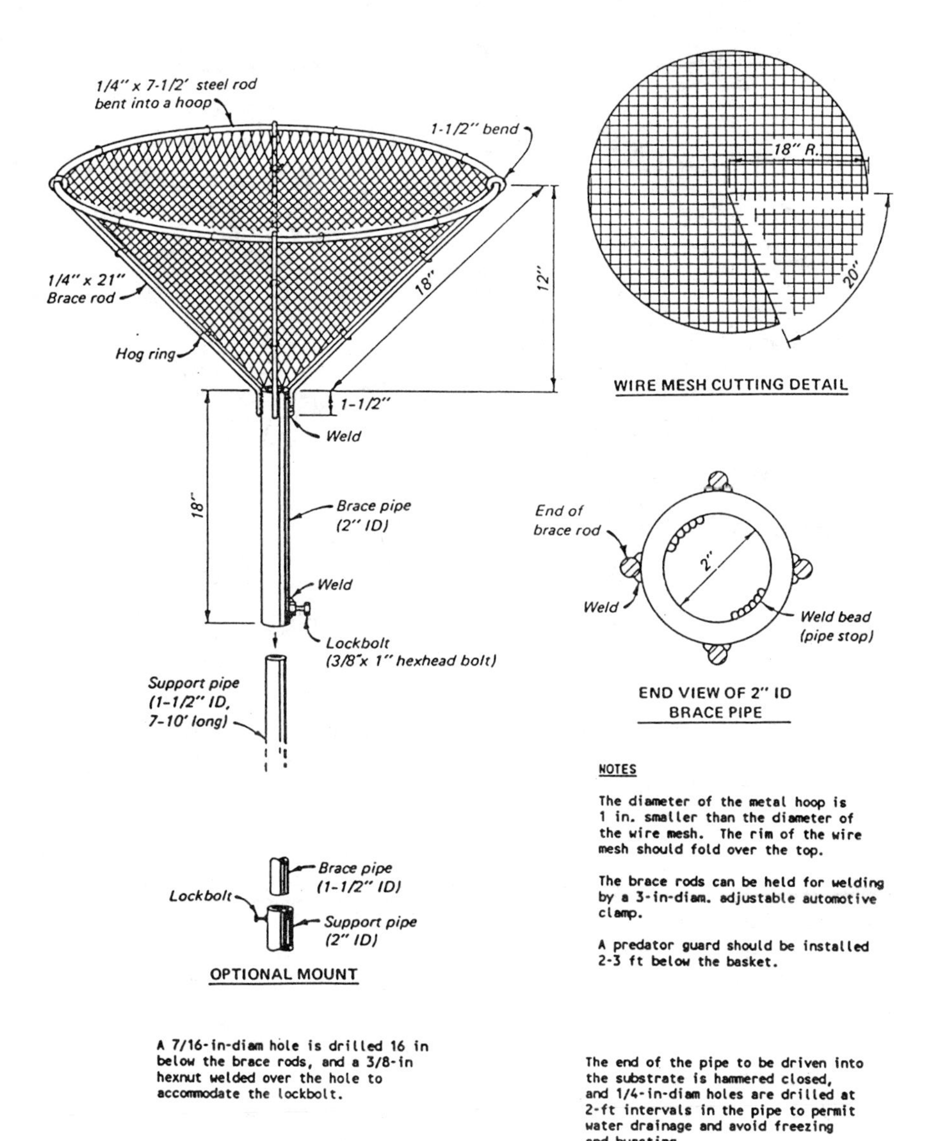

A 7/16-in-diam hole is drilled 16 in below the brace rods, and a 3/8-in hexnut welded over the hole to accommodate the lockbolt.

The end of the pipe to be driven into the substrate is hammered closed, and 1/4-in-diam holes are drilled at 2-ft intervals in the pipe to permit water drainage and avoid freezing and bursting.

Figure 6.13 Design of an open-basket nest structure for waterfowl (adapted from U.S. Fish and Wildlife Service 1970 *in* Marcy 1986). [For metric equivalence, see metric conversion table in the front of the book.]

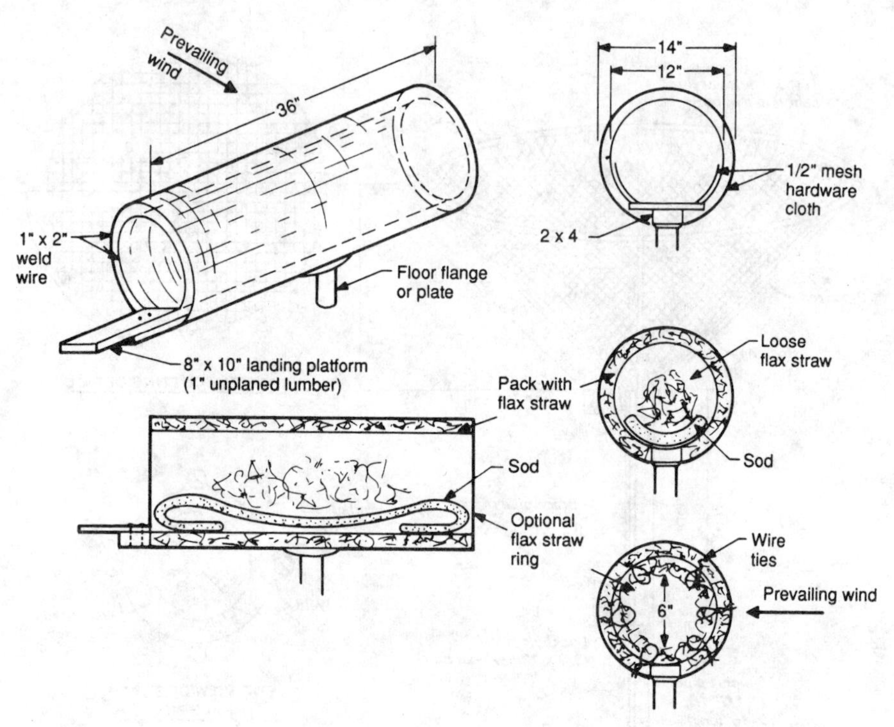

Figure 6.14 Horizontal (double-wire) cylinder nest structure for ducks (Ball et al. 1988). [For metric equivalence, see metric conversion table in the front of the book.]

loss of nest material by wind, but pioneering can be slow (Ball et al. 1988). With slow pioneering, installing open baskets and cylinders might expedite acceptance. A landing platform at one end (Fig. 6.14) might increase exploration by naive hens, but should be removed if used by avian predators. Annual maintenance is recommended until a nesting population is established, when a 2- to 3-year maintenance interval might suffice.

Rafts for nesting and loafing geese and ducks have serious drawbacks which largely can be overcome by using permanent structures (Ball et al. 1988), except in areas of severe ice action or water-level fluctuation. The best use of floating structures probably is in relatively small and intensive projects, because maintenance and tending are high. They last on average 6 years but are subject to waterlogging. Concurrent nesting of ducks and geese on one raft can be accomplished by providing an open nest compartment for geese and brush-covered boxes or cylinders for mallards. Grieb and Crawford (1967), Will and Crawford (1970), Young (1971), Fager and York (1975),

Brenner and Mondok (1979), Swift (1982), and Messmer et al. (1986) described various floating structures for geese and ducks.

Rafts can be hauled or dragged over ice to the site (Messmer et al. 1986). Rafts must be removed each fall to avoid ice damage, but anchors and at least 1.3-cm nylon anchor ropes can be left in place for one winter by attaching a short length of floating polypropylene rope as a marker (Ball et al. 1988). Earth anchors for mobile homes provide an excellent anchor system.

Messmer et al. (1986) described a raft for geese, with a wood nest box (Fig. 6.15) or a plastic nest tub. Brush attached along the edge seems

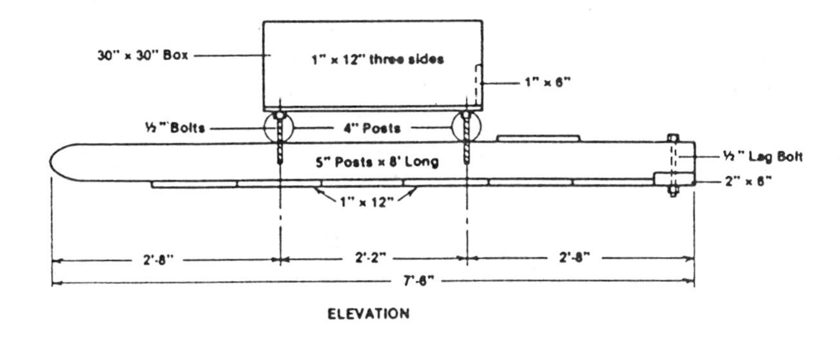

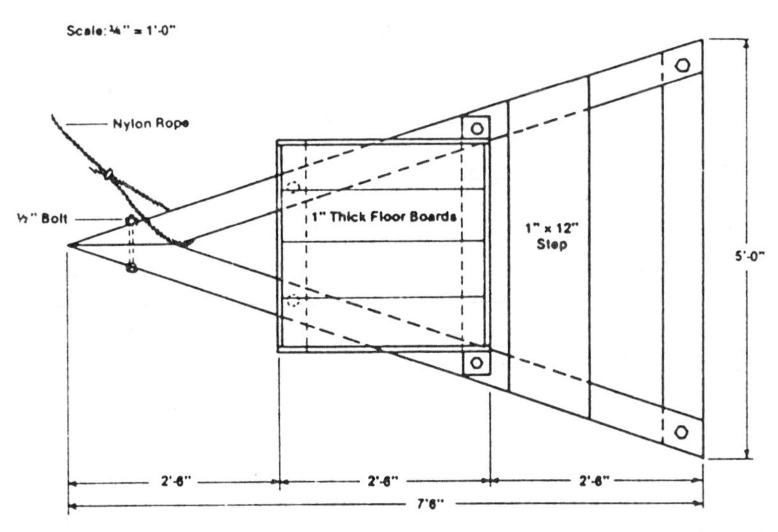

Figure 6.15 Floating nest platforms for Canada geese are most suited for areas with extreme water fluctuations (Messmer et al. 1986). [For metric equivalence, see metric conversion table in the front of the book.]

to discourage muskrats and might prevent avian predation of duck nests if arched over the nest box. Because wood is heavy and becomes waterlogged, 15- to 25-cm PVC pipe filled with styrofoam reduces structure weight and prevents muskrat damage to the styrofoam.

Fager and York (1975) described a "living" raft for ducks. Styrofoam is cut from shipping boxes for motorcycles and glued into a raft 1.2 by 1.2 m or 1.2 by 2.4 m and 10 cm thick. A 6- by 6-dm piece of styrofoam 5 cm thick is glued to the center. A shaped cover of chicken wire is placed over the styrofoam with willow or bulrush laced in the mesh, and grass is placed in the enclosure. (Mesh smaller or larger than the 2.5-cm chicken wire mesh is preferable.) To make a rooting zone for plants, a channel is cut around the raft 7.6 cm deep, 20 cm wide, and 2.5 cm from the outside edges. To allow water to seep in and keep root systems wet, 1.3-cm holes are drilled through the bottom of the channel. Soil and plants from the pond site are placed in the channels. To prevent coots from pecking holes in the styrofoam, the raft is framed with 20-cm-wide bark-out pine. A 6-dm square stabilizer is spliced into the anchor rope 9 dm from the raft to keep it from whipping in the wind. Grieb and Crawford (1967) described a floating goose nest structure with equalizer and anchor used for geese (Fig. 6.16).

Small loafing places can be made by anchoring logs or 1.2- by 1.2-m

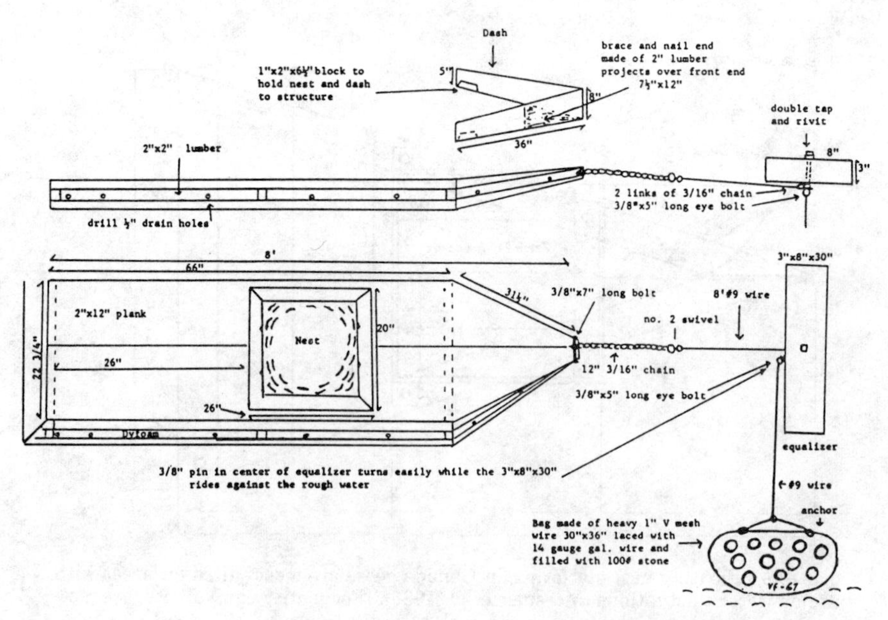

Figure 6.16 Floating nest structure for geese (Grieb and Crawford 1967). [For metric equivalence, see metric conversion table in the front of the book.]

rafts in open water (Atlantic Waterfowl Council 1972). The U.S. Soil Conservation Service recommends seven loafing sites per hectare.

Wood ducks. For cavity-nesting ducks such as the wood duck, chain saws can be used to create a cavity in a tree (Fig. 6.17) in about 8 min (Carey and Gill 1983). The chain saw must be modified with a 20-cm bar with a low-profile, antikick chain ("chisel chain") and kickback guard. Cavity sizes should resemble those of natural cavities. For wood ducks, natural cavities tend to be in trees with 35- to 40-cm dbh with a long life expectancy within 0.4 km of water, at least 2 m above the ground, with a 10-cm-diameter entrance (7.6 by 10 cm is better), 15 to 122 cm deep, and a bottom diameter of 25 to 28 cm (Sousa and Farmer 1983). Density averages three cavities per hectare (0.5 to 7.7 per hectare) (Semel et al. 1988a, b).

Nest boxes for wood ducks vary in design and materials. The best are a wooden box, a vertical metal box, a horizontal metal box, and a plastic pail. They can be mounted on pipes, posts, poles, or trees, but trees provide less security from climbing predators, including fire ants, than other structures placed in open water (Ridlehuber and Teaford 1986). Aspen trees should be rejected, as they are preferred and often cut down by beavers. Supports and boxes must be stable to

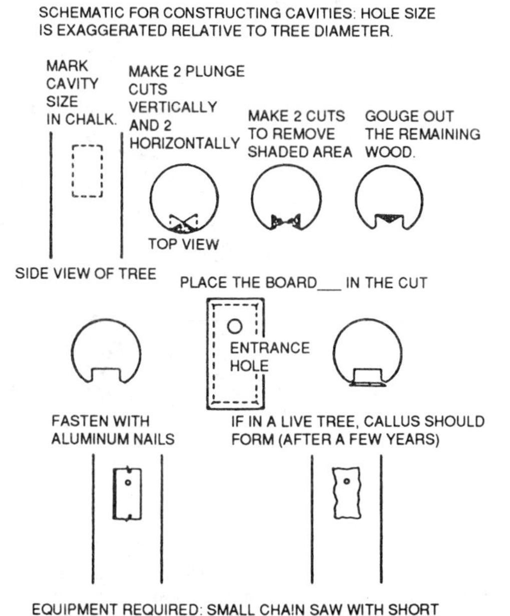

Figure 6.17 Constructing tree cavities with a chain saw (Carey and Gill 1983).

avoid rejection as nest sites. A cross brace on the post near the bottom can reduce sinking or leaning in soft marsh bottoms, but longer poles with fins are more stable.

A box is attached with a lag screw, a hanger bolt, a board and nails, or a bracket (Fig. 6.18). Lightweight, portable Swedish sectional tree-climbing ladders, available from forestry supply companies, are useful in mounting nest boxes.

Wooden boxes will last 15 to 25 years if built from decay-resistant lumber such as baldcypress, redwood, and western red cedar (Ridlehuber and Teaford 1986). Pressure-treated woods should be avoided. Rough-cut lumber is cheaper and provides toeholds for exiting ducklings. Smooth lumber requires a ladder of hardware cloth for exiting. A box with front opening (Table 6.3, Fig. 6.19) (Bellrose 1980) or side opening (U.S. Fish and Wildlife Service 1976b) facilitates cleaning and maintenance. A top-opened box (Bellrose 1980) facilitates banding of adults and ducklings.

Vertical metal boxes are less susceptible to predation by fox squirrels and are lightweight (Table 6.3, Fig. 6.20).

In areas where starlings compete with wood ducks for nest sites, the starling-proof horizontal metal box (U.S. Fish and Wildlife Service 1976b) is used (Fig. 6.21), although it is no deterrent to grackles and

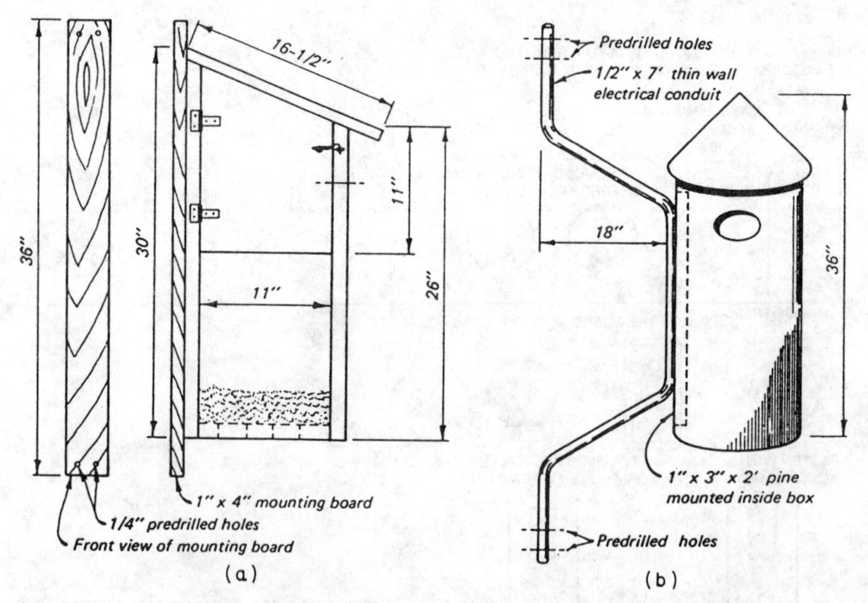

Figure 6.18 (*a*) Construction details for wood duck nest boxes, side view, for a wooden nest box with board attached for mounting; (*b*) method for attaching a metal nest box by using electrical conduit (Ridlehuber and Teaford 1986). [For metric equivalence, see metric conversion table in the front of the book.]

TABLE 6.3 Materials Needed to Build Wooden Nest Boxes and 12-in-Diameter Vertical Metal Boxes for Wood Ducks*

	Quantity	
Item	Per box	Per 100 boxes†
WOODEN BOX		
Lumber		
1 by 12 in	11 bd ft	1200 bd ft
Nails		
8-penny box, ring-shank or screw shank	40	35 lb
Staples, poultry netting, ¾ in	6	1½ lb
Hardware		
Hinges, 3-in T, light-duty	2	220
Hook and eye set, 2½ in	1 set	110 sets
Hardware cloth, ½- by ½-in mesh	64 in^2 (4 × 16 in)	17 linear ft of a 24-in-wide roll
Lag screws, ⅜ by 3½ in or hanger bolts (with nuts)	2	220
Flat washers, ⅜ by 2 in	2	220
METAL BOX		
Galvanized steel furnace pipe 26- to 28-gauge, 12-in diameter, 24 in long	1	100
Galvanized sheet metal, 26- to 28-ga		
Roof (16¼ by 31 in)	3.5 ft^2	250 linear ft of
Floor (13½ by 13½ in)	1.25 ft^2	a 20-in-wide roll
Hardware		
Hardware cloth, ½- by ½-in mesh	64 in^2 (4 × 16 in)	17 linear ft of a 24-in-wide roll
Lag screws, ⅜ by 3½ in or hanger bolts (with nuts)	1	110
Flat washers, ⅜ by 2 in	1	110
Sheet-metal screws, no. 6, ⅓ in‡	14	1500

*See metric conversion table in the front of the book.
†Quantities given assume a 10% loss or breakage rate.
‡Pop rivets can be substituted for sheet-metal screws.
SOURCE: Ridlehuber and Teaford (1986).

seems an attractive shooting target to vandals (Heusmann et al. 1977). The top front edge should be bent upward at a 45° angle to prevent injury to hens flying into the cylinder.

Plastic 19-L buckets modified for use as nest boxes (Fig. 6.22) are inexpensive and can last at least 20 years (Griffith and Fendley 1981).

Nest boxes have been developed from ammunition boxes and nail kegs (Bellrose 1953, Ridlehuber and Teaford 1986). Prefabricated wood duck nest boxes (e.g., Tom Tubbs box, Sonoco-designed box) are commercially available.

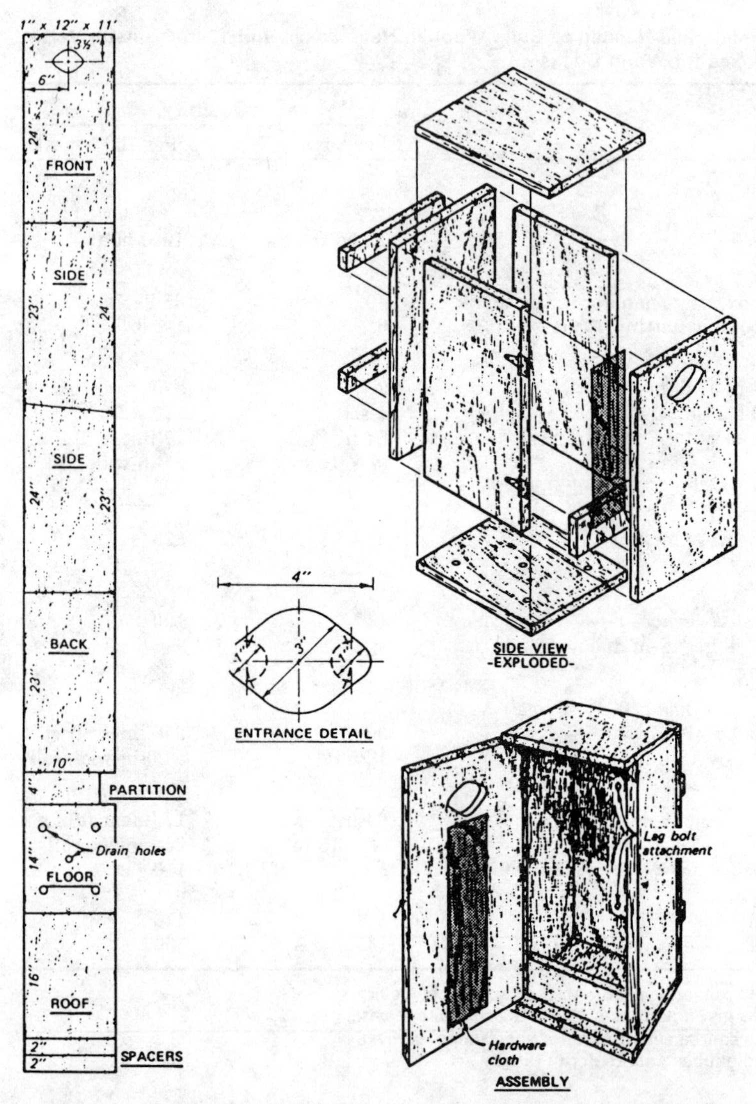

Figure 6.19 Construction details for a front-opening wooden nest box for wood ducks (entrance detail from Bellrose 1980 *in* Ridlehuber and Teaford 1986). [For metric equivalence, see metric conversion table in the front of the book.]

Other ducks. Hooded mergansers will use cylinders (Heusmann et al. 1977). Entrance holes of some wood duck nest boxes should be enlarged to a 12.7-cm circular opening to accommodate most other ducks (Bolen 1967, Bellrose 1980, Lumsden et al. 1980, 1986, Gauthier 1988, Savard 1988.) Buffleheads prefer relatively small

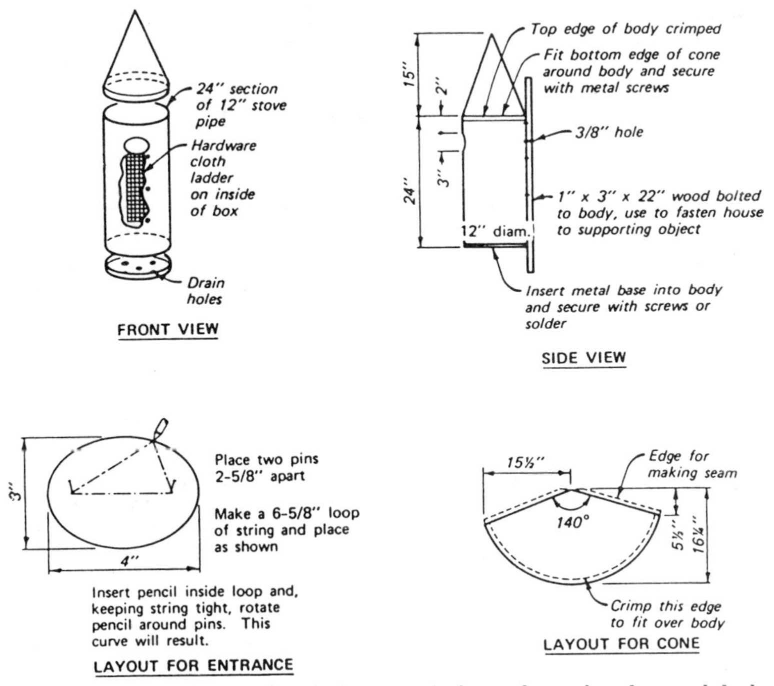

Figure 6.20 Construction details for a vertical metal nest box for wood ducks (after Bellrose 1980 *in* Ridlehuber and Teaford 1986). [For metric equivalence, see metric conversion table in the front of the book.]

boxes, 15 by 15 by 40 cm with entrance diameter of 6.5 cm (Gauthier 1988).

Nest shelters reduce predation by gulls on eggs of eider ducks (Clark et al. 1974). Wooden, lean-to shelters, 45 to 132 cm wide and 45 to 102 cm deep, should rest on a driftwood log at the rear and on 5- by 5-cm supports in front. The larger shelters often attract two or three nesting females. Another type of wooden nest box placed on the ground for eider ducks resembles the top portion of a wooden packing crate (Ducks Unlimited Canada 1989). It is about 18 cm high, 102 cm wide, and 89 cm long, with an open bottom, one wide side open, and a center division to accommodate two nests. Construction materials consist of one sheet of plywood 1.3 by 89 by 102 cm, one board 5 by 15 by 102 cm, three boards 5 by 15 by 89 cm, and nails.

Predator guards

Predators develop a search image for nest sites, so artificial nest structures need protection or they can become traps for waterfowl and

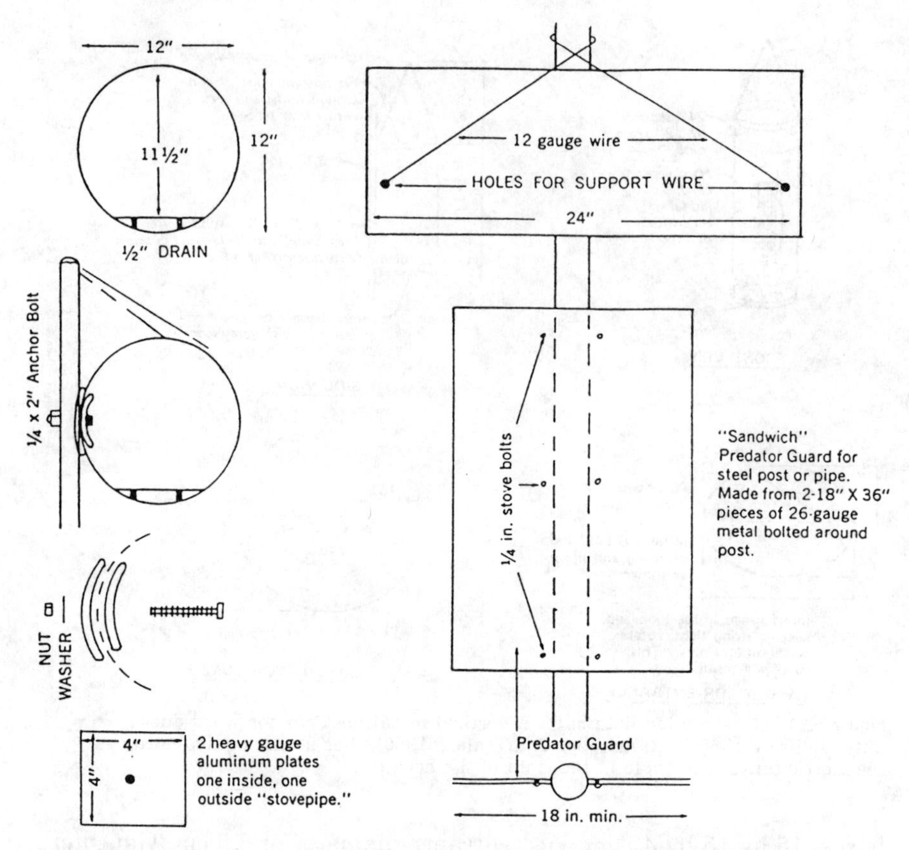

Figure 6.21 Stovepipe nesting shelter for wood ducks (U.S. Fish and Wildlife Service 1976b). [For metric equivalence, see metric conversion table in the front of the book.]

other birds and cause local population declines. The most common types of predator guards (Figs. 6.23 and 6.24) are (1) a metal cone and (2) a metal "sandwich" for wooden or metal posts, (3) a metal band for tree trunks, (4) a metal facing with entrance hole, and (5) a wooden tunnel for wooden boxes (Ridlehuber and Teaford 1986). The first three are generally effective against all climbing predators, the last two only against raccoons weighing at least 4.5 kg if the entrance is larger than that for wood ducks. None is effective against bears (Savard 1988). A pyramid shield (Fig. 6.25) can be substituted for the cone, which is more expensive and harder to make. Overhead branches must be removed to eliminate access from above by climbing predators. All guards should be tight to the support to deny access to snakes. A 25- to 30-cm band of sticky material (e.g., Tanglefoot, Tack-Trap) around the support prevents predation by snakes and fire ants

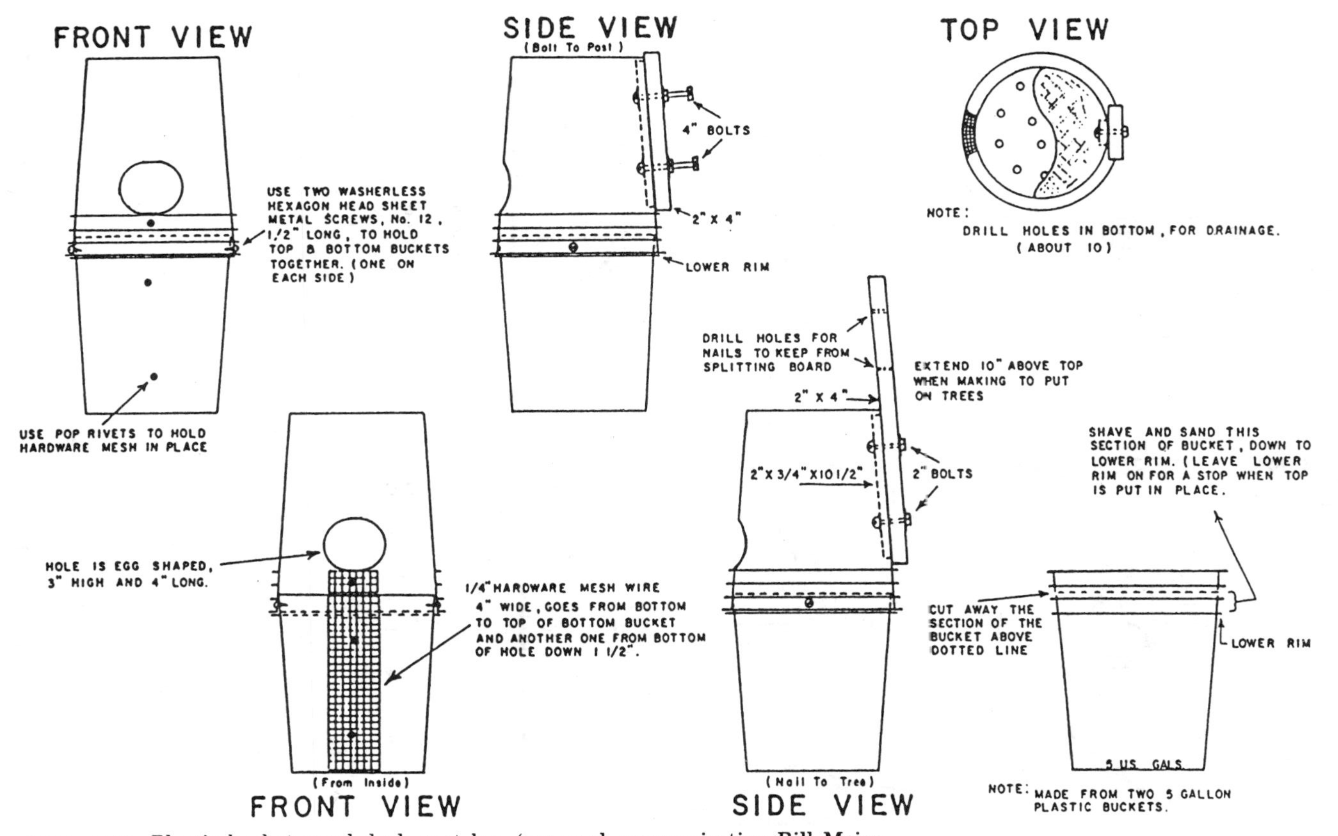

Figure 6.22 Plastic bucket wood duck nest box (personal communication Bill Meier, Wisconsin Department of Natural Resources). [For metric equivalence, see metric conversion table in the front of the book.]

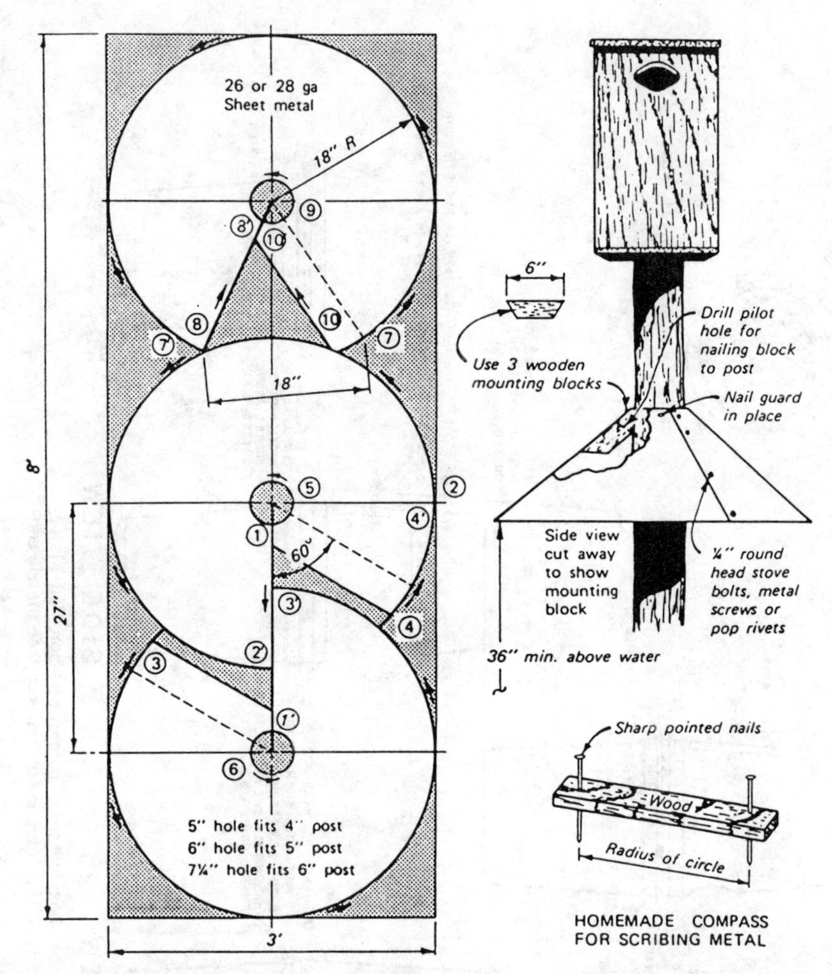

To facilitate cutting (on solid lines only) follow the sequence of numbers. Complete each cut before initiating the next (e.g. ① + ①' then ② + ②'). Make circular cuts in counterclockwise direction. To make initial cut at ① make slot with cold chisel. Cut complete circles at ⑤, ⑥, and ⑦. When installing guard, overlap the cut edge to the dashed line.

Figure 6.23 Construction details for a cone predator guard made from sheet metal (modified from U.S. Fish and Wildlife Service 1976b *in* Ridlehuber and Teaford 1986). [For metric equivalence, see metric conversion table in the front of the book.]

(Ridlehuber 1982). Spraying the interior with a disinfectant discourages wasps and bees (U.S. Fish and Wildlife Service 1976b).

Placement

Nest structures should be erected in sheltered sites in areas being used by waterfowl. Vehicles required for placement should enter the

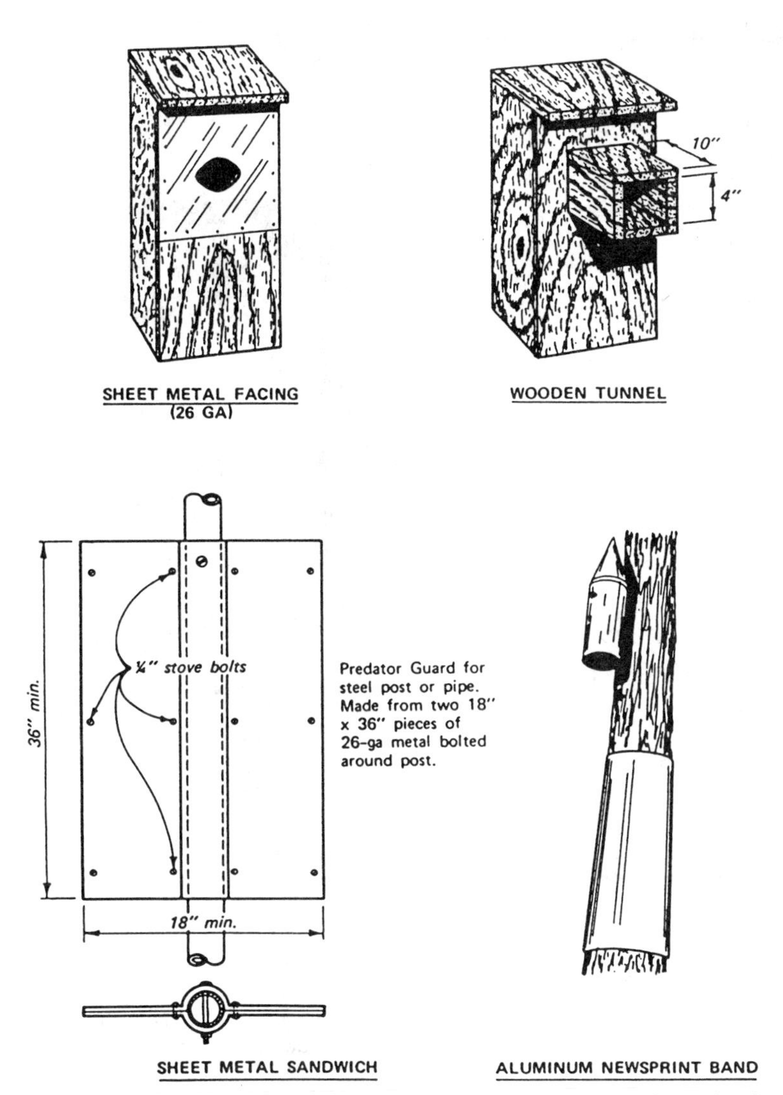

Figure 6.24 Four types of predator guards commonly used with artificial duck nests (Ridlehuber and Teaford 1986). [For metric equivalence, see metric conversion table in the front of the book.]

frozen or dried pond area from the shore opposite placement, to avoid leaving nearby tire tracks that persist through emergent vegetation to serve as travel lanes for predators (Ball et al. 1988). Water depth should be 4.5 to 6 or 9 dm. When possible, structures are placed in or near open stands of emergent vegetation for concealment or aesthetics, as far from shore as possible, on the leeward side of the wetland, protected somewhat from wind and ice action. Distances over 45 m

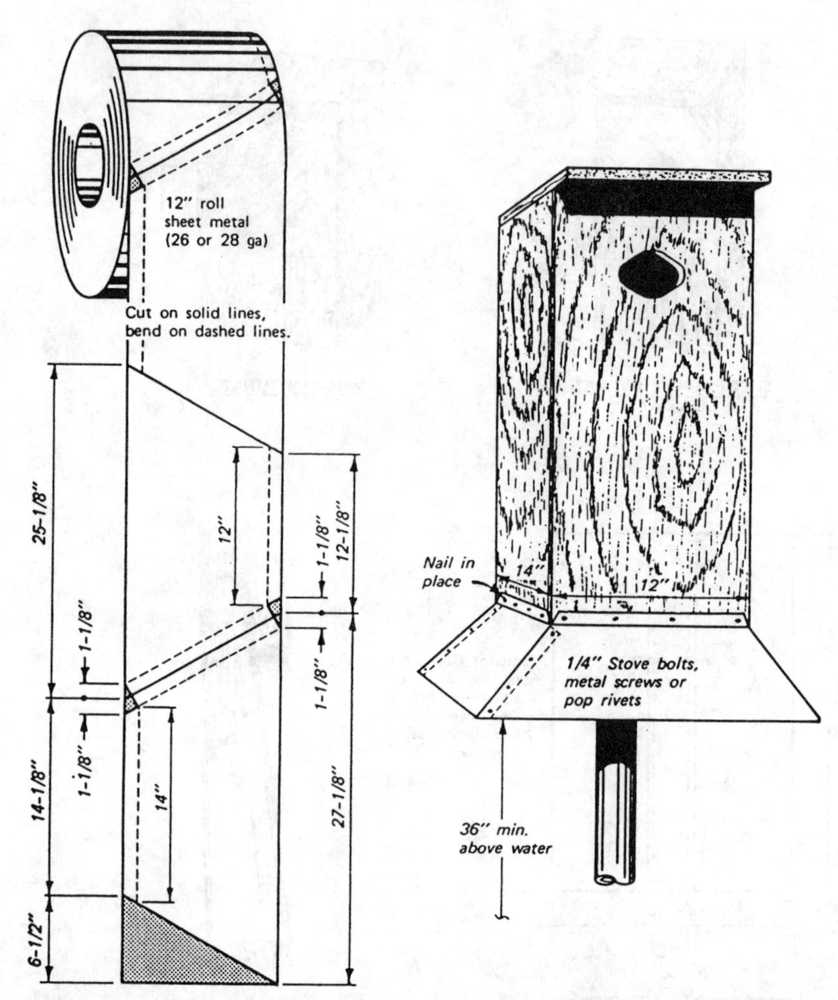

Figure 6.25 Construction details for a pyramid predator shield made from sheet metal (Ridlehuber and Teaford 1986). [For metric equivalence, see metric conversion table in the front of the book.]

from shore are best, but over 18 m might suffice with enough vegetative concealment.

Geese. Structures should be placed near brood habitat. Broods prefer green vegetation under 15 cm tall up to 27 m from water (Ball et al. 1988). Geese might not accept structures on land readily. Initially, a few structures can be placed 15 m offshore, then moved to 6 m offshore after 1 or 2 years of occupancy, and finally moved onto shore. Canada geese are adaptable. Density of structures should be one per hectare, spaced at least 91 m apart (Messmer et al. 1986, Ball et al. 1988).

Structures should be 9 to 12 dm over water to provide some security from predation, reduce unaesthetic impact, and reduce use as hunting perches by raptors. On land, structures 2.1 m high deter most leaping predators and prevent damage to nest material by livestock. Structures in trees should be above lower branches to reduce visual impact. Incubating geese on nests more than 4.6 to 6.1 m high seldom flush from passive human disturbance, but structures that high are more dangerous to install and maintain.

Mallards. Structures must not be placed on dry land. The predator guard should be at least 9 dm above the high-water level (Ridlehuber and Teaford 1986), and the upper surface of the nest structure should be slightly below the adjacent emergent vegetation (Ball et al. 1988). Structures should be placed in moderate stands of vegetation rather than dense stands, which mallards avoid. Mallards prefer the open water edge of shoreline stands of emergent vegetation and the edge of openings of at least 4.7 m^2 in emergents.

Structures can be spaced evenly around the perimeter of marshes, 9 to 91 m from shore in water 6 to 9 dm deep. Structures can be spaced one per hectare in wetlands under 2 ha or one per 2 ha on larger wetlands.

Wood ducks. Nest habitat is unproductive without nearby brood habitat. Optimum brood habitat contains an interspersion of 25 percent quiet water and 75 percent vegetative cover comprised of 30 to 50 percent shrubs, 40 to 70 percent herbaceous emergents, and 0 to 10 percent trees (McGilvrey 1968). Optimum distribution of water depths has 25 percent of the water area 0 to 3 dm deep, 50 percent 3 to 9 dm deep, and 25 percent 9 to 18 dm deep, with 25 to 30 loafing sites per hectare, such as small islands, muskrat houses, stumps, logs, tussocks of vegetation, and bare points of land. Areas less than 4 ha separated by more than 20 ha of land are marginal brood habitat. But complexes of small streamside areas and/or beaver ponds connected by water corridors are suitable.

Nest boxes should be installed over water preferably, with the predator guard at least 9 dm above the high-water level (Ridlehuber and Teaford 1986). Boxes can be placed in forest stands within 0.4 km of permanent water (Bellrose 1980) and at least 3 m above ground (U.S. Fish and Wildlife Service 1976b), but closer placement to water reduces predation of ducklings traveling from nest to water. The opening should face the water, at 90° to prevailing winds, with black interiors rather than unstained ones. Plastic nest boxes must be placed in full shade (Hartley and Hill 1988).

To reduce avian predation, interference from aggressive nest site competitors such as starlings, and dump nesting (brood parasitism) which reduces hatchability, nest structures are placed in habitats and at densities resembling the natural circumstances in which waterfowl evolved, i.e., three cavities per hectare (0.5 to 7.7 per hectare) (Semel et al. 1988a, b). Grouped nest boxes might help attract wood ducks in the first few years of a new nest box program and where they are visible to the public (e.g., at visitor centers).

Other ducks. Nest boxes for buffleheads should be installed at least 100 m apart along heavily forested coniferous shorelines, not in open aspen groves, which are the preferred nesting habitat of starlings which compete (Gauthier 1988). Boxes installed in spring will not be used by Barrow's goldeneyes until the next spring because most subadult females and unsuccessful females select nest sites the year before breeding (Eadie and Gauthier 1985). Common goldeneyes use entrances of 13 by 21 cm and 10 by 13 cm equally, but prefer black interiors over unstained ones; hooded mergansers prefer entrances 10 by 13 cm and boxes 33 cm deep (Lumsden et al. 1980). Nest boxes within 100 m of water had similar occupancy and success rates for black-bellied whistling ducks (McCamant and Bolen 1979), although 57 percent of known nest trees with natural cavities were more than 20 m from water, 30 percent were at least 500 m, and 5 percent were more than 1000 m, at a density of one per 7.7 ha (Delnicki and Bolen 1975). Suitable trees were 64-cm dbh and 52 cm at the cavity.

Common Loons

Where water fluctuations occur to cover shoreline nests of loons, rafts can improve production if floated in undisturbed areas of previously established territories (Sutcliffe 1979, Fair and Nielsen 1986). A raft is 1.5 m^2; framed with four 1.8- to 2.4-m, 20-cm-diameter cedar logs and one more in the middle; notched at both ends to overlap; and joined with 20-gauge galvanized spikes. Stapled to the bottom is 12.5-gauge galvanized wire fence with 5- by 10-cm mesh, covered with shoreline vegetation and duff. The raft should be floated within 2 weeks of ice-out, and anchored at opposite corners in 1 to 6 m of water 5 to 50 m from shore (Fig. 6.26) (Henderson undated).

Protected nest sites also can be improved by cutting away 3.3 to 9.3 m^2 of 40- to 60-cm thick free-floating pieces of bog with a hay saw and anchoring the nest site in open water (Mathisen 1969, McIntyre and Mathisen 1977). Plants typically include sedges (*Carex* spp.), leatherleaf, bog birch, and sphagnum moss. Four cedar logs, 7 to 10 cm in diameter, can be placed on the sides of the mat and wired at the corners.

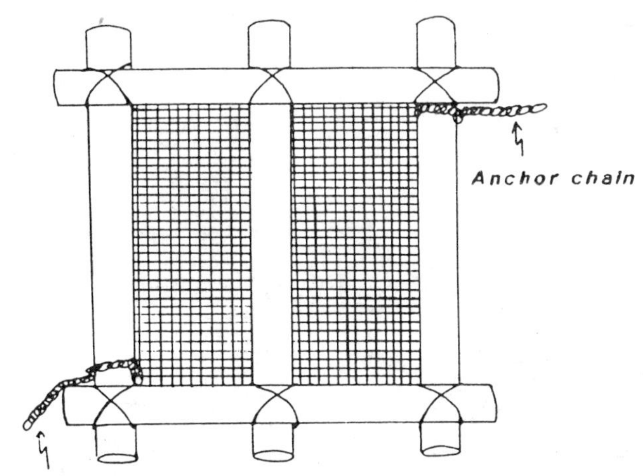

MATERIALS: Five 10" diameter cedar poles - 6' long
One 4'x4' welded wire screen
(2" x 4" mesh)

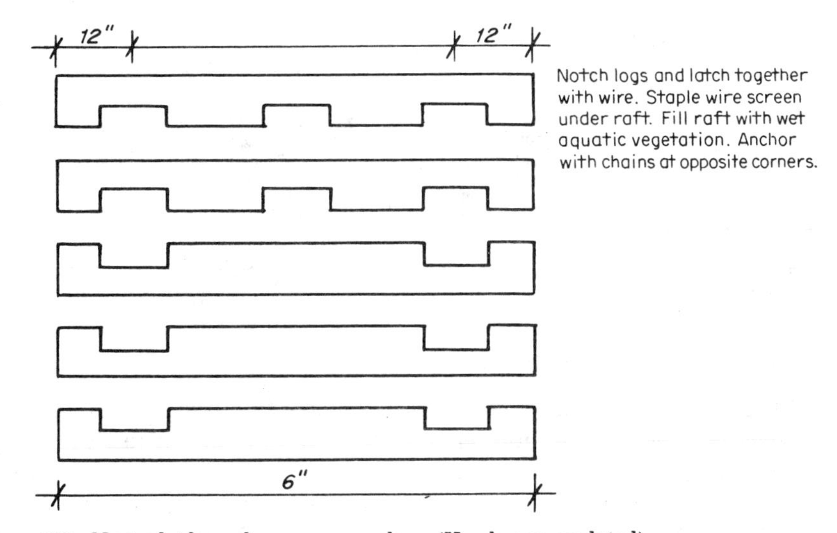

Figure 6.26 Nest platform for a common loon (Henderson undated). [For metric equivalence, see metric conversion table in the front of the book.]

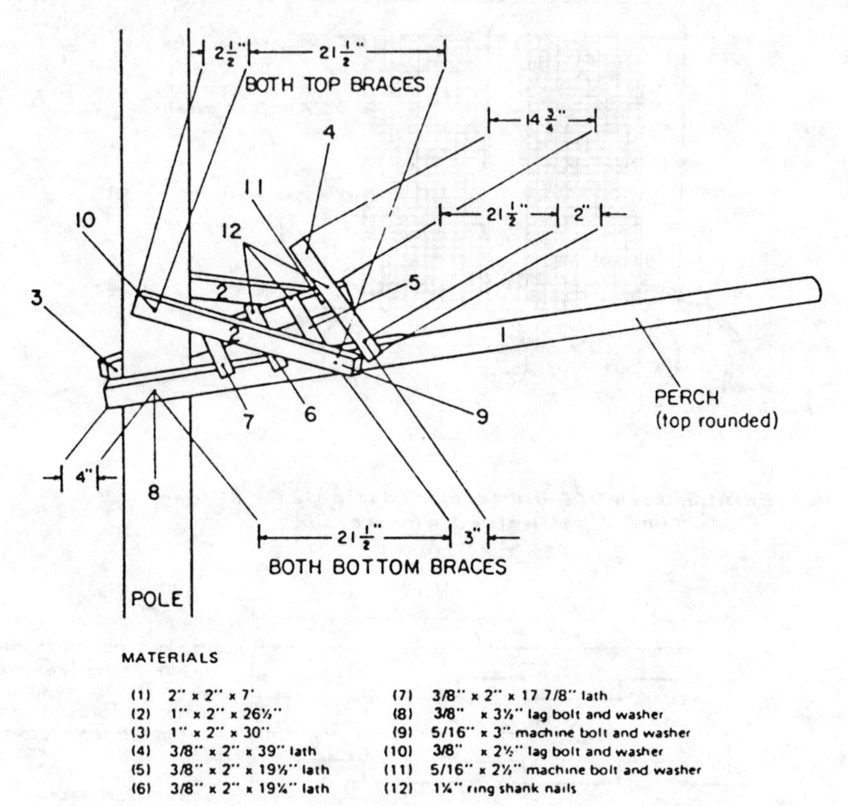

Figure 6.27 Nest platform for great blue herons, black-crowned night herons, and double-crested cormorants (Meier 1981). [For metric equivalence, see metric conversion table in the front of the book.]

Cormorants and Herons

Nest platforms for double-crested cormorants, great blue herons, and black-crowned night herons (Fig. 6.27) might prove useful to other similar colonial nesters when natural nesting trees are deteriorating or a new rookery is desired from a nearby source of birds. Three platforms with perch are attached to a 9-m treated pole with 20-cm base and 13-cm top diameters (Meier 1981). Poles should be within 7.6 m of natural nesting trees, if present, and 6 m apart, with additional poles added as expansion occurs. The platforms are made of southern yellow pine lumber treated with chromated copper arsenate under 0.40 retention which has relatively low toxicity and lasts 40 to 50 years. Platforms are placed at a 7° angle above horizontal to provide a nesting pocket, with three platforms per pole, 12 dm apart (9 dm for cormorants only), in a staggered 180° rotation. Platforms are erected 2.7 to 7.3 m above the water.

Poles are placed in areas with depths of at least 2 m of clay, 1.4 m or less of water, and little ice movement. Rock riprap is needed in less stable soils. Poles are installed in winter when ice will support the equipment (see Table 6.2). After a chain saw is used to cut a 4.5- by 4.5-dm hole through the ice, a truck-mounted hydraulic 20-cm auger is used to drill a hole 2 m deep into the bottom. A pole is positioned over the back of a bulldozer, with one end against the ice hole, and raised as the dozer backs up to a point where the pole drops into the hole. Then the dozer turns around, and a chain is attached to blade and pole, which is then hydraulically lowered. An extension ladder is used to reach the desired levels for attachment of platforms. Predator guards of 12-dm-long sheet metal might be needed around the middle of the pole. As natural nesting trees deteriorate within the rookery, additional perching structures on 9-m poles are needed to supplement the platform poles (Fig. 6.28) (Meier 1981).

Forster's Terns

To preserve a colony site of Forster's terns, 10 to 20 platforms should be placed in existing colonies in sheltered openings of cattails (Fig. 6.29) (Henderson undated). Steel posts 2.4 m long are driven into the

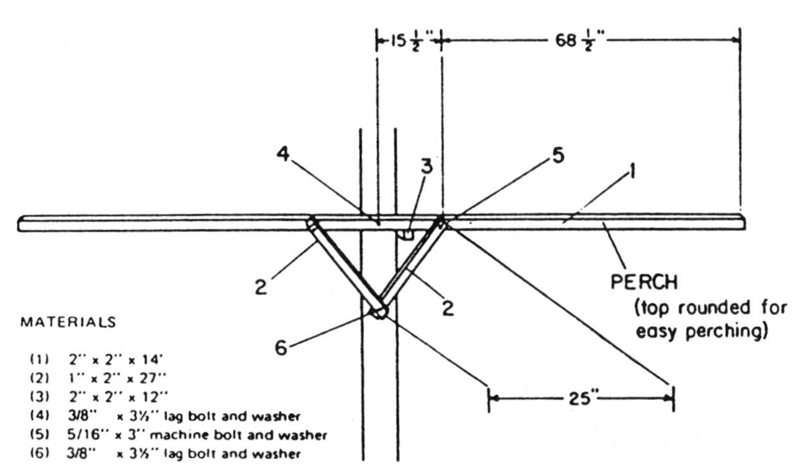

Figure 6.28 Perch for heron and cormorant rookeries (Meier 1981). [For metric equivalence, see metric conversion table in the front of the book.]

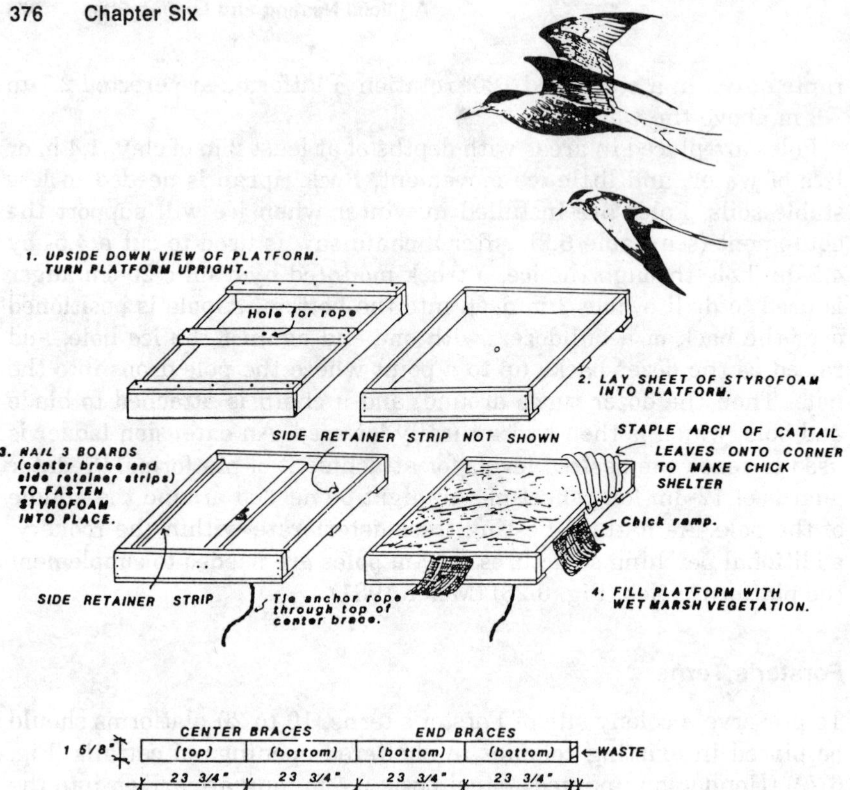

Figure 6.29 Nest platform for a Forster's tern (Henderson undated). [For metric equivalence, see metric conversion table in the front of the book.]

marsh bottom until the tops are near the high-water mark and are spaced about 7.6 m apart. Polypropylene rope, 0.6 cm, is tied between the posts 25 cm below the water surface. The fastening line from each platform is tied to the securing line 1.8 m apart, to prevent flooding from high water.

Other Colonial Waterbirds

To enhance habitat for establishing new colonies of species such as Atlantic puffins and Leach's storm petrels, artificial burrows have been used. In addition to habitat improvement, other solutions to the problems of establishing and maintaining colonial waterbirds include reducing contaminants, controlling disturbance, controlling pests, and improving the prey base, e.g., multispecies fishery management (Parnell et al. 1988).

Cliff Swallows

At sites where a colony of cliff swallows is eliminated through destruction of nests, such as bridge improvement or replacement, artificial nests will expedite recolonization (Henderson undated). An unused nest is coated three times with Blue Diamond Casting Plaster, removed with a putty knife, the mud is washed out, more plaster is added inside to approximate the inside dimensions, and the nest is filled with latex to set for about 10 min before excess latex is poured out to leave a thin sheet of latex over the inside of the mold. After hardening, the latex sheet is removed and placed on a flat surface, with the front facing upward. A collar of modeling clay is applied around the entrance, opening outward to a 5-cm diameter, and a 0.6-cm layer of wet casting plaster is applied over the model but not over the entrance, incorporating 2.5- by 15-cm strips of burlap for strength, with a flange of plaster on top and bottom for screws (Fig. 6.30). The plaster hardens in 15 min. Then the flexible latex mold is removed from inside. The waterproof qualities of the plaster are improved by sealing with acrylic or polyurethane spray paint before the outside of the nest and the inside of the entrance are coated with mud. Nests are attached horizontally on boards or in clusters on a plywood sheet. They last 2 to 3 years.

Phoebes, Barn Swallows, Songbirds, Bats

Platforms installed in culverts and bridges provide nesting for phoebes and barn swallows (Fig. 6.31) (Whitaker 1974). To improve concrete bridges for wildlife, design features include expanded beams for nesting, crevices for bats, roughened concrete to aid nest construction by some species of birds, wooden planks for nesting platforms, and bird boxes (Fig. 6.32) (Maser et al. 1979).

Ospreys

Artificial platforms for ospreys are used to provide nests where natural sites do not occur, to replace insecure natural nests, or to relocate nests away from excessive disturbance or hazardous areas (Martin et al. 1986). Sites should have an abundant fish population, clear and/or

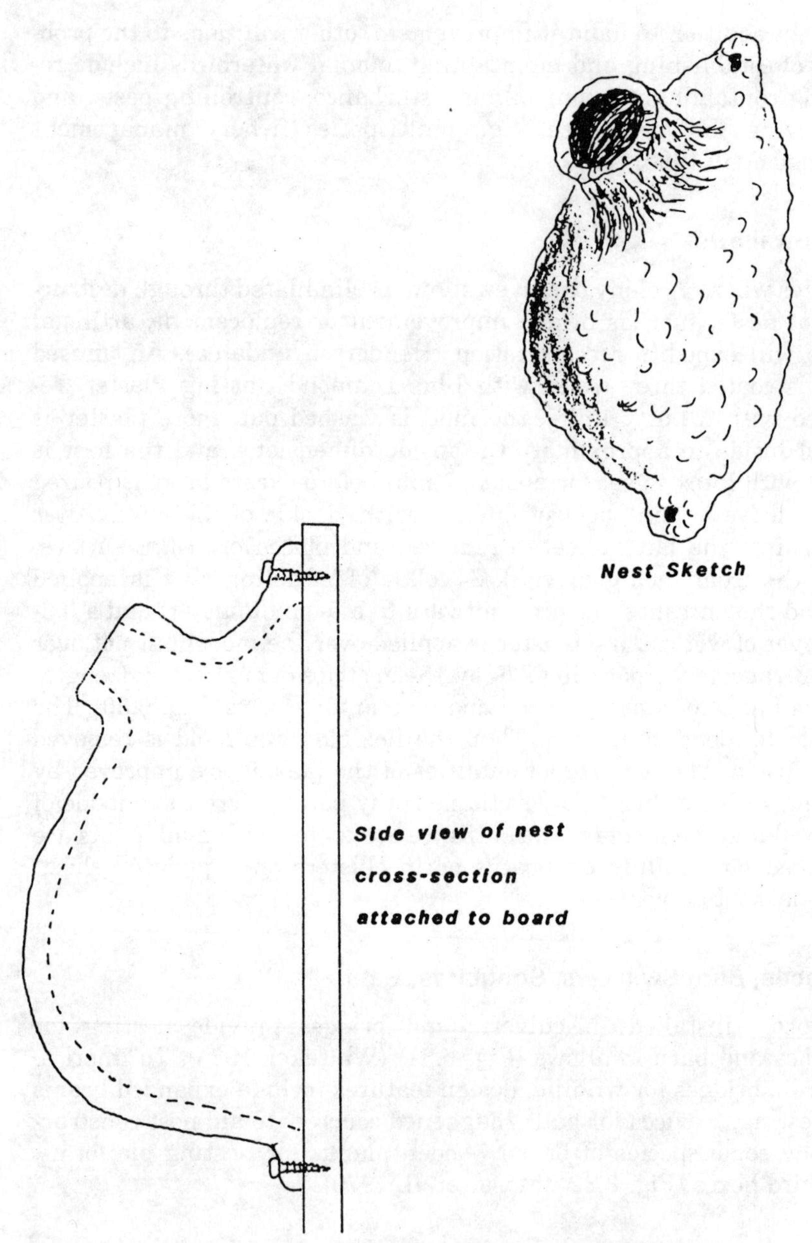

Figure 6.30 Artificial nest for cliff swallows (Henderson undated).

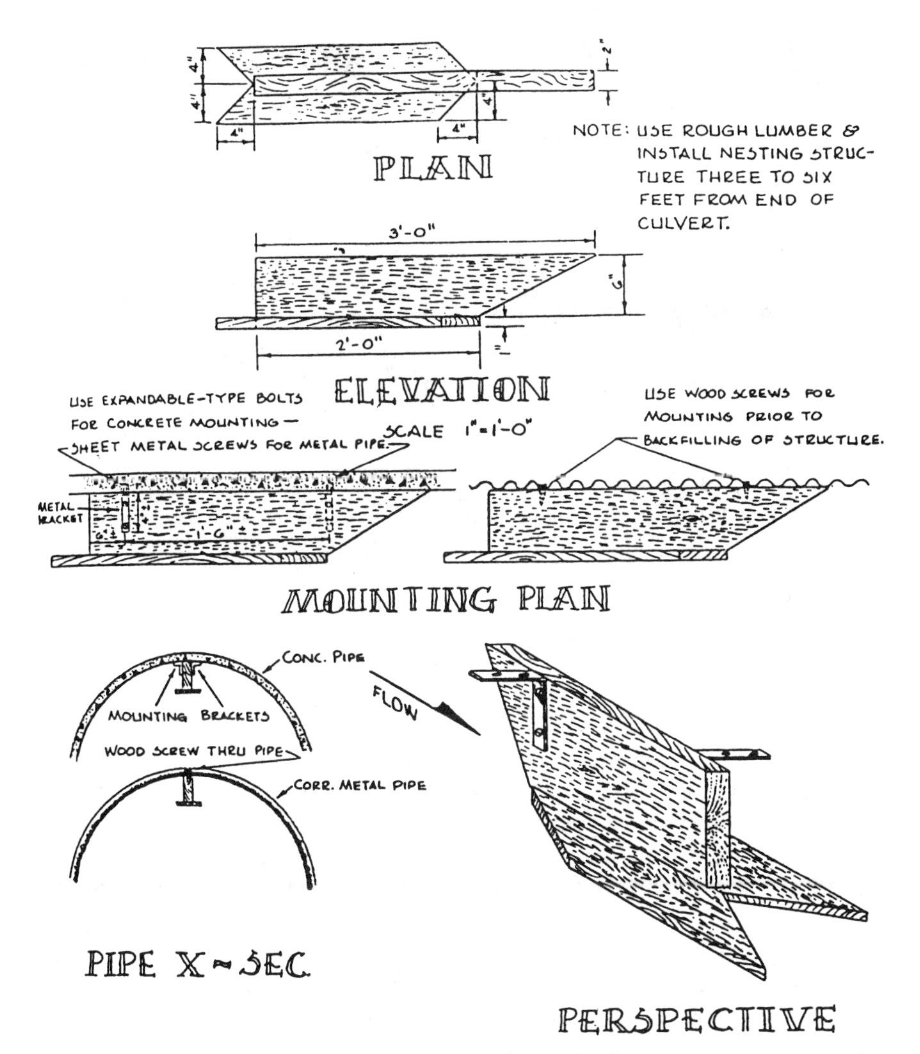

Figure 6.31 Construction and installation plans for artificial nesting structures on culverts, for use by phoebes and barn swallows (Whitaker 1974). [For metric equivalence, see metric conversion table in the front of the book.]

shallow good-quality water, relative isolation from human disturbance, and terrain and vegetation that are lower than the nest platform. Platforms should be at least 274 m apart; distant from fish hatcheries and nests or perches of other large raptors, crows, and ravens; and accessible for maintenance and data collection. The variety of platform designs and modifications available can be synthesized to three types of platform and three methods of support.

Figure 6.32 Design of concrete bridges to improve their potential as wildlife habitat: (1) Extended beams for nest construction, (2) artificial crevice in which bats can roost and rear young, (3) roughened concrete to aid nest construction by some species of birds, (4) wooden plank to create a platform on which birds can nest, and (5) bird boxes to enhance use by a variety of birds (Maser et al. 1979).

Platforms

The three basic designs of platform include frame (Fig. 6.33), solid-base (Fig. 6.34), and ring (Fig. 6.35). Frame platforms usually are mounted on single pole supports. Solid-base platforms can be attached to a pole or tree or attached to a tripod (Fig. 6.36). Ring platforms are used extensively in coastal areas and inland waterways and are attached to marine navigational aids. The Sanibel tripod is a lightweight, portable tripod-platform combination (Figs. 6.37 and 6.38), suitable for use in remote areas and in wet areas with soft substrates (Webb and Lloyd 1984, Martin et al. 1986). It might be useful if modified for bald eagles (compare Grubb 1980).

On power poles, a platform can be bolted onto 5- by 15-cm boards mounted on opposite sides of the pole and long enough to hold the nest at least 15 cm above the powerlines, or the attachment can be made with an angle-iron brace (Fig. 6.39) (Poole 1985). With power lines less than 1.8 m apart, nesting should be discouraged by bolting or nailing to the crossarms a series of 0.9-m-long pieces of 5- by 5-cm lumber 51 cm apart (Fig. 6.40) (Van Daele et al. 1980, Martin et al. 1986). Another method to discourage nesting is to bolt or screw two 5-cm by 6-dm by 51-cm or longer boards to the crossarms, one on each opposite end, and bolt or screw a piece of PVC plastic pipe, cut in half the length of the crossarms, to the top of the two boards (Fig. 6.40).

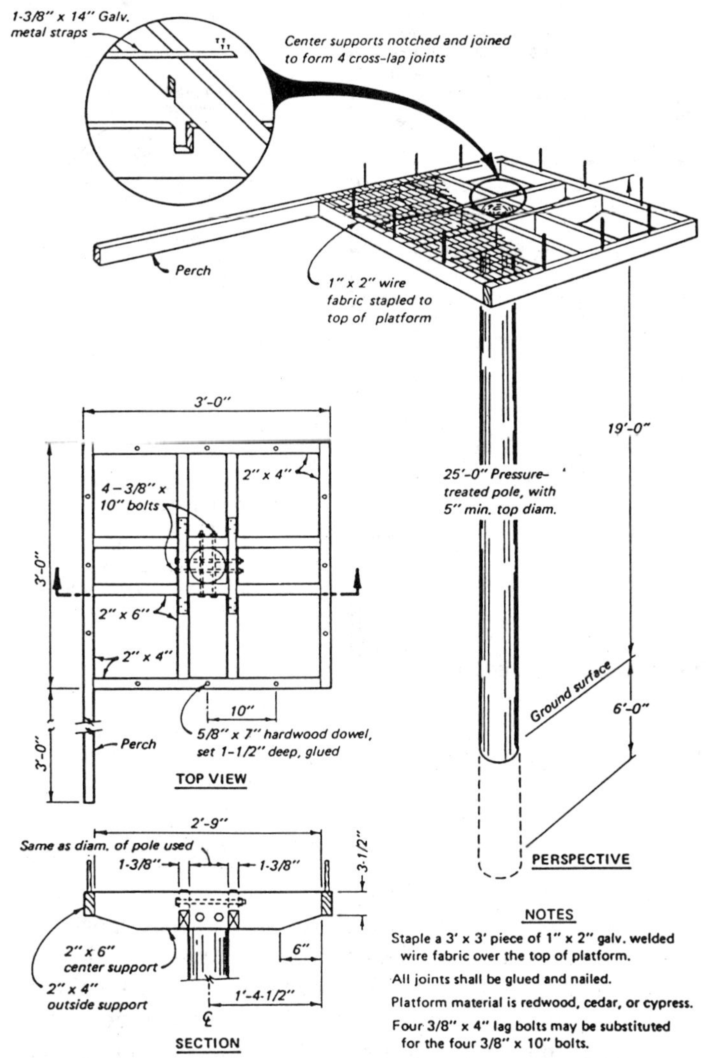

Figure 6.33 A frame nesting platform for ospreys (after guidelines provided by Bob Adair, U.S. Bureau of Reclamation *in* Martin et al. 1986). [For metric equivalence, see metric conversion table in the front of the book.]

Supports

The three basic methods of supporting platforms are top-mount (Figs. 6.33, 6.36, and 6.39), side-mount (Fig. 6.35), and tripod-mount (Figs. 6.36 and 6.37). Platform supports can be a snag, live-topped tree, pole, or tripod. Snags deteriorate most quickly, but others last 15 to 20

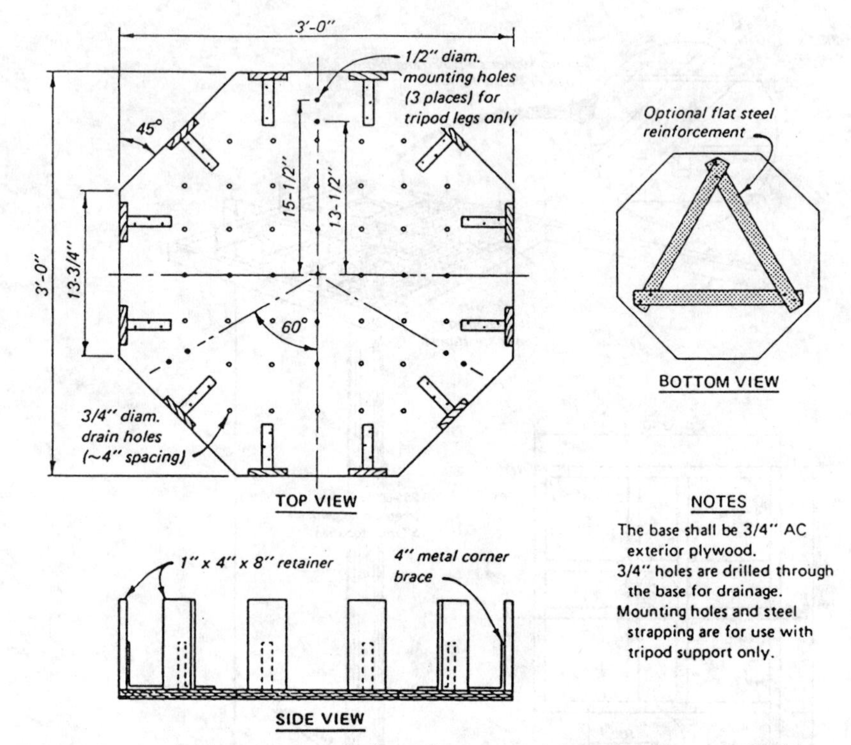

Figure 6.34 A solid-base nesting platform for ospreys (after guidelines provided by Thomas U. Fraser, Sr., Conservation for Survival *in* Martin et al. 1986). [For metric equivalence, see metric conversion table in the front of the book.]

years. Poles are pressure-treated, at least 7.6 m long for at least 1.8-m placement in ground, and at least 13-cm top diameter.

In clay soils, installation can occur in winter when ice supports the equipment. (See "Cormorants and Herons" in this chapter.) In less supportive substrates, installation should occur when water levels are lowest, usually in late summer or fall, so that the hole can be drilled and the pole set with a backhoe into dry soil, plumbed, and the soil tamped.

Snags and live trees are topped to at least 13-cm diameter of solid wood by a chain saw or primacord (200 gr), also called detonation cord, which is less hazardous than a chain saw. The Intermountain Region of the U.S. Forest Service reports that the primacord should be tightly wrapped just a few centimeters above a full whorl of limbs that will provide support for nest construction. The cap wire is attached to the primacord and to a lower limb with a slip knot and is dropped to the ground. The charge is detonated electronically at a safe distance (at least 20 m) from the tree. (See "Blasted Potholes" in Chap. 3.) The cap

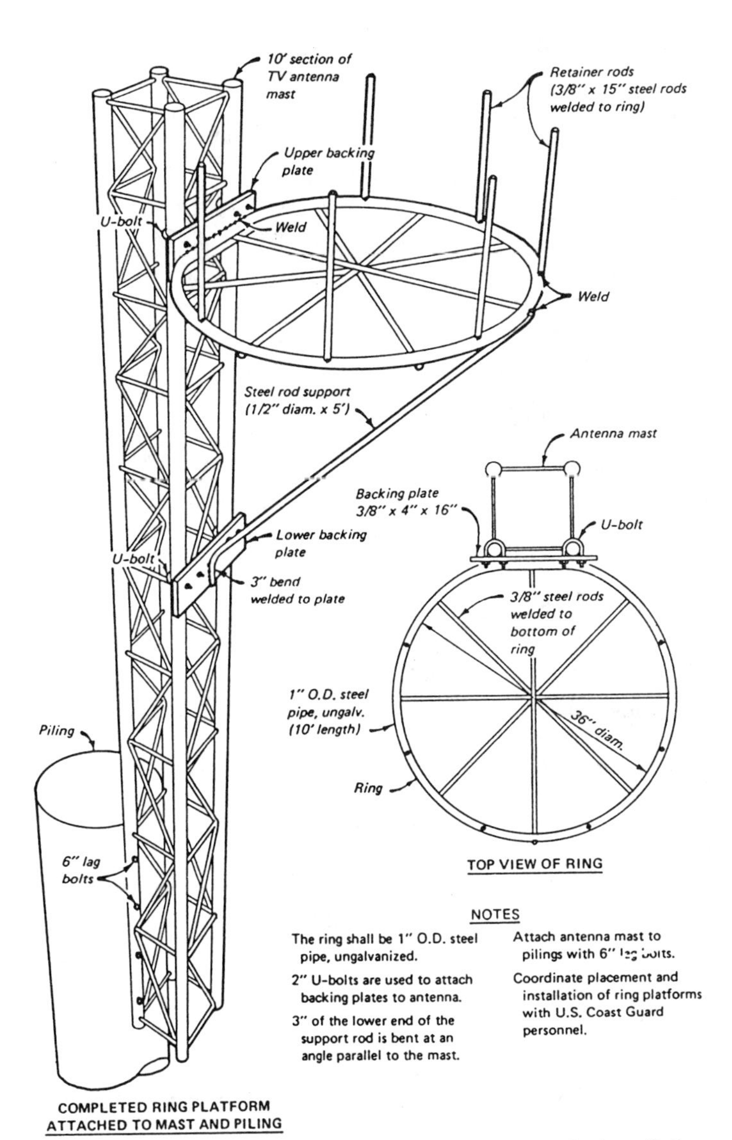

Figure 6.35 A ring nesting platform for ospreys that can be installed on marine navigation aids (Martin et al. 1986). [For metric equivalence, see metric conversion table in the front of the book.]

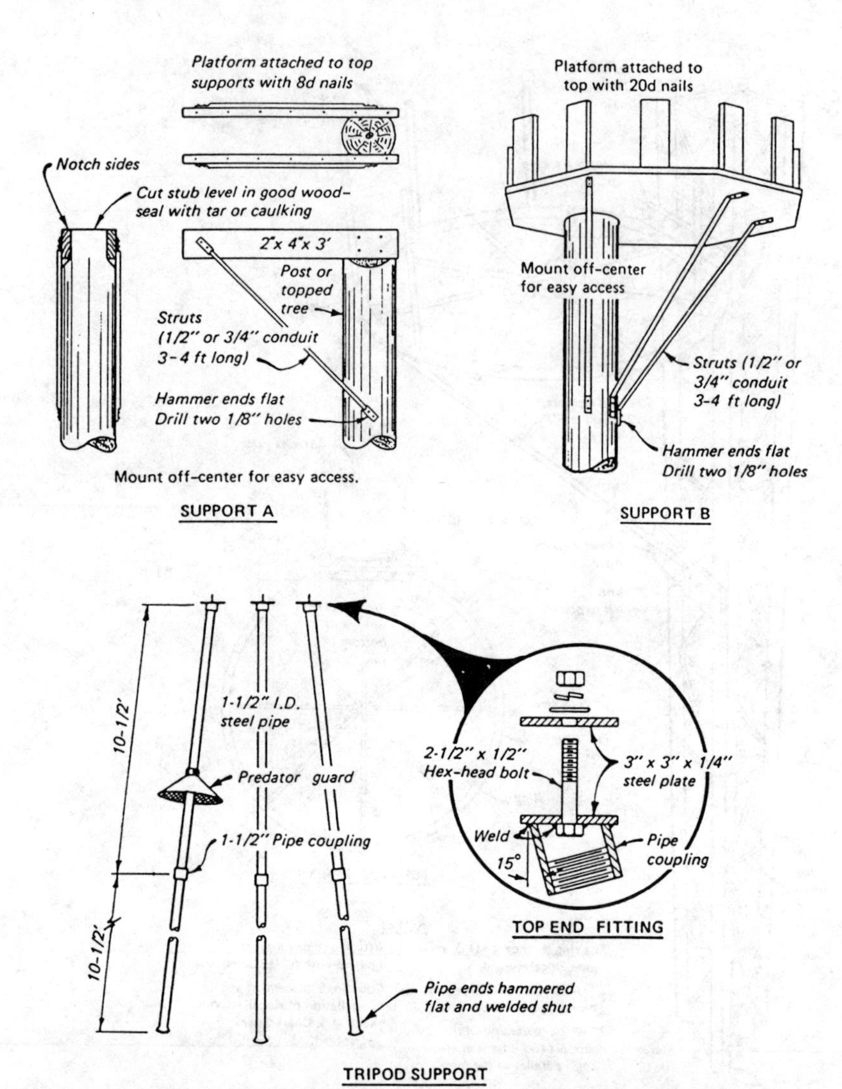

Figure 6.36 Supports used with a solid-base osprey platform (after guidelines provided by Thomas U. Fraser, Sr., Conservation for Survival *in* Martin et al. 1986). [For metric equivalence, see metric conversion table in the front of the book.]

wire is then pulled from the tree. The number of wraps is determined by experiment, but guidelines are as follows: seven wraps per 25-cm-diameter tree, five wraps per 15-cm tree, and four wraps per 10-cm tree, or about one wrap per 2.5 cm. After the tree is severed, any holes in the cut are sealed with tar or caulking. Predator guards of 1.2-m-long sheet metal might be needed around the middle of the pole or tree.

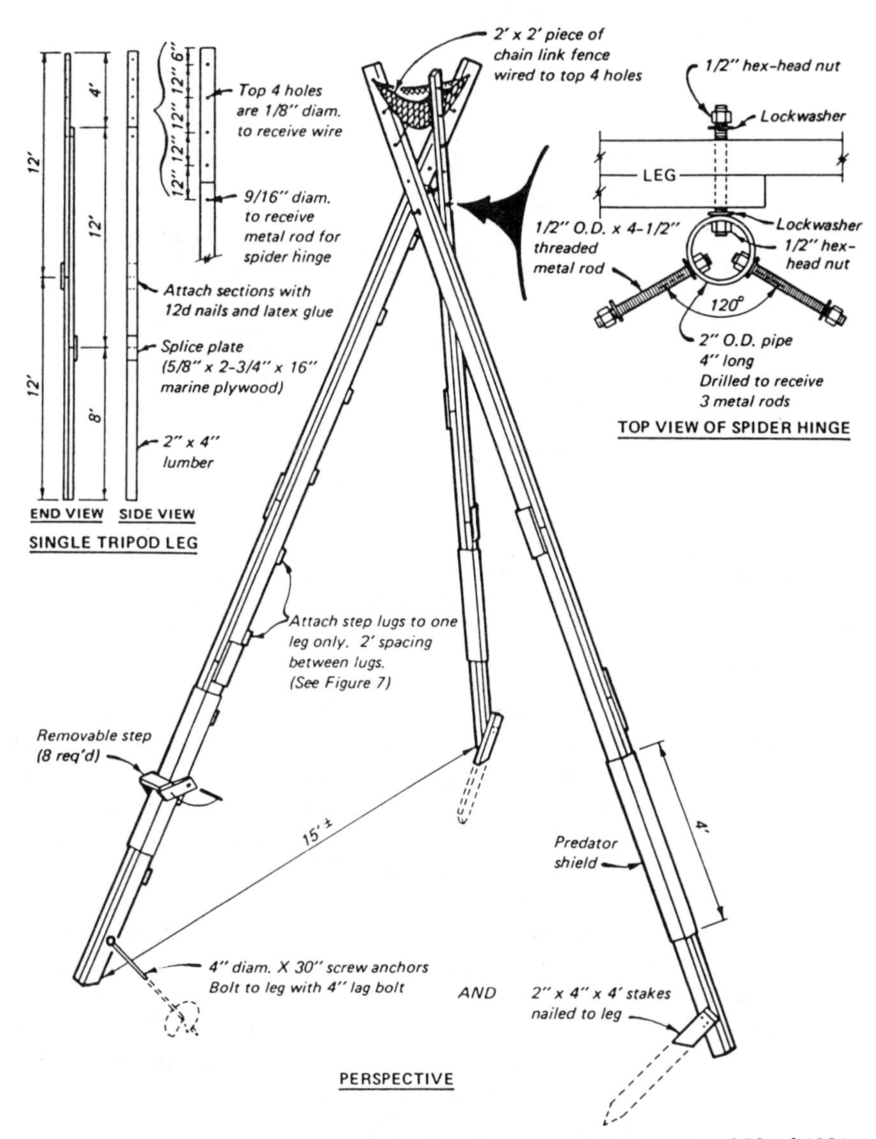

Figure 6.37 A Sanibel tripod nesting platform for ospreys (after Webb and Lloyd 1984 *in* Martin et al. 1986). [For metric equivalence, see metric conversion table in the front of the book.]

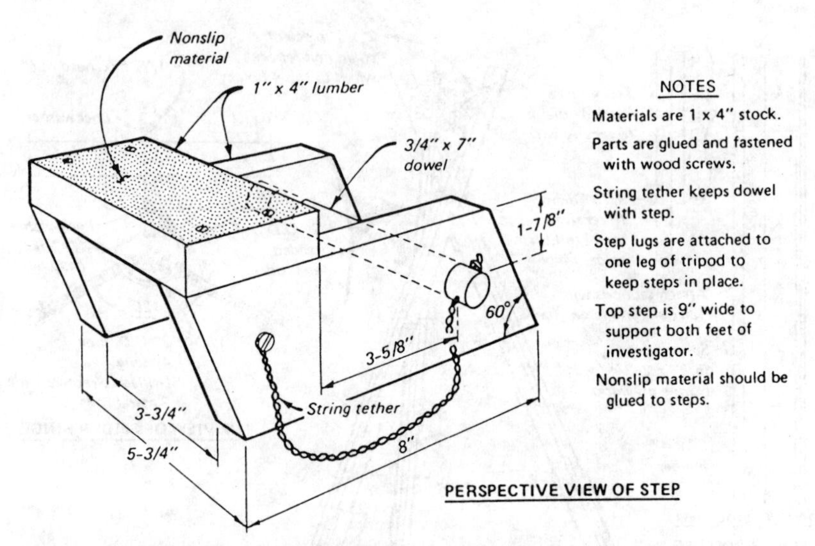

Figure 6.38 Material and details to construct removal steps attached to one leg of a Sanibel tripod (after Webb and Lloyd 1984 *in* Martin et al. 1986). [For metric equivalence, see metric conversion table in the front of the book.]

Everglade Kites

Nest structures for kites consist of a shallow basket 8 cm deep and with 55-cm inside diameter, of aircraft-grade 321-gauge stainless steel, galvanized steel, or aluminum sheet metal (Sykes and Chandler 1974). The outer tubular ring is 1 cm in diameter. Six concentric and 15 radial strips, each 1.5 cm wide, are riveted together. Three braces, woven into the basket, are riveted to a shaft of thin-walled metal tubing, 8 cm in diameter and 1.5 m long, left open at the bottom. The radial strips are cut from the upper end of the support shaft, leaving the lower ends attached. A wooden post or metal pipe, which is the size of the inside diameter of the support shaft, is driven into the ground and inserted into the support shaft. Nests are placed in tropical cattails usually, but to prevent predation by ants, no vegetation must touch the nest structure. A predator guard and sticky material (e.g., Tanglefoot, Tack-Trap) might be needed on the support shaft.

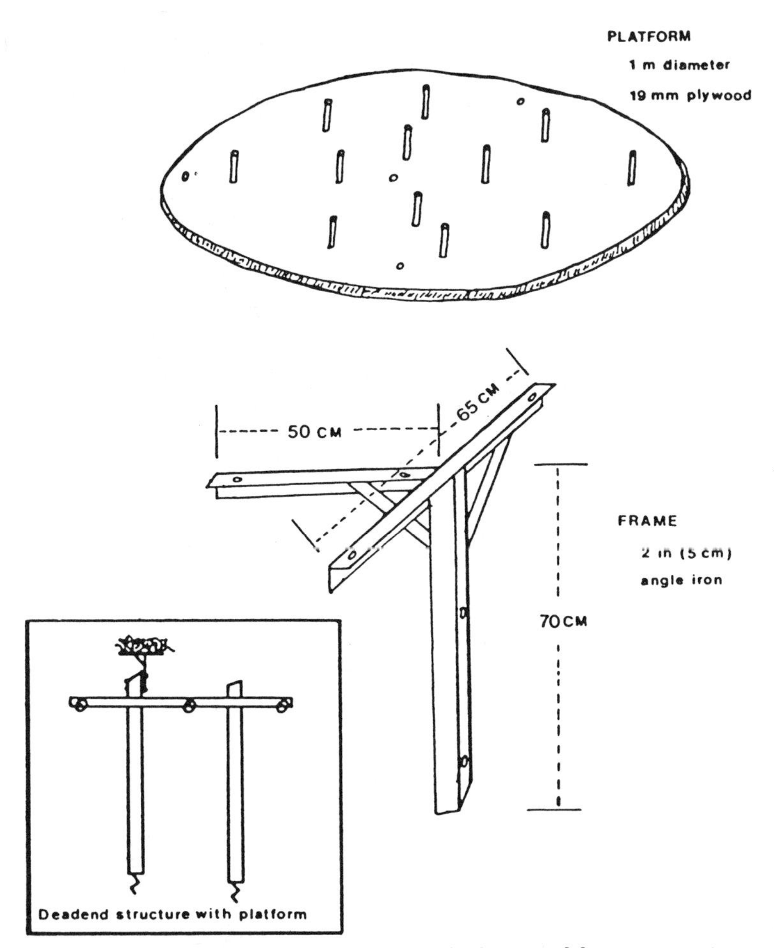

Figure 6.39 Angle-iron frame and wooden platform used for osprey nests on power poles (Poole 1985).

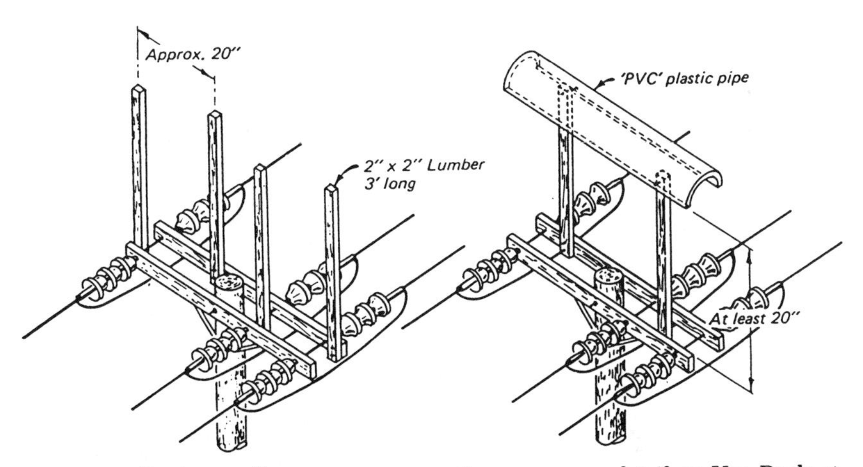

Figure 6.40 Devices to discourage osprey nesting on power poles (from Van Daele et al. 1980 *in* Martin et al. 1986). [For metric equivalence, see metric conversion table in the front of the book.]

References

Adamus, P. R., E. J. Clairain, Jr., R. D. Smith, and R. E. Young. 1987. Wetland evaluation technique (WET). Vol. 2: Methodology. Operational Draft Tech. Rep. Y-87-000. U.S. Army Eng. Waterways Exp. Stn., Vicksburg, MS. 206pp.

Addy, C. E., and L. G. MacNamara. 1948. Waterfowl management on small areas. Wildl. Manage. Inst., Washington. 84pp.

Ailstock, M. S. 1987. A review of beach prisms: their application for wetlands creation under moderate to high energy conditions. Proc. Annu. Conf. on Wetlands Restoration Creation 14:7–16.

Alberta Fish and Wildlife. Undated. Using grass-legume mixtures to improve wildlife habitat. Alberta Fish Wildl. Habitat Development Fact Sheet 2. 8pp.

Alberta Fish and Wildlife Division. Undated. Developing nesting islands to enhance waterfowl habitat. Alberta Energy Nat. Resour., Fish Wildl. Div. ENR Rep. 1/121-No.6. 4pp.

Allaire, P. N. 1979. Coal mining reclamation in Appalachia: low cost recommendations to improve bird/wildlife habitat. Pages 245–251 *in* G. A. Swanson, tech. coord. The mitigation symposium: a national workshop on mitigating losses of fish and wildlife habitats. U.S. For. Serv. Rocky Mountain For. Range Exp. Stn. Gen. Tech. Rep. RM-65.

Allen, A. W., and R. D. Hoffman. 1984. Habitat suitability index: muskrat. U.S. Fish Wildl. Serv. Rep. No. FWS/OBS-82/10.46. 27pp.

Allen, H. H., and C. V. Klimas. 1986. Reservoir shoreline revegetation guidelines. U.S. Army Eng. Waterways Exp. Stn. Tech. Rep. E-86-13. 87pp.

Allen, H. S. 1978. Role of wetland plants in erosion control of riparian shorelines. Pages 403–414 *in* P. E. Greeson, J. R. Clark, and J. E. Clark, eds. Wetland functions and values: the state of our understanding. Proc. Natl. Symp. on Wetland Functions and Values. Am. Water Resour. Assoc., Minneapolis.

Allen, K. O., and J. W. Hardy. 1980. Impacts of navigational dredging on fish and wildlife: a literature review. U.S. Fish Wildl. Serv. FWS/OBS-80/07. 81pp.

Allen, P. F. 1950. Ecological bases for land use planning in Gulf coast marshlands. J. Soil Water Conserv. 5:57–62, 85.

Allen, P. F., and W. L. Anderson. 1955. More wildlife from our marshes and wetlands. Pages 589–596 *in* U.S. Department of Agriculture. Water: the yearbook of agriculture. U.S. Dep. Agric., Washington.

Almeida, M. H. 1987. Nuisance furbearer damage control in urban and suburban areas. Pages 996–1006 *in* M. Novak, J. A. Baker, M. E. Obbard, and B. Malloch, eds. Wild furbearer management and conservation in North America. Ontario Min. Nat. Resour., Toronto.

Ambrose, R. E., C. R. Hinkle, and C. R. Wenzel. 1983. Practices for protecting and enhancing fish and wildlife on coal surface-mined land in the southcentral U.S. U.S. Fish Wildl. Serv. FWS/OBS-83/11. 229pp.

Anderson, B. W., J. Disano, D. L. Brooks, and R. D. Omart. 1984. Mortality and growth of cottonwood on dredge-fill. Pages 438–444 *in* R. E. Warner and K. M. Hendrix, eds. California riparian systems: ecology, conservation, and productive management. Univ. California Press, Berkeley.

Anderson, B. W., R. D. Ohmart, and J. Disano. 1978. Revegetating the riparian flood plain for wildlife. Pages 318–331 *in* R. R. Johnson and J. F. McCormick, tech. coord. Strategies for protection and management of floodplain wetlands and other riparian ecosystems. U.S. For. Serv. Gen. Tech. Rep. WO-12.

Anderson, G. R. 1985. Design and location of water impoundment structures. Pages 126–129 *in* M. D. Knighton, comp. Water impoundments for wildlife: a habitat management workshop. U.S. For. Serv. Gen. Tech. Rep. NC-100.

Anderson, J. W., and J. J. Cameron. 1980. The use of gabions to improve aquatic habitat. U.S. Bur. Land Manage. Tech. Note 432. 22pp.

Applied Biochemists, Inc. 1979. How to identify and control water weeds and algae. Applied Biochemists, Inc., Mequon, WI. 64pp.

Arner, D. H. 1963. Production of duck food in beaver ponds. J. Wildl. Manage. 27:76–81.

Aronson, J. G., and S. L. Ellis. 1979. Monitoring, maintenance, and enhancement of critical whooping crane habitat, Platte River, Nebraska. Pages 168–180 *in* G. A. Swanson, tech. coord. The mitigation symposium: a national workshop on mitigating losses of fish and wildlife habitats. U.S. For. Serv. Gen. Tech. Rep. RM-65.

Ash, A. N., C. B. McDonald, E. S. Kane, and C. A. Pories. 1983. Natural and modified pocosins: literature synthesis and management options. U.S. Fish Wildl. Serv. FWS/OBS-83/04. 156pp.

Atlantic Waterfowl Council. 1959. An illustrated small marsh construction manual based on standard designs. Vermont Fish Game Serv., Montpelier. 160pp.

Atlantic Waterfowl Council. 1972. Techniques handbook of waterfowl habitat development and management. 2d ed. Atlantic Waterfowl Counc., Bethany Beach, DE. 218pp.

Auble, G. T., D. B. Hamilton, J. E. Roelle, J. Clayton, and L. H. Fredrickson. 1988. A prototype expert system for moist soil management. Pages 137–143 *in* K. M. Mutz, D. J. Cooper, M. L. Scott, and L. K. Miller, tech. coord. Restoration, creation and management of wetland and riparian ecosystems in the American west. PIC Technologies, Denver.

Baldassarre, G. A., and E. G. Bolen. 1987. Management of waste corn for waterfowl wintering on the Texas high plains. Texas Tech Univ. Dep. Range Wildl. Manage. Note 13. 3pp.

Baldassare, G. A., R. J. Whyte, E. E. Quinlan, and E. G. Bolen. 1983. Dynamics and quality of waste corn available to postbreeding waterfowl in Texas. Wildl. Soc. Bull. 11:25–31.

Baldwin, W. P. 1968. Impoundments for waterfowl on south Atlantic and Gulf coastal marshes. Proc. Marsh Estuary Manage. Symp. 1:127–133.

Ball, I. J. 1990. Artificial nest structures for Canada geese. Pages 1–8 (13.2.12) *in* D. H. Cross, comp. Waterfowl management handbook. U.S. Fish Wildl. Serv. Fish Wildl. Leafl. 13.

Ball, I. J., S. K. Ball, and F. B. Lee. 1988. Artificial nest structures for mallards and Canada geese: a handbook. U.S. Fish Wildl. Serv. Tech. Rep. 86pp (draft).

Ball, I. J., R. D. Bauer, K. Vermeer, and M. J. Rabenberg. 1989. Northwest riverine and Pacific Coast. Pages 429–449 *in* L. M. Smith, R. L. Pederson, and R. M. Kaminski, eds. Habitat management for migrating and wintering waterfowl in North America. Texas Tech Univ. Press, Lubbock.

Ball, I. J., D. S. Gilmer, L. M. Cowardin, and J. H. Riechmann. 1975. Survival of wood duck and mallard broods in north-central Minnesota. J. Wildl. Manage. 38:776–780.

Ball, J. P., and T. D. Nudds. 1989. Mallard habitat selection: an experiment and implications for management. Pages 659–671 *in* R. R. Sharitz and J. W. Gibbons, ed. Freshwater wetlands and wildlife. CONF-8603101, DOE Symp. Ser. 61, U.S. Dep. Energy Off. Sci. Tech. Inf., Oak Ridge, TE.

Barfield, B. J., and S. C. Albrecht. 1982. Use of a vegetative filter zone to control fine-grained sediments from surface mines. Pages 481–490 *in* D. H. Graves, ed. Proceedings of a symposium on surface mining hydrology, sedimentation, and reclamation. Univ. Kentucky, Lexington.

Barnard, W. D. 1978. Prediction and control of dredged material dispersion around dredging and open-water pipeline disposal operations. U.S. Army Eng. Waterways Exp. Stn. Tech. Rep. DS-78-13. 114pp.

Barnard, W. D., and T. D. Hand. 1978. Treatment of contaminated material. U.S. Army Eng. Waterways Exp. Stn. Tech. Rep. DS-78-14. 46pp.

Bartos, M. J., Jr. 1977. Containment area management to promote natural dewatering of fine-grained dredged material. U.S. Army Eng. Waterways Exp. Stn. Tech. Rep. D-77-19. 86pp.

Bates, A. L., E. Pickard, and W. M. Dennis. 1978. Tree plantings—a diversified management tool for reservoir shorelines. Pages 190–194 *in* R. R. Johnson and J. F. McCormick, tech. coord. Strategies for protection and management of floodplain wetlands and other riparian ecosystems. U.S. For. Serv. Gen. Tech. Rep. WO-12.

Bates, G., G. L. Valentine, and F. H. Sprague. 1988. Waterfowl habitat created by floodwater-retarding structures in the southern United States. Pages 419–426 *in* M. W. Weller, ed. Waterfowl in winter. Univ. Minnesota Press, Minneapolis.

Beard, T. D. 1973. Overwinter drawdown impact on the aquatic vegetation in Murphy Flowage, Wisconsin. Wisconsin Dep. Nat. Resour. Tech. Bull. 61. 14pp.

Beauchamp, K. H. 1979. Structures. Pages 6-1–6-91 *in* Engineering field manual for conservation practices. U.S. Soil Conserv. Serv., Washington.

Beeman, S. 1983. Techniques for the creation and maintenance of intertidal saltmarsh wetlands for landscaping and shoreline protection. Proc. Annu. Conf. on Wetlands Restoration Prot. 10:33–43.

Beintema, A. J. 1982. Meadow birds in the Netherlands. Pages 83–91 *in* D. A. Scott, ed. Managing wetlands and their birds. Proc. 3d Tech. Meet. on West. Palearctic Migratory Bird Manage. Int. Waterfowl Res. Bur., Slimbridge, Glos., England.

Belanger, L., and R. Couture. 1988. Use of man-made ponds by dabbling duck broods. J. Wildl. Manage. 52:718–723.

Bellrose, F. C. 1953. Housing for wood ducks. Illinois Nat. Hist. Surv. Circ. 45. 47pp.

Bellrose, F. C. 1954. The value of waterfowl refuges in Illinois. J. Wildl. Manage. 18: 160–169.

Bellrose, F. C. 1980. Ducks, geese, and swans of North America. Stackpole Co., Harrisburg, PA. 540pp.

Bennett, F. D. 1974. Biological control. Pages 99–106 *in* D. S. Mitchell, ed. Aquatic vegetation and its use and control. U.N. Educ., Sci. and Cult. Organ., Paris.

Bennett, L. J. 1937. Grazing in relation to the nesting of the blue-winged teal. Trans. North Am. Wildl. Conf. 2:393–397.

Berger, J. J., ed. 1990. Environmental restoration: science and strategies for restoring the earth. Island Press, Washington. 398pp.

Beule, J. D. 1979. Control and management of cattails in southeastern Wisconsin wetlands. Wisconsin Dep. Nat. Resour. Tech. Bull. 112. 40pp.

Bishop, R. A., R. D. Andrews, and R. J. Bridges. 1979. Marsh management and its relationship to vegetation, waterfowl, and muskrats. Proc. Iowa Acad. Sci. 86:50–56.

Bolen, E. G. 1964. Plant ecology of spring-fed salt marshes in western Utah. Ecol. Monogr. 34:143–166.

Bolen, E. G. 1967. Nesting boxes for black-bellied tree ducks. J. Wildl. Manage. 31:794–797.

Bolen, E. G., G. A. Baldassarre, and F. S. Guthery. 1989. Playa lakes. Pages 341–365 *in* L. M. Smith, R. L. Pederson, and R. M. Kaminski, eds. Habitat management for migrating and wintering waterfowl in North America. Texas Tech Univ. Press, Lubbock.

Bookhout, T. A., K. E. Bednarik, and R. W. Kroll. 1989. The Great Lakes marshes. Pages 131–156 *in* L. M. Smith, R. L. Pederson, and R. M. Kaminski, eds. Habitat management for migrating and wintering waterfowl in North America. Texas Tech Univ. Press, Lubbock.

Born, S. M., T. L. Wirth, E. Brick, and J. O. Peterson. 1973. Restoring the recreational potential of small impoundments: the Marion Millpond experience. Wisconsin Dep. Nat. Resour. Tech. Bull. 71. 20pp.

Bossenmaier. E. F. 1964. Cows and cutter bars. Pages 627–634 *in* J. P. Linduska, ed. Waterfowl tomorrow. U.S. Fish Wildl. Serv., Washington.

Bradbury, H. M. 1938. Mosquito control operations on tide marshes in Massachusetts and their effect on shore birds and waterfowl. J. Wildl. Manage. 2:49–52.

Brannon, J. M. 1978. Evaluation of dredged material pollution potential. U.S. Army Eng. Waterways Exp. Stn. Tech. Rep. DS-78-6. 39pp.

Brenner, F. J., and J. J. Mondok. 1979. Waterfowl nesting rafts designed for fluctuating water levels. J. Wildl. Manage. 43:979–982.

British Columbia Ministry of Environment. 1980. Stream enhancement guide. British Columbia Ministry Environ., Fish. Oceans, Vancouver. 95pp.

Britton, C. M., and H. A. Wright. 1971. Correlations of weather and fuel variables to mesquite damage by fire. J. Range Manage. 23:136–141.

Britton, R. H. 1982. Managing the prey fauna. Pages 92–97 *in* D. A. Scott, ed. Managing wetlands and their birds. Proc. 3d Tech. Meet. on West. Palearctic Migratory Bird Manage. Int. Waterfowl Res. Bur., Slimbridge, Glos., England.

Broadfoot, W. M., and H. L. Williston. 1973. Flooding effects on southern forests. J. For. 71:584–587.

Broome, S. W. 1989. Creation and restoration of tidal wetlands of the southeastern United States. Pages 37–72 *in* J. A. Kusler and M. E. Kentula, eds. Wetland creation and restoration: the status of the science. Vol. 1: Regional reviews. U.S. Environ. Prot. Agency EPA/600/3-89/038a.

Broschart, M. R., and R. L. Linder. 1986. Aquatic invertebrates in level ditches and adjacent emergent marsh in a South Dakota wetland. Prairie Nat. 18:167–178.

Brown, D. 1977. Handbook of equipment for reclaiming strip-mined land. U.S. For. Serv. Equipment Develop. Cent., Missoula, MT. 58pp.

Brown, M., and J. J. Dinsmore. 1986. Implications of marsh size and isolation for marsh bird management. J. Wildl. Manage. 50:392–397.

Brown, P. W., and M. A. Brown. 1981. Nesting biology of the white-winged scoter. J. Wildl. Manage. 45:38–45.

Buckner, D. L. 1988. Construction of cattail wetlands along the east slope of the front range of Colorado. Pages 126–131 *in* K. M. Mutz, D. J. Cooper, M. L. Scott, and L. K. Miller, tech. coord. Restoration, creation and management of wetland and riparian ecosystems in the American west. PIC Technologies, Denver.

Bue, I. G., L. Blankenship, and W. H. Marshall. 1952. The relationship of grazing practices to waterfowl breeding populations and production of stock ponds in western South Dakota. Trans. North Am. Wildl. Conf. 17:396–414.

Bue, I. G., H. G. Uhlig, and J. D. Smith. 1964. Stock ponds and dugouts. Pages 391-398 *in* J. P. Linduska, ed. Waterfowl tomorrow. U.S. Fish Wildl. Serv., Washington.

Buech, R. R. 1985. Beaver in water impoundments: understanding a problem of water-level management. Pages 95–105 *in* M. D. Knighton, comp. Water impoundments for wildlife: a habitat management workshop. U.S. For. Serv. Gen. Tech. Rep. NC-100.

Bunting, S. C., and H. A. Wright. 1974. Ignition capabilities on nonflaming firebrands. J. For. 72:646–649.

Burger, G. V. 1973. Practical wildlife management. Winchester Press, New York. 218pp.

Burkhalter, A. P., L. M. Curtis, R. L. Lazar, M. L. Beach, and J. C. Hudson. 1974. Aquatic weed identification and control manual. Florida Dep. Nat. Resour. Bur. Aquatic Plant Res. Cont. Tallahassee. 100pp.

Burton, T. A., M. Moulton, and C. Kretsinger. 1989. Restoration of severely eroded streambanks using tree revetments and willow revegetation on Diamond Creek, Caribou County, Idaho. Pages 164–165 *in* R. E. Gresswell, B. A. Barton, and J. L. Kershner, eds. Practical approaches to riparian resource management: an educational workshop. U.S. Bur. Land Manage., Billings, MT.

Canada Soil Survey Committee. 1978. The Canadian system of soil classification. Can. Dep. Agric. Res. Branch Publ. 1646. 164pp.

Carangelo, P. D. 1988. Creation of sea grass habitat in Texas: results of research investigations and applied programs. Pages 286–300 *in* J. Zelazny and J. S. Feierabend, eds. Proceedings of a conference increasing our wetland resources. Natl. Wildl. Fed., Washington.

Carey, A. B., and J. D. Gill. 1983. Direct habitat improvement—some recent advances. Pages 80–87 *in* J. W. Davis, G. A. G. Goodwin, and R. A. Ockenfeis, eds. Snag habitat management: proceedings of symposium. U.S. For. Serv. Gen. Tech. Rep. RM-99.

Carey, R. C., P. S. Markovits, and J. B. Kirkwood, eds. 1981. Proceedings U.S. Fish and Wildlife Service workshop on coastal ecosystems of the southeastern United States. U.S. Fish Wildl. Serv. FWS/OBS-80/59. 257pp.

Carlton, R. L., and J. Jackson. 1984. Selected practices and plantings for wildlife. Univ. Georgia Coll. Agric. Bull. 733. 11pp.

Carothers, S. W., G. S. Mills, and R. R. Johnson. 1989. The creation and restoration of riparian habitat in southwestern and semi-arid regions. Pages 359–376 *in* J. A. Kusler and M. E. Kentula, eds. Wetland creation and restoration: the status of the science. Vol. 1: Regional reviews. U.S. Environ. Prot. Agency EPA/600/3-89/038a.

Carpenter, L. H., and G. L. Williams. 1972. A literature review on the role of mineral fertilizers in big game range management. Colorado Game, Fish, Parks Dep. Spec. Rep. 28. 25pp.

Cartwright, B. W. 1942. Regional burning as a marsh management technique. Trans. North Am. Wildl. Conf. 7:257–263.

Caslick, J. W., and D. J. Decker. 1981. Control of wildlife damage in homes and gardens. Cornell Univ. Coop. Ext. Serv. Inf. Bull. 176. 28pp.

Cely, J. E. 1979. The ecology and distribution of banana waterlily and its utilization by canvasback ducks. Proc. Annu. Conf. Southeast. Assoc. Fish Wildl. Agencies 33:43–47.

Chabreck, R. H. 1960. Coastal marsh impoundments for ducks in Louisiana. Proc. Annu. Conf. Southeast. Assoc. Game Fish Comm. 14:24–29.

Chabreck, R. H. 1968. Weirs, plugs and artificial potholes for the management of wildlife in coastal marshes. Proc. Marsh Estuary Manage. Symp. 1:178–192.

Chabreck, R. H., and C. M. Hoffpauer. 1962. The use of weirs in coastal marsh management in Louisiana. Proc. Annu. Conf. Southeast. Assoc. Game Fish Comm. 16:103–112.

Chabreck, R. H., T. Joanen, and S. L. Paulus. 1989. Southern coastal marshes and lakes. Pages 249–277 *in* L. M. Smith, R. L. Pederson, and R. M. Kaminski, eds. Habitat management for migrating and wintering waterfowl in North America. Texas Tech Univ. Press, Lubbock.

Chamberlain, E. B., Jr. 1960. Florida waterfowl populations, habitats and management. Florida Game Fresh Water Fish Comm. Tech. Bull. 7. 62pp.

Chapman, J. A., and G. A. Feldhamer, eds. 1982. Wild mammals of North America: biology, management, and economics. Johns Hopkins Univ. Press, Baltimore, MD. 1147pp.

Christiansen, J. E., and J. B. Low. 1970. Water requirements of waterfowl marshlands in northern Utah. Utah Div. Fish Game Publ. 69-12. 108pp.

Clark, S. H., H. L. Mendel, and W. Sarbello. 1974. Use of artificial nest shelters in eider management. Univ. Maine Res. in Life Sci. 22(2):1–15.

Clary, W. P., and B. F. Webster. 1989. Managing grazing of riparian areas in the Intermountain Region. U.S. For. Serv. Gen. Tech. Rep. INT-263. 11pp.

Clay, R. T., and M. Suprenant. 1987. Effects of burning and mechanical manipulation on *Phragmites australis* in Quebec. Final rep. Ducks Unlimited Canada, Montreal. 27pp.

Clayton, J. S., W. A. Ehrlich, D. B. Cann, J. H. Day, and I. B. Marshall. 1977. Soils of Canada. Agric. Can., Ottawa. 2 vols.

Clewell, A. F., and R. Lea. 1989. Creation and restoration of forested wetland vegetation in the southeastern United States. Pages 199–237 *in* J. A. Kusler and M. E. Kentula, eds. Wetland creation and restoration: the status of the science. Vol. 1: Regional reviews. U.S. Environ. Prot. Agency EPA/600/3-89/038a.

Coastal Zone Resources Division. 1978. Handbook for terrestrial wildlife habitat development on dredged material. U.S. Army Corps Eng. Tech. Rep. D-78-37. 388pp.

Condeletti, R. 1979. Naturalistic techniques for open marsh water management with the latest style rotary ditcher. Proc. New Jersey Mosquito Cont. Assoc. 66:74–78.

Cook, A. H. 1957. Control of muskrat burrow damage in earthen dikes. New York Fish Game J. 4:213–218.

Cook, A. H. 1964. Better living for ducks—through chemistry. Pages 569–578 *in* J. P. Linduska, ed. Waterfowl tomorrow. U.S. Fish Wildl. Serv., Washington.

Cooke, D. 1988. Lake and reservoir restoration and management techniques. Pages 6-1–6-38 *in* L. Moore and K. Thornton, eds. Lake and reservoir restoration guidance manual. North Am. Lake Manage. Soc., Environ. Res. Lab., Corvallis, OR.

Cooke, G. D., E. B. Welch, S. A. Peterson, and P. R. Newroth. 1986. Lake and reservoir restoration. Butterworth Publ., Boston. 392pp.

Cooperrider, A. Y., R. J. Boyd, and H. R. Stuart, eds. 1986. Inventory and monitoring of wildlife habitat. U.S. Bur. Land. Manage. Serv. Cent., Denver. 858pp.

Cottam, C. 1939. Food habits of North American diving ducks. U.S. Dep. Agric. Tech. Bull. 643. 140pp.

Countryman, C. M. 1964. Mass fires and fire behavior. U.S. For. Serv. Res. Pap. PSW-19. 53pp.

Countryman, C. M. 1971. Fire whorls...why, when, and where. U.S. For. Serv. Pac. Southwest For. Range Exp. Stn., Berkeley, CA. 11pp.

Cowan, W. F. 1982. Waterfowl production on zero-tillage farms. Wildl. Soc. Bull. 10: 305–308.

Cowardin, L. M. 1969. Use of flooded timber by waterfowl at the Montezuma National Wildlife Refuge. J. Wildl. Manage. 33:829–842.

Cowardin, L. M., V. Carter, F. C. Golet, and E. T. LaRoe. 1979. Classification of wetlands and deepwater habitats of the United States. U.S. Fish Wildl. Serv. Pub. FWS/OBS-79/31. 103pp.

Cranney, S., and V. Bachman. 1987. Explosives procedure handbook. Utah Div. Wildl. Resour., Salt Lake City. 63pp.

Crawford, H. S., Jr., and A. J. Bjugstad. 1967. Establishing grass range in the southwest Missouri Ozarks. U.S. For. Serv. Res. Note NC-22. 4pp.

Crawford, R. D., and J. A. Rossiter. 1982. General design considerations in creating artificial wetlands for wildlife. Pages 44–47 *in* W. D. Svedarsky and R. D. Crawford, eds. Wildlife values in gravel pits. Univ. Minnesota Agric. Exp. Stn. Misc. Publ. 17.

Cross, D. H., and K. L. Fleming. 1989. Control of Phragmites or common reed. Pages 1–5 (13.4.12) *in* D. H. Cross, comp. Waterfowl management handbook. U.S. Fish Wildl. Serv. Fish Wildl. Leafl. 13.

Currier, P. J., G. R. Lingle, and J. G. VanDerwalker. 1985. Migratory bird habitat on the Platte and North Platte rivers in Nebraska. The Platte River Whooping Crane Critical Habitat Maintenance Trust, Grand Island, NB. 177pp.

Cutshall, J. R. 1985. Vegetative establishment of smooth cordgrass (*Spartina alterniflora*) for shoreline erosion control. Proc. Coastal Marsh Estuary Manage. Symp. 4: 63–69.

Daiber, F. C. 1974. Salt marsh plants and future coastal salt marshes in relation to animals. Pages 475–508 in R. J. Reimold and W. H. Queens, eds. Ecology of halophytes. Academic Press, New York.

Daiber, F. C. 1986. Conservation of tidal marshes. Van Nostrand Reinhold, New York. 341pp.

Daiber, F. C. 1987. A brief history of tidal marsh mosquito control. Pages 234–252 *in* W. R. Whitman and W. H. Meredith, eds. Waterfowl and wetlands symposium: proceedings of a symposium on waterfowl and wetlands management in the coastal zone of the Atlantic Flyway. Delaware Dep. Nat. Resour. Environ. Cont., Dover.

Davis, D. J., D. W. Roberts, and K. M. Wicker. 1983. Components and controlling principles of coastal wetland management. Pages 41–65 *in* R. J. Varnell, ed. Proceedings of the water quality and wetland management conference. Louisiana Environmental Professionals Assoc., Metairie, LA.

Davis, G. A. 1977. Management alternatives for riparian habitat in the Southwest. Pages 59–67 *in* R. R. Johnson and D. A. Jones, tech. coord. Importance, preservation and management of riparian habitat: a symposium. U.S. For. Serv. Gen. Tech. Rep. RM-43.

Davis, J. W. 1982. Livestock vs. riparian habitat management—there are solutions. Pages 175–184 *in* J. M. Peek and P. D. Dalke, eds. Wildlife-livestock relationships symposium. Univ. Idaho For., Wildl., Range Exp. Stn., Moscow.

Davison, V. E., and W. W. Neely. 1959. Managing farm fields, wetlands, and waters for wild ducks in the south. U.S. Dep. Agric. Farmers' Bull. 2144. 14pp.

Davison, V. E., J. M. Lawrence, and L. V. Compton. 1962. Waterweed control on farms and ranches. U.S. Soil Conserv. Serv. Farmers' Bull. 2181. 22pp.

Dawson, F. H., and U. Kern-Hansen. 1979. The effect of natural and artificial shade on the macrophytes of lowland streams and the use of shade as a management technique. Int. Revue der Gesamter Hydrobiologie 64:437–455.

DeBano, L. F., and B. H. Heede. 1987. Enhancement of riparian ecosystems with channel structures. Water Resour. Bull. 23:463–470.

De la Cruz, A. A., and C. T. Hackney. 1980. The effects of winter fire and harvest on the vegetational structures and primary productivity of two tidal marsh communities in Mississippi. Mississippi-Alabama Sea Grant Consortium. Publ. MASGP-80-013. 115pp.

Delnicki, D. E., and E. G. Bolen. 1975. Natural nest site availability for black-bellied whistling ducks in south Texas. Southwest. Nat. 20:371–378.

Delnicki, D., and K. J. Reinecke. 1986. Midwinter food use and body weights of mallards and wood ducks in Mississippi. J. Wildl. Manage. 50:43–51.

Dennis, C. E. 1979. Elementary soil engineering. Pages 4-1–4-32 *in* Engineering field manual for conservation practices. U.S. Soil Conserv. Serv., Washington.

Derrenbacker, J., Jr., and R. R. Lewis, III. 1982. Seagrass habitat restoration, Lake Surprise, Florida Keys. Proc. Annu. Conf. on Wetlands Restoration Creation 9:132–154.

DeVaney, T. E. 1967. Chemical vegetation control manual for fish and wildlife management programs. U.S. Fish Wildl. Serv. Resour. Publ. 48. 42pp.

DeVoe, M. R., and D. S. Baughman. 1987. Coastal wetland impoundments: ecological characterization, management, status and use. Vol. 1: Executive summary. South Carolina Sea Grant Consortium Publ. SC-SG-TR-86-1. 42pp.

Dickerson, L. 1989. Beaver and beaver dam removal in Wisconsin trout streams. Proc. East. Wildl. Damage Cont. Conf. 4:135–141.

Dietz, R. H. 1967. Results of increasing waterfowl habitat and production by gull control. Trans. North Am. Wildl. Nat. Resour. Conf. 32:316–324.

Dillard, J. G. 1982. Missouri pond handbook. Missouri Dep. Conserv., Jefferson City. 61pp.

Dillon, O. W., Jr., W. W. Neely, V. E. Davison, and L. V. Compton. 1971. Warm-water fishponds. U.S. Soil Conserv. Serv. Farmers' Bull. 2250. 14pp.

Disano, J., B. W. Anderson, and R. D. Ohmart. 1984. Irrigation systems for riparian zone vegetation. Pages 471–476 *in* R. E. Warner and K. M. Hendrix, eds. California riparian systems: ecology, conservation, and productive management. Calif. Water Resour. Rep. 55, Univ. California Press, Berkeley.

Dobie, B. 1986. Private financing for wetland restoration. Pages 14–28 *in* J. L. Piehl, ed. Wetland restoration: a techniques workshop. Minnesota Chapter Wildl. Soc., Fergus Falls.

Doerr, T. B. 1986. Hydroseeders/mulchers. Section 8.4.7, U.S. Army Corps of Engineers wildlife resources management manual. U.S. Army Eng. Waterways Exp. Stn. Tech. Rep. EL–86–51. 8 pp.

Doran, W. L. 1957. Propagation of woody plants by cuttings. Univ. Massachusetts Coll. Agric. Exp. Stn. Bull. 491. 99pp.

Dore, W. G. 1969. Wild-rice. Can. Dep. Agric. Res. Branch Publ. 1393. 84pp.

Doty, H. A., and F. B. Lee. 1974. Homing to nest baskets by wild female mallards. J. Wildl. Manage. 38:714–719.

Doty, H. A., F. B. Lee, and A. D. Kruse. 1975. Use of elevated nest baskets by ducks. Wildl. Soc. Bull. 3:68–73.

Dozier, H. L. 1950. Muskrat trapping on the Montezuma National Wildlife Refuge, New York 1943–1948. J. Wildl. Manage. 14:403–412.

Dozier, H. L. 1953. Muskrat production and management. U.S. Fish Wildl. Serv. Circ. 18. 42pp.

Ducks Unlimited Canada. 1977. Construction 1976: a prediction come true. Ducks Unlimited 41(1):30–32, 41, 44–45.

Ducks Unlimited Canada. 1989. Focus on eiders. Conservator 10(1):6–9.

Duebbert, H. F. 1969. The ecology of Malheur Lake and management implications. U.S. Bur. Sport Fish. Wildl. Ref. Leafl. 412. 24pp.

Duebbert, H. F. 1982. Nesting of waterfowl on islands in Lake Audubon, North Dakota. Wildl. Soc. Bull. 10:232–237.

Duebbert, H. F., E. T. Jacobson, K. F. Higgins, and E. B. Podoll. 1981. Establishment of seeded grasslands for wildlife habitat in the prairie pothole region. U.S. Fish Wildl. Serv. Spec. Sci. Rep.-Wildl. 234. 21pp.

Duebbert, H. F., and H. A. Kantrud. 1987. Use of no-till winter wheat by nesting ducks in North Dakota. J. Soil Water Conserv. 42:50–53.

Duebbert, H. F., and J. T. Lokemoen. 1976. Duck nesting in fields of undisturbed grass-legume cover. J. Wildl. Manage. 40:39–49.

Duebbert, H. F., J. T. Lokemoen, and D. E. Sharp. 1986. Nest sites of ducks in grazed mixed-grass prairie in North Dakota. Prairie Nat. 18:99–108.
Duncan, D. C. 1986. Influences of vegetation on composition and density of island-nesting ducks. Wildl. Soc. Bull. 14:158–160.
Dunst, R. C. 1980. Sediment problems and lake restoration in Wisconsin. Pages 103–113 *in* S. A. Peterson and K. K. Randolph, eds. Management of bottom sediments containing toxic substances. U.S. Environ. Prot. Agency EPA 600/9-80-044.
Dyer, K. R. 1973. Estuaries, a physical introduction. Wiley, New York. 140pp.
Eadie, J. M., and G. Gauthier. 1985. Prospecting for nest sites by cavity-nesting ducks of the genus Bucephala. Condor 87:528–534.
Eckert, J. W., M. L. Giles, and G. M. Smith. 1978. Design concepts for in-water containment structures for marsh habitat development. U.S. Army Eng. Waterways Exp. Stn. Tech. Rep. D-78-31. 34pp.
Edminster, F. C. 1964. Farm ponds and waterfowl. Pages 399–407 *in* J. P. Linduska, ed. Waterfowl tomorrow. U.S. Fish Wildl. Serv., Washington.
Eng, R. L., J. D. Jones, and F. M. Gjersling. 1979. Construction and management of stockponds for waterfowl. U.S. Bur. Land Manage. Tech. Note 327. 39pp.
Engel, S. 1990. Ecosystem responses to growth and control of submerged macrophytes: a literature review. Wisconsin Dep. Nat. Resour. Tech. Bull. 170. 20pp.
Environmental Effects Laboratory. 1976. Ecological evaluation of proposed discharge of dredged or fill material into navigable waters. U.S. Army Eng. Waterways Exp. Stn. Misc. Pap. D-76-17. 83pp.
Erickson, R. C. 1964. Planting and misplanting. Pages 579–591 *in* J. P. Linduska, ed. Waterfowl tomorrow. U.S. Fish Wildl. Serv., Washington.
Ermacoff, N. 1968. Marsh and habitat management practices at the Mendota Wildlife Area. California Dep. Fish Game Leafl. 12. 10pp.
Errington, P. L. 1963. Muskrat populations. Iowa State Univ. Press, Ames. 665pp.
Erman, D. C., J. D. Newbold, and K. B. Roby. 1977. Evaluation of streamside buffer strips for protecting aquatic organisms. California Water Resour. Cent., Univ. California, Davis. 48pp.
Euliss, N. H., Jr., and G. Grodhaus. 1987. Management of midges and other invertebrates for waterfowl wintering in California. California Fish Game 73:238–243.
Evans, J. 1983. Nutria. Pages B61–70 *in* R. M. Timm, ed. Prevention and control of wildlife damage. Great Plains Agric. Counc. Wildl. Resour. Comm. and Nebraska Coop. Ext. Serv. Inst. Agric. Nat. Resour., Univ. Nebraska, Lincoln.
Evans, K. E., and R. R. Kerbs. 1977. Avian use of livestock watering ponds in western South Dakota. U.S. For. Serv. Gen. Tech. Rep. RM-35. 11pp.
Fager, L. F., and J. C. York. 1975. Floating islands for waterfowl in Arizona. Soil Conserv. 41(5):4–5.
Fair, J. S., and K. B. Nielsen. 1986. Use of artificial islands to enhance common loon nesting success in New Hampshire. Trans. Northeast Sect. Wildl. Soc. 43:73.
Falco, P. K., and F. J. Cali. 1977. Pregermination requirements and establishment techniques for salt marsh plants. U.S. Army Eng. Waterways Exp. Stn. Misc. Pap. D-77-1. 43pp.
Farmes, R. E. 1985. So you want to build a water impoundment. Pages 130–134 *in* M. D. Knighton, comp. Water impoundments for wildlife: a habitat management workshop. U.S. For. Serv. Gen. Tech. Rep. NC-100.
Federal Interagency Committee for Wetland Delineation. 1989. Federal manual for identifying and delineating jurisdictional wetlands. U.S. Army Corps Eng., U.S. Environ. Prot. Agency, U.S. Fish Wildl. Serv., U.S. Soil Conserv. Serv., Washington. 76pp.
Fellows, N. W., Jr. 1951. Results of a waterfowl food planting survival study in Maine. Maine Dep. Inland Fish Game, Portland. 23pp.
Florence, S. R. 1983. Prescribed burning for habitat improvement using the helitorch. Trans. Cal-Neva Wildl. 1983:162–167.
Folk, R. H., III, and C. W. Bales. 1982. An evaluation of wildlife mortality resulting from aerial ignition prescribed burning. Proc. Annu. Conf. Southeast. Assoc. Fish Wildl. Agencies 36:643–646.
Fonseca, M. S., W. J. Kenworthy, and G. W. Thayer. 1988. Restoration and management of seagrass systems: a review. Pages 353–368 *in* D. D. Hook, W. H. McKee, Jr.

H. K. Smith, J. Gregory, V. G. Burrell, Jr., M. R. DeVoe, R. E. Sojka, S. Gilbert, R. Banks, L. H. Stolzy, C. Brooks, T. D. Matthews, and T. H. Shear, eds. The ecology and management of wetlands. Vol. 2: Management, use and value of wetlands. Timber Press, Portland.

Foreman, H. J. 1979. Dikes and levees—wildlife wetland development. Pages 13-1–13-20 *in* Engineering field manual for conservation practices. U.S. Soil Conserv. Serv., Washington.

Foster, S. Q. 1986. Wetland values. Pages 177–214 *in* J. T. Windell, B. E. Willard, D. J. Cooper, S. Q. Foster, C. F. Knud-Hansen, L. P. Rink, and G. N. Kiladis, eds. An ecological characterization of Rocky Mountain montane and subalpine wetlands. U.S. Fish Wildl. Serv. Biol. Rep. 86(11).

Fowler, D. K., and D. A. Hammer. 1976. Techniques for establishing vegetation on reservoir inundation zones. J. Soil Water Conserv. 31:116–118.

Fraser, J. E., and D. L. Britt. 1982. Liming of acidified waters: a review of methods and effects on aquatic ecosystems. U.S. Fish Wildl. Serv. FWS/OBS-80/40.13. 189pp.

Fredette, T. J., M. S. Fonseca, W. J. Kenworthy, and G. W. Thayer. 1985. Seagrass transplanting: 10 years of U.S. Army Corps of Engineers research. Proc. Annu. Conf. on Wetlands Restoration Creation 12:121–134.

Fredrickson, L. H. 1978. Lowland hardwood wetlands: current status and habitat values for wildlife. Pages 296–306 *in* P. E. Greeson, J. R. Clark, and J. E. Clark, eds. Wetland functions and values: the state of our understanding. Proc. Natl. Symp. on Wetland Functions Values. Am. Water Resour. Assoc., Minneapolis.

Fredrickson, L. H. 1985. Managed wetland habitats for wildlife: why are they important? Pages 1–8 *in* M. D. Knighton, comp. Water impoundments for wildlife: a habitat management workshop. U.S. For. Serv. Gen. Tech. Rep. NC-100.

Fredrickson, L. H., and R. D. Drobney. 1979. Habitat utilization by postbreeding waterfowl. Pages 119–131 in T. A. Bookhout, ed. Waterfowl and wetlands—An integrated review. North Cent. Sect. Wildl. Soc., Madison, WI.

Fredrickson, L. H., and F. A. Reid. 1986. Wetland and riparian habitats: a nongame management overview. Pages 59–96 *in* J. B. Hale, L. B. Best, and R. L. Clawson, eds. Management of nongame wildlife in the midwest: a developing art. North Cent. Sect. Wildl. Soc., Chelsea, MI.

Fredrickson, L. H., and F. A. Reid. 1988a. Invertebrate response to wetland management. Pages 1–5 (13.3.1) *in* D. H. Cross, comp. Waterfowl management handbook. U.S. Fish Wildl. Serv. Fish Wildl. Leafl. 13.

Fredrickson, L. H., and F. A. Reid. 1988b. Waterfowl use of wetland complexes. Pages 1–6 (13.2.1) *in* D. H. Cross, comp. Waterfowl management handbook. U.S. Fish Wildl. Serv. Fish Wildl. Leafl. 13.

Fredrickson, L. H., and T. S. Taylor. 1982. Management of seasonally flooded impoundments for wildlife. U.S. Fish Wildl. Serv. Res. Publ. 148. 29pp.

French, R. H. 1985. Open channel hydraulics. McGraw-Hill, New York. 705pp.

Frentress, C. 1989. An improved drain for beaver ponds. U.S. Army Corps Eng. Wildl. Resour. Notes 7(1):6–7.

Fulton, G. W., J. L. Richardson, and W. T. Barker. 1986. Wetland soils and vegetation. North Dakota State Univ. Agric. Exp. Stn. Rep. 106. 15pp.

Gaby, R. 1986. Mechanical harvesting of seagrasses for mitigation projects. Proc. Annu. Conf. on Wetlands Restoration Creation 13:87–93.

Gangstad, E. O. 1986. Freshwater vegetation management. Thomas Publ., Fresno, CA. 380pp.

Garbisch, E. W. 1986. Highway and wetlands: compensating wetland losses. U.S. Fed. Highway Admin. Rep. FHWA-1P-86-22. 60pp.

Gauthier, G. 1988. Factors affecting nest-box use by buffleheads and other cavity-nesting birds. Wildl. Soc. Bull. 16:132–141.

Gavin, A. 1964. Ducks unlimited. Pages 545–553 *in* J. P. Linduska, ed. Waterfowl tomorrow. U.S. Fish Wildl. Serv., Washington.

George, H. A. 1963. Planting alkali bulrush for waterfowl food. Calif. Dep. Fish Game Manage. Leafl. 9. 9pp.

Gilbert, F. F., and D. D. Dodds. 1987. The philosophy and practice of wildlife management. Robert E. Krieger Publ. Co., Malabar, FL. 279pp.

Giroux, J. 1981. Use of artificial islands by nesting waterfowl in southeastern Alberta. J. Wildl. Manage. 45:669–679.

Giroux, J., D. E. Jelinski, and R. W. Boychuk. 1983. Use of rock islands and round straw bales by nesting Canada geese. Wildl. Soc. Bull. 11:172–178.

Givens, L. S. 1962. Use of fire on southeastern wildlife refuges. Proc. Annu. Tall Timbers Fire Ecol. Conf. 1:121–126.

Givens, L. S., M. C. Nelson, and V. Ekedahl. 1964. Farming for waterfowl. Pages 599–610 *in* J. P. Linduska, ed. Waterfowl tomorrow. U.S. Fish Wildl. Serv., Washington.

Gnann, J. W. 1985. Aerial ignition (ping-pong balls). Pages 87–93 *in* D. D. Wade, comp. Prescribed fire and smoke management in the South: conference proceedings. U.S. For. Serv. Southeast. For. Exp. Stn., Ashville, NC.

Godshalk, G. L., and R. G. Wetzel. 1978. Decomposition in the littoral zone of lakes. Pages 131–143 *in* R. E. Good, D. F. Whigham, and R. L. Simpson, eds. Freshwater wetlands, ecological processes and management potential. Academic Press, New York.

Goldsmith, A., and E. H. Clark II. 1990. Nonregulatory programs promoting wetlands protection. Pages 75–110 *in* G. Bingham, E. H. Clark II, L. V. Haygood, and M. Leslie, eds. Issues in wetlands protection: background papers prepared for the National Wetlands Policy forum. Conserv. Found., Washington.

Golet, F. C. 1979. Rating the wildlife value of northeastern fresh water wetlands. Pages 63–73 *in* P. E. Greeson, J. R. Clark, and J. E. Clark, eds. Wetland functions and values: the state of our understanding. Proc. Natl. Symp. on Wetlands. Am. Water Resour. Assoc., Minneapolis.

Golet, F. C., and J. S. Larson. 1974. Classification of freshwater wetlands in the glaciated northeast. U.S. Fish Wildl. Serv. Res. Publ. 116. 56pp.

Gordon, D. H., B. T. Gray, R. D. Perry, M. B. Prevost, T. H. Strange, and R. K. Williams. 1989. South Atlantic coastal wetlands. Pages 57–92 *in* L. M. Smith, R. L. Pederson, and R. M. Kaminski, eds. Habitat management for migrating and wintering waterfowl in North America. Texas Tech Univ. Press, Lubbock.

Green, C. E., and A. A. Rula. 1977. Low-ground-pressure construction equipment for use in dredged material containment area operation and maintenance—equipment inventory. U.S. Army Corps Eng. Waterways Exp. Stn. Tech. Rep. D-77-1. 126pp.

Green, J. E., and R. E. Salter. 1987. Methods for reclamation of wildlife habitat in the Canadian prairie provinces. Environ. Can., Edmonton, Alta. 114pp.

Green, W. E., L. G. MacNamara, and F. M. Uhler. 1964. Water off and on. Pages 557–568 *in* J. P. Linduska, ed. Waterfowl tomorrow. U.S. Fish Wildl. Serv., Washington.

Greenwood, R. J., A. B. Sargeant, D. H. Johnson, L. M. Cowardin, and T. L. Sheffer. 1987. Mallard nest success and recruitment in prairie Canada. Trans. North Am. Wildl. Nat. Resour. Conf. 52:298–309.

Grieb, J. R., and G. I. Crawford. 1967. Nesting structures for Canada geese. Colorado Game, Fish, Parks Dep. Game Inf. Leafl. 48. 4pp.

Griffith, M. A., and T. T. Fendley. 1981. Five-gallon plastic bucket: an inexpensive wood duck nesting structure. J. Wildl. Manage. 45:281–284.

Griffith, R. 1948. Improving waterfowl habitat. Trans. North Am. Wildl. Conf. 13:609–617.

Grim, E. C., and R. D. Hill. 1974. Environmental protection in surface mining of coal. U.S. Environ. Prot. Agency EPA/670/2-74-093. 292pp.

Grubb, T. G. 1980. An artificial bald eagle nest structure. U.S. For. Serv. Res. Note RM-383. 4 pp.

Guthery, F. S., and F. C. Bryant. 1982. Status of playas in the southern Great Plains. Wildl. Soc. Bull. 10:309–317.

Guthery, F. S., and F. A. Stormer. 1984a. Managing playas for wildlife in the southern high plains of Texas. Texas Tech Univ. Range Wildl. Manage. Note 4. 5pp.

Guthery, F. S., and F. A. Stormer. 1984b. Playa management. Pages 177B–182B *in* F. R. Henderson, ed. Guidelines for increasing wildlife on farms and ranches. Kansas State Univ. Coop. Ext. Serv., Manhattan.

Hackney, C. T., and A. A. de la Cruz. 1978. The effects of fire on the productivity and species composition of two St. Louis Bay, Mississippi tidal marshes dominated by *Juncus roemerianus* and *Spartina cynosuroides*, respectively. J. Mississippi Acad. Sci. 23(suppl.):109.

Hackney. C. T., and A. A. de la Cruz. 1981. Effects of fire on brackish marsh communities: management implications. Wetlands 1:75–86.
Hagan, P. D. 1980. A guide to managing marsh impoundments and farm areas on Mattamuskeet National Wildlife Refuge. U.S. Fish Wildl. Serv. Mattamuskeet Natl. Wildl. Ref., New Holland, NC. 44pp.
Hair, J. D., G. T. Hepp, L. M. Luckett, K. P. Reese, and D. K. Woodward. 1978. Beaver pond ecosystems and their relationships to multi-use natural resource management. Pages 80–92 *in* R. R. Johnson and J. F. McCormick, tech. coord. Strategies for protection and management of floodplain wetlands and other riparian ecosystems. U.S. For. Serv. Gen. Tech. Rep. WO-12.
Haliburton, T. A. 1978. Guidelines for dewatering/densifying confined dredged material. U.S. Army Eng. Waterways Exp. Stn. Tech. Rep. DS-78-11. 128pp.
Hall, D. L. 1962. Food utilization by waterfowl in green timber reservoirs at Noxubee National Wildlife Refuge. Proc. Southeast. Assoc. Game Fish Comm. 16:184–199.
Hammer, D. P., and E. D. Blackburn. 1977. Design and construction of retaining dikes for containment of dredged material. U.S. Army Eng. Waterways Exp. Stn. Tech. Rep. D-77-9. 200pp.
Hammond, M. C., and G. E. Mann. 1956. Waterfowl nesting islands. J. Wildl. Manage. 20:345–352.
Hamor, W. H., H. G. Uhlig, and L. V. Compton. 1968. Ponds and marshes for wild ducks on farms and ranches in the northern plains. U.S. Soil Conserv. Serv. Farmers' Bull. 2234. 16pp.
Hansen, G. W., F. E. Oliver, and N. E. Otto. 1984. Herbicide manual. U.S. Bur. Reclam., Denver. 346pp.
Harris, S. W., and W. H. Marshall. 1963. Ecology of water-level manipulations on a northern marsh. Ecology 44:331–343.
Hartley, D. R., and E. P. Hill. 1988. Effects of high temperatures on wood duck production in plastic nest boxes. U.S. Army Corps Eng. Wildl. Resour. Notes 6(1):2–3.
Hartman, H. T., and E. E. Kester. 1975. Plant propagation principles and practices. 3d ed. Prentice-Hall, Englewood Cliffs, NJ. 662pp.
Harvey, H. T., P. Williams, and J. Haltiner. 1983. Guidelines for enhancement and restoration of diked historic baylands. San Francisco Bay Conserv. Develop. Comm., San Francisco. 38pp.
Heitmeyer, M. E., D. P. Connelly, and R. L. Pederson. 1989. The Central, Imperial, and Coachella valleys of California. Pages 475–505 *in* L. M. Smith, R. L. Pederson, and R. M. Kaminski, eds. Habitat management for migrating and wintering waterfowl in North America. Texas Tech Univ. Press, Lubbock.
Henderson, C. L. Undated. Woodworking for wildlife. Minnesota Dep. Nat. Resour., St. Paul. 47pp.
Herricks, E. E. 1982. Development of aquatic habitat potential of gravel pits. Pages 196–207 *in* W. D. Svedarsky and R. D. Crawford, eds. Wildlife values of gravel pits. Univ. Minnesota Agric. Exp. Stn. Misc. Publ. 17.
Herricks, E. E., A. J. Krzysik, R. E. Szafoni, and D. J. Tazik. 1982. Best current practices for fish and wildlife on surface-mined lands in the eastern interior coal region. U.S. Fish Wildl. Serv. FWS/OBS-80/68. 212pp.
Heusmann, H. W., W. W. Blandin, and R. E. Turner. 1977. Starling-deterrent nesting cylinders in wood duck management. Wildl. Soc. Bull. 5:14–18.
Higgins, K. F. 1986. Further evaluation of duck nesting on small man-made islands in North Dakota. Wildl. Soc. Bull. 14:155–157.
Higgins, K. F., and W. T. Barker. 1982. Changes in vegetation structure in seeded nesting cover in the prairie pothole region. U.S. Fish Wildl. Serv. Spec. Sci. Rep.-Wildl. 242. 26pp.
Higgins, K. F., H. W. Miller, and L. M. Kirsch. 1986. Waterfowl nesting on an earth-filled cement culvert. Prairie Nat. 18:115–116.
Higgins, K. F., A. D. Kruse, and J. L. Piehl. 1989. Prescribed burning guidelines in the Northern Great Plains. EC 760. Wildl. Fish Sci. Dep., South Dakota State Univ., Brookings, 36pp.
Higgins, K. F., D. P. Fellows, J. M. Callow, A. D. Kruse, and J. L. Piehl. Undated. Annotated bibliography of fire literature relative to northern grasslands in south-

central Canada and north-central United States. U.S. Fish Wildl. Serv. and Coop. Ext. Serv., South Dakota State Univ., Brookings, 20pp.

Higgins, K. F., A. D. Kruse, and J. L. Piehl. Undated. Effects of fire in the Northern Great Plains. EC 761. Wildl. Fish. Sci. Dep., South Dakota State Univ., Brookings, 47pp.

Hill, T. K. 1990. Control of cattail in ponds. Univ. Tennessee Dep. For., Wildl. Fish. Renewable Resour.—Timely Tips 5(1):1.

Hindman, L. J., and V. D. Stotts. 1989. Chesapeake Bay and North Carolina sounds. Pages 27–55 *in* L. M. Smith, R. L. Pederson, and R. M. Kaminski, eds. Habitat management for migrating and wintering waterfowl in North America. Texas Tech Univ. Press, Lubbock.

Hirsch, N. D., L. H. DiSalvo, and R. Peddicord. 1978. Effects of dredging on aquatic organisms. U.S. Army Eng. Waterways Exp. Stn. Tech. Rep. DS-78-5. 41pp.

Hobaugh, W. C. 1984. Habital use by snow geese wintering in southeast Texas. J. Wildl. Manage. 48:1085–1096.

Hobaugh, W. C., C. D. Stutzenbaker, and E. L. Flickinger. 1989. The rice prairies. Pages 367–383 *in* L. M. Smith, R. L. Pederson, and R. M. Kaminski, eds. Habitat management for migrating and wintering waterfowl in North America. Texas Tech Univ. Press, Lubbock.

Hobaugh, W. C., and J. G. Teer. 1981. Waterfowl use characteristics of flood-prevention lakes in north-central Texas. J. Wildl. Manage. 45:16–26.

Hoeger, S. 1988. Schwimmkampen: Germany's artificial floating islands. J. Soil Water Conserv. 43:304–306.

Hoffman, R. D. 1988. Ducks Unlimited's United States construction program for enhancing waterfowl production. Pages 109–113 *in* J. Zelazny and J. S. Feierabend, eds. Proceedings of a conference increasing our wetland resources. Natl. Wildl. Fed., Washington.

Hoffpauer, C. M. 1968. Burning for coastal marsh management. Proc. Marsh Estuary Manage. Symp. 1:134–139.

Holsapple, L., and J. Lott. 1979. Blasting waterfowl potholes. U.S. For. Serv. Equipment Develop. Cent., Missoula, MT. 12pp.

Hopkins, R. C. 1962. Drawdown for ducks. Wis. Conserv. Bull. 27(4):18–19.

Hopper, R. M. 1971. Use of ammonium nitrate–fuel oil mixtures in blasting potholes for wildlife. Colorado Dep. Nat. Resour. Game Inf. Leafl. 85. 4pp.

Hopper, R. M. 1972. Waterfowl use in relation to size and cost of potholes. J. Wildl. Manage. 36:459–468.

Hopper, R. M. 1978. Evaluation of pothole blasting for waterfowl in Colorado. Colorado Div. Wildl. Spec. Rep. 44. 21pp.

Horstman, L. P., and J. R. Gunson, comps. 1983. Prevention and control of wildlife damage in Alberta. Alberta Energy Nat. Resour., Edmonton.

Howard, R., D. G. Rhodes, and J. W. Simmers. 1978. A review of the biology and potential control techniques for *Phragmites australis*. U.S. Army Eng. Waterways Exp. Stn. Tech. Rep. D-78-26. 80pp.

Howard, R. J., and J. A. Allen. 1989. Streamside habitats in southern forested wetlands: their role and implications for management. Pages 97–106 *in* D. D. Hook and R. Lea, eds. Proceedings of the symposium: the forested wetlands of the southern United States. U.S. For. Serv. Gen. Tech. Rep. SE-50.

Hubert, W. A., and J. N. Krull. 1973. Seasonal fluctuations of aquatic macroinvertebrates in Oakwood Bottoms Greentree Reservoir. Am. Midl. Nat. 90:177–185.

Hudson, M. S. 1983. Waterfowl production on three age-classes of stock ponds in Montana. J. Wildl. Manage. 47:112–117.

Hughes, J. H., and E. L. Young, Jr. 1982. Autumn foods of dabbling ducks in southeastern Alaska. J. Wildl. Manage. 46:259–263.

Hunt, L. J., A. W. Ford, M. C. Landin, and B. R. Wells. 1978. Upland habitat development with dredged material: engineering and plant propagation. U.S. Army Corps Eng. Tech. Rep. DS-78-17. 160pp.

Hunter, C. G. 1978. Managing green tree reservoirs for waterfowl. Int. Waterfowl Symp. 3:217–223.

Hunter, M. L., Jr. 1990. Wildlife, forests, and forestry principles of managing forests for biological diversity. Prentice-Hall, Englewood Cliffs, NJ. 370pp.

Hynson, J., P. Adamus, S. Tibbetts, and R. Darnell. 1982. Handbook for protection of fish and wildlife from construction of farm and forest roads. U.S. Fish Wildl. Serv. FWS/OBS-82/18. 153pp.

Illinois Department of Conservation. 1986. Management of small lakes and ponds in Illinois. Illinois Dep. Conserv., Springfield. 82pp.

Jensen, S. E., and W. S. Platts. 1989. Restoration of degraded riverine/riparian habitat in the Great Basin and Snake River regions. Pages 377–415 *in* J. A. Kusler and M. E. Kentula, eds. Wetland creation and restoration: the status of the science. Vol. 1: Regional reviews. U.S. Environ. Prot. Agency EPA/600/3-89/038a.

Johnson, F. A., and F. Montalbano. 1989. Southern reservoirs and lakes. Pages 93–116 *in* L. M. Smith, R. L. Pederson, and R. M. Kaminski, eds. Habitat management for migrating and wintering waterfowl in North America. Texas Tech Univ. Press, Lubbock.

Johnson, L. E., and W. V. McGuinness, Jr. 1975. Guidelines for material placement in marsh creation. U.S. Army Eng. Waterways Exp. Stn. Rep. D-75-2. 230pp.

Johnson, N. F. 1984. Muskrat. Pages 156C–160C *in* F. R. Henderson, ed. Guidelines for increasing wildlife on farms and ranches. Kansas State Univ. Coop. Ext. Serv., Manhattan.

Johnson, R. F., R. O. Woodward, and L. M. Kirsch. 1978. Waterfowl nesting on small man-made islands in the prairie wetlands. Wildl. Soc. Bull. 6:240–243.

Johnson, R. R., and J. J. Dinsmore. 1986. Habitat use by breeding Virginia rails and soras. J. Wildl. Manage. 50:387–392.

Johnson, T. R. 1983. Wildlife watering holes: their construction, value and use by amphibians and management. Missouri Dep. Conserv., Jefferson City. 4pp.

Johnson, W. W., and M. T. Finley. 1980. Handbook of acute toxicity of chemicals to fish and aquatic invertebrates. U.S. Fish Wildl. Serv. Resour. Publ. 137. 98pp.

Jones, J. D. 1975. Waterfowl nesting island development. U.S. Bur. Land Manage. Tech. Note 260. 17pp.

Jones, W. L., and W. C. Lehman. 1987. Phragmites control and revegetation following aerial applications of glyphosate in Delaware. Pages 185–196 *in* W. R. Whitman and W. H. Meredith, eds. Waterfowl and wetlands symposium: proceedings of a symposium on waterfowl and wetlands management in the coastal zone of the Atlantic Flyway. Delaware Dep. Nat. Resour. Environ. Cont., Dover.

Jorde, D. G., G. L. Krapu, and R. D. Crawford. 1983. Feeding ecology of mallards wintering in Nebraska. J. Wildl. Manage. 47:1044–1053.

Josselyn, M., and J. Buccholz. 1984. Marsh restoration in San Francisco Bay: a guide to design and planning. Tech. Rep. 3. Tiburon Cent. for Environ. Studies, San Francisco State Univ., San Francisco. 103pp.

Josselyn, M., J. Zedler, and T. Griswold. 1989. Wetland mitigation along the Pacific Coast of the United States. Pages 1–35 *in* J. A. Kusler and M. E. Kentula, eds. Wetland creation and restoration: the status of the science. Vol. 1: Regional reviews. U.S. Environ. Prot. Agency EPA/600/3-89/038a.

Kadlec, J. A. 1960. The effect of a drawdown on the ecology of a waterfowl impoundment. Michigan Dep. Conserv. Rep. 2276. 181pp.

Kadlec, J. A., and L. M. Smith. 1984. Marsh plant establishment on newly flooded salt flats. Wildl. Soc. Bull. 12:388–394.

Kadlec, J. A., and L. M. Smith. 1989. The Great Basin marshes. Pages 451–474 *in* L. M. Smith, R. L. Pederson, and R. M. Kaminski, eds. Habitat management for migrating and wintering waterfowl in North America. Texas Tech Univ. Press, Lubbock.

Kadlec, J. A., and L. M. Smith. In press. Habitat management for waterfowl breeding areas. *In* B. D. J. Batt, A. D. Afton, M. G. Anderson, C. D. Ankney, D. H. Johnson, J. D. Kadlec, and G. L. Krapu, eds. Ecology and management of breeding waterfowl. Univ. Minnesota Press, Minneapolis.

Kadlec, J. A., and W. A. Wentz. 1974. State-of-the-art survey and evaluation of marsh plant establishment techniques: induced and natural. Vol. 1: Report of research. U.S. Army Corps Eng. Rep. D-74-9. 266pp.

Kahl, R. 1991. Restoration of canvasback migrational staging habitat in Wisconsin: a research plan with implications for shallow lake management. Wisconsin Dep. Nat. Resour. Tech. Bull. 172. 47pp.

Kaminski, R. M., H. R. Murkin, and C. E. Smith. 1985. Control of cattail and bulrush by cutting and flooding. Pages 253–262 *in* H. H. Prince and F. M. D'Itri, eds. Coastal wetlands. Lewis Publ., Inc., Chelsea, MI.

Kaminski, R. M., and H. H. Prince. 1981. Dabbling duck and aquatic macroinvertebrate responses to manipulated wetland habitat. J. Wildl. Manage. 45:1–15.

Kantrud, H. A. 1986a. Effects of vegetation manipulation on breeding waterfowl in prairie wetlands: a literature review. U.S. Fish Wildl. Serv. Tech. Rep. 3. 15pp.

Kantrud, H. A. 1986b. Western Stump Lake, a major canvasback staging area in eastern North Dakota. Prairie Nat. 18:247–253.

Keith, L. B. 1961. A study of waterfowl ecology on small impoundments in southeastern Alberta. Wildl. Monogr. 6. 88pp.

Kent, K. M., and W. A. Styner. 1979. Estimating runoff. Pages 2-1–2-76 *in* Engineering field manual for conservation practices. U.S. Soil Conserv. Serv., Washington.

Kirby, R. E., S. J. Lewis, and T. N. Sexson. 1988. Fire in North American wetland ecosystems and fire-wildlife relations: an annotated bibliography. U.S. Fish Wildl. Serv. Biol. Rep. 88(1). 146pp.

Kierstead, M. W. Undated. Wetlands creation and management. Pages 1–57 (chap. 8) *in* Community wildlife involvement program field manual. Ontario Min. Nat. Resour., Toronto.

Kirsch, L. M., and K. F. Higgins. 1976. Upland sandpiper nesting and management in North Dakota. Wildl. Soc. Bull. 4:16–20.

Kirsch, L. M., H. F. Duebbert, and A. D. Kruse. 1978. Grazing and haying effects of upland nesting birds. Trans. North Am. Wildl. Nat. Resour. Conf. 43:486–497.

Kjellsen, M. L., and K. F. Higgins. 1990. Grasslands: benefits of management by fire. FS 857. U.S. Fish Wildl. Serv. and Coop. Ext. Serv., South Dakota State Univ., Brookings. 4pp.

Klett, A. T., H. F. Duebbert, and G. L. Heismeyer. 1984. Use of seeded native grasses as nesting cover by ducks. Wildl. Soc. Bull. 12:134–138.

Knighton, M. D. 1985. Vegetation management in water impoundments: water-level control. Pages 39–50 *in* M. D. Knighton, comp. Water impoundments for wildlife: a habitat management workshop. U.S. For. Serv. Gen. Tech. Rep. NC-100.

Knighton, M. D., and E. S. Verry. 1983. How to evaluate water impoundment sites. U.S. For. Serv. North Cent. For. Exp. Stn. HT-58. 6pp.

Knutson, P. L., 1978. Planting guidelines for dune creation and stabilization. U.S. Army Corps Eng. Coastal Eng. Res. Cent. Rep. CERC-REPRINT-78-12. 20pp.

Knutson, P. L., and W. W. Woodhouse, Jr. 1982. Pacific coastal marshes. Pages 111–130 *in* R. R. Lewis, III, ed. Creation and restoration of coastal plant communities. CRC Press, Boca Raton, FL.

Knutson, P. L., and W. W. Woodhouse, Jr. 1983. Shore stabilization with salt marsh vegetation. U.S. Army Corps Eng. Coastal Eng. Res. Cent. Spec. Rep. 9. 95pp.

Koegel, R. G., D. F. Livermore, and H. D. Bruhn. 1974. Evaluation of large scale mechanical management of aquatic plants in waters of Dane County, Wisconsin. Univ. Wisconsin Water Resour. Cent. Tech. Rep. WIS WRC 74-08. 36pp.

Koegel, R. G., D. F. Livermore, and H. D. Bruhn. 1978. Improvement and evaluation of techniques for the mechanical removal and utilization of excess aquatic vegetation. Univ. Wisconsin Water Resour. Cent. Tech. Rep. WIS WRC 78-02. 68pp.

Korschgen, C. E. 1989. Riverine and deepwater habitats for diving ducks. Pages 157–180 *in* L. M. Smith, R. L. Pederson, and R. M. Kaminski, eds. Habitat management for migrating and wintering waterfowl in North America. Texas Tech Univ. Press, Lubbock.

Korschgen, C. E., and W. L. Green. 1988. American wildcelery (*Vallisnaria americana*): ecological considerations for restoration. U.S. Fish Wildl. Serv. Tech. Rep. 19. 24pp.

Kroll, R. W., and R. L. Meeks. 1985. Muskrat population recovery following habitat reestablishment near southwestern Lake Erie. Wildl. Soc. Bull. 13:483–486.

Kruczynski, W. L. 1989. Options to be considered in preparation and evaluation of mitigation plans. Pages 143–158 *in* J. A. Kusler and M. E. Kentula, eds. Wetland creation and restoration: the status of the science. Vol. 2: Perspectives. U.S. Environ. Prot. Agency EPA 60/3-89/0038b.

Kruczynski, W. L., R. T. Huffman, and M. K. Vincent. 1978. Habitat development field investigations, Apalachicola Bay marsh development site, Apalachicola Bay, Florida; summary report. U.S. Army Corps Eng. Tech. Rep. D-78-32. 39pp.

Krueger, H. O., and S. H. Anderson. 1985. The use of cattle as a management tool for wildlife in shrub-willow riparian systems. Pages 300–304 *in* R. R. Johnson, C. D. Ziebell, D. R. Patton, P. F. Ffolliott, and R. H. Hamre, tech. coord. Riparian ecosystems and their management: reconciling conflicting uses. U.S. For. Serv. Gen. Tech. Rep. RM-120.

Landers, J. L., A. S. Johnson, P. H. Morgan, and W. P. Baldwin. 1976. Duck foods in managed tidal impoundments in South Carolina. J. Wildl. Manage. 40:721–728.

Laramie, H. A., Jr. 1963. A device for control of problem beavers. J. Wildl. Manage. 27:471–476.

Laramie, H. A, Jr. 1978. Water level control in beaver ponds and culverts. New Hampshire Fish Game Dep., Concord. 5pp.

Larrick, W. J., Jr., and R. H. Chabreck. 1976. The effects of weirs on aquatic vegetation along the Louisiana coast. Proc. Annu. Conf. Southeast. Assoc. Game Fish Comm. 30:581–589.

Larson, J. E. 1980. Revegetation equipment catalog. U.S. For. Serv. Equipment Develop. Cent., Missoula, MT. 198pp.

Larson, J. S., M. S. Bedinger, C. F. Bryan, S. Brown, R. T. Huffman, E. L. Miller, D. G. Rhodes, and B. A. Touchet. 1981. Transition from wetlands to uplands in southeastern bottomland hardwood forests. Pages 225–273 *in* J. R. Clark and J. Benforado, eds. Wetlands of bottomland hardwood forests. Elsevier Sci. Publ., New York.

Larsson, T. 1982. Restoration of lakes and other wetlands in Sweden. Pages 107–122 *in* D. A. Scott, ed. Managing wetlands and their birds. Proc. 3d Tech. Meet. on West. Palearctic Migratory Bird Manage. Int. Waterfowl Res. Bur., Slimbridge, Glos., England.

Lea, R. 1988. Management of eastern United States bottomland hardwood forests. Pages 185–194 *in* D. D. Hook, W. H. McKee, Jr., H. K. Smith, J. Gregory, V. G. Burrell, Jr., M. R. DeVoe, R. E. Sojka, S. Gilbert, R. Banks, L. H. Stolzy, C. Brooks, T. D. Matthews, and T. H. Shears, eds. The ecology and management of wetlands. Vol. 2: Management, use and value of wetlands. Timber Press, Portland, OR.

Leedy, D. L., and L. W. Adams. 1982. Wildlife considerations in planning and managing highway corridors. U.S. Fed. Highway Admin. Rep. FHWA-TS-82-212. 93pp.

Leege, T. A., and M. C. Fultz. 1972. Aerial ignition of Idaho elk ranges. J. Wildl. Manage. 36:1332–1336.

Lejcher, T. 1986. Hydrology and watershed considerations. Pages 4–9 *in* J. L. Piehl, ed. Wetland restoration: a techniques workshop. Minnesota chapter Wildl. Soc., Fergus Falls.

Leslie, A. J., Jr. 1988. Literature review of drawdown for aquatic plant control. Aquatics 10(1):12–18.

Leslie, M., E. H. Clark II, and R. B. Reed. 1990. Overview of existing regulatory programs. Pages 141–172 *in* G. Bingham, E. H. Clark II, L. V. Haygood, and M. Leslie, eds. Issues in wetlands protection: background papers prepared for the National Wetlands Policy forum. Conserv. Found., Washington.

Lewis, J. C., and E. W. Bunce, eds. 1980. Rehabilitation and creation of selected coastal habitats: proceedings of a workshop. U.S. Fish Wildl. Serv. FWS/OBS-80/27. 162pp.

Lewis, R. R., III, ed. 1982a. Creation and restoration of coastal plant communities. CRC Press, Boca Raton, FL. 218pp.

Lewis, R. R., III, ed. 1982b. Mangrove forests. Pages 153–171 *in* R. R. Lewis, III, ed. Creation and restoration of coastal plant communities. CRC Press, Boca Raton, FL.

Linde, A. F. 1969. Techniques for wetland management. Wisconsin Dep. Nat. Resour. Rep. 45. 156pp.

Linde, A. F. 1985. Vegetation management in water impoundments: alternatives and supplements to water-level control. Pages 51–60 *in* M. D. Knighton, comp. Water impoundments for wildlife: a habitat management workshop. U.S. For. Serv. Gen. Tech. Rep. NC-100.

Linde, A. F., T. Janisch, and D. Smith. 1976. Cattail—the significance of its growth, phenology and carbohydrate storage to its control and management. Wisconsin Dep. Nat. Resour. Tech. Bull. 94. 26pp.

Lindenmuth, A. W., Jr., and J. R. Davis. 1973. Predicting fire spread in Arizona oak chapparrel. U.S. For. Serv. Res. Pap. RM-101. 11pp.

Lokemoen, J. T. 1973. Waterfowl production on stock-watering ponds in the Northern Plains. J. Range Manage. 26:179–184.

Lokemoen, J. T., F. B. Lee, H. F. Duebbert, and G. A. Swanson. 1984. Aquatic habitats—waterfowl. Pages 161B–176B *in* F. R. Henderson, ed. Guidelines for increasing wildlife on farms and ranches. Kansas State Univ. Coop. Ext. Serv., Manhattan.

Long, S. G., J. K. Burrell, N. F. Laurenson, and J. H. Nyenhuis. 1984. Manual of revegetation techniques. U.S. For. Serv. Equipment Develop. Cent., Missoula, MT. 145pp.

Lopinot, A. C. 1986. Aquatic weeds, their identification and methods of control. Illinois Dep. Conserv. Fish. Bull. 4. 54pp.

Lott, J. R. 1977. Forest Service blasters manual. U.S. For. Serv. Equipment Develop. Cent., Missoula, MT. 193pp.

Low, J. B. 1945. Ecology and management of the redhead (*Nyroca americana*) in Iowa. Ecol. Monogr. 15:35–69.

Lumsden, H. G., R. E. Page, and M. Gauthier. 1980. Choice of nest boxes by common goldeneyes in Ontario. Wilson Bull. 92:497–505.

Lumsden, H. G., J. Robinson, and R. Hartford. 1986. Choice of nest boxes by cavity-nesting ducks. Wilson Bull. 98:167–168.

Lunz, J. D., R. J. Diaz, and R. A. Cole. 1978. Upland and wetland habitat development with dredged material: ecological considerations. U.S. Army Eng. Waterways Exp. Stn. Tech. Rep. DS-78-15. 50pp.

Lynch, J. J. 1941. The place of burning in management of the Gulf Coast wildlife refuges. J. Wildl. Manage. 5:454–457.

Lynch, J. J., T. O'Neil, and D. W. Lay. 1947. Management significance of damage by geese and muskrats to Gulf Coast marshes. J. Wildl. Manage. 11:50–76.

Lyon, L. J., H. S. Crawford, E. Czuhai, R. L. Fredriksen, R. F. Harlow, L. J. Metz, and H. A. Pearson. 1978. Effects of fire on fauna: a state-of-knowledge review. U.S. For. Serv. Gen. Tech. Rep. WO-6. 22pp.

Mackey, D. L., W. C. Matthews, Jr., and I. J. Ball. 1988. Elevated nest structures for Canada geese. Wildl. Soc. Bull. 16:362–367.

Maguire, J. D., and G. A. Heuterman. 1978. Influence of pregermination conditions on the viability of selected marsh plants. U.S. Army Corps Eng. Tech. Rep. D-78-51. 103pp.

Mahler, D., and J. Walther. 1990. Habitat restoration on a central Texas office building site. Pages 160–170 *in* J. J. Berger, ed. Environmental restoration: science and strategies for restoring the earth. Island Press, Washington.

Mallik, A. V., and R. W. Wein. 1986. Response of a *Typha* marsh community to draining, flooding, and seasonal burning. Can. J. Bot. 64:2136–2143.

Marcy, L. E. 1986. Waterfowl nest baskets. Sect. 5.1.3, U.S. Army Corps of Engineers wildlife resources management manual. U.S. Army Eng. Waterways Exp. Stn. Tech. Rep. EL-86-15. 16pp.

Markell, L. 1986. Design and planning of a wetland restoration project. Pages 1–3 *in* J. L. Piehl, ed. Wetland restoration: a techniques workshop. Minnesota chap. Wildl. Soc., Fergus Falls.

Martin, A. C., N. Hotchkiss, F. M. Uhler, and W. S. Bourn. 1953. Classification of wetlands of the United States. U.S. Fish Wildl. Serv. Spec. Sci. Rep.-Wildl. 20. 14pp.

Martin, A. C., and F. M. Uhler. 1939. Food of game ducks in the United States and Canada. U.S. Dep. Agric. Tech. Bull. 634. 308pp.

Martin, A. C., H. S. Zim, and A. L. Nelson. 1951. American wildlife and plants, a guide to wildlife food habits. McGraw-Hill, New York. 500pp.

Martin, C. O., and L. E. Marcy. 1989. Artificial potholes—blasting techniques. Sect. 5.5.4, U.S. Army Corps of Engineers wildlife resources management manual. U.S. Army Eng. Waterways Exp. Stn. Tech. Rep. EL-89-14. 45pp.

Martin, C. O., W. A. Mitchell, and D. A. Hammer. 1986. Osprey nest platforms. Sect. 5.1.6, U.S. Army Corps of Engineers wildlife resources management manual. U.S. Army Eng. Waterways Exp. Stn. Tech. Rep. EL-86-21. 31pp.

Martin, R. E., S. E. Coleman, and A. H. Johnson. 1977. Wetline technique for prescribed burning firelines in rangelands. U.S. For. Serv. Res. Note PNW-292. 6 pp.

Maser, C., J. W. Thomas, I. D. Luman, and R. Anderson. 1979. Wildlife habitats in managed rangelands—the Great Basin of southeastern Oregon, manmade habitats. U.S. For. Serv. Gen. Tech. Rep. PNW-86. 40pp.

Mathiak, H. A. 1965. Pothole blasting for wildlife. Wisconsin Conserv. Dep. Publ. 352. 31pp.
Mathiak, H. A., and A. F. Linde. 1956. Studies on level ditching for marsh management. Wisconsin Conserv. Dep. Tech. Wildl. Bull. 12. 49pp.
Mathisen, J. E. 1969. Use of man-made islands as nesting sites of the common loon. Wilson Bull. 81:331.
Mathisen, J. E. 1985. Wildlife impoundments in the north central states: why do we need them? Pages 23–30 *in* M. D. Knighton, comp. Water impoundments for wildlife: a habitat management workshop. U.S. For. Serv. Gen. Tech. Rep. NC-100.
Mathisen, J. E., J. Byelich, and R. Radtke. 1964. The use of ammonium nitrate for marsh blasting. Trans. North Am. Wildl. Nat. Resour. Conf. 29:143–150.
Matter, W. J., and R. W. Mannan. 1988. Sand and gravel pits as fish and wildlife habitat in the Southwest. U.S. Fish Wildl. Serv. Resour. Publ. 171. 11pp.
May, B. E., and B. Davis. 1982. Practices for livestock grazing and aquatic habitat protection on western rangelands. Pages 271–278 *in* J. M. Peek and P. D. Dalke, eds. Wildlife-livestock relationships symposium. Univ. Idaho For. Wildl., Range Exp. Stn., Moscow.
Mayer, F. L., Jr., and M. R. Ellersieck. 1986. Manual of acute toxicity: interpretation and data base for 410 chemicals and 66 species of freshwater animals. U.S. Fish Wildl. Serv. Resour. Publ. 160. 506pp.
Mayhew, J. K., and S. T. Runkel. 1962. The control of nuisance aquatic vegetation with black polyethylene plastic. Proc. Iowa Acad. Sci. 69:302–307.
Maynord, S. T. 1978. Practical riprap design. U.S. Army Eng. Waterways Exp. Stn. Misc. Pap. H-78-7. 66pp.
Mazzoni, J., M. Barber, R. Critchlow, W. Miller, and G. Studenski. 1983. Livestock and waterfowl. Pages 87–94 *in* J. W. Menke, ed. Proceedings of the workshop on livestock and wildlife-fisheries relationships in the Great Basin. Univ. California Div. Agric. Sci., Berkeley.
McAtee, W. L. 1939. Wildlife food plants, their value, propagation and management. Collegiate Press, Ames, IA. 141pp.
McBride, J. R., and J. Strahan. 1984. Establishment and survival of woody riparian species on gravel bars of an intermittent stream. Am. Midl. Nat. 112:235–245.
McCabe, T. R. 1982. Muskrat population levels and vegetation utilization: a basis for an index. Ph.D. diss., Utah State Univ., Logan. 111pp.
McCamant, R. E., and E. G. Bolen. 1979. A 12-year study of nest box utilization by black-bellied whistling ducks. J. Wildl. Manage. 43:936–943.
McCluskey, D. C. V., J. Brown, D. Bornholdt, D. A. Duff, and A. H. Winword. 1983. Willow planting for riparian habitat improvement. U.S. Bur. Land Manage. Tech. Note 363. 21pp.
McDonald, M. E. 1955. Causes and effects of a dieoff of emergent vegetation. J. Wildl. Manage. 19:24–35.
McGilvrey, F. D., comp. 1968. A guide to wood duck production habitat requirements. U.S. Fish Wildl. Serv. Resour. Publ. 60. 32pp.
McGinn, L. R., and L. L. Glascow. 1963. Loss of waterfowl foods in ricefields in southwest Louisiana. Proc. Annu. Conf. Southeast. Assoc. Game Fish Comm. 17:69–79.
McIntyre, J. W., and J. E. Mathisen. 1977. Artificial islands as nest sites for common loons. J. Wildl. Manage. 41:317–319.
McKendrick, J. D. 1987. *Arctophila fulva* for revegetating arctic wetlands. Restor. Manage. Notes 5(2):93.
McPherson, G. R., G. A. Rasmussen, H. A. Wright, and C. M. Britton. 1986. Getting started in prescribed burning. Texas Tech Univ. Dep. Range Wildl. Manage. Note 9. 5pp.
McQuilkin, R. A., and R. A. Musbach. 1977. Pin oak acorn production on green tree reservoirs in southeastern Missouri. J. Wildl. Manage. 41:218–225.
McRoy, C. P., and C. Helfferich. 1980. Applied aspects of seagrasses. Pages 298–343 *in* R. C. Phillips and C. P. McRoy, eds. Handbook of seagrass biology: an ecosystem perspective. Garland STPM Press, New York.
Meeker, F. J., and S. Nielsen. 1986. Observations concerning the establishment of marsh vegetation (principally *Spartina alterniflora*) by use of the plant roll technique. Proc. Annu. Conf. on Wetlands Restoration Creation 13:134–144.

Meeks, R. L. 1969. The effect of drawdown date on wetland plant succession. J. Wildl. Manage. 33:817–821.
Meier, T. I. 1981. Artificial nesting structures for the double-crested cormorant. Wisconsin Dep. Nat. Resour. Tech. Bull. 126. 13pp.
Melton, B. L., R. L. Hoover, R. L. Moore, and D. J. Pfankuch. 1987. Aquatic and riparian wildlife. Pages 260–301 *in* R. L. Hoover and D. L. Wills, eds. Managing forested lands for wildlife. Colorado Div. Wildl. and U.S. For. Serv., Denver.
Meredith, W. H., D. E. Saveikis, and C. J. Stachecki. 1983. Delaware's open marsh management research program: an overview and update. Proc. New Jersey Mosquito Cont. Assoc. 70:42–47.
Meredith, W. H., D. E. Saveikis, and C. J. Stachecki. 1985. Guidelines for "open marsh water management" in Delaware's salt marshes—objectives, system designs, and installation procedures. Wetlands 5:119–133.
Merendino, M. T., and L. M. Smith. 1991. Influence of drawdown date and reflood depth on wetland vegetation establishment. Wildl. Soc. Bull. 19:143–150.
Merendino, M. T., L. M. Smith, H. R. Murkin, and R. L. Pederson. 1990. The response of prairie wetland vegetation to seasonality of drawdown. Wildl. Soc. Bull. 18:245–251.
Merila, E., and P. Vikberg. 1980. Artificial raising of the nests of *Anatidae* and *Laridae*. Suomen Riista 28:118–122.
Messmer, T. A., M. A. Johnson, and F. B. Lee. 1986. Homemade nest sites for giant Canada geese. North Dakota State Univ. Coop. Ext. Serv., Fargo. 15pp.
Meyer, M. I. 1987. Planting grasslands for wildlife habitat. U.S. Fish Wildl. Serv. North. Prairie Wildl. Res. Cent., Jamestown, ND. 12pp.
Miglarese, J. V., and P. A. Sandifer, eds. 1982. An ecological characterization of South Carolina wetland impoundments. South Carolina Wildl. and Marine Resour. Dep. Tech. Bull. 51. 132pp.
Millar, J. B. 1973. Vegetation changes in shallow marsh wetlands under improving moisture regime. Can. J. Bot. 51:1443–1457.
Miller, A. W. 1962. Waterfowl habitat improvement in California. Proc. West. Assoc. State Game Fish Comm. 42:112–116.
Miller, J. E. 1983a. Beavers. Pages B1–B11 *in* R. M. Timm, ed. Prevention and control of wildlife damage. Great Plains Agric. Counc. Wildl. Resour. Comm. and Nebraska Coop. Ext. Serv. Inst. Agric. Nat. Resour., Univ. Nebraska, Lincoln.
Miller, J. E. 1983b. Muskrats. Pages B51–B59 *in* R. M. Timm, ed. Prevention and control of wildlife damage. Great Plains Agric. Counc. Wildl. Resour. Comm. and Nebraska Coop. Ext. Serv. Inst. Agric. Nat. Resour., Univ. Nebraska, Lincoln.
Miller, R., and G. E. Pope. 1984. An effective technique for planting trees in riparian habitat. U.S. Army Corps Eng. Wildl. Resour. Notes 2(4):1–2.
Mirov, N. T., and C. J. Kraebel. 1939. Collecting and handling seeds of wild plants. U.S. For. Serv. Civ. Conserv. Corps, California For. Range Exp. Stn. For. Publ. 5. 42pp.
Mitchell, W. A. 1989. Japanese millet (*Echinochloa crusgalli* var. *frumentacea*). Sect. 7.1.6, U.S. Army Corps of Engineers wildlife resources management manual. U.S. Army Eng. Waterways Exp. Stn. Tech. Rep. EL-89-13. 18pp.
Mitchell, W. A., and C. O. Martin. 1986. Chufa (*Cyperus esculentus*). Sect. 7.4.1, U.S. Army Corps of Engineers wildlife resources manual. U.S. Army Eng. Waterways Exp. Stn. Tech. Rep. EL-86-22. 14pp.
Mitchell, W. A., and C. J. Newling. 1986. Greentree reservoirs. Sect. 5.5.3, U.S. Army Corps of Engineers wildlife resources management manual. U.S. Army Eng. Waterways Exp. Stn. Tech. Rep. EL-86-9. 22pp.
Mitchell, W. A., and W. H. Tomlinson, Jr. 1989. Browntop millet (*Panicum ramosum*). Sect. 7.1.5, U.S. Army Corps of Engineers wildlife resources management manual. U.S. Army Eng. Waterways Exp. Stn. Tech. Rep. EL-89-12. 14pp.
Mitsch, W. J., and J. G. Gosselink. 1986. Wetlands. Van Nostrand Reinhold, New York. 539pp.
Mobley, H. E., R. S. Jackson, W. E. Blamer, W. E. Ruziska, and W. A. Hough. 1973 (rev. 1978). A guide for prescribed fire in southern forests. U.S. For. Serv. South. Reg., Atlanta. 40pp.
Molini, W. A. 1977. Livestock interactions with upland game, nongame, and waterfowl in the Great Basin. A workshop synopsis. Trans. Cal.-Neva. Wildl. 1977:97–103.

Monda, M. J., and J. T. Ratti. 1988. Niche overlap and habitat use by sympatric duck broods in eastern Washington. J. Wildl. Manage. 52:95–103.
Montgomery, R. L., A. W. Ford, M. E. Poindexter, and M. J. Bartos. 1978. Guidelines for dredged material disposal area reuse management. U.S. Army Eng. Waterways Exp. Stn. Tech. Rep. DS-78-12. 110pp.
More Game Birds in America. 1936. Waterfowl food plants. More Game Birds in America, New York. 45pp.
Morgan, J. P. 1973. Impact of subsidence and erosion on Louisiana coastal marshes and estuaries. Proc. Coastal Marsh Estuary Manage. Symp. 2:217–233.
Morgan, P. H., A. S. Johnson, W. P. Baldwin, and J. L. Landers. 1975. Characteristics and management of tidal impoundments for wildlife in a South Carolina estuary. Proc. Annu. Conf. Southeast. Assoc. Game Fish Comm. 29:526–539.
Morrison, D. G. 1982. Principles of revegetating mined lands. Pages 51–58 *in* W. D. Svedarsky and R. D. Crawford, eds. Wildlife values in gravel pits. Univ. Minnesota Agric. Exp. Stn. Misc. Publ. 17.
Morton, J. W. 1977. Ecological effects of dredging and dredge spoil disposal: a literature review. U.S. Fish Wildl. Serv. Tech. Pap. 94. 33pp.
Mosby, H. S. 1980. Reconnaissance mapping and map use. Pages 277–290 *in* S. D. Schemnitz, ed. Wildlife management techniques manual. Wildl. Soc., Washington.
Moyle, J. B., ed. 1964. Ducks and land use in Minnesota. Minnesota Div. Game Fish Tech. Bull. 8. 140pp.
Moyle, J. B., and N. Hotchkiss. 1945. The aquatic and marsh vegetation of Minnesota and its value to waterfowl. Minnesota Fish. Res. Lab. Tech. Bull. 3. 122pp.
Moyle, J. B., and J. H. Kuehn. 1964. Carp, a sometimes villain. Pages 635–642 *in* J. P. Linduska, ed. Waterfowl tomorrow. U.S. Fish Wildl. Serv., Washington.
Mundinger, J. G. 1976. Waterfowl response to rest-rotation grazing. J. Wildl. Manage. 40:60–68.
Munro, D. A. 1963. Ducks and the Great Plains wetlands. Can. Audubon 25:105–111.
Murkin, H. R., and P. Ward. 1980. Early spring cutting to control cattail in a northern marsh. Wildl. Soc. Bull. 8:254–256.
Murphy, W. L., and T. W. Zeigler. 1974. Practices and problems in the confinement of dredged material in Corps of Engineers projects. U.S. Army Eng. Waterways Exp. Stn. Tech. Rep. D-74-2. 187pp.
Myers, L. H. 1989. Grazing and riparian management in southwestern Montana. Pages 117–120 *in* R. E. Gresswell, B. A. Barton, and J. L. Kershner, eds. Practical approaches to riparian resource management: an educational workshop. U.S. Bur. Land Manage., Billings, MT.
National Wetlands Working Group Canada Committee on Ecological Land Classification. 1987. The Canadian wetland classification system. Can. Wildl. Serv. Land Classification Ser. 21. 18pp.
National Wetlands Working Group Canada Committee on Ecological Land Classification. 1988. Wetlands of Canada. Ecological Land Classification Ser. 24. Sustainable Develop. Branch, Ottawa, and Polysci. Pub. Inc., Montreal. 452pp.
Neal, T. J. 1968. A comparison of two muskrat populations. Iowa State J. Sci. 43:193–210.
Neckles, H. A., J. W. Nelson, and R. L. Pederson. 1985. Management of whitetop (*Scolochloa festucacea*) marshes for livestock forage and wildlife. Delta Waterfowl and Wetlands Res. Stn. Tech. Bull. 1. Portage la Prairie, Manit. 12pp.
Neely, W. W. 1959. Snipe field management in the southeastern states. Proc. Annu. Conf. Southeast. Assoc. Game Fish Comm. 13:288–291.
Neely, W. W. 1960. Managing *Scirpus robustus* for ducks. Proc. Annu. Conf. Southeast. Assoc. Game Fish Comm. 14:30–34.
Neely, W. W. 1962. Saline soils and brackish waters in management of wildlife, fish and shrimp. Trans. North Am. Wildl. Nat. Resour. Conf. 27:321–335.
Neely, W. W. 1968. Planting, disking, mowing, and grazing. Proc. Marsh Estuary Manage. Symp. 1:212–221.
Neely, W. W., and V. E. Davison. 1971. Wild ducks on farmland in the south. U.S. Dep. Agric. Farmers' Bull. 2218. 14pp.
Nelson, H. K., and H. F. Duebbert. 1974. New concepts regarding the production of

waterfowl and other game birds in areas of diversified agriculture. Int. Cong. Game Biol. 11:385–394.

Nelson, N. F., and R. H. Dietz. 1966. Cattail control methods in Utah. Utah Dep. Fish Game Publ. 66-2. 31pp.

Nelson, R. W., G. C. Horah, and J. E. Olson. 1978. Western reservoir and stream habitat improvements handbook. U.S. Fish Wildl. Serv. FWS/OBS-78/56. 250pp.

New, J. Undated. Marsh development for wildlife. Indiana Dep. Nat. Resour. Manage. Ser. 8. 9pp.

Nichols, S. A. 1974. Mechanical and habitat manipulation for aquatic plant management. Wisconsin Dep. Nat. Resour. Tech. Bull. 77. 34pp.

North Dakota State Highway Department. 1978. Wetland developments within the right-of-way and their biological significance. North Dakota State Highway Dep. Programming and Surveys Div., Bismarck. 15pp.

Novak, M., J. A. Baker, M. E. Obbard, and B. Malloch, eds. 1987. Wild furbearer management and conservation in North America. Ontario Min. Nat. Resour., Toronto. 1150pp.

Oetting, R. B. 1982. Management of waterfowl habitat on public rights-of-way in North Dakota, U.S.A. Pages 163–171 *in* D. A. Scott, ed. Managing wetlands and their birds. Proc. 3d Tech. Meet. on West. Palearctic Migratory Bird Manage. Int. Waterfowl Res. Bur., Slimbridge, Glos., England.

Office, Chief of Engineers. 1952. Soil mechanics design, seepage control. U.S. Dep. Army Eng. Manual EM 1110-2-1901. 40pp.

Office, Chief of Engineers. 1953. Soil mechanics design, settlement analysis. U.S. Dep. Army Eng. Manual EM 1110-2-1904. 36pp.

Office, Chief of Engineers. 1958. Bearing capacity of soils. U.S. Dep. Army Eng. Manual EM 1110-2-1903. 36pp.

Office of Technology Assessment. 1984. Wetlands: their use and regulation. Off. Technol. Assessment, Washington. 208p.

Ohlsson, K. E., A. E. Robb, Jr., C. E. Guindon, Jr., D. E. Samuel, and R. L. Smith. 1982. Best current practices for fish and wildlife on surface-mined land in the northern Appalachian coal region. U.S. Fish Wildl. Serv. FWS/OBS-81/45. 305pp.

O'Leary, W. G., W. D. Klimstra, and J. R. Nawrot. 1984. Waterfowl habitats on reclaimed surface mined lands in southwestern Illinois. Pages 377–382 *in* D. H. Graves, ed. Proceedings of a symposium on surface mining hydrology, sedimentation, and reclamation. Univ. Kentucky, Lexington.

O'Neil, T. 1949. The muskrat in the Louisiana Coastal marshes. Louisiana Dep. Wildl. Fish., New Orleans. 152pp.

Opler, P. A. 1981. Management of prairie habitats for insect conservation. J. Nat. Areas Assoc. 1:3–6.

Opler, P. A., P. H. Eschmeyer, C. H. Halvorson, and J. L. Fenner. 1989. Fisheries and wildlife research and development 1987/1988. U.S. Fish Wildl. Serv., Denver. 90pp.

Orth, R. J., and K. A. Moore. 1982. The effect of fertilizers on transplanted eelgrass, *Zostera marina* L. Annu. Conf. on Wetland Restoration Creation 9:104–131.

Owen, M. 1975. Cutting and fertilizing grassland for winter goose management. J. Wildl. Manage. 39:163–167.

Owensby, C. 1984. Prescribed or controlled burning. Pages 47B–54B *in* F. R. Henderson, ed. Guidelines for increasing wildlife on farms and ranches. Kansas State Univ. Coop. Ext. Serv., Manhattan.

Palermo, M. R., R. L. Montgomery, and M. E. Poindexter. 1978. Guidelines for designing, operating, and managing dredged material containment areas. U.S. Army Eng. Waterways Exp. Stn. Tech. Rep. DS-78-10. 156pp.

Palmisano, A. W. 1972. The effects of salinity on the germination and growth of plants important to wildlife in the Gulf Coast marshes. Proc. Annu. Conf. Southeast. Assoc. Game Fish Comm. 25:215–223.

Parnell, J. F., D. G. Ainley, H. Blokpoel, B. Cain, T. W. Custer, J. L. Dusi, S. Kress, J. A. Kushlan, W. E. Southern, L. E. Stenzel, and B. C. Thompson. 1988. Colonial waterbird management in North America. Colonial Waterbirds 11:129–169.

Payne, N. F., and F. Copes. 1986. Wildlife and fisheries habitat improvement handbook. U.S. For. Serv., Washington. 402pp.

Payne, N. F., B. S. McGinnes, and H. S. Mosby. 1987. Capture rates of cottontails, opossums, and other wildlife in unbaited wooden box traps. Virginia J. Sci. 38:23–26.

Pederson, R. L., D. G. Jorde, and S. G. Simpson. 1989. Northern Great Plains. Pages 281–310 *in* L. M. Smith, R. L. Pederson, and R. M. Kaminski, eds. Habitat management for migrating and wintering waterfowl in North America. Texas Tech Univ. Press, Lubbock.

Pederson, R. L., and L. M. Smith. 1988. Implications of wetland seed bank research: a review of Great Basin and prairie marsh studies. Pages 81–98 *in* D. A. Wilcox, ed. Interdisciplinary approaches to freshwater research. Michigan State Univ. Press, East Lansing.

Peppard, W. 1971. Improve your land for waterfowl. Maine Dep. Inland Fish. Game. 3pp.

Perkins, C. J. 1968. Controlled burning in the management of muskrats and waterfowl in Louisiana coastal marshes. Proc. Annu. Tall Timbers Fire Ecol. Conf. 8:269–280.

Perkins, M. A. 1984. An evaluation of pigmented nylon film for use in aquatic plant management. Pages 467–471 *in* J. Taggart, L. Moore, and K. M. Mackenthun, eds. Lake and reservoir management. U.S. Environ. Prot. Agency EPA 440/5/84-001.

Perkins, M. A., H. L. Boston, and E. F. Curren. 1980. The use of fiberglass screens for control of Eurasian watermilfoil. J. Aquat. Plant Manage. 18:13–19.

Petersen, L. R., M. A. Martin, J. M. Cole, J. R. March, and C. M. Pils. 1982. Evaluation of waterfowl production areas in Wisconsin. Wisconsin Dep. Nat. Resour. Tech. Bull. 135. 32pp.

Peterson, J. O., J. P. Wall, T. L. Wirth, and S. M. Born. 1973. Eutrophication control: nutrient inactivation by chemical precipitation at Horseshoe Lake, Wisconsin. Wisconsin Dep. Nat. Resour. Tech. Bull. 62. 20pp.

Peterson, L. P., H. R. Murkin, and D. A. Wrubleski. 1989. Waterfowl predation on benthic macroinvertebrates during fall drawdown of a northern prairie marsh. Pages 681–689 *in* R. R. Sharitz and J. W. Gibbons, eds. Freshwater wetlands and wildlife. CONF-8603101, DOE Symp. Ser. 61, U.S. Dep. Energy Off. Sci. Tech. Inf., Oak Ridge, TN.

Phillips, R. C. 1980a. Creation of seagrass beds. Pages 91–104 *in* J. C. Lewis and E. W. Bunce, eds. Rehabilitation and creation of selected coastal habitats: proceedings of a workshop. U.S. Fish Wildl. Serv. FWS/OBS-80/27.

Phillips, R. C. 1980b. Planting guidelines for seagrasses. U.S. Army Corps Eng. Coastal Eng. Res. Cent. CETA 80-2. 28pp.

Piehl, J. 1986. Restoration of drained wetlands. Pages 33–37 *in* J. L. Piehl, ed. Wetland restoration: a techniques workshop. Minnesota Chap. Wildl. Soc., Fergus Falls.

Pierce, N. D. 1970. Inland lake dredging evaluation. Wisconsin Dep. Nat. Resour. Tech. Bull. 46. 68pp.

Platts, W. S. 1989. Compatibility of livestock grazing strategies with fisheries. Pages 103–110 *in* R. E. Gresswell, B. A. Barton, and J. L. Kershner, eds. Practical approaches to riparian resource management: an educational workshop. U.S. Bur. Land Manage., Billings, MT.

Platts, W. S., C. Armour, G. D. Booth, M. Bryant, J. L. Bufford, P. Cuplin, S. Jensen, G. W. Lienkaemper, G. W. Minshall, S. B. Monsen, R. L. Nelson, J. R. Sedell, and J. S. Tuhy. 1987. Methods for evaluating riparian habitats with applications to management. U.S. For. Serv. Gen. Tech. Rep. INT-221. 177pp.

Plummer, A. P., D. R. Christensen, and S. B. Monsen. 1968. Restoring big game range in Utah. Utah Div. Fish Game Publ. 68-3. 183pp.

Poff, R. J. 1985. Managing waterfowl impoundments for fisheries. Pages 106–109 *in* M. D. Knighton, comp. Water impoundments for wildlife: a habitat management workshop. U.S. For. Serv. Gen. Tech. Rep. NC-100.

Poole, K. G. 1985. Artificial nesting platforms for ospreys near Yellowknife, Northwest Territories. Northwest Territories Wildl. Serv. File Rep. 50. 25pp.

Poston, H. J., and R. K. Schmidt. 1981. Wildlife habitat: a handbook for Canada's prairies and parklands. Can. Wildl. Serv., Edmonton, Alta. 51pp.

Potter, M. J. 1988. Tree shelters improve survival and increase early growth rates. J. For. 86:39–41.

Prevost, M. B. 1987. Management of plant communities for waterfowl in coastal South Carolina. Pages 167–183 *in* W. R. Whitman and W. H. Meredith, eds. Waterfowl and wetlands symposium: proceedings of a symposium on waterfowl and wetlands

mangement in the coastal zone of the Atlantic Flyway. Delaware Coastal Manage. Prog., Dep. Nat. Resour. Environ. Cont., Dover.
Proctor, B. R., R. W. Thompson, J. E. Bunin, K. W. Fucik, G. R. Tamm, and E. G. Wolf. 1983a. Practices for protecting and enhancing fish and wildlife on coal mined land in the Uinta-southwestern Utah region. U.S. Fish Wildl. Serv. FWS/OBS-83/12. 250pp.
Proctor, B. R., R. W. Thompson, J. E. Bunin, K. W. Fucik, G. R. Tamm, and E. G. Wolf. 1983b. Practices for protecting and enhancing fish and wildlife on coal surface-mined land in the Green River-Hams Fork region. U.S. Fish Wildl. Serv. FWS/OBS-83/09. 242pp.
Proctor, B. R., R. W. Thompson, J. E. Bunin, K. W. Fucik, G. R. Tamm, and E. G. Wolf. 1983c. Practices for protecting and enhancing fish and wildlife on coal surface-mined land in the Powder River-Fort Union region. U.S. Fish Wildl. Serv. FWS/OBS-83/10. 246pp.
Provost, M. W. 1968. Managing impounded salt marsh for mosquito control and estuarine resource conservation. Proc. Marsh Estuary Manage. Symp. 1:163–171.
Rakstad, D., and J. Probst. 1985. Wildlife occurrence in water impoundments. Pages 80–94 *in* M. D. Knighton, comp. Water impoundments for wildlife: a habitat management workshop. U.S. For. Serv. Gen. Tech. Rep. NC-100.
Range Seeding and Equipment Committee. 1970. Range seeding and equipment handbook. U.S. Dep. Agric. and Dep. Inter., Washington. 156pp.
Rasmussen, G. A., G. R. McPherson, and H. A. Wright. 1986. Prescribed burning juniper communities in Texas. Texas Tech Univ. Dep. Range Wildl. Manage. Note 10. 5pp.
Reed, P. B., Jr. 1988. National list of plant species that occur in wetlands: national summary. U.S. Fish Wildl. Serv. Biol. Rep. 88(24). 244pp.
Rees, J. R. 1982. Potential uses of cattle grazing to manage waterfowl nesting cover on Turnbill National Wildlife Refuge. Pages 86–93 *in* J. M. Peek and P. D. Dalke, eds. Wildlife-livestock relationships symposium. Univ. Idaho For., Wildl., Range Exp. Stn., Moscow.
Reid, F. A. 1985. Wetland invertebrates in relation to hydrology and water chemistry. Pages 72–79 *in* M. D. Knighton, ed. Water impoundments for wildlife: a habitat management workshop. U.S. For. Serv. Gen. Tech. Rep. NC-100.
Reid, F. A., J. R. Kelley, Jr., T. S. Taylor, and L. H. Fredrickson. 1989. Upper Mississippi Valley wetlands—refuges and moist-soil impoundments. Pages 181–202 *in* L. M. Smith, R. L. Pederson, and R. M. Kaminski, eds. Habitat management for migrating and wintering waterfowl in North America. Texas Tech Univ. Press, Lubbock.
Reinecke, K. J. 1977. The importance of freshwater invertebrates and female energy reserves for black ducks breeding in Maine. Ph.D. diss. Univ. Maine, Orono. 113pp.
Reinecke, K. J., R. M. Kaminski, D. J. Moorhead, J. D. Hodges, and J. R. Nassar. 1989. Mississippi alluvial valley. Pages 203–247 *in* L. M. Smith, R. L. Pederson, and R. M. Kaminski, eds. Habitat management for migrating and wintering waterfowl in North America. Texas Tech Univ. Press, Lubbock.
Renfro, G. M. 1979. Ponds and reservoirs. Pages 11-1–11-60 *in* Engineering field manual for conservation practices. U.S. Soil Conserv. Serv., Washington.
Ridlehuber, K. T. 1982. Fire ant predation on wood duck ducklings and pipped eggs. Southwest. Nat. 27:222.
Ridlehuber, K. T., and J. W. Teaford. 1986. Wood duck nest boxes. Sect. 5.1.2, U.S. Army Corps of Engineers wildlife resources management manual. U.S. Army Eng. Waterways Exp. Stn. Tech. Rep. EL-86-12. 21pp.
Ringelman, J. K. 1990. Habitat mangagement for molting waterfowl. Pages 1–6 (13.4.4) *in* D. H. Cross, comp. Waterfowl management handbook. U.S. Fish Wildl. Serv. Fish Wildl. Leafl. 13.
Ringelman, J. K., W. R. Eddleman, and H. W. Miller. 1989. High plains reservoirs and sloughs. Pages 311–340 *in* L. M. Smith, R. L. Pederson, and R. M. Kaminski, eds. Habitat management for migrating and wintering waterfowl in North America. Texas Tech Univ. Press, Lubbock.
Robson, T. O. 1974. Mechanical control. Pages 72–84 *in* D. S. Mitchell, ed. Aquatic vegetation and its use and control. U.N. Educ., Sci. Cult. Organ., Paris.
Roby, G. A., and L. R. Green. 1976. Mechanical methods of chaparral modification. U. S. For. Serv. Agric. Handb. 487. 46pp.

Rogalsky, J. R., K. W. Clark, and J. M. Stewart. 1971. Wild rice paddy production in Manitoba. Manitoba Dep. Agric. Publ. 527.

Rogers, B. D., and W. H. Herke. 1985. Estuarine-dependent fish and crustacean movements and weir management. Proc. Coastal Marsh Estuary Manage. Symp. 4:201–219.

Rogers, B. D., W. H. Herke, and E. E. Knudson. 1987. Investigation of a weir-design alternative for coastal fisheries benefit. Louisiana State Univ. Agric. Cent., Baton Rouge. 98pp.

Rollings, C. T., and R. L. Warden. 1964. Weedkillers and waterfowl. Pages 593–598 *in* J. P. Linduska, ed. Waterfowl tomorrow. U.S. Fish Wildl. Serv., Washington.

Rollins, G. L. 1981. A guide to waterfowl habitat management in Suisun Marsh. California Dep. Fish Game, Sacramento. 109pp.

Ross, D., C. Kocur, and W. Jurgens. 1985. Wetlands creation techniques for heavy construction equipment. Proc. Annu. Conf. on Wetlands Restoration Creation 12:210–220.

Rudolph, D. C., and J. G. Dickson. 1990. Streamside zone width and amphibian and reptile abundance. Southwest. Nat. 35:472–476.

Rudolph, R. R., and C. G. Hunter. 1964. Green trees and greenheads. Pages 611–618 *in* J. P. Linduska, ed. Waterfowl tomorrow. U.S. Fish Wildl. Serv., Washington.

Rumble, M. A. 1989. Surface mine impoundments as wildlife and fish habitat. U.S. For. Serv. Gen. Tech. Rep. RM-183. 6pp.

Rumble, M. A., and L. D. Flake. 1983. Management considerations to enhance use of stock ponds by waterfowl broods. J. Range Manage. 36:691–694.

Rundle, W. D., and L. H. Fredrickson. 1981. Managing seasonally flooded impoundments for migrant rails and shorebirds. Wildl. Soc. Bull. 9:80–87.

Rutherford, W. H., and W. D. Snyder. 1983. Guidelines for habitat modification to benefit wildlife. Colorado Div. Wildl., Denver. 194pp.

Sale, P. J. M., and R. G. Wetzel. 1983. Growth and metabolism of *Typha* species in relation to cutting treatments. South. Aquat. Bot. 15:321–334.

Sanders, W. D., J. B. Burley, and C. A. Churchward. 1982. A study for vegetation and wildlife habitat on Lower Grey Cloud Island. Pages 102–108 *in* W. D. Svedarsky and R. D. Crawford, eds. Wildlife values in gravel pits. Univ. Minnesota Agric. Exp. Stn. Misc. Publ. 17.

Sanderson, G. C. 1980. Conservation of waterfowl. Pages 43–63 *in* F. C. Bellrose, ed. Ducks, geese, and swans of North America. Stackpole Books, Harrisburg, PA.

Sanderson, G. C., and F. C. Bellrose. 1969. Wildlife habitat management of wetlands. Suppl. An. Acad. Bras. Cienc. 41:153–204.

Sather, J. H., and R. D. Smith. 1984. An overview of major wetland functions and values. U.S. Fish Wildl. Serv. FWS/OBS-84/18. 68pp.

Savard, J. L. 1988. Use of nest boxes by Barrow's goldeneyes: nesting success and effect on the breeding population. Wildl. Soc. Bull. 16:125–132.

Schitoskey, F., Jr., and R. L. Linder. 1978. Use of wetlands by upland wildlife. Pages 307–311 *in* P. E. Greeson, J. R. Clark, and J. E. Clark, eds. Wetland functions and values: the state of our understanding. Proc. Natl. Symp. on Wetland Functions Values. Am. Water Resour. Assoc., Minneapolis.

Schlichtemeier, G. 1967. Marsh burning for waterfowl. Proc. Annu. Tall Timbers Fire Ecol. Conf. 6:40–46.

Schmutz, J. K., W. D. Wishart, J. Allen, R. Bjorge, and D. A. Moore. 1988. Dual use of nest platforms by hawks and Canada geese. Wildl. Soc. Bull. 16:141–145.

Schneller-McDonald, K., L. S. Ischinger, and G. T. Auble. 1990. Wetland creation and restoration: description and summary of the literature. U.S. Fish Wildl. Serv. Biol. Rep. 90(3). 198pp.

Schnick, R. A., J. A. Morton, J. C. Mochalski, and J. T. Beall. 1982. Mitigation and enhancement techniques for the upper Mississippi River system and other large river systems. U.S. Fish Wildl. Serv. Resour. Publ. 149. 714pp.

Schroeder, L. D. 1973. A literature review on the role of invertebrates in waterfowl management. Colorado Div. Wildl. Spec. Rep. 29. 13pp.

Scifres, C. J. 1980. Brush management. Texas A & M Univ. Press, College Station. 360pp.

Seehorn, M. E. 1985. Fish habitat improvement handbook. U.S. For. Serv. Tech. Publ. R8-TP7. 21pp.

Semel, B., P. W. Sherman, and S. M. Byers. 1988a. Effects of brood parasitism and nest box placement on wood duck breeding ecology. Condor 90:920–930.
Semel, B., P. W. Sherman, and S. M. Byers. 1988b. Nest boxes and brood parasitism in wood ducks: a management dilemma. Proc. North Am. Wood Duck Symp. 1:163–170.
Seneca, E. D. 1980. Dune community creation along the Atlantic coast. Pages 58–62 *in* J. C. Lewis and E. W. Bunce, eds. Rehabilitation and creation of selected coastal habitats: proceedings of a workshop. U.S. Fish Wildl. Serv. FWS/OBS-80/27.
Shaw, S. P., and C. G. Fredine. 1956. Wetlands of the United States. U.S. Fish Wildl. Serv. Circ. 39. 67pp.
Singleton, J. R. 1951. Production and utilization of waterfowl food plants on the east Texas Gulf coast. J. Wildl. Manage. 15:46–56.
Singleton, J. R., 1965. Waterfowl habitat management in Texas. Texas Parks Wildl. Dep. Bull. 47. 68pp.
Sipple, W. S. 1979. A review of the biology, ecology, and management of *Scirpus olneyi*. Vol. 2: a synthesis of selected references. Maryland Dep. Nat. Resour. Wetland Publ. 4. 85pp.
Sipple, W. S. 1987a. Wetland identification and delineation manual. Vol. 1: Rationale, wetland parameters, and overview of jurisdictional approach. U.S. Environ. Prot. Agency Off. of Wetlands Prot., Washington. 28pp.
Sipple, W. S. 1987b. Wetland identification and delineation manual. Vol. 2: Field methodology. U.S. Environ. Prot. Agency Off. of Wetlands Prot., Washington. 29pp.
Skinner, T. 1982. Cebolla marsh pothole development—1981 followup. Southwest Habitation 3(2):1–2.
Smith, E. R. 1960. Evaluation of a leveed Louisiana marsh. Trans. North Am. Wildl. Nat. Resour. Conf. 35:265–275.
Smith, G. J. 1987. Pesticide use and toxicology in relation to wildlife: organophosphorus and carbamate compounds. U.S. Fish Wildl. Serv. Resour. Publ. 170. 171pp.
Smith, H. K. 1978. An introduction to habitat development on dredged material. U.S. Army Eng. Waterways Exp. Stn. Tech. Rep. DS-78-19. 40pp.
Smith, L. M., and J. A. Kadlec. 1983. Seed banks and their role during drawdown of a North American marsh. J. Appl. Ecol. 20:673–684.
Smith, L. M., and J. A. Kadlec. 1985. The effects of disturbance on marsh seed banks. Can. J. Bot. 63:2133–2137.
Smith, L. M., and J. A. Kadlec. 1986. Habitat management for wildlife in marshes of Great Salt Lake. Trans. North Am. Wildl. Nat. Resour. Conf. 51:222–231.
Smith, R. H. 1942. Management of salt marshes on the Atlantic coast of the United States. Trans. North Am. Wildl. Conf. 7:272–277.
Snyder, B. D., and J. L. Snyder. 1984. Feasibility of using oil shale wastewater for waterfowl wetlands. U.S. Fish Wildl. Serv. FWS/OBS-84/01. 290pp.
Snyder, W. D. 1988. Stem cutting propagation of woody phreatophytes in eastern Colorado. Pages 151–156 *in* K. M. Mutz, D. J. Cooper, M. L. Scott, and L. K. Miller, tech. coord. Restoration, creation and management of wetland and riparian ecosystems in the American west. PIC Technologies, Denver.
Soots, R. F., Jr., and M. C. Landin. 1978. Development and management of avian habitat on dredged material islands. U.S. Army Corps Eng. Tech. Rep. DS-78-18. 123pp.
Soots, R. F., Jr., and J. F. Parnell. 1975. Introduction to the nature of dredge islands and their wildlife in North Carolina and recommendations for management. Pages 1–34 *in* J. F. Parnell and R. F. Soots, eds. Proceedings of a conference on management of dredge islands in North Carolina estuaries. Univ. North Carolina Sea Grant Coll. Prog. Publ. UNC-SG-75-01.
Sousa, P. J. 1987. Habitat management models for selected wildlife management practices in the northern Great Plains. U.S. Bur. Reclam. Rep. REC-ERC 87-11.
Sousa, P. J., and A. H. Farmer. 1983. Habitat suitability index models: wood duck. U.S. Fish Wildl. Serv. FWS/OBS-82/10.43. 27pp.
Spaine, P. A., J. L. Llopis, and E. R. Perrier. 1978. Guidance and land improvement using dredged material. U.S. Army Eng. Waterways Exp. Stn. Tech. Rep. DS-78-21. 112pp.
Spiller, S. F., and R. H. Chabreck. 1975. Wildlife populations in coastal marshes influenced by weirs. Proc. Annu. Conf. Southeast. Assoc. Game Fish Comm. 29:518–525.

Stabb, M. 1989. Wetlands conservation on a small scale: the beaver baffler and other macro-management projects on private lands. Pages 87–96 *in* M. J. Bardecki and N. Patterson, eds. Wetlands: inertia or momentum. Fed. Ontario Nat., Don Mills, Ont.

Stakhiv, A., ed. Undated. Dredging is for the birds. U.S. Army Corps Eng., Ft. Belvoir, VA. 21pp.

Stanley, W. R. 1979. Preparation of engineering plans. Pages 5-2–5-23 *in* Engineering field manual for conservation practices. U.S. Soil Conserv. Serv., Washington.

Stanton, F. W. 1957. Planting food for waterfowl. Oregon Game Comm. Misc. Wildl. Publ. 1. 17pp.

Steenis, J. H., E. W. Ball, V. D. Stotts, and C. K. Rawls. 1968. Pest plant control with herbicides. Proc. Marsh Estuary Manage. Symp. 1:140–148.

Stevens, G. E. 1985. Aerial ignition flying drip torch. Pages 95–100 *in* D. D. Wade, comp. Prescribed fire and smoke management in the South: conference proceedings. U.S. For. Serv. Southeast. For. Exp. Stn., Ashville, NC.

Stewart, R. E., and H. Gratkowski. 1976. Aerial application equipment for herbicidal drift reduction. U.S. For. Serv. Gen. Tech. Rep. PNW-54. 21pp.

Stewart, R. E., and H. A. Kantrud. 1971. Classification of natural ponds and lakes in the glaciated prairie region. U.S. Fish Wildl. Serv. Res. Publ. 92. 57pp.

Stewart, R. E., and H. A. Kantrud. 1972. Vegetation of prairie potholes, North Dakota, in relation to quality of water and other environmental factors. U.S. Geol. Surv. Prof. Pap. 585-D. 36pp.

Stoecker, R. E. 1982. Creating small islands for wildlife and visual enhancement. Pages 48–50 *in* W. D. Svedarsky and R. D. Crawford, eds. Wildlife values of gravel pits. Univ. Minnesota Agric. Exp. Stn. Misc. Publ. 17.

Stormer, F. A., E. G. Bolen, and C. D. Simpson. 1981. Management of playas for migratory birds—information needs. Pages 52–61 *in* J. S. Barclay and W. V. White, eds. Playa lakes symposium proceedings. U.S. Fish Wildl. Serv. FWS/OBS-81/07.

Stotts, V. D. 1971. Improving (changing) wetlands for waterfowl and other wildlife. Proc. Maryland Mosquito Wildl. Conf. Annapolis. 8pp.

Stoudt, J. H. 1971. Ecological factors affecting waterfowl production in the Saskatchewan parklands. U.S. Fish Wildl. Serv. Res. Publ. 99. 58pp.

Stoudt, J. H. 1982. Habitat use and productivity of canvasbacks in southwestern Manitoba, 1961-72. U.S. Fish Wildl. Serv. Spec. Sci. Rep. Wildl. 248. 31pp.

Street, M. 1982. The use of waste straw to promote the production of invertebrate food for waterfowl in man-made wetlands. Pages 98–103 *in* D. A. Scott, ed. Managing wetlands and their birds. Proc. 3d Tech. Meet. on West. Palearctic Migratory Bird Manage. Int. Waterfowl Res. Bur., Slimbridge, Glos., England.

Stutzenbaker, C. D., and M. W. Weller. 1989. The Texas coast. Pages 385–405 *in* L. M. Smith, R. L. Pederson, and R. M. Kaminski, eds. Habitat management for migrating and wintering waterfowl in North America. Texas Tech Univ. Press, Lubbock.

Summers, M. W. 1984. Managing Louisiana fish ponds. Louisiana Dep. Wildl. Fish. Tech. Bull. 4. 49pp.

Sutcliffe, S. A. 1979. Artificial common loon nesting constructions, placement and utilization in New Hampshire. Proc. North Am. Conf. on Common Loon Res. Manage. 2:147–152.

Swenson, E. A. 1988. Progress in the understanding of how to reestablish native riparian plants in New Mexico. Pages 144–150 *in* K. M. Mutz, D. J. Cooper, M. L. Scott, and L. K. Miller, tech. coord. Restoration, creation and management of wetland and riparian ecosystems in the American west. PIC Technologies, Denver.

Swenson, E. A., and C. L. Mullins. 1985. Revegetating riparian trees in southwestern floodplains. Pages 135–138 *in* R. R. Johnson, C. D. Ziebell, D. R. Patton, P. F. Ffolliott, and R. H. Hamre, tech. coord. Riparian ecosystems and their management: reconciling conflicting uses. U.S. For. Serv. Gen. Tech. Rep. RM-120.

Swickhard, D. K. 1974. An evaluation of two artificial least tern nesting sites. California Fish Game 60:88–90.

Swiderek, P. K., A. S. Johnson, P. E. Hale, and R. L. Joyner. 1988. Production, management, and waterfowl use of sea purslane, Gulf Coast muskgrass, and widgeongrass in brackish impoundments. Pages 441–457 *in* M. W. Weller, ed. Waterfowl in winter. Univ. Minnesota Press, Minneapolis.

Swift, B. L. 1984. Status of riparian ecosystems in the United States. Water Resour. Bull. 20:223–228.

Swift, J. A. 1982. Construction of rafts and islands. Pages 200–203 *in* D. A. Scott, ed. Managing wetlands and their birds. Proc. 3d Tech. Meet. on West. Paleartic Migratory Bird Manage. Int. Waterfowl Res. Bur., Slimbridge, Glos., England.

Sykes, P. W., and R. Chandler. 1974. Use of artificial nest structures by Everglade kites. Wilson Bull. 86:282–284.

Sypulski, J. L. 1943. The Seny bulrush picker. J. Wildl. Manage. 7:230–231.

Teaford, J. W. 1986. Beaver pond management. Sec. 5.5.2, U.S. Army Corps of Engineers wildlife resources management manual. U.S. Army Eng. Waterways Exp. Stn. Tech. Rep. EL-86-10. 10pp.

Teas, H. J. 1980. Mangrove swamp creation. Pages 63–90 *in* J. C. Lewis and E. W. Bunce, eds. Rehabilitation and creation of selected coastal habitats: proceedings of a workshop. U.S. Fish Wildl. Serv. FWS/OBS-80/27.

Teas, H. J. 1981. Restoration of mangrove ecosystems. Pages 95–103 *in* R. C. Carey, P. S. Markovits, and J. B. Kirkwood, eds. Proceedings: U.S. Fish and Wildlife workshop on coastal ecosystems of the southeastern United States. U.S. Fish Wildl. Serv. FWS/OBS-80/59.

Ternyik, W. E. 1980. Sand dune habitat creation on the Pacific coast. Pages 55–57 *in* J. C. Lewis and E. W. Bunce, eds. Rehabilitation and creation of selected coastal habitats: proceedings of a workshop. U.S. Fish Wildl. Serv. FWS/OBS-80/27.

Teskey, R. O., and T. M. Hinckley. 1977a. Impact of water level changes on woody riparian and wetland communities. Vol. 2: The southern forest region. U.S. Fish Wildl. Serv. FWS/OBS-77/59. 46pp.

Teskey, R. O., and T. M. Hinckley. 1977b. Impact of water level changes on woody riparian and wetland communities. Vol. 3: The central forest region. U.S. Fish Wildl. Serv. FWS/OBS-77/60. 36pp.

Tester, J. R., and W. H. Marshall. 1962. Minnesota prairie management techniques and their wildlife implications. Trans. North Am. Wildl. Nat. Resour. Conf. 27:267–287.

Thomas, J. W., C. Maser, and J. E. Rodiek. 1979. Riparian zones. Pages 40–47 *in* J. W. Thomas, ed. Wildlife habitats in managed forests: the Blue Mountains of Oregon and Washington. U.S. For. Serv., Portland, OR.

Thompson, D. Q. 1989. Control of purple loosestrife. Pages 1–6 (13.4.11) *in* D. H. Cross, comp. Waterfowl management handbook. U.S. Fish Wildl. Serv. Fish Wildl. Leafl. 13.

Thorhaug, A. 1980. Techniques for creating seagrass meadows in damaged areas along the east coast of the U.S.A. Pages 105–116 *in* J. C. Lewis and E. W. Bunce, eds. Rehabilitation and creation of selected coastal habitats: proceedings of a workshop. U.S. Fish Wildl. Serv. FWS/OBS-8/27.

Thorhaug, A. 1990. Restoration of mangroves and seagrasses—economic benefits for fisheries and mariculture. Pages 265–281 *in* J. J. Berger, ed. Environmental restoration: science and strategies for restoring the earth. Island Press, Washington.

Timm, R. M., ed. 1983. Prevention and control of wildlife damage. Great Plains Agric. Counc. Wildl. Resour. Comm. and Nebraska Coop. Ext. Serv. Inst. Agric. and Nat. Resour., Univ. Nebraska, Lincoln.

Tisdal, S. L., and W. L. Nelson. 1975. Soil fertility and fertilizers. MacMillan., New York. 694pp.

Townsend, F. C. 1979. Use of lime in levee restoration. U.S. Army Eng. Waterways Exp. Stn. Tech. Rep. GL-79-12. 96pp.

Tunberg, L. D. 1966. The use of bentonite for sealing farm ponds and small reservoirs. Colorado Game, Fish Parks Dep. Fish. Inf. Leafl. 1. 2pp.

Uhler, F. M. 1944. Control of undesirable plants in waterfowl habitats. Trans. North Am. Wildl. Conf. 9:295–303.

Uresk, D. W., and K. Severson. 1988. Waterfowl and shorebird use of surface-mined and livestock water impoundments on the northern Great Plains. Great Basin Nat. 48: 353–357.

Urquhart, W. J. 1979. Hydraulics. Pages 3-1–3-112 *in* Engineering field manual for conservation practices. U.S. Soil Conserv. Serv., Washington.

U.S. Army. 1986. Explosives and demolitions. Field Man. 5-25. Headquarters, U.S. Dep. Army, Washington. 186pp.

U.S. Army Coastal Engineering Research Center. 1977. Shore protection manual. Stock No. 008-022-00113-1. U.S. Gov. Print. Off., Washington. 3 vols.

U.S. Army Engineer Waterways Experiment Station. 1978. Wetland habitat development with dredged material: engineering and plant propagation. U.S. Army Eng. Waterways Exp. Stn. Tech. Rep. DS-78-16. 158pp.

U.S. Department of Agriculture. 1961. Seeds. U.S. Dep. Agric. Washington. 791pp.

U.S. Department of Commerce. 1986. Climatological data, California. Vol. 90. Natl. Oceanic Atmos. Admin., Natl. Climatic Cent., Asheville, NC. 10pp.

U.S. Department of Commerce, National Oceanic and Atmospheric Administration. 1980. The relationship between the upper limits of coastal wetlands and tidal datums along the Pacific Coast. U.S. Dep. Commerce Natl. Oceanic Atmospheric Admin., Rockville, MD.

U.S. Environmental Protection Agency. 1974 plus updates. EPA compendium of registered pesticides. Vol. 1: Herbicides and plant regulators. U.S. Environ. Prot. Agency, Washington. 630pp.

U.S. Fish and Wildlife Service. 1970. New homes for prairie ducks. U.S. Fish Wildl. Serv., Washington. 5 pp.

U.S. Fish and Wildlife Service. 1976a. Selected list of federal laws and treaties relating to sport fish and wildlife. U.S. Fish Wildl. Serv., Washington. 19pp.

U.S. Fish and Wildlife Service. 1976b. Nest boxes for wood ducks. U.S. Fish Wildl. Serv. Wildl. Leafl. 510. 14pp.

U.S. Fish and Wildlife Service. 1979. Aquatic plant control herbicide handbook. U.S. Fish Wildl. Serv., Washington. 266pp.

U.S. Fish and Wildlife Service. 1980. Habitat evaluation procedures (HEP). U.S. Fish Wildl. Serv. Div. Ecol. Serv. Man. 102 ESM.

U.S. Fish and Wildlife Service. 1985. Evaluation of mitigation wetlands: U.S. Highway 83 between Bismark and Minot. U.S. Fish Wildl. Serv. unpubl. rep. Bismark, ND. 10pp.

U.S. Fish and Wildlife Service. 1989. National wetlands priority conservation plan. U.S. Fish Wildl. Serv., Washington. 58pp.

U.S. Forest Service. 1969. Land treatment measures handbook. U.S. For. Serv. Handb. FSH 2509.11. 124pp.

U.S. Forest Service. 1974a. Seeds of woody plants in the United States. U.S. For. Serv. Handb. 450. 883pp.

U.S. Forest Service. 1974b. Watershed structural measures handbook. U.S. For. Serv. Handb. FSH 2509.12. 103pp.

U.S. Forest Service. 1978. Animal damage control handbook. U.S. For. Serv. Reg. 6 FSH 2609.22.

U.S. Forest Service. 1979. User guide to vegetation. U.S. For. Serv. Gen. Tech. Rep. INT-64. 85pp.

U.S. Forest Service. 1980. Blaster's handbook. U.S. For. Serv. FSH 7109.51. 146pp.

U.S. Soil Conservation Service. 1976. National range handbook. U.S. Soil Conserv. Serv., Washington. 143pp.

U.S. Soil Conservation Service. 1977. Gulf Coast wetlands handbook. Alexandria, LA. 89pp.

U.S. Soil Conservation Service. 1978a. Land resources regions and major land resources areas of the United States. U.S. Soil Conserv Serv. Agric. Handb. 296. 82pp.

U.S. Soil Conservation Service. 1978b. Technical guide; standards and specifications, North Dakota. U.S. Soil Conserv. Serv., Bismarck, ND.

U.S. Soil Conservation Service. 1979. Engineering field manual for conservation practices. U.S. Soil Conserv. Serv., Washington.

U.S. Soil Conservation Service. 1982. Ponds—Planning, design, construction. U.S. Soil Conserv. Serv. Agric. Handb. 590. 51pp.

Vallentine, J. F. 1983. The application and use of herbicides for range plant control. Pages 39–48 *in* S. B. Monsen and N. Shaw, eds. Managing intermountain rangelands—improvement of range and wildlife habitats. U.S. For. Serv. Gen. Tech. Rep. INT-157.

Vallentine, J. F. 1989. Range development and improvement. 3d ed. Brigham Young Univ. Press, Provo, UT. 524pp.

Van Daele, L. J., H. A. Van Daele, and D. R. Johnson. 1980. The status and management of ospreys nesting in Long Valley, Idaho. Univ. Idaho, Moscow. 49pp.
Vaughn, A. 1976. Doughnuts for ducks. Soil Conserv. 4(11):18.
Verry, E. S. 1985a. Selection of water impoundment sites in the Lake States. Pages 31–38 *in* M. D. Knighton, comp. Water impoundments for wildlife: a habitat management workshop. U.S. For. Serv. Gen. Tech. Rep. NC-100.
Verry, E. S. 1985b. Water quality and nutrient dynamics in shallow water impoundments. Pages 61–71 *in* M. D. Knighton, comp. Water impoundments for wildlife: a habitat management workshop. U.S. For. Serv. Gen. Tech. Rep. NC-100.
Verry, E. S. 1989. Selection and management of shallow water impoundments for wildlife. Pages 1177–1193 *in* R. R. Sharitz and J. W. Gibbons, eds. Freshwater wetlands and wildlife. CONF-8603101, DOE Symp. Ser. 61, U.S. Dep. Energy Off. Sci. Tech. Inf., Oak Ridge, TN.
Voorhees, L. D., and J. F. Cassel. 1980. Highway right-of-way: mowing versus succession as related to duck nesting. J. Wildl. Manage. 44:155–163.
Wade, D. D., and J. D. Lunsford. 1989. A guide for prescribed fire in southern forests. U.S. For. Serv. Tech. Publ. R8-TP11. 56pp.
Walsh, M. R., and M. D. Malkasian. 1978. Productive land use of dredged material containment areas: planning and implementation considerations. U.S. Army Eng. Waterways Exp. Stn. Tech. Rep. DS-78-20. 116pp.
Walski, T. M., and P. R. Schroeder. 1978. Weir design to maintain effluent quality from dredged material containment areas. U.S. Army Eng. Waterways Exp. Stn. Tech. Rep. D-78-18. 94pp.
Ward, E. 1942. Phragmites management. Trans. North Am. Wildl. Conf. 7:294–298.
Ward, P. 1968. Fire in relation to waterfowl habitat of the Delta Marshes. Proc. Annu. Tall Timbers Fire Ecol. Conf. 8:255–268.
Warren, J., and D. Bandel. 1968. Pothole blasting in Maryland wetlands. Proc. Annu. Conf. Southeast. Game Fish Comm. 22:58–68.
Webb, W. L., and A. H. Lloyd. 1984. Design and use of tripods as osprey nest platforms. Pages 99–107 *in* M. A. Westall, ed. Proceedings of the southeastern U.S. and Caribbean osprey symposium. Int. Osprey Found. Sanibel Island, FL.
Weed Science Society of America. 1979. Herbicide handbook of the Weed Science Society of America. 4th ed. Weed Sci. Soc. Am., Champaign, IL. 479pp.
Weed Science Society of America. 1989. Herbicide handbook of the Weed Science Society of America. 6th ed. Weed Sci. Soc. Am., Champaign, IL. 301pp.
Wellborn, T. L., A. J. Herring, and R. Callahan. 1984. Farm pond management. Mississippi State Univ. Coop. Ext. Serv. Publ. 1428. 15pp.
Weller, M. W. 1975. Studies of cattail in relation to management for marsh wildlife. Iowa State J. Res. 49:383–412.
Weller, M. W. 1978. Management of freshwater marshes for wildlife. Pages 267–284 *in* R. E. Good, D. F. Whigham, and R. L. Simpson, eds. Freshwater wetlands: ecological processes and management potential. Academic Press, New York.
Weller, M. W. 1986. Marshes. Pages 201–224 *in* A. Y. Cooperrider, R. J. Boyd, and H. R. Stuart, eds. Inventory and monitoring of wildlife habitat. U.S. Bur. Land Manage. Serv. Cent., Denver.
Weller, M. W. 1987. Freshwater marshes ecology and wildlife management. 2d ed. Univ. Minnesota Press, Minneapolis. 150pp.
Weller, M. W., and C. S. Spatcher. 1965. Role of habitat in the distribution and abundance of marsh birds. Iowa State Univ. Agric. Home Econ. Exp. Stn. Spec. Rep. 43. 31pp.
Weller, M. W. , and D. K. Voights. 1983. Changes in the vegetation and wildlife use of a small prairie wetland following drought. Proc. Iowa. Acad. Sci. 90:50–54.
Welsh, R. G., and D. Müller-Schwarze. 1989. Experimental habitat scenting inhibits colonization by beaver, *Castor canadensis*. J. Chem. Ecol. 15:887–893.
Wenger, K. E. 1984. Forestry handbook. 2d ed. Wiley, New York. 1335pp.
Wentz, W. A. 1981. Wetlands values and management. U.S. Fish Wildl. Serv. and U.S. Environ. Prot. Agency, Washington. 25pp.
Wesley, D. E. 1979. Serving up the rice. Ducks Unlimited 43(4):32–33, 35, 37, 39.
Wetzel, R. G. 1975. Limnology. W. B. Saunders, Philadelphia. 743pp.

Whitaker, G. A. 1974. Phoebe and barn swallow structures. Northeast Fish Wildl. Conf. 31:57–62.
White, W. M., and G. W. Malaher. 1964. Reservoirs. Pages 381–389 *in* J. P. Linduska, ed. Waterfowl tomorrow. U.S. Fish Wildl. Serv., Washington.
Whitman, W. R. 1974. The response of macro-invertebrates to experimental marsh management. Ph.D. diss. Univ. Maine, Orono. 114pp.
Whitman, W. R. 1982. Construction of impoundments and ponds at Tintamarre National Wildlife Area, Canada. Pages 156–162 *in* D. A. Scott, ed. Managing wetlands and their birds. Proc. 3d Tech. Meet. on West. Palearctic Migratory Bird Manage., Int. Waterfowl Res. Bur., Slimbridge, Glos., England.
Whitman, W. R., and R. V. Cole. 1987. Ecological conditions and implications for waterfowl management in selected coastal impoundments in Delaware. Pages 99–119 *in* W. R. Whitman and W. H. Meredith, eds. Waterfowl and wetlands symposium: proceedings of a symposium on waterfowl and wetlands management in the coastal zone of the Atlantic Flyway. Delaware Dep. Nat. Resour. Environ. Cont., Dover.
Wicker, K. M., D. Davis, and D. Roberts. 1983. Rockefeller State Wildlife Refuge and Game Preserve: evaluation of wetland management techniques. Louisiana Dep. Nat. Resour., Baton Rouge.
Wigley, T. B., and T. H. Filer, Jr. 1989. Greentree reservoirs: current management, use and problems. Wildl. Soc. Bull. 17:136–142.
Wiley, J. E. III. 1988. Wildlife guidelines for the public reserved lands of Maine. Maine Dep. Conserv., Augusta. 71pp.
Wiley, M. J., P. P. Tazik, and S. T. Sobaski. 1987. Controlling aquatic vegetation with triploid grass carp. Illinois Nat. Hist. Surv. Circ. 57. 16pp.
Will, G. C., and G. I. Crawford. 1970. Elevated and floating nest structures for Canada geese. J. Wildl. Manage. 34:583–586.
Willard, D., M. Leslie, and R. B. Reed. 1990. Defining and delineating wetlands. Pages 111–140 *in* G. Bingham, E. H. Clark II, L. V. Haygood, and M. Leslie, eds. Issues in wetlands protection: background papers prepared for the National Wetlands Policy forum. The Conserv. Found., Washington.
Williams, G. L. 1985. Classifying wetlands according to relative wildlife value: application to water impoundments. Pages 110–119 *in* M. D. Knighton, comp. Water impoundments for wildlife: a habitat management workshop. U.S. For. Serv. Gen. Tech. Rep. NC-100.
Williams, R. D., and S. H. Hanks. 1976. Hardwood nurseryman's guide. U.S. For. Serv. Agric. Handb. 473. 78pp.
Williams, R. K. 1987. Construction, maintenance, and water control structures of tidal impoundments in South Carolina. Pages 139–166 *in* W. R. Whitman and W. H. Meredith, eds. Waterfowl and wetlands symposium: proceedings of a symposium on waterfowl and wetlands management in the coastal zone of the Atlantic Flyway. Delaware Dep. Nat. Resour. Environ. Cont., Dover.
Willoughby, W. E. 1978. Assessment of low-ground-pressure equipment for use in containment area operation and maintenance. U.S. Army Corps Eng. Waterways Exp. Stn. Tech. Rep. DS-78-9. 106pp.
Wilson, K. A. 1968. Fur production on southeastern coastal marshes. Proc. Marsh Estuary Manage. Symp. 1:149–162.
Wisconsin Department of Natural Resources. 1990. Prescribed burn handbook. Wisconsin Dep. Nat. Resour., Madison. 47pp.
Woehler, E. E. 1987. Use of herbicides to control woody nuisance plants on public lands. Wisconsin Dep. Nat. Resour. Res. Manage. Findings 5. 4pp.
Woodhouse, W. W., Jr. 1978. Dune building and stabilization with vegetation. U.S. Army Corps Eng. Coastal Eng. Res. Cent. Spec. Rep. 3. 112pp.
Woodhouse, W. W., Jr. 1979. Building salt marshes along the coasts of the continental United States. U.S. Army Corps Eng. Coastal Eng. Res. Cent. Spec. Rep. 4. 96pp.
Woodhouse, W. W., Jr. 1982. Coastal sand dunes of the U.S. Pages 1–44 *in* R. R. Lewis, III, ed. Creation and restoration of coastal plant communities. CRC Press, Boca Raton, FL.
Woodhouse, W. W., Jr., and P. L. Knutson. 1982. Atlantic coastal marshes. Pages 45–70

in R. R. Lewis, III, ed. Creation and restoration of coastal plant communities. CRC Press, Boca Raton, FL.

Woodhouse, W. W., Jr., E. D. Seneca, and S. W. Broome. 1972. Marsh building with dredge spoil in North Carolina. North Carolina State Univ. Agric. Exp. Stn. Bull. 445. 28pp.

Woodward-Clyde Consultants. 1980. Gravel removal guidelines manual for arctic and subarctic floodplains. U.S. Fish Wildl. Serv. FWS/OBS-80/09. 169pp.

Wright, H. A., and A. W. Bailey. 1982. Fire ecology: United States and southern Canada. Wiley, New York. 501pp.

Wright, T. D. 1978. Aquatic dredged material disposal impacts. U.S. Army Eng. Waterways Exp. Stn. Tech. Rep. DS-78-1. 57pp.

Yancey, R. K. 1964. Matches and marshes. Pages 619–626 *in* J. P. Linduska, ed. Waterfowl tomorrow. U.S. Fish Wildl. Serv., Washington.

Yoakum, J., W. P. Dasmann, H. R. Sanderson, C. M. Nixon, and H. S. Crawford. 1980. Habitat improvement techniques. Pages 329–403 *in* S. D. Schemnitz, ed. Wildlife management techniques manual. 4th ed. Wildl. Soc., Washington.

York, J. C. 1985. Dormant stub planting techniques. Pages 513–514 *in* R. R. Johnson, C. D. Ziebell, D. R. Patton, P. F. Ffolliott, and R. H. Hamre, tech. coord. Riparian ecosystems and their management: reconciling conflicting uses. U.S. For. Serv. Gen. Tech. Rep. RM-120.

Young, C. M. 1971. A nesting raft for ducks. Can. Field-Nat. 85:179–181.

Zedler, J. B. 1984. Salt marsh restoration a guidebook for southern California. San Diego State Univ. California Sea Grant Rep. T-CSGCP-009. 46pp.

Zedler, P. H. 1987. The ecology of southern California vernal pools: a community profile. U.S. Fish Wildl. Serv. Biol. Rep. 85(7.11). 136pp.

Zurbuch, P. E. 1984. Neutralization of acidified streams in West Virginia. Fisheries 9: 42–47.

Appendix

Wetland Types

The Canadian wetland classification system contains five wetland classes: bog, fen, marsh, swamp, and shallow water (National Wetlands Working Group Canada Committee on Ecological Land Classification 1987, 1988). These five classes are divided like a taxonomic key into 70 wetland forms based on surface morphology, surface pattern, water type, and morphology of underlying mineral soil (Table A.1). Wetland forms can be divided into 16 wetland types based on the general physiognomy of the vegetative cover.

Cowardin et al. (1979) divided wetlands and deepwater habitats of North America into five general systems: marine (over continental shelf), estuarine (brackish tidewater areas), riverine (rivers and streams), lacustrine (lakes and river reservoirs over 8 ha without woody and persistent herbaceous vegetation), palustrine (nontidal marshes and other wetlands of woody or persistent herbaceous growth, or less than 8 ha of surface water, or water depth less than 2 m, or tidewater areas with salinity less than 0.5 ppt). Like a taxonomic key, these are subdivided into 11 subsystems of 55 classes (Table A.2). The name of the class reflects either the dominant vegetation, when over 30 percent of the cover is vegetation, or the type of substrate. Further definition of the class is obtained from subclasses, modifiers, and dominance types, which include information about the chemical and physical environment. Such characteristics are difficult to determine in a large-scale wetland inventory. Much of the wetland legislation passed in the United States refers to Circular 39 (Shaw and Fredine 1956), published originally in Martin et al. (1953). Although it is very generalized, some states find it more practical to use because it is the most widely known wetland classification system with wildlife management emphasis (Knighton 1985, Mitsch and Gosselink 1986). Circular 39 divides wetlands into four areas of 20

TABLE A.1 Classes, Forms, and Types of Wetlands in the Canadian Wetlands Classification System

Wetland Classes and Forms		
Bog	Fen	Marsh
Palsa	Northern ribbed	Estuarine high
Peat mound	Atlantic ribbed	Estuarine low
Mound	Ladder	Coastal high
Domed	Net	Coastal low
Polygonal	Floating	Tidal freshwater
Lowland polygon	Stream	Floodplain
Peat plateau	Shore	Stream
Northern plateau	Collapse scar	Channel
Atlantic plateau	Palsa	Active delta
Collapse scar	Snowpatch	Inactive delta
Floating	Spring	Terminal basin
Shore	Feather	Shallow basin
Basin	Slope	Kettle
Flat	Lowland polygon	Seepage track
String	Horizontal	Shore
Blanket	Channel	
Slope	Basin	
Veneer		

Wetland classes and forms		
Swamp	Shallow water	Wetland type
Stream	Stream	Treed
Shore	Channel	Coniferous
Peat margin	Oxbow	Hardwood
Basin	Delta	Shrub
Spring	Terminal basin	Tall
Flat	Shallow basin	Low
Floodplain	Kettle	Mixed
	Tundra pool	Forb
	Thermokarst	Graminoid
	Shore	Grass
	Estuarine	Reed
	Tidal	Tall rush
	Nontidal	Low rush
		Sedge
		Moss
		Lichen
		Aquatic
		Floating
		Submerged
		Nonvegetal

SOURCE: National Wetlands Working Group Canada Committee on Ecological Land Classification (1987, 1988).

TABLE A.2 Wetland Classification System Used in the United States

- A. Marine
  - 1. Subtidal
    - *a.* Rock bottom
    - *b.* Unconsolidated bottom
    - *c.* Aquatic bed
    - *d.* Reef
  - 2. Intertidal
    - *a.* Aquatic bed
    - *b.* Reef
    - *c.* Rocky shore
    - *d.* Unconsolidated shore
- B. Estuarine
  - 1. Subtidal
    - *a.* Rock bottom
    - *b.* Unconsolidated bottom
    - *c.* Aquatic bed
    - *d.* Reef
- C. Riverine
  - 1. Tidal
    - *a.* Rock bottom
    - *b.* Unconsolidated bottom
    - *c.* Aquatic bed
    - *d.* Rocky shore
    - *e.* Unconsolidated shore
    - *f.* Emergent wetland
  - 2. Lower perennial
    - *a.* Rock bottom
    - *b.* Unconsolidated bottom
    - *c.* Aquatic bed
    - *d.* Rocky shore
    - *e.* Unconsolidated shore
    - *f.* Emergent wetland
  - 3. Upper perennial
    - *a.* Rock bottom
    - *b.* Unconsolidated bottom
    - *c.* Aquatic bed
    - *d.* Rocky shore
    - *e.* Unconsolidated shore
  - 4. Intermittent
    - *a.* Streambed
- D. Lacustrine
  - 1. Limnetic
    - *a.* Rock bottom
    - *b.* Unconsolidated bottom
    - *c.* Aquatic bed
  - 2. Littoral
    - *a.* Rock bottom
    - *b.* Unconsolidated bottom
    - *c.* Aquatic bed
    - *d.* Rocky shore
    - *e.* Unconsolidated shore
    - *f.* Emergent wetland
- E. Palustrine
  - 1. Palustrine
    - *a.* Rock bottom
    - *b.* Unconsolidated bottom
    - *c.* Aquatic bed
    - *d.* Unconsolidated shore
    - *e.* Moss-lichen wetland
    - *f.* Emergent wetland
    - *g.* Scrub-shrub wetland
    - *h.* Forested wetland

SOURCE: Cowardin et al. (1979).

types by life-forms of vegetation and depth of flooding (Table A.3). Cowardin et al. (1979) compared their system with Circular 39 (Table A.4). Verry (1985a, 1989) presented a brief comparison of pertinent wetland types between Shaw and Fredine (1956) and Cowardin et al. (1979) (see Table. 2.4).

Golet and Larson (1974) and Golet (1979) described a classification system for freshwater wetlands of the glaciated northeastern United States similar to Types 1 through 8 of Circular 39 (Shaw and Fredine 1956), with 8 classes and 26 subclasses (Table A.5). Further definition

TABLE A.3 Early Wetland Classification by U.S. Fish and Wildlife Service

Type number	Wetland type	Site characteristics
	Inland fresh areas	
1	Seasonally flooded	Soil in upland depressions and bottomlands covered with water or waterlogged during variable periods, but well drained during much of the growing season, with bottomland hardwoods and herbaceous plants.
2	Fresh meadows	Waterlogged to within a few centimeters of surface, but without standing water during growing season; herbaceous plants.
3	Shallow fresh marshes	Soil waterlogged and often covered with ≥15 cm of water; emergents during growing season.
4	Deep fresh marshes	Soil covered with ≥15 cm to 0.9 m of water during growing season; submergents.
5	Open fresh water	Water <3 m deep; submergents, fringed with emergents.
6	Shrub swamps	Soil waterlogged during growing season, often covered with ≥15 cm of water; swamp shrubs.
7	Wooded swamps	Soil waterlogged during growing season, often covered with 30 cm of water; along sluggish streams, flat uplands, shallow lake basins; swamp trees.
8	Bogs	Soil waterlogged; spongy covering of mosses, with other herbaceous and woody plants.
	Inland saline areas	
9	Saline flats	Flooded after periods of heavy precipitation; waterlogged within a few centimeters of surface during the growing season; salt-tolerant herbs.
10	Saline marshes	Soil waterlogged during growing season; often covered with 0.7 to 1 m of water; shallow lake basins; alkali or hardstem bulrush, sago, and widgeongrass.
	Coastal fresh areas	
11	Open saline water	Permanent areas of shallow saline water of variable depth; submergents.
12	Shallow fresh marshes	Soil waterlogged during growing season; at high tide ≤15 cm of water; on landward side, deep marshes along tidal rivers, sounds, deltas; grasses and emergents.
13	Deep fresh marshes	At high tide covered with 15 cm to 0.9 m of water during growing season; along tidal rivers and bays; emergents and often submergents.
14	Open fresh water	Shallow portions of open water along fresh tidal rivers and sounds; plants absent or emergents in water <1.8 m.
	Coastal saline areas	
15	Salt flats	Soil waterlogged during growing season; sites occasionally to fairly regularly covered by high tide; landward sides or islands within salt meadows and marshes; sparse grasses.

TABLE A.3 **Early Wetland Classification by U.S. Fish and Wildlife Service** *(Continued)*

Type number	Wetland type	Site characteristics
16	Salt meadows	Soil waterlogged during growing season; rarely covered by tide water; landward side of salt marshes; grasses and sedges.
17	Irregularly flooded salt marshes	Covered by wind tides at irregular intervals during growing season; along shores of nearly enclosed bays, sounds, etc.; needlerush.
18	Regularly flooded salt marshes	Covered at average high tide with ≥15 cm of water; along open ocean and sounds; saltmarsh cordgrass on Atlantic, alkali bulrush on Pacific.
19	Sounds and bays	Portions of saltwater sounds and bays shallow enough to be diked and filled; all water landward from average low-tide line; submergents.
20	Mangrove swamps	Soil covered at average high tide with 15 cm to 0.9 m of water; along coast of southern Florida; mangroves.

SOURCE: Shaw and Fredine (1956).

includes six surrounding habitat types, three vegetative interspersion types, eight cover types, seven site types, and five wetland size categories (very small: less than 4 ha; small: 4 to 20 ha; medium: 21 to 40 ha; large: 41 to 202 ha; very large: more than 202 ha).

Stewart and Kantrud (1971) defined seven classes of wetlands in the glaciated prairie region of North America, based on vegetation zones that occupied the deepest part of each pothole basin and its hydrologic characteristics. Cowardin et al. (1979) compared these vegetative zones with their water regime modifiers (Table A.6). Weller (1987) compared those classes to the wetland types of Shaw and Fredine (1956): Class I, ephemeral ponds (Type 1); Class II, temporary ponds (Type 2); Class III, seasonal ponds and lakes (Type 3); Class IV, semipermanent ponds and lakes (Type 4); Class V, permanent ponds and lakes (Type 5); Class VI, alkali ponds and lakes (Type 10); and Class VII, fen ponds (Type 8). Verry (1989) indicated that Types 2, 3, 4, and 6 (Shaw and Fredine 1956) are best suited to water impoundments (See Table 2.4).

Mitsch and Gosselink (1986) divided wetlands into seven types because they include most of the wetlands in North America, they are distinct in form and function, and they are commonly distinguished in the literature:

Coastal wetland ecosystems	Inland wetland ecosystems
Tidal salt marshes	Freshwater marshes
Tidal freshwater marshes	Northern peatlands and bogs
Mangrove swamps	Southern deepwater swamps
	Riparian wetlands

TABLE A.4 Comparison of Early and Recent Classification of Wetlands in the United States

Shaw and Fredine (1956) with examples of typical vegetation	Cowardin et al. (1979) Classes	Water regimes	Water chemistry
Type 1—Seasonally flooded basins or flats			
Wet meadow	Emergent wetland	Temporarily flooded	Fresh
Bottomland hardwoods Shallow freshwater swamps	Forested wetland	Intermittently flooded	Mixosaline
Type 2—Inland fresh meadows			
Fen, northern sedge meadow	Emergent wetland	Saturated	Fresh Mixosaline
Type 3—Inland shallow fresh marshes			
Shallow marsh	Emergent wetland	Semipermanently flooded Seasonally flooded	Fresh Mixosaline
Type 4—Inland deep fresh marshes			
Deep marsh	Emergent wetland Aquatic bed	Permanently flooded Intermittently exposed Semipermanently flooded	Fresh Mixosaline
Type 5—Inland open fresh water			
Open water Submerged aquatic	Aquatic bed Unconsolidated bottom	Permanently flooded Intermittently exposed	Fresh Mixosaline
Type 6—Shrub swamps			
Shrub swamp Shrub-carr, alder thicket	Scrub-shrub wetland	All nontidal regimes except permanently flooded	Fresh
Type 7—Wooded swamps			
Wooded swamp	Forested wetland	All nontidal regimes except permanently flooded	Fresh
Type 8—Bogs			
Bog	Scrub-shrub wetland	Saturated	Fresh (acid only)
Pocosin	Forested wetland Moss-lichen wetland		
Type 9—Inland saline flats			
Intermittent alkali zone	Unconsolidated shore	Seasonally flooded Temporarily flooded Intermittently flooded	Eusaline Hypersaline
Type 10—Inland saline marshes			
Inland salt marshes	Emergent wetland	Semipermanently flooded Seasonally flooded	Eusaline
Type 11—Inland open saline water			
Inland saline lake community	Unconsolidated bottom	Permanently flooded Intermittently exposed	Eusaline

TABLE A.4 Comparison of Early and Recent Classification of Wetlands in the United States (*Continued*)

	Cowardin et al. (1979)		
Shaw and Fredine (1956) with examples of typical vegetation	Classes	Water regimes	Water chemistry
Type 12—Coastal shallow fresh marshes			
Marsh Estuarine bay marshes, estuarine river marshes Fresh and intermediate marshes	Emergent wetland	Regularly flooded Irregularly flooded Semipermanently flooded—tidal	Mixohaline Fresh
Type 13—Coastal deep fresh marshes			
Marsh Estuarine bay marshes, estuarine river marshes Fresh and intermediate marshes	Emergent wetland	Regularly flooded Semipermanently flooded—tidal	Mixohaline Fresh
Type 14—Coastal open fresh water			
Estuarine bays	Aquatic bed Unconsolidated bottom	Subtidal Permanently flooded—tidal	Mixohaline Fresh
Type 15—Coastal salt flats			
Panne, slough marsh Marsh pans	Unconsolidated shore	Regularly flooded Irregularly flooded	Hyperhaline Euhaline
Type 16—Coastal salt meadows			
Salt marsh	Emergent wetland	Irregularly flooded	Euhaline Mixohaline
Type 17—Irregularly flooded salt marshes			
Salt marsh Saline, brackish, and intermediate marsh	Emergent wetland	Irregularly flooded	Euhaline Mixohaline
Type 18—Regularly flooded salt marshes			
Salt marsh	Emergent wetland	Regularly flooded	Euhaline Mixohaline
Type 19—Sounds and bays			
Kelp beds, temperate grass flats Tropical marine meadows Eelgrass beds	Unconsolidated bottom Aquatic bed Unconsolidated shore	Subtidal Irregularly exposed Regularly flooded Irregularly flooded	Euhaline Mixohaline
Type 20—Mangrove swamps			
Mangrove swamp systems	Scrub-shrub wetland Forested wetland	Irregularly exposed Regularly flooded Irregularly flooded	Hyperhaline Euhaline Mixohaline Fresh

SOURCE: Cowardin et al. (1979).

TABLE A.5 Classification of Freshwater Wetlands in the Glaciated Northern United States

Open water
- (OW-1) Vegetated
- (OW-2) Nonvegetated
- (OW-3) Shallow vegetated

Deep marsh
- (DM-1) Dead woody
- (DM-2) Shrub
- (DM-3) Subshrub
- (DM-4) Robust
- (DM-5) Narrow-leaved
- (DM-6) Broad-leaved

Shallow marsh
- (SM-1) Robust
- (SM-2) Narrow-leaved
- (SM-3) Floating-leaved

Meadow
- (M-1) Ungrazed
- (M-2) Grazed

Shrub swamp
- (SS-1) Deciduous sapling
- (SS-2) Bushy
- (SS-3) Compact
- (SS-4) Aquatic
- (SS-5) Evergreen sapling

Forested swamp
- (WS-1) Deciduous
- (WS-2) Evergreen

Fen (alkaline bog)
- (F-1) Emergent
- (F-2) Low shrub

Bog (acidic bog)
- (BG-1) Emergent
- (BG-2) Shrub
- (BG-3) Forested

SOURCE: Golet and Larson (1974), Golet (1979).

TABLE A.6 Comparison of the Classes and Zones of Stewart and Kantrud (1979) with the Water Regime Modifiers of Cowardin et al. (1979)

Class	Zone	Water regime modifier	
I	Ephemeral pond	Wetland low prairie	Nonwetland
II	Temporary pond	Wet meadow	Temporarily flooded
III	Seasonal ponds and lakes	Shallow marsh	Seasonally flooded
IV	Semipermanent ponds and lakes	Deep marsh	Semipermanently flooded, intermittently exposed
V	Permanent ponds and lakes	Permanent open water	Permanently flooded
VI	Alkali ponds and lakes	Intermittent alkali	Intermittently flooded
VII	Fen (alkaline bog) ponds	Fen	Saturated

SOURCE: Cowardin et al. (1979).

Willard et al. (1990) summarized various definitions of wetlands. Leslie et al. (1990) presented regulations protecting wetlands, and Goldsmith and Clark (1990) discussed nonregulatory programs promoting wetlands protection.

Tidal salt marshes surround the coastline of North America intermittently. In the United States, 61 percent occur along the Gulf Coast, 32 percent along the south Atlantic Coast, and 7 percent along the

West Coast (Mitsch and Gosselink 1986). Salt marshes cover less than 5 percent of Canada's arctic coast (National Wetlands Working Group Canada Committee on Ecological Land Classification 1988). Tidal freshwater marshes occur mainly along the Atlantic Coast, mostly in New Jersey, and along the Gulf Coast, mostly in Louisiana. Mangrove swamps occur mainly along the Atlantic and Gulf coasts of Florida. Inland freshwater marshes occur throughout North America. Northern peatlands and bogs occur mainly in Canada, Alaska, and north-central United States. Similar peat deposits called *pocosins* occur on the coastal plain of southeastern United States, mainly in North Carolina. Southern deepwater swamps occur in the southeastern United States, coinciding with the distribution of cypress. Riparian wetlands, called bottomland hardwood forests in the southeastern United States, occur throughout North America along rivers and streams.

Coastal marshes in the United States and Canada mainly consist of tidal marshes and those along the Great Lakes (Woodhouse 1979). Vegetation along East Coast marshes contains few dominant species. Smooth cordgrass occupies the intertidal zone to Texas. Saltmeadow cordgrass and saltgrass usually dominate the zone immediately above high tide, with two rushes on slightly higher sites—black-grass north of the Virginia Capes and black needlerush southward. South of Chesapeake Bay, marshes of the south Atlantic and Gulf coasts form behind barrier beaches and in estuaries where rivers deposit much silt. South of Daytona Beach, Florida, mangrove trees dominate many of the south Atlantic marshes. Due to limited rainfall and evaporation from high temperatures, hypersaline conditions in the Laguna Madre largely exclude marshes from the south Texas coast. Swamps of black mangroves occupy occasional coastal areas south of Galveston and offshore islands north. Gulf cordgrass replaces saltmeadow cordgrass.

Vegetation on the West Coast is more varied than on the East Coast. Pacific cordgrass, the West Coast counterpart of smooth cordgrass, occurs along the coasts only of southern and central California. Sedges, tufted hairgrass, and seaside arrowgrass occur north of the range of Pacific cordgrass.

Vegetation along the arctic and subarctic coasts is dominated by goose grass (National Wetlands Working Group Canada Committee on Ecological Land Classification 1988). Marshes of the Great Lakes are limited to protected shores of bays and inlets, occurring mostly in Lake Michigan and Lake Huron. Freshwater marsh plants dominate.

Salt marshes can be classified as lower (intertidal), upper (high) (Mitsch and Gosselink 1986, National Wetlands Working Group Canada Committee on Ecological Land Classification 1988), or interior (Kadlec and Wentz 1974). Interior salt marshes, mostly in the western United States, have vegetation similar to that in coastal salt marshes.

The low marsh is flooded almost daily, with less than 10 days of continuous exposure, resulting in generally higher salinity than in the high marsh. The high marsh is above high-tide levels, flooded only during highest tides or storm surges. Low and high marshes can be classified as coastal or estuarine marsh. Coastal marshes occur on marine terraces, flats, embayments, or lagoons behind barrier beaches, remote from estuaries, with periodic inundation by tidal brackish or salt water, including salt spray. Estuarine marshes occur in river estuaries or connecting bays where tidal flats, channels, and pools are inundated periodically by tidal water of varying salinity.

Management of wetlands generally involves swamps, dominated by shrubs and trees, and marshes. Marshes can be classified by water salinity as fresh (less than 1 ppt), intermediate (1 to 5 ppt), brackish (5 to 20 ppt), brackish-salt (20 to 30 ppt), and salt (30 to 35 ppt) (Gordon et al. 1989). Fresh marshes usually occur next to uplands and intermediate marshes between fresh and brackish marshes; brackish and brackish-salt marshes serve as a buffer between intermediate and saline marshes which occur next to the coast. Swamps usually are associated only with fresh water, except for mangrove swamps (U.S. Soil Conservation Service 1977).

Regionally, wetlands can be identified in the United States as follows (Sanderson and Bellrose 1969, Sanderson 1980):

1. Tundra and northern forests
   a. Tundra
   b. Open boreal forest
   c. Closed boreal forest
   d. Mixed forest
2. Great Plains (prairie pothole region)
   a. Aspen parkland
   b. Grasslands
3. Interior wetlands (between Appalachians and Rockies)
   a. Western Great Plains (including playas of west Texas)
   b. Interlake area (includes most of Great Lakes area)
      (1) Northeast pines
      (2) Northeast hardwoods
   c. Mississippi River basin
      (1) Bottomland swamps (subject to seasonal flooding)
      (2) Sump swamps (collecting water from local drainage)
4. Coastal wetlands
   a. Atlantic coast
      (1) New England states
      (2) Coastal plain (between Appalachians and Atlantic and Gulf including all Florida and pocosins in North Carolina)

(3) Coastal marshes and estuaries (south of Long Island, New York) (includes Everglades and mangrove swamps)

b. Gulf coast

c. Pacific coast

5. Intermountain wetlands (between ranges of Rockies)

In Canada, wetland regions are as follows (National Wetlands Working Group Canada Committee on Ecological Land Classification 1988):

1. Arctic
   a. High arctic
   b. Mid arctic
   c. Low arctic
2. Subarctic
   a. High subarctic
   b. Low subarctic
   c. Atlantic subarctic
3. Boreal
   a. High boreal
   b. Midboreal
   c. Low boreal
   d. Atlantic boreal
4. Temperate
   a. Eastern temperate
   b. Pacific temperate
5. Mountain
   a. Coastal mountain
   b. Interior mountain
   c. Rocky mountain
   d. Eastern mountain
6. Prairie
   a. Continental prairie
   b. Intermountain prairie
7. Oceanic
   a. Atlantic oceanic
   b. Pacific oceanic

TABLE A.7 Some United States and Canadian Federal Legislation Affecting Wildlife Management

United States
Agriculture Appropriation Act of 1907
Anadromous Fish Conservation Act of 1965
Bald Eagle Protection Act of 1972
Classification and Multiple Use Act of 1964
Clean Air Act of 1977
Clean Air Act of 1990
Clean Water Act of 1977
Coastal Zone Management Act of 1972
Comprehensive Environmental Response, Compensation, and Liability Act of 1980
Conservation Reserve Program of Federal Food Security Act (Farm Bill) of 1985
Cooperative Forestry Assistance Act of 1978
Domestic Water Supply Act of 1930
Emergency Wetlands Loan Act of 1961
Emergency Wetlands Resources Act of 1986
Endangered Species Act of 1969
Endangered Species Conservation Act of 1969
Endangered Species Preservation Act of 1966
Environmental Quality Improvement Act of 1970
Estuarine Areas Act
Federal Aid in Wildlife Restoration Act of 1937 (Pittman-Robertson Act)
Federal Environmental Pesticide Control Act of 1972
Federal Insecticide, Fungicide and Rodenticide Act of 1947
Federal Land Policy and Management Act of 1976
Federal Power Act of 1970
Federal Water Pollution Control Act of 1972
Federal Water Project Recreation Act of 1965
Federal Water Quality Act of 1965
Fish and Wildlife Conservation Act of 1980
Fish and Wildlife Coordination Act of 1956
Fish Restoration and Management Act of 1950 (Dingell-Johnson Act)
Flood Control Act of 1944
Flood Disaster Protection Act of 1973
Food and Agriculture Act of 1962
Food Security Act (Swampbuster) of 1985
Forest and Rangelands Renewable Resources Planning Act of 1974
Forest Reserve Act of 1891
Forest Reserve Transfer Act of 1905
Fur Seal Act
Insecticide Act of 1910
Knutson-Vandenburg Act
Lacey Act of 1900
Land and Water Conservation Fund Act of 1965
Lea Act of 1948
Marine Mammal Protection Act of 1972
Marine Protection, Research, and Sanctuaries Act of 1972
Migratory Bird Act of 1913
Migratory Bird Conservation Act of 1929
Migratory Bird Hunting Stamp Act of 1934
Migratory Bird Treaty Act of 1918
Multiple Use-Sustained Yield Act of 1960
National Environmental Policy Act of 1969
National Flood Insurance Act of 1968
National Forest Management Act of 1976
National Forests Organic Act of 1897
National Park Service Act
National Wildlife Refuge System Administration Act of 1966
Nonindigenous Aquatic Nuisance Protection and Control Act of 1990
North American Wetlands Conservation Act of 1989
Oregon and California (O & C Railroad) Act of 1937

Organic Administration Act of 1897	Sustained Yield Forest Management Act of 1944
Protection of Wild Horses and Burros Act of 1971	Taylor Grazing Act of 1964
Refuge Recreation Act of 1962	Toxic Substances Control Act of 1976
Refuge Revenue Sharing Act (amended 1974)	Water Bank Act of 1970
Resource Conservation and Recovery Act of 1976	Water Resources Development Act of 1976
River and Harbor Act of 1899	Water Resources Planning Act of 1965
Rivers and Harbors Act of 1988	Watershed Protection and Flood Prevention Act of 1954
Rivers and Harbors and Flood Control Acts of 1970	Wetlands Loan Act
Shorelines Management Act of 1971	Whaling Convention Act of 1949
Sikes Act of 1960	Wild and Scenic Rivers Act of 1976
Sikes Act Extension	Wild Free-Roaming Horses and Burros Act
Small Watersheds Act	Wilderness Act of 1964
Soil and Water Resources Conservation Act of 1977	
Canada	
Agriculture Rehabilitation and Development Act	Fisheries Act
Animal Contagious Disease Act	Forest Development and Research Act
Canada Water Act	Game Export Act
Canada Wildlife Act	Indian Act
Canada Wildlife Week Act	Maritime Marshland Reclamation Act
Customs Act	Migratory Bird Convention Act
Environmental Contaminants Act	National Parks Act
Export and Import Permits Act	Plant Quarantine Act
Export, Import and Interprovincial Transport of Wildlife Act*	Prairie Farmer Rehabilitation Act

*Will replace Export and Import Permits Act and Game Export Act.
SOURCE: Adapted from Gilbert and Dodds (1987).

TABLE A.8 Relationship of Certain Aquatic and Marsh Plants to Various Physical and Chemical Conditions

Species	Regions*	Depth range,† cm	Salinity‡	Alkalinity§	Bottom type¶	pH
Acorus calamus	1–3	—	F	20.0–202.5	—	5.9–8.8
Alisma planatago-aquatica	1–6	< 15	F	31.8–297.5	—	7.0–8.8
Alternanthera philoxeroides	1	—	F	—	—	7.4–8.0
Aster tenuifolius	—	Mean 80 above MSL	B	—	Si-C, Sa	—
Atriplex patula	—	80 above MSL	B-S	65	—	7.5
Avicennia nitida	1	− 15 − + 2.5	B	—	—	5.7–7.7
Baccharis halmifolia	—	80–100 above MSL	F-S	—	—	4.0–7.5
Bidens spp.	1–6	—	—	—	O, Sa, Sie	—
Borrichia frutescens	—	Mean 90 above MSL	B-S	93.3–161.1	—	6.1–8.0
Brasenia schreberi	1–3, 6	< 180	F	32.5–144.0	OO, O, L	4.9–8.8
Carex rostrata	2–6	0–90	F	45–160	—	5.1–7.8
Carex spp.	1–6	< 15	F	45–160	P, O, BO	4.5–7.8
Cephalanthus occidentalis	1–6	—	F	4.2–127.2	—	4.9–8.9
Ceratophyllum demersum	1–6	30–150	F-B	8.5–376.0	O	5.4–9.1
Chara spp.	1–6	30–800	F-B	70–256.6	O, Ma, Sa, L, Si-L, OO	5.9–9.5
Cladium jamaicense	1	− 15− + 100	F-B	—	P, O	4.5–7.5
Cyperus spp.	1–6	< 30	F	—	Sa, C, Si-L	—
Decodon verticillatus	1–3	—	F	—	—	4.5–5.3
Distichlis spicata	1–6	30 below MSL to 150 above MSL	B-S	170–8600	Si-C, Sa, C, O	4.1–9.5
Echinochloa spp.	1–6	30	F	17.0–127.2	Sa	6.2–8.7
Eichornia crassipes	1, 4, 5	—	F	4.2–182.3	—	4.5–9.1
Elodea canadensis	1–6	30–300	F	35.3–297.5	OO, BO, C	5.4–8.8
Eleocharis aricularis	1–6	< 120	F	18.7–376.0	P, O, Sa, Si	7.0–8.0
Eleocharis equisetoides	1–3	—	F	4.2–46.6	O	4.4–7.6
Eleocharis palustris	1–6	< 50	F	0.5–220.0	Sa, Si	5.9–9.0
Eleocharis parvula	1–6	Wet soil	F-B	—	—	3.7–6.7

Eleocharis robbinsii	1–3	—	—	—	O	—
Equisetum fluviatile	2–4, 6	—	—	7.5–297.5	—	6.8–8.8
Eriocaulon septangulare	1–3	—	F	10.0–44.3	P	6.7–7.8
Fimbristylis castanea	1–4	Mean 90 above MSL	B	—	—	5.4–7.8
Frankenia grandifolia	5	60 above MSL to MHW	—	—	C	—
Glyceria borealis	2–4, 6	—	F	8.0–187.5	—	5.9–8.8
Glyceria grandis	2–4, 6	—	F	8.0–245.0	—	5.9–8.8
Halodule wrightii	1	40–60 below MLW	B-S	34.0–195.0	Sa, Si-Sa	7.3–9.2
Heteranthera dubia	1–6	—	—	22.5–245.0	O	7.6–9.0
Hibiscus palustris	1–3	—	B	—	P-Sa	—
Hippuris vulgaris	3, 4, 6	—	—	30.0–297.0	—	6.8–8.8
Isoetes braunii	2–4, 6	—	F	8.0–45.0	Sa, Si, P, O	7.0–8.0
Isoetes lacustris	—	—	—	—	C, Sa, Si	6.0–7.9
Juncus roemerianus	1	Mean 70 above MSL	F-B-S	8.5–243.8	Si-C, Sa, L	4.3–9.5
Jaumea carnosa	5–6	50–90 above MSL	—	—	C	—
Leersia oryzoides	1–6	Wet soil	F	30.4–277.0	—	7.2–9.0
Lemna minor	1–6	—	F	41.2–262.5	—	5.9–9.0
Lemna trisulca	1–6	—	F	41.2–297.5	—	4.9–8.8
Limonium carolinianum	—	Mean 70 above MSL	B	—	Si-C, Sa	—
Littorella uniflora	3	20–60	F	—	Sa, Si	5.5–8.0
Lobelia dortmanna	2–4, 6	10–240	F	12.5–42.0	Si, BO, Sa, P	5.5–8.0
Megalodonta beckii	2–4	—	F	31.8–190.0	—	7.0–8.8
Menyanthes trifoliata	—	—	F	80–90	C	4.5–7.3
Myriophyllum alterniflorum	3	—	—	"Soft"	All	—
Myriophyllum exalbescens	1–6	—	F-B	22.5–376.0	O, Sa	7.2–8.9
Myriophyllum heterophyllum	1–4	—	F	4.2–46.6	O	5.2–7.2
Myriophyllum pinnatum	—	—	—	—	O, Sa	—
Myriophyllum spicatum	—	30–270	F-B	—	C, all	5.8–9.5
Najas flexilis	1–4, 6	30–800	F-B	18.7–307.7	BIO, OO, Sa, Si	6.9–9.0

TABLE A.8 Relationship of Certain Aquatic and Marsh Plants to Various Physical and Chemical Conditions (*Continued*)

Species	Regions*	Depth range,† cm	Salinity‡	Alkalinity§	Bottom type¶	pH
Najas guadalupensis	1–6	—	F	27.6–237.4	—	6.2–9.1
Najas marina	1–6	—	F-B	146.8–376.0	OO	8.2–9.0
Nelumbo lutea	—	30–150	F	—	Si, C-O	5.3–6.2
Nitella opaca	—	90–300	F	—	BlO, Si, Sa	5.5–7.2
Nuphar advena	1–4	< 300	F	2.1–139.9	OO	3.7–8.9
Nuphar microphyllum	2–3	< 300	F	7.5–41.1	O	6.8–7.3
Nuphar rubrodiscum	2–3	—	—	0.5–31.8	O	6.8–7.3
Nuphar variegatum	2–4	100–300	F	7.5–220.0	O, Sa, Si	5.8–8.6
Nymphaea alba	—	90–160	—	—	Si, C, BO	5.3–6.2
Nymphaea odorata	1–3	< 300	F	2.1–297.0	OO, O, P, Sa, Si	4.4–9.1
Nymphoides aquaticum	1	—	F	4.2–50.9	O	4.9–8.2
Phalaris arundinacea	1–6	—	F	22.5–160	—	6.9–8.8
Phragmites australis	1–6	– 100– + 200	F-B	0.5–297.5	BO, Sa, Si, C, P-Sa	3.7–9.0
Potamogeton alpinus	2–6	—	F	12.5–113.7	BO, O, Si, OO	5.4–8.6
Potamogeton amplifolius	1–3, 5, 6	100–800	F	2.5–208.7	O, Sa, Si	7.1–8.8
Potamogeton angustifolius	1–6	—	F	121.5–307.5	—	7.0–9.0
Potamogeton crispus	1–6	—	F	113.2–262.5	—	7.6–8.4
Potamogeton epihydrus	2–6	100–300	F	10.0–113.7	Sa, Si, O, OO	6.7–8.6
Potamogeton foliosus	1–6	60–180	F-B	37.5–228.7	—	5.9–8.8
Potamogeton friesii	2–4	—	F	71.6–376.1	—	7.7–8.8
Potamogeton gramineus	3, 4, 6	0–800	F	0.5–226	Sa, Si, O, G	5.9–8.8
Potamogeton illinoensis	1–6	—	F	8.5–164.0	—	5.5–9.1
Potamogeton natans	2–6	90–200	F	18.7–307.7	BlO, P, Sa, Si OO, C	5.9–9.0
Potamogeton nodosus	1–6	—	F	41.2–312.0	—	7.3–8.5
Potamogeton obtusifolius	3–6	—	—	30.4–70.0	—	7.0–7.9
Potamogeton pectinatus	1–6	5–300	F-B	31.8–376.0	Si, Sa, O	5.9–9.0
Potamogeton perfoliatus	1–6	60–240	F-B	—	C, Si, BlO, BO, Sa	5.5–7.2
Potamogeton pusillus	1–6	60–800	F-B	31.5–187.8	Sa, Si, O	7.0–8.8
Potamogeton praelongus	1–4, 6	160–280	F	12.5–307.7	Si, Sa, O	7.1–9.0
Potamogeton richardsonii	1–4, 6	0–250 below MLW	F-B	31.8–376.0	Si-O	7.0–9.1

Potamogeton robbinsii	2–4, 6	100–800	F	32.5–173	Sa, Si, O	7.2–8.6
Potamogeton spirillus	1–6	0–300	F	18.7–46.0	Sa, Si, O	7.2–8.6
Potamogeton strictifolius	2–4	—	—	31.8–262.5	—	7.4–9.0
Potamogeton vaginatus	3, 6	—	—	107.5–307.7	—	8.0–9.0
Potamogeton zosteriformis	3, 4, 6	800	F	18.0–245.0	O	6.9–9.0
Paspalum lividum	—	—	B-S	—	—	—
Pontederia cordata	1–3	< 120	F	4.2–144.2	O	4.9–8.9
Polygonum amphibium	2–6	< 300	F	30.0–260.0	Sa, Si, O	5.4–8.8
Polygonum coccineum	2–6	—	—	75.0–208.7	—	7.7–8.8
Polygonum spp.	1–6	Wet soil	F	45–160	Si, Sa	5.1–7.8
Ranunculus longirostris	2–4	—	—	113.2–144.0	—	7.9–8.4
Ranunculus trichophyllus	2–6	—	—	12.5–297.5	—	7.1–8.8
Rhizophora mangle	1	—	F-S	106.0–233.2	—	7.7–9.1
Ruppia maritima	1–6	30–300 below MLW	F-B-S	33.9–284.0	Si, Sa, O	3.7–9.5
Sagittaria cristata	2–4	—	F	18.7–183.0	—	7.2–8.8
Sagittaria cuneata	2–4	—	F	20.0–376.0	OO	7.3–9.0
Sagittaria lancifolia	1–2	—	F-B	2.1–136.0	—	4.8–8.9
Sagittaria latifolia	1–6	< 30	F	0.5–297.5	O	5.9–8.8
Sagittaria platyphylla	1	< 30	F-B	—	—	—
Sagittaria rigida	1–4	—	F	32.5–397.5	—	7.4–8.8
Salicornia ambigua	—	30 above MSL to MHW	S	—	C	—
Salicornia bigelovii	5	—	B-S	—	—	6.6–8.5
Salicornia europea	—	Mean 70 above MSL	B-S	—	Si-C, Sa, P	—
Salicornia perennis	—	Mean 80 above MSL	S	—	Si-C, Sa	—
Salicornia virginica	—	Mean 60 above MSL	S	—	P	—
Salix spp.	1–6	—	F	4.2–237.4	—	4.5–8.3
Scirpus acutus	2–6	< 150	F-B	17.1–220.0	Sa, all even hard	6.7–9.1
Scirpus americanus	1–6	< 60	F-B	12.7–277.0	Sa, C	6.7–8.9
Scirpus californicus	1, 4, 5	< 180	F-B	6.4–144.2	Si, Sa	4.1–6.2
Scirpus fluviatilis	1–6	< 50	F-B	30.4–220.0	—	7.0–9.1
Scirpus heterochaetus	2–4, 6	—	—	41.2–198.7	—	7.3–8.6
Scirpus olneyi	1, 4, 5	− 7– + 120	B-S	175–630	O, C	3.7–8.0
Scirpus paludosus	4–6	—	F-B	146.8–197.5	—	8.4–9.0
Scirpus robustus	1–6	− 15– + 120	F-B	140–890	O, C	4.0–6.9
Scirpus subterminalis	2–4, 6	—	F	8.0–42.5	OO	6.8–7.5

TABLE A.8 Relationship of Certain Aquatic and Marsh Plants to Various Physical and Chemical Conditions (*Continued*)

Species	Regions*	Depth range,† cm	Salinity‡	Alkalinity§	Bottom type¶	pH
Scirpus validus	1–6	< 120	F	115	Sa, C, Ma	5.3–7.8
Sparganium americanum	1–4	< 30	—	—	—	—
Sparganium chlorocarpum	2–4	—	F	16.5–160	—	7.0–8.4
Sparganium eurycarpum	1–6	< 120	F	35.3–376.0	—	6.7–8.8
Sparganium fluctuans	2–3	< 180	F	20.0–45.0	OO	7.0–7.3
Sparganium minimum	2–4, 6	—	F	86–115	BO, P, BlO	5.4–7.8
Spartina alterniflora	1	Slightly below to more than MSL	B-S	33.9–555.4	Si-C, Sa, P	4.5–8.5
Spartina cynosuroides	—	High marsh	B-S	—	—	4.3–6.9
Spartina foliosa	5	0–100 over MSL	—	—	C	—
Spartina patens	—	30 below MSL to +40 over MHW	B-S	170–8600	Si-C, Sa, P, Sa-P, O	3.7–7.9
Spartina spartinae	—	− 10– + 5	F-S	4.2–131.4	—	4.9–8.5
Spirodela polyrhiza	1–6	—	F	49.0–297.5	—	5.9–8.8
Syringodium filiforme	1	45–60 below LW	S	—	Sa, Si-Sa	—
Thalassia testudinum	1	Below LW to 800 above	B-S	—	P, Sa, Si-Sa	4.9–7.2
Typha angustifolia	1–6	60–90 above MSL, 100	F-B	86–115	O, BlO	3.7–8.5
Typha domingensis	1, 4, 5	—	F-B	12.7–148.1	—	6.0–8.5
Typha glauca	—	< 60	F-B	45–160	O	4.5–7.8
Typha latifolia	1–6	< 30	F-B	10.0–376.0	P, O	4.5–9.0
Utricularia intermedia	2–4, 6	—	F	8.0–245.0	—	5.1–8.6
Utricularia vulgaris	1–6	—	F-B	16.5–287.5	—	6.7–8.9
Vallisneria americana	1–3	30–300	F-B	18.7–277.0	Sa-O, Si, Sa, O	5.9–9.1
Wolffia columbiana	1–2	—	F	85.0–220.0	—	6.4–8.4
Zannichellia palustris	1–6	30–150	F-B	75.0–337.5	Si	7.6–9.0
Zizania aquatica	1–3	5–180	F	8.0–297.5	O, C	6.2–8.8
Zizaniopsis miliacea	—	—	F-B	6.4–63.6	—	6.0–7.4
Zostera marina	1, 3, 5, 6	< LW 30–180	B-S	—	Sandy mud, Sa, G	—

*See Fig. A.1.

†MLW = mean low water, LW = low water, MSL = mean sea level, MHW = mean high water.

‡F = fresh, 0 to 5 ppt; B = brackish, 6 to 25 ppt; S = saline, 25+ ppt. §ppm as $CaCO_3$.

¶C = clay, Si = silt, Sa = sand, BO = brown mud, P = peat, PO = peaty mud, O = organic, BlO = black mud, OO = ooze, Ma = marl, L = loam, G = gravel.

SOURCE: Kadlec and Wentz (1974).

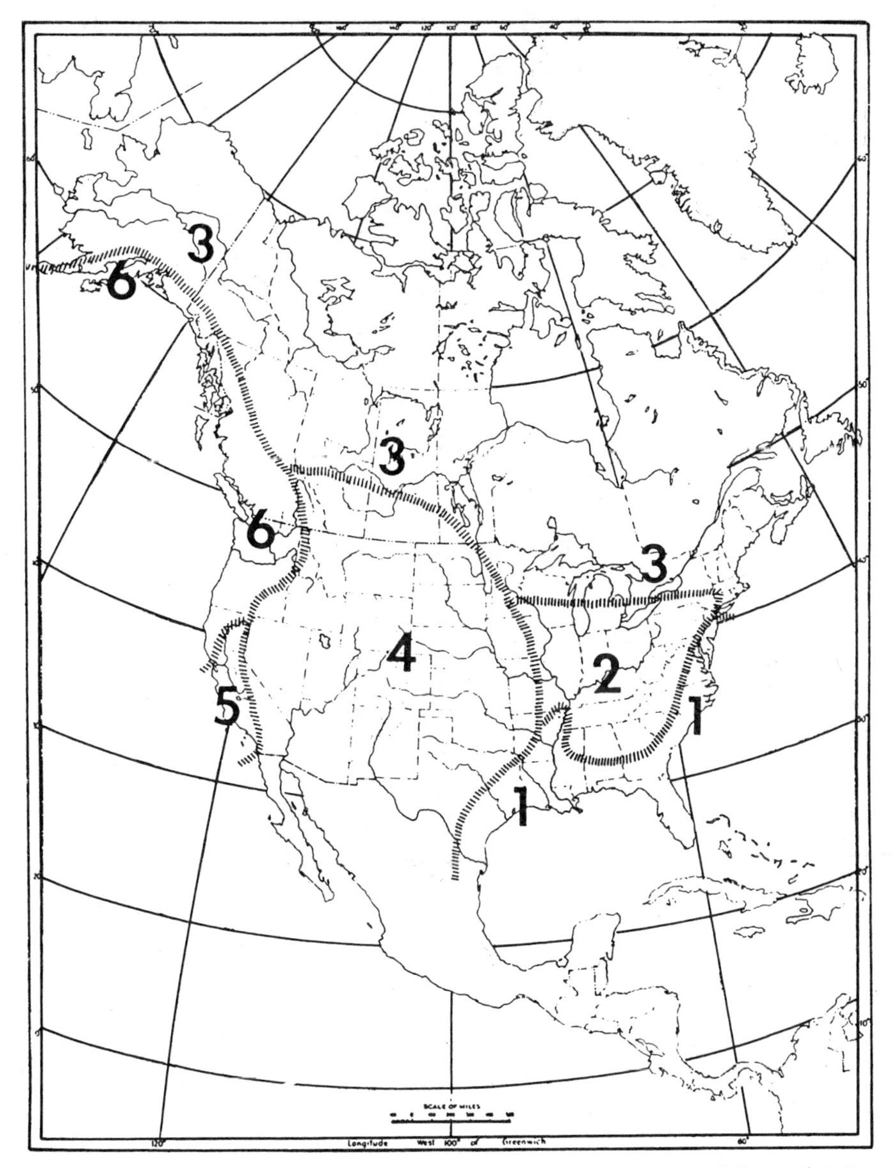

Figure A.1 Geographic regions of North America based on vegetational characteristics (*Kadlec and Wentz 1974*).

TABLE A.9 Marsh and Aquatic Plants Apparently Intolerant of Pollution, Turbidity, and Related Factors

Submergents and floating-leaved	Emergents
Megalodonta beckii	*Carex aquatilis*
Myriophyllum alterniflorum	*Equisetum fluviatile*
Najas flexilis	*Hibiscus militaris*
Najas gracillima	*Justicia americana*
Najas guadalupensis	*Lippia lanceolata*
Potamogeton amplifolius	*Rumex verticillatus*
Potamogeton filiformis	*Sagittaria rigida*
Potamogeton friesii	*Saururus cernuus*
Potamogeton gramineus	*Scirpus americanus*
Potamogeton praelongus	*Scirpus expansus*
Potamogeton richardsonii	
Potamogeton zonsteriformis	

SOURCE: Kadlec and Wentz (1974).

TABLE A.10 Marsh and Aquatic Plants Apparently Tolerant of Moderate Pollution, Turbidity, and Related Factors

Submergents and floating-leaved	Emergents
Alisma plantago-aquatica	*Butomus umbellatus*
Ceratophyllum demersum	*Polygonum hydropiper*
Elodea spp.	*Polygonum lapathifolium*
Heteranthera dubia	*Polygonum pensylvanicum*
Lemna minor	*Polygonum punctatum*
Myriophyllum exalbescens	*Sagittaria latifolia*
Myriophyllum verticillatum	*Sagittaria sagittifolia*
Najas minor	*Sparganium eurycarpum*
Nuphar lutea	*Typha angustifolia*
Potamogeton crispus	*Typha latifolia*
Potamogeton pectinatus	
Riccia fluitans	
Ricciocarpus natans	
Spirodela polyrhiza	
Utricularia vulgaris	
Vallisneria americana	
Zannichellia palustris	

SOURCE: Kadlec and Wentz (1974).

TABLE A.11 Evaluation of Aquatic and Marsh Plants for Wildlife, Substrate Stabilization, and Potential Nuisance*

Species	Waterfowl foods		Waterfowl cover	Birds other than waterfowl		Muskrat food	Substrate stabilization	Potential nuisance
	Part consumed	Value rating		Food	Cover			
Acnida cannabina	a	g						
Acorus calamus			x					x
Alisma plantago-aquatica	a	f						x
Alternanthera philoxeroides								x
Aneilema keisak	a	e						
Aster spp.			x					
Atriplex patula	a, c	g						
Avicennia spp.							x	
Azolla caroliniana								x
Baccharis halimifolia								x
Bacopa spp.	a, c	p						
Beckmannia spp.	a	f						
Bidens spp.	a	p	x					
Boltonia asteroides			x					
Brasenia schreberi	a	g						x
Butomus umbellatus	a	f						
Cabomba caroliniana								x
Calamagrostis canadensis			x					
Carex spp.	a	f	x					
Cephalanthus occidentalis	a	g	x		x			
Ceratophyllum demersum	a, c	f		a				x
Chara spp.	c	g						
Cladium jamaicense	a	f	x	a	x	x		x
Cyperus esculentus	a, b	e						
Cyperus odoratus	c	f						
Cyperus spp.	a	f		a				
Damasomium californicum	a	f						
Decodon verticillatus	a	p						
Deschampia spp.	a	f						
Distichlis spicata	a	f		a			x	
Echinochloa spp.	a	e	x	a	x	x		
Eichornia crassipes								x
Eleocharis acicularis								x
Eleocharis palustris			x					
Eleocharis spp.	b	q		a	x	x		
Elodea canadensis	c	f						x
Equisetum spp.	c	p				x		
Eragrostis spp.	a	p						
Glyceria maxima							x	x
Glyceria striata	a	f	x					
Heliotropium spp.	a	p						
Heteranthera dubia	a	p						x
Hippuris vulgaris	a	p						
Hydrilla verticillata								x
Hydrochloa carolinensis	a, c	f						
Hydrocotyl spp.	a	p						x
Iris versicolor			x					
Iva frutescens								x
Juncus roemerianus				a	x			x

TABLE A.11 Evaluation of Aquatic and Marsh Plants for Wildlife, Substrate Stabilization, and Potential Nuisance* (*Continued*)

Species	Waterfowl foods		Waterfowl cover	Birds other than waterfowl		Muskrat food	Substrate stabilization	Potential nuisance
	Part consumed	Value rating		Food	Cover			
Jussiaea spp.	a	g						
Leersia oryzoides	a, b	g	x	a	x	x		
Lemnaceae	c	g		c				x
Leptochloa fasicularis	a	g						
Limnobium spongia	a	f						x
Lophotocarpus calycinus	a	g						
Ludwigia peruviana	a	f						x
Marasilea vestita	a	f						
Menyanthes trifoliata	a	f						
Myrica spp.	a	p						
Myriophyllum brasiliense								x
Myriophyllum spicatum	a, c	p						x
Myriophyllum spp.	a, c	p		a				
Najas flexilis	a, c	e						
Najas guadalupensis	a, c	e						x
Najas marina	a, c	f						x
Najas spp.				a				
Nasturtium officinale	c	f	x					x
Nelumbo lutea								x
Nuphar luteum				a				x
Nuphar mexicana				a				x
Nuphar microphyllum	a	f		a				
Nuphar variegatum	a	f		a				
Nymphaea odorata	a	f						x
Nymphaea spp.	a	p		a, b, c		x		x
Nymphaea tuberosa	a	f						
Panicum dichotomiflorum	a	g						
Panicum hemitomum								x
Panicum purpurascens	a	f						x
Panicum repens	a	p						x
Paspalum boscianum	a	g						
Paspalum distichum	a	f						
Paspalum fruitans	a	p						x
Paspalum vaginatum	a	f						
Peltandra virginica	a	p	x	a				
Phragmites australis			x		x		x	x
Pistia stratiotes								x
Planera aquatica	a	f						
Polygonum amphibium	a	e						
Polygonum aviculare	a	g						
Polygonum coccineum	a	g						
Polygonum densiflorum	a	g						
Polygonum hydropiper	a	g						
Polygonum hydropiperoides	a	g						
Polygonum lapathifolium	a	e						
Polygonum muhlenbergii	a	e						
Polygonum natans	a	g						
Polygonum pensylvanicum	a	e						
Polygonum persicaria	a	e						

TABLE A.11 Evaluation of Aquatic and Marsh Plants for Wildlife, Substrate Stabilization, and Potential Nuisance* (*Continued*)

Species	Waterfowl foods		Water-fowl cover	Birds other than waterfowl		Musk-rat food	Sub-strate stabi-lization	Po-tential nui-sance
	Part consumed	Value rating		Food	Cover			
Polygonum portoricense	a	e						
Polygonum punctatum	a	e						
Polygonum ramoisissimum	a	g						
Polygonum sagitatum	a	f						
Polygonum spp.			x	a	x	x		
Pontederia cordata	a	p	x			x		x
Porserpinaca palustris	a	p						
Potamogeton amplifolius	a	f						
Potamogeton capillaceus	a	f						
Potamogeton compressus	a, b, c	g						
Potamogeton crispus	a, b	p						x
Potamogeton diversifolius	a	f						
Potamogeton epihydrus	a, b, c	g						
Potamogeton foliosus	a, b, c	g						
Potamogeton friesii	a, c	g						
Potamogeton gramineus	a, b	g						
Potamogeton heterophyllus	a, b, c	g						
Potamogeton illinoensis	a	f						x
Potamogeton natans	a, b	g					x	
Potamogeton nodosus	a	g						
Potamogeton perfoliatus	a, b, c	g						
Potamogeton praelongus	a, b, c	f						
Potamogeton pusillus	a, b, c	g						
Potamogeton richardsonii	a, b, c	g						
Potamogeton robinsii								x
Potamogeton spirillus	a	f						
Potamogeton strictifolius	a	g						
Potamogeton zosteriformes	a	f						
Ranunculus spp.	a, c	p						
Raphanus sativus	a	p						
Rhizophora mangle							x	
Rhynchospora spp.	a	f						
Rumex spp.	a	p						
Ruppia maritima	a, b, c	e		a				
Sagittaria cuneata	a, b	f						
Sagittaria heterophylla	a, b	f						
Sagittaria platyphylla	a, b	e						
Sagittaria spp.			x	a	x	x		
Salicornia virginica	a, c	f						
Salix interior								x
Salix spp.							x	
Salvinia rotundifolia								x
Saururus cernuus	a	p						
Schoenoplectus spp.								x
Scirpus acutus	a	e					x	
Scirpus americanus	a	g					x	
Scirpus californicus	a	f						
Scirpus campestris	a	g						
Scirpus cyperinus								x

TABLE A.11 Evaluation of Aquatic and Marsh Plants for Wildlife, Substrate Stabilization, and Potential Nuisance* (*Continued*)

Species	Waterfowl foods		Water-fowl cover	Birds other than waterfowl		Musk-rat food	Sub-strate stabi-lization	Po-tential nui-sance
	Part consumed	Value rating		Food	Cover			
Scirpus fluviatilis	a	p						
Scirpus heterochaetus	a	g						
Scirpus olneyi	a	e						
Scirpus paludosus	a	e					x	
Scirpus robustus	a	e					x	
Scirpus spp.			x	a, b	x	x		
Scirpus validus							x	
Scolochloa festucacea	a	g						
Sesuvium portulascastrum	a	g						
Setaria lutescens	a	g						
Setaria magna	a	f						
Sparganium americanum	a	f						
Sparganium eurycarpum	a	f						
Sparganium spp.								
Spartina alterniflora	a, b	f	x				x	
Spartina bakeri	a	f						
Spartina cynosuroides	a	f	x				x	x
Spartina foliosa	a	f						
Spartina gracillus	a	f						
Spartina patens	a	f					x	
Spartina pectinata								x
Spartina spp.				a	x	x		
Thalassia testudinum	c	f					x	
Thalia divarcata	a	f						
Torestria acuminata	a	f						
Trapa natans								x
Triglochin maritima	a	g						
Typha spp.	b, c	p	x	a	x	x		x
Utricularia spp.	c	p						x
Zanichellia palustris	a, c	g		a				
Zizania aquatica	a	e	x	a	x			
Zizaniopsis miliacea								x
Zostera marina	a, b, c	e		a			x	

*a = seeds or comparable structures, b = tubers and roots, c = foliage and stems; e = excellent, g = good, f = fair, p = poor; x = indicates plant is functional in specified category; a blank space indicates that plant is not functional in specified category or information is lacking.

SOURCE: Kadlec and Wentz (1974).

TABLE A.12 Regional and Seasonal Wetland Food Plants Preferred by Waterfowl*

Plant species	Preference by region†					Preference by waterfowl group and season‡															
						1		2		3		4		5		6		7		8	
	NE	SE	PR	MT	PC	N	MW	N	MW	N	MW	N	MW	N	MW	N	MW	N	MW	N	MW
Submergent																					
Aneilema keisak		*					*														
Brasenia schreberi	*	*					*				*		*	*							
Ceratophyllum demersum	*	*	*	*	*						*		*								
Chara spp.		*	*	*	*		*						*								
Halodule spp.		*									*		*								
Lemna spp.	*	*	*	*	*				*		*										
Najas (except *marina*)	*	*					*				*										
Nuphar spp.	*	*			*					*											
Nymphaea spp.	*	*							*				*								
Polygonum spp.	*	*	*	*	*	*	*	*	*		*										
Potamogeton spp.	*	*	*	*	*	*	*	*	*	*	*	*	*	*	*	*	*				
Rorippa spp.	*	*					*										*				
Ruppia maritima	*	*	*	*	*	*	*						*	*		*	*				
Spirodela spp.	*	*	*	*	*				*		*										
Vallisneria spp.	*	*											*	*			*				
Wolffia spp.	*	*	*	*	*				*		*										
Zanichellia palustris		*	*	*		*	*						*	*							
Zostera marina	*				*						*		*				*	*			
Emergent																					
Acnida cannabinus	*	*					*														
Atriplex patula		*	*				*														
Carex spp.	*	*	*	*	*	*	*			*	*										

TABLE A.12 Regional and Seasonal Wetland Food Plants Preferred by Waterfowl* (*Continued*)

	Preference by region†					Preference by waterfowl group and season‡															
						1		2		3		4		5		6		7		8	
Plant species	NE	SE	PR	MT	PC	N	MW	N	MW	N	MW	N	MW	N	MW	N	MW	N	MW	N	MW
Cladium jamaicense		*					*										*				
Distichlis spicata	*	*	*	*	*												*				
Echinochloa spp.		*	*		*		*														
Eleocharis spp.	*	*	*	*	*	*	*				*						*				
Equisetum spp.	*		*	*	*	*	*										*				
Juncus spp.	*	*	*	*	*	*															
Jussiaea spp.		*					*														
Leersia oryzoides	*	*	*				*														
Leptochloa fasicularis	*	*	*				*														
Lophotocarpus calycinus	*	*	*				*														
Oryza spp.		*				*	*						*		*		*			*	*
Panicum spp.	*	*	*			*	*														
Paspalum boscianum	*	*				*	*														
Peltandra virginica	*	*						*	*			*									
Sagittaria platyphylla		*					*														
Salicornia virginica	*	*	*	*	*												*				
Scirpus spp.	*	*	*	*	*	*	*		*	*	*	*	*								
Scolochloa festucacea		*	*				*														
Sesuvium portulascastrum		*	*																		
Setaria spp.	*		*				*														
Sparganium spp.	*		*	*	*	*	*		*								*				
Spartina spp.	*	*							*												
Triglochin maritima					*																
Zizania aquatica	*	*					*	*	*	*	*	*	*	*							

*All plants are of above-average attractiveness to ≥1 waterfowl group in ≥1 region as indicated by the asterisks in the body of the table. Plants not listed are seldom preferred as food; nevertheless, they sometimes might be valuable as cover, nesting material, or as food when preferred plants are locally scarce. This table reflects the attractiveness of plants to waterfowl and not necessarily their nutritive values. Nuts, mast, and fruits of woody species might be important locally but are not considered here. The presence of adequate cover and dense concentrations of aquatic invertebrates might be at least as important as the presence of preferred plants to some groups at some seasons. This is particularly true for groups 1 through 5 and group 8 during the breeding season.

†Regions in which the plant is preferred (combining all waterfowl species and seasons):

Northeast (NE): ME, NH, VT, MA, CT, RI, NJ, NY, PA, DE, MD, WV, OH, IN, MI, WI, KY, west NC, east TN, south IL, east MN, west VA.

Southeast (SE): SC, GA, FL, AL, MS, AK, LA, east OK, east TX, south MO, west TN, east NC, south VA.

Prairie (PR): IA, IL, KS, NE, SD, ND, east MT, east WY, east CO, east MN, west OK, west NM, north MO, central TX.

Mountains (MT): AZ, UT, NV, ID, west NM, west CO, west WY, west MT, east OR, east WA, southeast CA.

Pacific (PC): CA, west OR, west WA. ‡Plants preferred by waterfowl species during particular periods of the year. The periods of the year are abbreviated: N = nesting and brood rearing; MW = migration and winter. The waterfowl groups are defined below.

Group 1: Prairie dabblers; Group 2: Wood duck and black duck; Group 3: Goldeneye and bufflehead; Group 4: Canvasback, redhead, ruddy duck, and ring-necked duck; Group 5: Greater scaup and lesser scaup; Group 6: Inland geese and swans; Group 7: Brant; Group 8: Whistling ducks

SOURCE: Adamus et al. (1987).

TABLE A.13 General Growth Requirements and Characteristics of Selected Marsh Plants*

	†Region						Soil conditions							
							pH			Salinity			Texture	
Species	SA	NA	PF	GC	WC	I	Acid	Neutral	Alkaline	Fresh	Brackish	Saline	Fine	Coarse
Alkali bulrush (*Scirpus peludosus*)		*			*	*		*	*	*	*		*	*
Arrow arum (*Peltandra virginica*)	*	*	*	*		*	*	*	*	*			*	
Beak rush (*Rynchospora tracyi*)	*	*	*	*		*	*	*	*	*	*		*	*
Beggar's ticks (*Bidens* spp.)	*	*	*	*	*	*	*	*	*	*	*		*	*
Big cordgrass (*Spartina cynosuroides*)	*	*	*	*			*	*	*	*	*		*	*
Bigelow's glasswort (*Salicornia bigelovii*)	*	*	*	*			*	*	*		*	*	*	*
Black mangrove (*Avicennia nitida*)			*	*			*	*	*		*	*	*	*
Black needlerush (*Juncus roemerianus*)	*		*	*			*	*	*	*	*		*	*
Bladderworts (*Utricularia* spp.)	*	*	*	*	*	*	*	*	*	*			*	*
Broadleaf arrowhead (*Sagittaria latifolia*)	*	*	*	*	*	*	*	*	*	*			*	
Broadleaf cattail (*Typha latifolia*)	*	*	*	*	*	*	*	*	*	*			*	*
Bulrushes (*Scirpus* spp.)	*	*	*	*	*	*	*	*	*	*	*	*	*	*
Burreed (*Sparganium americanum*)	*	*	*	*	*	*	*	*	*	*			*	*
Buttercups (*Ranunculus* spp.)	*	*	*	*	*	*	*	*	*	*			*	*
Buttonbush (*Cephalanthus occidentalis*)	*		*	*		*	*	*	*	*			*	*
Chufa (*Cyperus esculentus*)	*	*	*	*	*	*	*	*	*	*			*	
Common reed (*Phragmites australis*)	*	*	*	*	*	*	*	*	*	*	*		*	*
Common threesquare (*Scirpus americanus*)	*	*	*	*	*	*	*	*	*	*	*		*	*
Delta duckpotato (*Sagittaria platyphylla*)			*	*			*	*	*	*			*	*
Docks (*Rumex* spp.)	*	*	*	*	*	*	*	*	*	*			*	*
Dotted smartweed (*Polygonum punctatum*)	*	*	*	*	*	*	*	*	*	*			*	*
Duckpotato (*Sagittaria cuneata*)		*			*	*	*	*					*	
Duckweeds (*Lemna* spp.)	*	*	*	*	*	*	*	*		*			*	
Eel grass (*Zostera marina*)	*	*			*		*	*	*		*	*	*	*
European glasswort (*Salicornia europaea*)	*	*	*	*			*	*	*			*	*	*
Fimbristylis (*Fimbristylis castanea*)	*	*	*	*			*	*	*	*	*		*	*
Foxtail grasses (*Setaria* spp.)	*	*	*	*	*	*	*	*	*	*			*	*
Frankenia (*Frankenia grandifolia*)					*		*	*	*		*		*	*
Frog bit (*Limnobium spongia*)	*	*	*	*		*	*	*	*	*			*	*

*An asterisk (*) in the table body indicates positive occurrence tolerance, value, or characteristic. No asterisk indicates negative or doubtful information.

Marsh moisture conditions												
	Tidal		Interior			Wildlife value			Morphology			
Standing water	Low	High	Low fresh	High fresh	Brackish	Food	Cover	Nesting/ breeding	Perennial	Annual	Potential nuisance	Soil stabilizer
		*		*	*	*	*		*			*
				*		*	*	*	*		*	*
		*		*		*	*		*			*
		*		*		*	*		*	*	*	*
	*	*				*	*	*	*		*	*
	*	*					*		*			*
	*						*	*	*			*
		*					*		*		*	*
*			*			*	*		*		*	
*			*	*		*	*	*	*			*
*			*			*	*	*	*		*	*
	*	*	*	*	*	*	*	*	*		*	*
				*		*	*		*			*
*			*				*		*		*	
*			*	*		*	*	*	*			*
				*		*	*		*			*
		*		*	*		*	*	*		*	*
	*	*			*	*	*	*	*			*
			*	*		*	*		*			*
				*		*	*		*		*	*
				*		*	*		*	*		*
			*	*		*	*		*			*
*						*	*		*		*	
*	*					*	*		*			*
	*						*	*	*			*
		*		*	*	*	*		*			*
				*		*	*	*	*	*	*	*
	*	*					*		*			
			*	*		*	*		*	*		*

†SA = south Atlantic; NA = north Atlantic; PF = peninsula Florida; GC = gulf coasts; WC = west coast; I = interior.

SOURCE: U.S. Army Engineer Waterways Experiment Station (1978).

TABLE A.13 General Growth Requirements and Characteristics of Selected Marsh Plants* ***(Continued)***

	†Region						Soil conditions							
							pH			Salinity			Texture	
Species	SA	NA	PF	GC	WC	I	Acid	Neutral	Alkaline	Fresh	Brackish	Saline	Fine	Coarse
Giant reed (*Arundo donax*)	*		*	*		*	*	*	*	*			*	*
Groundsel tree (*Baccharis halimifolia*)	*	*	*	*			*	*	*	*	*		*	*
Hardstem bulrush (*Scirpus acutus*)	*	*	*	*	*	*	*	*	*	*	*		*	*
Horned pondweed (*Zannichellia palustris*)	*	*	*	*	*	*	*	*	*	*	*		*	*
Horsetails (*Equisetum* spp.)	*	*	*	*	*	*	*	*	*	*	*		*	*
Japanese millet (*Echinochloa crusgalli*)	*	*	*	*	*	*	*	*	*	*			*	*
Ladysthumb (*Polygonum persicaria*)	*	*	*	*	*	*	*	*	*	*			*	*
Lizard's tail (*Saururus cernuus*)	*	*	*	*		*	*	*		*			*	
Lobelia (*Lobelia dortmanna*)		*			*	*	*	*	*	*			*	*
Lotus (*Nelumbo lutea*)	*	*	*	*	*	*	*	*		*			*	*
Lyngbye's sedge (*Carex lyngbyei*)					*		*	*		*			*	*
Manna grass (*Glyceria acutiflora*)	*	*		*		*	*	*	*	*			*	*
Manna grass (*Glyceria fluitans*)	*	*	*	*	*	*	*	*	*	*			*	*
Marsh elder (*Iva frutescens*)	*	*	*	*			*	*	*	*	*		*	*
Marsh hibiscus (*Hibiscus moscheutos*)	*		*	*			*	*		*			*	*
Marsh pepper (*Polygonum hydropiper*)	*	*	*	*	*	*	*	*		*		*	*	
Marsh smartweed (*Polygonum hydropiperoides*)	*	*	*	*		*	*	*		*			*	*
Mud plantain (*Heteranthera reniformis*)	*	*	*	*		*	*	*		*			*	
Nodding smartweed (*Polygonum lapathifolium*)	*	*	*	*	*	*	*	*		*			*	*
Nutsedges (*Cyperus* spp.)	*	*	*	*	*	*	*	*	*	*			*	*
Olney's threesquare (*Scirpus olneyi*)	*			*	*		*	*		*	*		*	
Orache (*Atriplex patula*)	*	*		*			*	*	*	*	*			*
Pacific cordgrass (*Spartina foliosa*)					*		*	*			*	*	*	*
Pacific glasswort (*Salicornia pacifica*)					*		*	*	*		*	*	*	*
Pacific sedge (*Carex obnupta*)					*		*	*		*	*		*	*
Panic grasses (*Panicum* spp.)	*	*	*	*	*	*	*	*	*	*	*	*	*	*
Paspalum grasses (*Paspalum* spp.)	*	*	*	*	*	*	*	*	*	*	*		*	*
Pennsylvania smartweed (*Polygonum pensylvanicum*)	*	*	*	*		*	*	*		*			*	*
Pennyworts (*Hydrocotyle* spp.)	*	*	*	*	*	*	*	*	*	*	*		*	

*An asterisk (*) in the table body indicates positive occurrence tolerance, value, or characteristic. No asterisk indicates negative or doubtful information.

Marsh moisture conditions												
	Tidal		Interior			Wildlife value			Morphology			
Standing water	Low	High	Low fresh	High fresh	Brackish	Food	Cover	Nesting/ breeding	Perennial	Annual	Potential nuisance	Soil stabilizer
			*	*		*	*	*	*			*
		*					*	*	*			*
		*	*	*	*	*	*		*			*
*	*		*		*	*	*		*			*
		*	*	*	*		*	*	*		*	*
				*		*	*			*		*
				*			*		*			*
			*	*		*	*		*			*
				*			*					
*			*			*	*		*		*	*
				*		*	*		*			*
*			*			*	*	*	*			*
			*	*		*	*	*	*			*
		*					*	*	*			*
			*	*		*	*	*	*			*
			*	*		*	*		*	*	*	*
			*	*		*	*		*		*	*
				*			*		*			*
			*	*		*	*	*	*		*	*
			*	*		*	*	*	*		*	*
		*			*	*	*		*			*
		*			*		*		*			*
	*					*	*		*			*
	*	*					*		*			*
		*				*	*	*	*			*
*		*	*	*	*	*	*	*	*	*		*
		*		*	*	*	*	*	*	*		*
			*	*		*	*		*			*
*		*	*	*	*		*	*	*		*	*

†SA = south Atlantic; NA = north Atlantic; PF = peninsula Florida; GC = gulf coasts; WC = west coast; I = interior.

SOURCE: U.S. Army Engineer Waterways Experiment Station (1978).

TABLE A.13 General Growth Requirements and Characteristics of Selected Marsh Plants* (*Continued*)

	†Region						Soil conditions							
							pH			Salinity			Texture	
Species	SA	NA	PF	GC	WC	I	Acid	Neutral	Alkaline	Fresh	Brackish	Saline	Fine	Coarse
Pickerelweed (*Pontederia cordata*)	*	*	*	*		*	*	*		*			*	
Pondweeds (*Potomogeton* spp.)	*	*	*	*	*	*	*	*	*	*			*	*
Prairie cordgrass (*Spartina pectinata*)	*	*	*	*		*	*	*	*	*	*		*	*
Red mangrove (*Rhizophora mangle*)			*				*	*	*			*	*	*
Reed canary grass (*Phalaris arundinacea*)	*	*	*	*	*	*	*	*	*	*			*	*
Reed grass (*Calamagrostis canadensis*)	*				*	*	*	*	*	*			*	*
Reed manna grass (*Glyceria grandis*)	*	*		*	*	*	*	*	*	*			*	*
Rice cutgrass (*Leersia oryzoides*)	*	*	*	*	*	*	*	*		*			*	*
River bulrush (*Scirpus fluviatilis*)		*			*	*	*	*	*	*			*	*
Rushes (*Juncus* spp.)	*	*	*	*	*	*	*	*	*	*	*		*	*
Saltgrass (*Distichlis spicata*)	*	*	*	*		*	*	*	*		*	*	*	*
Saltmarsh aster (*Aster tenuifolius*)	*	*	*	*			*	*	*		*		*	*
Saltmarsh bulrush (*Scirpus robustus*)	*	*	*	*			*	*	*		*	*	*	
Saltmarsh cattail (*Typha angustifolia*)	*	*	*	*			*	*		*	*		*	*
Saltmarsh jaumea (*Jaumea carnosa*)					*		*	*					*	*
Saltmeadow cordgrass (*Spartina patens*)	*	*	*	*			*	*	*		*		*	*
Sawgrass (*Cladium jamaicense*)			*				*	*	*	*			*	*
Sea lavender (*Limonium carolinianum*)	*	*	*	*			*	*	*	*	*		*	*
Sea lavender (*Limonium vulgare*)	*	*	*	*	*		*	*	*	*	*		*	*
Sea ox-eye (*Borrichia frutescens*)	*		*	*			*	*	*			*	*	*
Sea purslane (*Sesuvium portulacastrum*)	*	*	*	*			*	*	*	*	*		*	*
Seaside arrowgrass (*Triglochin maritima*)		*			*			*	*	*	*	*	*	*
Sedges (*Carex* spp.)	*	*	*	*	*	*	*	*	*	*	*		*	*
Shoal grass (*Halodule wrightii*)	*		*	*				*	*			*	*	*
Slough grass (*Backmannia syzigachne*)		*		*	*		*	*		*			*	*
Slough sedge (*Carex obnupta*)		*			*	*	*	*		*			*	*
Smartweeds (*Polygonum* spp.)	*	*	*	*	*	*	*	*	*	*			*	*
Smooth cordgrass (*Spartina alterniflora*)	*	*	*	*	*		*	*	*		*	*	*	*
Soft rush (*Juncus effusus*)	*		*	*		*	*	*			*		*	

*An asterisk (*) in the table body indicates positive occurrence tolerance, value, or characteristic. No asterisk indicates negative or doubtful information.

Marsh moisture conditions						Wildlife value			Morphology			
	Tidal		Interior									
Standing water	Low	High	Low fresh	High fresh	Brackish	Food	Cover	Nesting/ breeding	Perennial	Annual	Potential nuisance	Soil stabilizer
			*			*	*	*	*		*	*
*						*	*		*		*	
	*	*			*	*	*	*	*			*
*	*						*	*	*			*
				*		*	*	*	*			*
				*		*	*	*	*			*
				*		*	*	*	*			*
*			*			*	*		*			*
*			*	*		*	*	*	*			*
		*		*	*	*	*	*	*		*	*
	*	*				*	*	*	*			*
		*					*		*			*
		*			*	*	*	*	*			*
		*		*	*		*	*	*			*
		*					*		*			*
		*				*	*	*	*			*
*			*	*		*	*	*	*			*
		*					*		*			*
		*					*	*	*			*
	*	*					*	*	*			*
		*					*	*	*			*
	*	*	*	*	*		*		*			*
		*	*	*	*	*	*	*	*			*
*						*	*		*			*
				*			*		*			*
			*	*		*	*	*	*			*
*			*	*		*	*	*	*		*	*
	*					*	*	*	*			*
				*		*	*	*	*		*	*

†SA = south Atlantic; NA = north Atlantic; PF = peninsula Florida; GC = gulf coasts; WC = west coast; I = interior.

SOURCE: U.S. Army Engineer Waterways Experiment Station (1978).

TABLE A.13 General Growth Requirements and Characteristics of Selected Marsh Plants* (*Continued*)

	†Region						Soil conditions							
							pH			Salinity			Texture	
Species	SA	NA	PF	GC	WC	I	Acid	Neu-tral	Alka-line	Fresh	Brack-ish	Sa-line	Fine	Coarse
Softstem bulrush (*Scirpus validus*)	*	*	*	*	*	*	*	*			*		*	
Southern bulrush (*Scirpus californicus*)	*		*	*	*	*	*	*			*		*	*
Southern smartweed (*Polygonum densiflorum*)	*		*	*		*	*	*			*		*	*
Southern cutgrass (*Zizaniopsis miliacea*)	*		*	*		*	*	*	*		*		*	
Spatterdock (*Nuphar luteum*)	*	*	*	*	*	*	*	*			*		*	
Spikerushes (*Eleocharis* spp.)	*	*	*	*	*	*	*	*	*	*	*		*	*
Spirodella (*Spirodella polyrhiza*)	*	*	*	*	*	*	*	*		*			*	
Sprangletop (*Leptochloa fascicularis*)	*			*			*	*		*			*	*
Sweet flag (*Acorus calamus*)	*					*	*	*		*			*	
Tufted hairgrass (*Deschampsia caespitosa*)		*			*	*	*	*	*	*	*		*	*
Turtle grass (*Thalassia testudinum*)	*		*		*			*	*			*	*	*
Walter's millet (*Ecbinochloa walteri*)	*	*	*	*		*	*	*	*	*			*	*
Water hemp (*Acnida cannabina*)	*		*				*	*		*			*	*
Water hyssop (*Bacopa caroliniana*)	*		*	*		*	*	*	*	*	*	*	*	*
Water lilies (*Nymphaea* spp.)	*	*	*	*	*	*	*	*		*			*	*
Watermilfoils (*Myriophyllum* spp.)	*	*	*	*	*	*	*	*		*			*	*
Water nymph (*Najas* spp.)	*	*	*	*	*	*	*	*			*		*	*
Water plantain (*Alisma plantago-aquatica*)	*	*		*		*	*	*		*			*	*
Water shield (*Brasenia schreberi*)	*		*				*	*			*		*	*
Water smartweed (*Polygonum amphibium*)	*	*	*	*	*	*	*	*		*			*	*
Water willow (*Decodon verticillatus*)	*	*	*	*			*	*		*			*	*
White mangrove (*Laguncularia racemosa*)			*				*	*	*		*	*	*	*
Widgeongrass (*Ruppia maritima*)	*	*	*	*	*	*		*	*		*		*	*
Wild celery (*Vallisneria americana*)	*	*	*	*	*	*	*	*		*	*		*	*
Wild rice (*Zizania aquatica*)	*	*	*	*		*	*	*		*			*	*
Willows (*Salix* spp.)	*	*	*	*	*	*	*	*	*	*			*	*
Wolffias (*Wolffia* spp.)	*	*	*	*		*	*	*	*	*			*	
Yellow flag (*Iris pseudacorus*)		*				*	*	*		*			*	*

*An asterisk (*) in the table body indicates positive occurrence tolerance, value, or characteristic. No asterisk indicates negative or doubtful information.

Marsh moisture conditions						Wildlife value			Morphology			
	Tidal		Interior									
Standing water	Low	High	Low fresh	High fresh	Brackish	Food	Cover	Nesting/ breeding	Perennial	Annual	Potential nuisance	Soil stabilizer
			*	*		*	*	*	*			*
			*	*		*	*	*	*			*
			*	*		*	*		*			*
			*			*	*	*	*			*
			*	*		*	*		*			*
		*	*	*	*	*	*		*			*
*						*	*		*		*	*
				*		*	*	*	*			*
				*			*		*			*
	*	*	*	*			*	*	*			*
*						*	*		*			*
				*		*	*			*		*
			*	*		*	*		*			*
		*		*			*		*			*
*							*		*		*	
*						*	*		*		*	
*							*		*		*	
			*	*		*	*		*			*
*							*		*		*	
*			*	*		*	*		*		*	*
				*			*		*			*
*	*						*	*	*			*
*	*	*				*	*		*			
*			*			*	*		*		*	
			*	*		*	*		*			*
*			*	*		*	*	*	*		*	*
*						*	*		*			
				*			*		*			*

†SA = south Atlantic; NA = north Atlantic; PF = peninsula Florida; GC = gulf coasts; WC = west coast; I = interior.

SOURCE: U.S. Army Engineer Waterways Experiment Station (1978).

TABLE A.14 Plants Recommended for Planting on Dredged Material Islands As Nesting Habitat for Colonial Waterbirds

Species	Occurrence range*	Habitat category†							
		BS	SH	MH	DH	HS	ST	SF	F
Saltmeadow cordgrass *Spartina patens*	6		x						
Seaside paspalum *Paspalum virginatum*	2		x						
Saltgrass *Distichlis stricta*	6		x						
Evening primrose *Oenothera humifusa*	5, 6		x						
Camphorweed *Heterotheca subaxillaris*	2, 5		x						
Horseweed *Erigeron canadensis*	2, 5		x						
Beach pea *Strophostyles helvola*	2		x	x					
Sedge *Carex* spp.	5, 6		x	x					
Rush *Juncus* spp.	5, 6		x	x					
Smartweed *Polygonum* spp.	5, 6		x	x					
Fescue *Festuca* spp.	5, 6		x	x	x	x			
Knotweed *Polygonum* spp.	5, 6		x						
Spurge *Euphorbia polygonifolia*	5, 6		x	x					
Sea ox-eye *Borrichia frutescens*	2		x						
Sea blite *Suaeda maritima*	2		x						
Dog fennel *Eupatorium capillifolium*	5, 6			x	x	x			
Scotch broom *Cytisus scoparium*	4, 5			x	x	x			
Broomsedge *Andropogon* spp.	5, 6			x	x	x			
American beachgrass *Ammophila breviligulata*	6			x	x	x			
Wild rye *Elymus virginicus*	5, 6			x	x	x			
Sea oats *Uniola paniculata*	2			x	x	x			
Pepper grass *Lipidium virginicum*	5, 6		x	x	x	x			
Croton *Croton punctatus*	5, 6		x	x	x	x			
Purple top *Tripalsis purpurea*	5, 6			x	x	x			
Beach panic grass *Panicum anceps*	6			x	x	x			
Reed canarygrass *Phalaris arundinacea*	5, 6			x	x	x			
Goldenrod *Solidago* spp.	5, 6				x	x			

TABLE A.14 Plants Recommended for Planting on Dredged Material Islands As Nesting Habitat for Colonial Waterbirds (*Continued*)

Species	Occurrence range*	Habitat category†							
		BS	SH	MH	DH	HS	ST	SF	F
Ragweed *Ambrosia* spp.	5, 6				x	x			
Switchgrass *Panicum virgatum*	2, 5				x	x			
Marsh elder *Iva frutescens*	2					x	x	x	
Groundsel tree *Baccharis halimifolia*	2					x	x	x	
Wax myrtle *Myrica cerifera*	2					x	x	x	x
Bayberry *Myrica pennsylvanica*	3					x	x	x	
Shrub verbena *Lantan camara*	2, 5					x	x	x	
Wild indigo *Baptisia leucophaea*	2, 5					x	x	x	
Yaupon *Ilex vomitoria*	2					x	x	x	x
Huisache tree *Acacia smallii*	4, 5					x	x	x	x
Brazilian pepper *Schinus terebinthifolius*	1					x	x	x	x
White mangrove *Laguncularia racemosa*	1					x	x	x	x
Red mangrove *Rhisophora mangle*	1					x	x	x	x
Black mangrove *Avicennia germinans*	1					x	x	x	x
Oleander *Nerium oleander*	5, 6					x	x	x	
Eastern red cedar *Juniperus virginiana*	2, 5					x	x	x	x
Live oak *Quercus virginiana*	2							x	x
Saltcedar *Tamarix chinensis*	2, 4						x	x	x
Sand pine *Pinus clausa*	2							x	x
Loblolly pine *Pinus taeda*	2, 5							x	x
Hackberry *Celtis occidentalis*	5, 6							x	x
Australian pine *Casuarina equisetifolia*	1							x	x
Eastern cottonwood *Populus deltoides*	5, 6							x	x
Peachleaf willow *Salix amygdaloides*	3, 5					x	x	x	

*1 = extreme southern United States (freeze-intolerant), 2 = midsouth (south of Virginia), 3 = northern United States only, 4 = western United States only, 5 = freshwater conditions only, 6 = entire United States.

†BS = bare substrate, SH = sparse herb, MH = medium herb, DH = dense herb, HS = herb-shrub, ST = shrub thicket, SF = shrub-forest, F = forest.

SOURCE: Soots and Landin (1978).

TABLE A.15 Propagules and Planting Techniques Recommended for Selected Marsh Plants

Species	Recommended propagules	General collection, handling, and planting techniques	Remarks
Alkali bulrush *Scirpus paludosus*	Transplants,* tubers	Dig plants; divide; replant on site at same depth or pot for holding in nursery or greenhouse.	Seeds often eaten by waterfowl and other birds; used for soil stabilization; prefers fine soils.
Arrow arum† *Peltandra virginica*	Transplants, seeds	Dig plants; separate; replant at same depth on site or pot for holding. Gather seeds when mature; store in freshwater at 1–3°C; broadcast on site and rake into soil.	Mainly a good soil stabilizer, although seeds are infrequently eaten by waterfowl and muskrats use it for lodge material. Potential pest plant.
Beak rush† *Rynchospora tracyi*	Seeds	Gather seeds when mature (Jul–Sep); store in freshwater at 5°C; broadcast on site and rake into soil.	Seeds eaten by waterfowl mainly.
Beggar's ticks† *Bidens* spp.	Seeds	Gather seeds when mature (Jul–Sep); store dry at 5°C; broadcast on site and rake into soil.	Good food source for songbirds, game birds, and chicks. Potential pest.
Big cordgrass† *Spartina cynosuroides*	Transplants, seedlings	Dig young plants from natural stands; separate; replant on site at same depth or pot for holding. Germinate seeds and grow seedlings until ready for planting (3–6 mo).	Excellent soil stabilizer in low, brackish marshes. Salinity prevents this species from competing with smooth cordgrass. Seeds eaten by many birds; rodents eat young tender foliage. Potential pest.
Bigelow's glasswort† *Salicornia bigelovii*	Cuttings, rootstock	Collect 5- to 15-cm cuttings of top shoots and broadcast in wet area on site. If cuttings must be stored, keep moist. Dig rootstock; replant on site at same depth.	Low tidal area soil stabilizer. Tolerates fairly high salinities. Easily propagated. Poor source of wildlife foods. Occasionally used by nesting colonial seabirds.
Black mangrove† *Avicennia germinans*	Seeds, seedlings	Collect seed pods when mature (summer, fall); plant whole pod upright in soil with stem end up and out of soil. Dig seedlings from natural stand or grow from seed pods.	Excellent soil stabilizer in south Florida. Often occurs on dredged material islands and used by colonial nesting wading birds. Tolerates to 40-ppt salinity.

Black needlerush† *Juncus roemerianus*	Transplants	Dig clumps; divide into sections with cutting device; replant on site at same depth or pot for holding.	Good high marsh soil stabilizer. Will not tolerate extended inundation and naturally occurs on tidal creek banks and high spots in marsh. Seeds eaten by birds and small mammals.
Bladderworts *Utricularia* spp.	Cuttings	Collect quantities of cuttings in buckets of water by scooping plants out of natural stands (in water); transfer to standing water on site.	Good waterfowl food source, especially for dabbling ducks. Potential pest plant in reservoirs.
Broadleaf arrowhead† *Sagittaria latifolia*	Transplants	Dig clumps; separate individuals; replant on site or pot for holding.	Good waterfowl food source; good cover for wildlife; muskrat food.
Bulrushes† *Scirpus* spp.	Transplants, tubers	Dig plants; divide; replant on site or pot for holding. Dig tubers; separate; cut off top shoots if present; replant on site or pot for holding.	Excellent waterfowl and songbird food (seeds); foliage eaten by muskrats; used for cover and breeding and nesting by many species.
Burreed† *Sparganium americanum*	Transplants	Dig plants; divide; replant on site or pot for holding.	Seeds infrequent source of wildlife food.
Buttercups *Ranunculus* spp.	Cuttings	Collect quantities of cuttings in buckets of water by scooping plants out of natural stand (in water); transfer to standing water on site.	Good waterfowl food source. Potential pest plant in reservoirs.
Buttonbush† *Cephalanthus occidentalis*	Transplants, seeds	Dig small plants (large seedlings); transplant to site or pot for holding. Collect seeds Aug–Sep; store seeds in fresh water at 5°C.	Seeds good source of food for waterfowl and other birds, insects, beaver, and muskrats. Provides cover and nesting habitat for birds.
Chufa†,‡ *Cyperus esculentus*	Tubers	Dig tubers when mature (Jul–Sep); separate from other plant material; store moist but not wet at 5°C; broadcast on site and rake into soil. Tubers are very small and can be treated as seeds.	Excellent food source for waterfowl, turkeys, deer, wild boar, songbirds; highly productive plants can produce hundreds of tubers per plant. Seeds, tubers, foliage all relished.

TABLE A.15 Propagules and Planting Techniques Recommended for Selected Marsh Plants (*Continued*)

Species	Recommended propagules	General collection, handling, and planting techniques	Remarks
Common reed† *Phragmites australis*	Transplants, rootstock	Dig plants; divide; replant on site or pot for holding. Dig rootstock; separate into sections with at least one growth point; plant on site.	Used for nesting by songbirds, marsh birds, and waterbirds. Stabilizes soil; rapid growth with tall rank form. Definite pest plant.
Common threesquare† *Scirpus americana*	Transplants, tubers	Dig plants, divide, replant on site at same depth or pot for holding. Dig tubers; divide; cut off top shoots if present; replant on site.	Good source of food for waterfowl, muskrats, and nutria. Used for soil stabilization.
Delta duckpotato†,‡ *Sagittaria platyphylla*	Transplants	Dig plants, separate individuals; replant on site at same depth or pot for holding.	Excellent waterfowl food source; good soil stabilizer; grows well only on fine textured soils.
Dock† *Rumex* spp.	Seeds	Collect seeds when mature (May–Jul); store dry at room temperature or less; broadcast on site and rake into soil.	Good food source for songbirds. Hardy species and good soil stabilizer.
Dotted smartweed† *Polygonum punctatum*	Seeds, cuttings	Collect seeds; store dry at room temperature or less; broadcast on site and rake into soil. Take cuttings from natural stand; broadcast on wet area on site (not standing water).	Good soil stabilizer; good cover for ducklings; seeds eaten by waterfowl, muskrats, deer.
Duckpotato† *Sagittaria cuneata*	Transplants	Dig plants; separate individuals; replant on site or pot for holding.	Excellent food source for waterfowl.
Duckweed† *Lemna* spp.	Whole plants	Collect buckets of plants from natural stand in water; place whole plants in standing permanent water on site.	Excellent food source for waterfowl, especially wood ducks. Good cover. In deep south can be pest in standing water that should be kept open.
Eelgrass† *Zostera marina*	Transplants	Dig clumps with coring device; replant in shallow seawater with a minimum of current and wave action.	Good soil stabilizer; food source for diving ducks; provides cover for marine organisms.

European glasswort† *Salicornia europea*	Cuttings, rootstock	Take 5- to 15-cm cuttings from top shoots; broadcast on wet area of site. Dig rootstock; divide into clumps; replant on site at same depth.	Used mainly for soil stabilization. Poor wildlife food use; occasionally used by nesting colonial seabirds.
Fimbristylis† *Fimbristylis castanea*	Transplants, seeds	Dig plants; separate individuals; replant on site at same depth or pot for holding. Collect seeds when mature (Jul–Sep); store dry; broadcast on site and rake into soil.	Fair food source for songbirds and occasionally waterfowl.
Foxtail grasses† *Setaria* spp.	Sprigs, seeds	Dig young plants; replant as sprigs on site at same depth or pot for holding as transplants. Collect seeds when mature (Jun–Oct, depending upon species); store dry at 5°C; broadcast on site.	Good source of food for most birds, browsers and grazers, rodents. Cover for many wildlife species.
Frankenia *Frankenia grandifolia*	Transplants	Dig plants; separate individuals; replant on site at same depth or pot for holding.	Soil stabilizer; poor source of food but some use as cover by wildlife.
Frog bit† *Limnobium spongia*	Seeds	Collect seeds when mature (Jul–Sep); store dry at room temperatures or less; broadcast on site and rake into soil.	Good seed source for songbirds; cover for birds and other small animals; some use for stabilization.
Giant reed† *Arundo donax*	Seeds, transplants	Collect seeds when mature; store dry at room temperatures or less; broadcast on site and rake into soil. Dig plants; divide; replant on site or pot for holding.	Hardy plant; good seed source for wildlife; used for soil stabilization.
Groundsel tree† *Baccharis halimifolia*	Seedlings	Dig seedlings ≥3 to 5 dm high in natural stands; replant on site at same depth or pot for holding.	Excellent for cover nesting/breeding; used often by colonial nesting wading birds on dredged material islands. Poor food source.
Hardstem bulrush† *Scirpus acutus*	Transplants, tubers	Dig plants; divide; replant on site or pot for holding. Dig tubers, divide from other plant materials; cut off top shoots if present; plant on site at same depth.	Excellent seed source for birds; hardy species; used by muskrats and for soil stabilization..

TABLE A.15 Propagules and Planting Techniques Recommended for Selected Marsh Plants (*Continued*)

Species	Recommended propagules	General collection, handling, and planting techniques	Remarks
Horned pondweed *Zannichellia palustris*	Cuttings, rootstock	Gather plant material from standing water; place on site in permanent standing water areas. Dig rootstock from shallow water areas where possible; plant intact on site.	Fair food source for waterfowl, especially dabbling ducks; good sediment stabilizer.
Horsetails† *Equisteum* spp.	Transplants	Dig plants; separate individuals; replant on site or pot for holding.	Poor food source; used for soil stabilization.
Japanese millet†,‡ *Echinochloa crusgalli frumentacea*	Seeds	Buy seeds from commercial seed source.	Excellent upland and marsh bird food; relished by waterfowl; eaten by turkeys, raccoons and other small animals, deer; used in game management as food plot source.
Ladysthumb† *Polygonum persicaria*	Cuttings, seeds	Take cuttings 5 to 15 cm from top shoots broadcast on wet area of site; rake into soil. Collect seeds when mature; store in fresh water; broadcast on site and rake into soil.	Excellent source of food for waterfowl and upland game and songbirds.
Lizard's tail† *Saururus cernuus*	Transplants, seeds	Dig plants; separate individuals; replant on site or pot for holding. Collect seeds when mature (Jun–Aug); store in fresh water; broadcast on site and rake into soil.	Fair food source; used for stabilization in intermittent pond area.
Lobelia *Lobelia dortmanna*	Transplants	Dig plants; separate individuals; replant on site or pot for holding.	Fair food source; possibly used for soil stablization.
Lotus *Lelumbo lutea*	Seeds, rootstock	Collect seeds when mature (Aug–Oct); remove from pods; store in fresh water at 5°C; broadcast in shallow water on site. Dig rootstock when water is very low (late summer, fall); plant in shallow water on site.	Fair food source for waterfowl; relished by wild boar (roots); excellent cover for ducklings; potential pest in standing water and shallow reservoirs.

Lyngby's sedge† *Carex lyngbyei*	Transplants, seeds	Dig plants, separate individuals; replant on site or pot for holding. Collect seeds when mature (Jul–Sep); store dry at room temperature; broadcast on site.	Good food source for waterfowl and other birds; good cover for many species.
Mannagrass† *Glyceria acutiflora*	Seeds, sprigs	Collect seeds when mature; store dry at room temperature or less; broadcast on site. Dig young plants for sprigs; replant on site or pot for holding as transplants.	Excellent seed source for many bird species; foliage eaten by small and large animals; good cover.
Mannagrass† *Glyceria fluitans*	Seeds, sprigs	Same as *Glyceria acutiflora*.	Excellent seed source for many bird species and other wildlife; good cover. Grows in wetter areas than *Glyceria acutiflora*.
Marsh elder† *Iva frutescens*	Seedlings	Dig seedlings in natural stands near parent plants; separate individuals; replant on site or pot for holding. Seedlings should be ≥0.3 m tall.	Excellent cover species for birds and small mammals and herps; used by colonial nesting wading birds for nesting substrate. Potential pest plant.
Marsh hibiscus† *Hibiscus moscheutos*	Seeds, transplants	Collect seeds when mature (Aug–Oct); store dry at 5°C; plant on site ≥3 to 5 cm deep. Dig plants, replant on site or pot for holding.	Good cover for birds, sunning turtles; grows on banks of streams and ponds, in ditches; good soil stabilizer.
Marsh pepper† *Polygonum hydropiper*	Cuttings, rootstock	Take 5- to 15-cm cuttings from top shoots; broadcast on wet area of site; rake into soil. Dig rootstock; divide into sections; plant in wet area of site.	Excellent seed source for waterfowl and other birds; foliage bitter to browsers; good cover and soil stabilizer.
Marsh smartweed† *Polygonum hydropiperoides*	Cuttings, seeds	Take 5- to 15-cm cuttings from top shoots; broadcast on wet area of site; rake into soil. Collect seeds when mature (Jun–Sep); store or plant immediately on site; rake in soil.	Excellent seed source for waterfowl and other birds; good cover for many wildlife species.
Mud plantain† *Heteranthera reniformis*	Cuttings	Take 5- to 15-cm sections from top shoots; replant in mud and wet areas on site, taking care to bury portions of cuttings in soil.	Good soil stabilizer in intermittent ponds and streams.

TABLE A.15 Propagules and Planting Techniques Recommended for Selected Marsh Plants (*Continued*)

Species	Recommended propagules	General collection, handling, and planting techniques	Remarks
Nodding smartweed† *Polygonum lapathifolium*	Seeds	Collect seeds when mature (Jun–Sep); store in fresh water at 5°C; broadcast on site; rake into soil.	Abundant seed source for upland birds and waterfowl; grows in drier soils than most smartweeds; potential pest.
Nutsedges† *Cyperus* spp.	Tubers, rootstock	Dig tubers in late summer and fall; divide; plant on site or pot for using as transplants. Dig rootstock; divide into sections; plant on site, same depth.	Excellent food source for most wildlife, especially chufa and red-rooted sedge; commercially available; potential pest in agronomic areas.
Olney's threesquare† *Scirpus olneyi*	Transplants, tubers	Dig plants, separate individuals; plant on site or pot for holding. Dig tubers; separate; plant on site at same depth.	Excellent food source for waterfowl, muskrats, nutria, small animals. Good soil stabilizer.
Orache† *Atriplex patula*	Seeds	Collect seeds when mature; store dry at room temperature or less; broadcast on site; rake into soil.	Good source of seeds for birds and rodents; good soil stabilizer.
Pacific cordgrass† *Spartina foliosa*	Transplants, sprigs	Dig young plants from edge of marsh; plant at same depth immediately as sprigs, or grow in pots and transplant into site as larger plants. Growing from seeds not recommended, as seeds have very low viability.	Only low-marsh soil stabilizer on West Coast that tolerates both high salinities and strong tidal action. Good soil stabilizer; good cover; very slow growth.
Red mangrove† *Rhizophora mangle*	Seeds, seedlings	Collect seed pods when mature; plant whole pod upright in soil with stem end up and out of soil. Dig seedlings from natural stand or grow from seed pods.	Excellent soil stabilizer in south Florida. Often occurs on dredged material islands and used by colonial nesting wading birds for nesting. Tolerates sea-strength salinities.
Reed canarygrass†,‡ *Phalaris arundinacea*	Seeds	Buy seeds from commercial seed source.	Excellent soil stabilizer; seeds good wildlife food source; used to dewater and filter wastewater.

Reed grass† *Calamogrostis canadensis*	Seeds, sprigs	Collect seeds when mature (Jul–Sep); store dry at 5°C; broadcast on site. Dig young plants to use for sprigs; separate individuals; plant on site or pot for growing as transplants.	Excellent seed source for birds; grazed heavily by rodents and other mammals. Good soil stabilizer.
Reed mannagrass† *Glyceria grandis*	Seeds, sprigs	Same procedures as for reed grass.	Same value as for reed grass.
Rice cutgrass† *Leersia oryzoides*	Seeds, sprigs	Collect seeds when mature (May–Jul); store in fresh water at 5°C; broadcast on site and rake into soil (in wet area). Dig young plants; separate individuals; plant on site at same depth in wet areas.	Good seed and foliage food source for many wildlife species, especially waterfowl and marsh birds. Good soil stabilizer of banks.
River bulrush† *Scirpus fluviatilis*	Rootstock, transplants	Dig rootstock, divide into sections; plant at same depth on site. Dig plants; separate individuals; transplant to site or pot for holding.	Used often by nesting waterfowl and marsh birds; seed good food source for many wildlife species. Good soil stabilizer.
Rushes† *Juncus* spp.	Transplants, rootstock, seeds	Dig plants; separate individuals; transplant to site or pot for holding. Dig rootstock; divide into sections; plant at same depth on site. Collect seeds when mature (Jul–Oct); store in fresh water at 5°C; broadcast on site; rake into soil.	This group of plant species excellent for waterfowl, other birds, small mammals. Used as nesting substrate by waterfowl and marsh birds; good soil stabilizers; hardy plants.
Saltgrass† *Distichlis spicata*	Sprigs, rhizomes	Dig young plants; divide into sections; plant on site or pot for holding. Dig roots; divide rhizomes into small sections; plant on site; rake into soil.	Excellent soil stabilizer; grows well in high brackish marshes; used as lodge material by muskrats; seeds fair food source, but foliage poor source.
Saltmarsh aster† *Aster tenuifolius*	Seeds	Collect seeds when mature (Jul–Sep); store dry at room temperature or less; broadcast on site; rake into soil.	Good soil stabilizer in high coastal marshes.
Saltmarsh bulrush† *Scirpus robustus*	Transplants, tubers	Dig plants; divide; plant on site at same depth or pot for holding; Dig tubers; separate tubers; cut off top shoots if present; plant on site at same depth.	Excellent food source for waterfowl muskrats, nutria, other small animals. Good cover; good soil stabilizer; used by muskrats for lodge material.

TABLE A.15 Propagules and Planting Techniques Recommended for Selected Marsh Plants (*Continued*)

Species	Recommended propagules	General collection, handling, and planting techniques	Remarks
Saltmarsh cattail† *Typha angustifolia*	Transplants, rootstock	Dig plants; separate individuals; plant on site at same depth. Dig roots; separate; cut off top shoots if present; plant on site.	Good soil stabilizer in brackish soils. Occurs in ditches, intermittent ponds, primarily on coasts. Low food value; fair cover.
Saltmarsh jaumea *Jaumea carnosa*	Transplants	Dig plants, separate individuals; plant on site at same depth or pot for holding.	Fair soil stabilizer on West Coast in high brackish marshes.
Saltmeadow cordgrass† *Spartina patens*	Transplants, sprigs	Dig plants; divide into clumps; plant on site at same depth or pot for holding. Dig young plants; separate; plant on site at same depth.	Excellent soil stabilizer in brackish marshes; also used to stabilize dunes on Atlantic coast. Seed production often poor; low food value; some cover value.
Sawgrass† *Cladium jamaicense*	Sprigs, seeds	Dig young plants; separate individuals; plant on site or pot for holding. Collect seeds when mature (Jul–Sep); store in fresh water at 5°C; broadcast on site; rake into soil.	Species very site specific; occurs only in south Florida. Will not tolerate high nutrient levels. Good soil stabilizer; good cover; seeds eaten by some wildlife.
Sea lavender† *Limonium carolinianum*	Seeds	Collect seeds when mature (Jul–Aug); store dry at 5°C; broadcast on site; rake into soil.	Fair soil stabilizer; cover. Low food value. Some nesting substrate value.
Sea lavender† *Limonium vulgare*	Seeds	Same procedures as for *Limonium carolinianum.*	Same values as *Limonium carolinianum.*
Sea ox-eye† *Borrichia frutescens*	Transplants, seeds	Dig plants; separate individuals; plant on site at same depth or pot for holding. Collect seed heads when mature (Jul–Oct); store seeds in fresh water at 5°C; plant on site; rake into soil.	Excellent soil stabilizer; grows in high brackish marshes and on shores. Low food value; some cover and nesting value.
Sea purslane† *Sesuvium portolacastrum*	Seeds	Collect seeds when mature; store dry at room temperature or less; plant on site; rake into soil.	Fair soil stabilizer value; low food value; some seed value as food. Some cover use.

Seaside arrowgrass† *Triglochin maritima*	Transplants	Dig plants, divide into individuals or clumps; plant on site at same depth or pot for holding.	Excellent soil stabilizer in brackish tidal marshes in Pacific northwest; some cover value; low food value.
Sedges† *Carex* spp.	Transplants, seeds	Dig plants; separate into clumps or individuals; plant on site or pot for holding. Collect seeds when mature (Jun–Sep); store dry at 5°C; broadcast on site; rake into soil.	Group of species far-ranging and widely varied. Usually excellent seed value for wildlife; also good cover. Prolific plants.
Shoalgrass† *Halodule wrightii*	Transplants	Dig plugs with coring device in water at low tide; plant at site immediately at same depth.	Propagules must be stabilized to prevent tidal scour. Good cover value for marine organisms; good sediment stabilizer.
Slough grass† *Beckmannia syzigachne*	Transplants, seeds	Dig plants; divide into clumps or individuals; plant at same depth on site or pot for holding. Collect seeds when mature (Jul–Sep); store in fresh water at 5°C; broadcast on wet site area.	Good food value for waterfowl and other seed-eating birds; foliage eaten by small animals. Good soil stabilizer.
Slough sedge† *Carex trichocarpa*	Transplants, seeds	Dig plants; separate into clumps; plant on site at same depth or pot for holding. Collect seeds when mature (Jul–Oct); store in fresh water at 5°C; broadcast on wet site; rake into soil if needed.	Excellent wildlife seed source; foliage also eaten. Good soil stabilizer.
Smartweeds† *Polygonum* spp.	Cuttings, seeds	Take 5- to 15-cm cuttings from top shoots; broadcast on site; rake into soil and cover parts of cuttings (site should be wet). Collect seeds, store in fresh water or dry depending on species; broadcast on site; rake into soil.	Excellent group of plants for wildlife value; seeds readily consumed by waterfowl and many other birds and small mammals. Good soil stabilizers.
Smooth cordgrass†,‡ *Spartina alterniflora*	Sprigs, transplants	Dig young plants, separate individuals; plant as sprigs on site or pot to hold as transplants. Dig transplants from natural marsh or grow from seeds; plant on site, covering all roots.	Best soil stabilizer of low salt marshes on East and Gulf coasts. Used extensively for stabilization and marsh creation projects. Good cover value; good food value. Tolerant of tidal inundation for long periods.

TABLE A.15 Propagules and Planting Techniques Recommended for Selected Marsh Plants (*Continued*)

Species	Recommended propagules	General collection, handling, and planting techniques	Remarks
Soft rush† *Juncus effusus*	Transplants	Dig clumps; divide into sections with cutting device; plant on site at same depth or pot for holding.	Persistent high marsh species; good cover value. Some seed value, but foliage inedible. Known pest in pastoral areas.
Softstem bulrush† *Scirpus validus*	Rhizomes, transplants	Dig roots; divide rhizomes leaving ≥1 growth point on each; plant on site 2 to 3 cm deep. Dig plants; divide into sections; plant on site or pot for holding.	Excellent soil stabilizer of freshwater coastal and interior marshes. Good seed value for wildlife. Used as cover and nesting material by waterfowl and other wildlife.
Southern bulrush *Scirpus californicus*	Rhizomes, transplants	Same procedures as for softstem bulrush.	Same values as for softstem bulrush, but does not occur as extensively and grows to be much larger and more robust.
Southern smartweed† *Polygonum densiflorum*	Cuttings, seeds	Take 5- to 15-cm cuttings from top shoots; broadcast in wet area on site; rake or place cuttings into soil. Collect seeds when mature (Jul–Oct); store in fresh water at 5°C; broadcast on site; rake into soil.	Excellent food source for waterfowl and marsh birds. Prolific growth habits; forms dense tall stands. Good cover value.
Southern cutgrass† *Zizaniopsis mileacea*	Seeds, sprigs	Collect seeds when mature (May–Jul); store in fresh water at 5°C; broadcast on wet site, rake into soil if needed.	Excellent seed value for waterfowl and other birds; tender, young foliage eaten by small animals and grazers. Good soil stabilizer.
Spatterdock† *Nyphar lutea*	Transplants	Dig plants; separate individuals; plant on site at same depth or pot for holding.	Good waterfowl food; good soil stabilizer.
Spikerushes† *Eleocharis* spp.	Transplants	Dig plants; divide into clumps; plant on site at same depth or pot for holding.	Excellent soil stabilizer; fair waterfowl food.
Spirodella† *Spirodella polyrhiza*	Whole plants	Scoop buckets of plants from standing water; transfer to standing water on site.	Good waterfowl food, especially for wood ducks.

Sprangletop† *Leptochloa fascicularis*	Seeds, sprigs	Collect seeds when mature (summer, fall); store dry at room temperature or less; broadcast on site; rake into soil. Dig young plants; plant on site as sprigs.	Excellent seed source for wildlife; good soil stabilizer; used for cover.
Sweet flag *Acorus calamus*	Transplants	Dig plants; divide individuals; plant on site in high marsh at same depth.	Good soil stabilizer; fair wildlife value; potential pest plant.
Tufted hairgrass† *Deschampsia caespitosa*	Transplants, sprigs	Dig plants; divide individuals; plant on site or pot for holding. Dig young plants; plant as sprigs on site.	Excellent low-marsh species for Pacific northwest; prolific growth; good cover and fair food value. Good soil stabilizer.
Turtle grass† *Thalassia testudium*	Transplants	Dig clumps with coring device from water at low tide; clumps need ≥1 growth point; plant on site in water.	Excellent cover and wildlife value; good cover for marine organisms. Species susceptible to environmental changes; rare in some areas.
Walters millet†,‡ *Echinochloa walteri*	Seeds	Buy from commercial seed source.	Excellent food value for waterfowl and other wildlife such as raccoons, turkeys, deer, muskrats. Good temporary soil stabilizer.
Water hemp† *Acnida cannabina*	Seeds	Collect seeds when mature; store in fresh water at 5°C; broadcast in wet area on site; rake into soil if needed.	Good seed source for wildlife; fair soil stabilizer.
Water hyssop *Bacopa caroliniana*	Cuttings, sprigs	Take 5- to 15-cm cuttings from top shoots; plant in mud on site. Dig young plants; divide; plant on site in wet area.	Good soil stabilizer; fair wildlife food.
Water lilies†,‡ *Nymphaea* spp.	Rootstock	Dig rootstock in late summer and fall when water levels are low; transplant to shallow water on site.	Good cover for ducklings; some food value. Excellent sediment stabilizer; potential pest.
Water milfoils *Nyriophyllum* spp.	Cuttings	Remove buckets of segments of plants from standing water; transfer to standing water on site.	Excellent dabbling duck food; good cover. Potential pest in standing water and reservoirs.
Water nymphs *Najas* spp.	Cuttings	Same procedures as for water milfoils.	Same value as for water milfoils.

TABLE A.15 Propagules and Planting Techniques Recommended for Selected Marsh Plants (*Continued*)

Species	Recommended propagules	General collection, handling, and planting techniques	Remarks
Water plantain† *Alisma plantago-aquatica*	Transplants	Dig plants; divide individuals; plant on site at same depth.	Good food source for wildlife; fair soil stabilizer.
Water shield *Brasenia schreberi*	Rootstock	Dig roots in shallow water in late summer and fall; transfer to standing shallow water on site.	Good cover value, good sediment stabilizer.
Water smartweed† *Polygonum amphibium*	Cuttings, seeds	Take 5- to 15-cm cuttings from top shoots; plant on site in wet area by burying part of cutting. Collect seeds when mature (Jul–Sept); store in fresh water at 5°C; broadcast on wet site.	Excellent waterfowl food; good cover. Excellent sediment and soil stabilizer.
Water willow *Decodon verticillatus*	Transplants	Dig plants; divide individuals; plant on site at same depth.	Fair soil stabilizer; low wildlife value.
White mangrove† *Laguncularia racemosa*	Seeds, seedlings	Collect seeds when mature; plant immediately on site. Dig seedlings from natural stand; plant on site.	Excellent soil stabilizer; good cover; low food value; used by nesting birds.
Widgeongrass† *Ruppia maritima*	Cuttings	Remove buckets of segments of plants from standing water; transfer to standing water on site.	Excellent waterfowl food; grown by waterfowl managers for attracting waterfowl.
Wildcelery *Vallisneria americana*	Whole plants	Remove whole plants from standing water; transfer to standing water on site.	Excellent cover value; harbors many invertebrates fed on by wildlife. Shades out aquatic plants; pest in Florida and deep south in some areas.
Wildrice† *Zizania aquatica*	Sprigs, seeds	Dig young plants, divide individuals; plant in shallow water on site. Collect seeds when mature; plant on wet site.	Low tolerance for pollution; must have fine-textured soils in slow-moving water. Excellent wildlife food, good soil stabilizer.

Willows† *Salix* spp.	Cuttings	Take 10- to 30-cm cuttings from dormant trees (winter months, early spring); plant on site with butt end two-thirds in soil.	Excellent soil stabilizer of stream and pond banks. Good cover and food value for songbirds. Very fast growing, potential pest.
Wolffias *Wolffia* spp.	Whole plants	Remove buckets of plants from standing water; transfer to standing water on site.	Excellent waterfowl food; good cover value.
Yellow flag *Iris pseudacorus*	Transplants, rhizomes	Dig plants; divide individuals; plant in high marsh on site. Dig rhizomes; divide, with 1 growth point on each rhizome; plant shallowly on site.	Good soil stabilizer, low wildlife value; showy flowers.

*Transplants include plugs, groups of individuals, very large seedlings, and large whole plants.
†Known to occur on dredged material.
‡Commercially available.
SOURCE: U.S. Army Engineer Waterways Experiment Station (1978).

TABLE A.16 Seed Production, Germination, and Storage of Some Marsh and Aquatic Plants*

Species	Seed production, kg/ha†	Dormancy or after-ripening	Ripening dates	Storage
Alisma plantago-aquatica	144,000/plant	Yes		
Cephalanthus occidentalis	1200			
Cladium jamaicense	592			
Cyperus erythrorhizos	674			
Cyperus esculentus	125	Yes		
Cyperus strigosus	370			
Distichlis spicata		Yes	Early Sep–Oct	Dry, room temp.
Echinochloa crusgalli	1681–2925			
Echinochloa crusgalli frumentaceae	3276			
Echinochloa walteri	912–1500			
Echinodorus cordifoliosis	26,900/plant			
Eichornia crassipes	≤ 500/plant			
Eleocharis palustris	15–40/spike	Yes		
Eleocharis quadrangulata	9			
Eragrostis hypnoides	396			
Eragrostis pectinacea	155			
Juncus filiformis	75–150/capsule			
Juncus macer	100–300/capsule		Jul–Sep	
Leersia oryzoides	154			
Limonium vulgare	5/spike		Late Sep	Dry, room temp
Lobelia dortmanna	118/capsule	No	Jul–Oct	Wet, 1 to 3°C
Lophotocarpus calycinus	401		Sep	
Myriophyllum spicatum		Yes	Nov	
Najas guadalupensis	78,500/plant			
Nelumbo lutea	26/pod			Dry, room temp.
Phragmites australis	≤ 1000/panicle		Nov	
Polygonum hydropiper	300–900/plant	Yes	Aug	
Polygonum lapathifolium	772	Yes	Late summer	
Polygonum muhlenbergii	49	Yes		
Polygonum pensylvanicum	1000	Yes	Sep	
Polygonum persicaria	200–800/plant	Yes	Jun	
Polygonum punctatum	150			
Pontederia cordata	388–722		Sep	Wet, 1 to 3°C
Potamogeton crispus	960/plant			
Potamogeton foliosus	≤ 124,950/plant			
Potamogeton nodosus	959	Yes	Sep	Wet, 1 to 3°C
Potamogeton pectinatus	63,000/plant	Yes	Aug	Wet, 1 to 3°C
Potamogeton richardsonii	114–29,200/plant			
Potamogeton vaginatus	90/plant			
Ranunculus aquatilis	150/plant			

TABLE A.16 Seed Production, Germination, and Storage of Some Marsh and Aquatic Plants* (*Continued*)

Species	Seed production, kg/ha†	Dormancy or after-ripening	Ripening dates	Storage
Ruppia maritima	15			
Rhynchospora corniculata	1021			
Sagittaria latifolia	313	Yes	Sep	Wet, 1 to 3°C
Scirpus americana	240	Yes	Sep	Dry, room temp.
Scirpus californicus	49	Yes		Dry, room temp.
Scirpus olneyi	27	Yes		Dry, cool, or room temp.
Scirpus robustus	335	Yes		Dry, room temp.
Scolochloa festuceae		Yes	Jul	
Sparganium eurycarpum	930		Oct	
Spartina alterniflora	40–210	Yes	Sep	Wet, 1 to 3°C
Spartina foliosa		Yes		
Suaeda maritima	≤ 1000/plant	Yes	Oct	
Subularia aquatica	8–25/plant	No		
Thalassia testudinum		No	Jun (Fla.)	
Typha latifolia	222,000/18-cm spike	Yes	Late Aug	Dry, room temp.
Vallisneria americana	Several hundred/pod		Sep	Cool, wet
Zannichellia palustris	2,080,000/plant			
Zizania aquatica	28–45	Yes	Sep	Wet, 1 to 3°C
Zostera marina		No	Aug	

*Omission indicates no data available.
†Unless otherwise noted.
SOURCE: Kadlec and Wentz (1974).

TABLE A.17 Some Native Grasslike Plants Recommended for Planting in Riparian Sites

Species	Areas*	Habitat	Abundance	Rooting habit	Comments
Carex aquatilis Water sedge	Asp.-SF	Wet meadows	Abundant	Caespitose, long rhizomes	Excellent streambank stability, highly palatable. Principal species for revegetation.
Carex aurea Golden sedge	Val.-SF	Marsh, wet meadows	Frequent	Caespitose, long rootstocks	Widely distributed, good ground cover.
Carex disperma Softleaved sedge	Asp.-Alp.	Swamps, meadows	Frequent	Caespitose, long rhizomes	Shady areas, solid mat, moderate vigor.
Carex douglasii Douglas sedge	PJ-Asp.	Dry meadows, alkali tolerant	Abundant	Creeping rootstocks, long culms	Adapted to compact soils, low palatability, increases under grazing.
Carex elynoides Black sedge-root	Alp.	Open, dry meadows	Common	Caespitose	Vigorous, abundant.
Carex hoodii Hood sedge	Mtn.B.-SF	Open parks, drainage ways, bottoms	Abundant	Densely caespitose	Excellent ground cover, useful forage species.
Carex lanuginosa Woolly sedge	Val.-SF	Dry to wet meadows	Abundant	Caespitose, long rootstocks	Very robust, principal species for streambank stabilization.
Carex lenticularis Kellogg sedge	Mtn.B.-SF	Wet meadows, marshes	Abundant	Caespitose, long rootstocks	Pioneer species, invades water's edge.
Carex microptera Smallwing sedge	Mtn.B.-Asp.	Meadow edges	Abundant	Densely caespitose	Good cover for streambank, palatable, spreads by seeds, widely distributed.
Carex nardina Hepburn sedge	Alp.	Open meadows	Abundant	Densely caespitose	Short stature, open cover.
Carex nebrascensis Nebraska sedge	Val.-Asp.	Marshes and meadows, alkali-tolerant	Common	Strongly rhizomatous	Excellent soil stabilizer, palatable, widely distributed.
Carex nigricans Black alpine sedge	SF-Alp.	Well-drained meadows	Frequent	Creeping rootstock	Good cover for wet areas.
Carex praegracilis Slim sedge	Val.-Asp.	Dry to moist, alkali bottomlands	Abundant	Long, creeping rootstocks	Large plant, dense, persistent, moderately palatable.

Carex rostrata Beaked sedge	Val.-SF	Streams, water's edge, standing water	Abundant	Culms from stout, long rhizomes	**Principal species for streambank stabilization, low palatability, fluctuating water level, wide elevational range.**
Carex rupestris Rock sedge	Alp.	Dry slopes and meadows	Abundant	Short rhizomes	**Vigorous, spreads rapidly, limited distribution.**
Carex saxatillis	LPP-SF	Water's edge	Abundant	Culms from long, creeping rootstocks	**Excellent streambank cover, limited distribution.**
Carex scirpoidea Downy sedge	Alp.	Dry and wet meadows	Abundant	Rhizomatous	**Vigorous, spreads rapidly.**
Carex simulata Analogne sedge	PP-SF	Bogs and wet meadows, calcareous soils	Frequent	Long, creeping rootstocks	**Excellent cover, widely distributed.**
Carex vallicola Valley sedge	Sage-Asp.	Dry slopes	Abundant	Caespitose	**Spreads onto dry grass-sage sites.**
Eleocharis palustris Spikerush	Val.-SF	Wet meadows and streams, alkali-tolerant	Abundant	Rhizomatous	**Spreads rapidly, low palatability, wide elevational range.**
Juncus arcticus balticus Baltic rush	Val.-Asp.	Wet and semiwet meadows	Abundant	Rhizomatous	**Principal species for stabilization. Uses adapted ecotypes, spreads aggressively, persists with grazing.**
Juncus drummondii Drummond rush	LPP-Alp.	Wet and dry meadows	Common	Caespitose	**Spreads after disturbance, occupies infertile soil.**
Juncus ensifolius Swordleaf rush	Sage-SF	Streams, wet meadows, seeps	Abundant	Strongly rhizomatous	**Moderately palatable, wide elevational range.**
Juncus longistylis Longstyle rush	Sage-SF	Wet meadows, streams	Common	Rhizomatous	**Moderately palatable.**
Juncus torreyi Torrey rush	Val.-PJ	Streams, wet meadows, seeps, alkali-tolerant	Common	Strongly rhizomatous	**Spreads onto disturbances.**
Scirpus acutus Tule bulrush	Val.-Mtn.B.	Lake edge	Abundant	Rhizomatous	**Tall, rank, dense patches, restricted to water's edge.**
Scirpus maritimus Saltmarsh bulrush	Mtn.B.	Lake edge, stream bank, alkali sites	Abundant	Rhizomatous	**Dense patches, spreads rapidly.**

*Areas: Alp. = alpine, SF = spruce-fir, Asp. = aspen, LPP = lodgepole pine, PP = ponderosa pine, Mtn.B. = mountain brush, PJ = pinyon-juniper, Sage = big sagebrush, Val. = valley.

SOURCE: Platts et al. (1987).

TABLE A.18 Grasses Recommended for Direct Seeding and Transplanting in Riparian Sites

Species	Areas of adaptation*	Origin	Seeding trait	Trans-plant capability	Growth rate	Rooting habit	Salinity tolerance†	Flooding tolerance	Palatability	Spreadability
Agropyron elongatum										
Tall wheatgrass	Mtn.B.-V	Introduced	Excellent	Good	Rapid	Large clump	MT	Moderate	Fair	Good
Agropyron repens										
Quackgrass	Asp.-V	Introduced	Fair	Excellent	Slow	Rhizomatous	MT	Moderate	Good	Excellent
Agropyron smithii										
Western wheatgrass	PP-SDS	Native	Poor	Excellent	Slow	Rhizomatous	MS	Moderate	Good	Good
Agropyron trachycaulum										
Slender wheatgrass	SF-PJ	Native	Excellent	Excellent	Rapid	Rhizomatous	MS	Sensitive	Excellent	Good
Agrostis stolonifera										
Redtop	Salp.-SF	Introduced	Fair	Good	Moderate	Rhizomatous	MS	Moderate	Good	Excellent
Alopecurus pratensis										
Meadow foxtail	Alp.-Mtn.B.	Introduced	Excellent	Good	Rapid	Rhizomatous	MT	Tolerant	Good	Excellent
Bromus carinatus										
Mountain brome	Alp.-PJ	Native	Excellent	Excellent	Rapid	Rhizomatous	MT	Moderate	Good	Good
Bromus erectus										
Meadow brome	Alp.-PJ	Introduced	Excellent	Excellent	Moderate	Rhizomatous	MT	Moderate	Good	Excellent
Bromus inermis										
Smooth brome	Alp.-Mtn.B.	Introduced	Good	Excellent	Moderate	Rhizomatous	MT	Moderate	Good	Excellent
Calamagrostis canadensis										
Bluejoint reedgrass	SF-Sage	Native	Good	Excellent	Moderate	Rhizomatous	MT	Tolerant	Good	Excellent
Calamagrostis epigeois										
Chee reedgrass	Alp.-PJ	Introduced	Poor	Good	Slow	Rhizomatous	MT	Tolerant	Good	Good
Dactylis glomerata										
Orchardgrass	Alp.-Sage	Introduced	Good	Good	Rapid	Bunch	MS	Sensitive	Excellent	Fair
Deschampsia caespitosa										
Tufted hairgrass	Alp.-SF	Native	Poor	Fair	Slow	Bunch	MT	Tolerant	Fair	Poor
Distichlis spicata										
Saltgrass	V	Native	Poor	Excellent	Slow	Rhizomatous	T	Tolerant	Fair	Excellent
Elymus cinereus										
Great Basin wildrye	Mtn.B.-V	Native	Good	Good	Moderate	Large clump	T	Moderate	Good	Fair

Elymus giganteus										
Mammoth wildrye	Mtn.B.-Sage	Introduced	Fair	Good	Moderate	Rhizomatous	T	Tolerant	Good	Good
Elymus junceus										
Russian wildrye	Mtn.B.-V	Introduced	Fair	Good	Moderate	Bunch	T	Moderate	Excellent	Fair
Elymus triticoides										
Creeping wildrye	JP-V	Introduced	Good	Excellent	Moderate	Rhizomatous	T	Tolerant	Poor	Good
Festuca arundinacea										
Reed fescue (alta or tall)	Asp.-SDS	Introduced	Excellent	Excellent	Rapid	Rhizomatous	T	Tolerant	Good	Excellent
Hordeum brachyantherum										
Meadow barley	Alp.-Asp.	Native	Excellent	Excellent	Moderate	Bunch	T	Tolerant	Fair	Good
Lolium perenne										
Perennial ryegrass	SF-PP	Introduced	Excellent	Good	Rapid	Small bunch	MT	Sensitive	Good	Good
Phalaris arundinacea										
Reed canarygrass	Asp.-V	Native	Poor	Excellent	Slow	Rhizomatous	T	Tolerant	Fair	Excellent
Phleum pratense										
Timothy	Asp.-Mtn.B.	Introduced	Good	Good	Rapid	Bunch	MS	Moderate	Good	Good
Poa pratensis										
Kentucky bluegrass	Asp.-PJ	Introduced	Fair	Good	Slow	Rhizomatous	MT	Moderate	Good	Excellent
Poa secunda										
Sandberg bluegrass	Mtn.B.-Sage	Native	Fair	Good	Slow	Bunch	MT	Moderate	Good	Fair
Sitanion hystrix										
Bottlebrush squirreltail	Mtn.B.-SDS	Native	Good	Fair	Moderate	Bunch	MT	Moderate	Good	Good
Sporobolus airoides										
Alkali sacaton		Native	Fair	Good	Slow	Bunch	MT	Moderate	Good	Excellent

*Areas of adaptation: Alp = alpine; SF = spruce-fir; Asp. = aspen; Mtn.B. = mountain brush; PJ = pinyon-juniper; PP = ponderosa pine; Sage = big sagebrush; Salp. = subalpine; SDS = salt desert shrub; V = valley bottom.

†Salinity tolerance: S = sensitive; MS = moderately sensitive; MT = moderately tolerant; T = tolerant.

SOURCE: Platts et al. (1987).

TABLE A.19 Forbs Recommended for Planting in Riparian Sites

Species	Areas of adaptation*	Origin	Seeding trait	Transplant capability	Growth rate	Salinity tolerance†	Flooding tolerance	Palatability	Spreadability
Achillea millefolium lanulosa									
Western yarrow	Alp.-V	Native	Excellent	Excellent	Rapid	MS	Moderate	Poor	Excellent
Artemisia ludoviciana ludoviciana									
Louisiana sagewort	Alp.-Sage	Native	Excellent	Excellent	Rapid	MS	Moderate	Poor	Excellent
Aster chilensis adscendens									
Pacific aster	Asp.-V	Native	Poor	Excellent	Moderate	MS	Moderate	Excellent	Excellent
Bassia hyssopifolia									
Fivehook bassia	PJ-SDS	Native	Excellent	Good	Rapid	T	Tolerant	Good	Good
Coronilla varia									
Crownvetch	PJ-Mtn.B.	Introduced	Good	Excellent	Rapid	MS	Moderate	Good	Good
Epilobium angustifolium									
Fireweed	Asp.-Mtn.B.	Native	Excellent	Good	Rapid	S	Moderate	Fair	Excellent
Heracleum lanatum									
Common cowparsnip	Alp.-Mtn.B	Native	Poor	Poor	Poor	S	Sensitive	Excellent	Fair
Linum lewisii									
Lewis flax	Asp.-Sage	Native	Excellent	Good	Moderate	S	Sensitive	Good	Good
Medicago lupulina									
Black medic	Asp.-Sage	Introduced	Excellent	Good	Moderate	MT	Moderate	Good	Good
Medicago sativa									
Alfalfa	Asp.-Sage	Introduced	Excellent	Good	Rapid	MT	Moderate	Excellent	Fair
Melilotus officinalis									
Yellow sweetclover	Asp.-Sage	Introduced	Excellent	Poor	Rapid	MT	Moderate	Good	Excellent
Potentilla glandulosa glandulosa									
Gland cinquefoil	Asp.-PP	Native	Good	Excellent	Moderate	S	Moderate	Fair	Good
Senecio serra									
Butterweed groundsel	Asp.-PP	Native	Good	Excellent	Moderate	S	Moderate	Good	Good
Sidalcea oregana									
Oregon checkermallow	Asp.-Mtn.B.	Native	Good	Good	Moderate	S	Moderate	Fair	Good

Smilacina racemosa amplexicaulis									
Western Solomons-seal	Asp.-Mtn.B.	Native	Poor	Fair	Slow	S	Moderate	Excellent	Fair
Trifolium fragiferum									
Strawberry clover	V	Introduced	Good	Fair	Moderate	MT	Moderate	Excellent	Excellent
Trifolium hybridum									
Alsike clover	Asp.-Mtn.B.	Introduced	Good	Fair	Moderate	S	Moderate	Good	Good
Valeriana edulis									
Edible valerian	Asp.-Mtn.B.	Native	Poor	Fair	Slow	S	Moderate	Fair	Fair

*Areas of adaptation: Alp. = alpine; Asp. = aspen; PP = ponderosa pine; Mtn.B. = mountain brush; PJ = pinyon-juniper; Sage = sagebrush; SDS = salt desert shrub; V = valley bottoms.

†Salinity tolerance: S = sensitive; MS = moderately sensitive; MT = moderately tolerant; T = tolerant.

SOURCE: Platts et al. (1987).

TABLE A.20 Trees and Shrubs Recommended for Planting in Disturbed Riparian Sites

	Areas of occurrence				Establishment traits			
Species	Zones*	Habitat	Adaptation to disturbed sites	Methods of culture†	Seedling establishment	Growth rates	Soil stability value	Comments
Alnus tenuifolia Thinleaf alder	SF-Mtn.B.	Stream edge and well-drained soils.	Excellent	NS, CS, DS	Excellent	Rapid	Excellent	Easily established, adapted to harsh sites, grows rapidly.
Amelanchier alnifolia Saskatoon serviceberry	Asp.-Mtn.B.	Well-drained soils, occasional seeps.	Good	NS, CS	Fair	Slow	Good	Slow to establish, sensitive to understory competition.
Artemisia cana viscidula Silver sagebrush	Asp.-Sage	Well-drained and moist soils, valley bottoms.	Fair	DS, NS, CS	Good	Rapid	Fair	Well adapted to exposed moist soils able to tolerate flooding for short time.
Artemisia tridentata tridentata Basin big sagebrush	Mtn.B.-SDS	Deep, well-drained soils, occasional flooding.	Excellent	DS, NS, CS	Good	Rapid	Fair	Useful for planting extremely disturbed and well-drained soils.
Artemisia tridentata vaseyana Mountain big sagebrush	Asp.-Mtn.B.	Well-drained soils, moist sites.	Excellent	DS, NS. CS	Good	Rapid	Fair	Adapted to disturbed sites, suited to moist but not saturated soils.
Artemisia tripartita Tall threetip sagebrush	Asp.-Mtn.B.	Well-drained soils, moist sites.	Excellent	DS, NS, CS	Excellent	Rapid	Fair	Well suited to eroded exposed soils, spreads quickly.
Atriplex canescens Fourwing saltbush	Mtn.B.-V	Well-drained soils, frequent flooding and shallow water table.	Good	DS, NS	Excellent	Rapid	Good	Useful for well-drained and disturbed soils.

Atriplex gardneri Gardner saltbush	SDS-V	Semiarid deserts. Withstands seasonal flooding, and alternating wet/dry period.	Fair	DS, NS, CS	Fair	Moderate	Fair	Adapted to arid sites subjected to seasonal saturated soils.
Betula occidentalis occidentalis Water birch	SF-Mtn.B.	Stream edges.	Good	NS	Excellent	Rapid	Excellent	Establishes well by transplanting, adapted to streambanks and bogs.
Ceanothus sanguineus Redstem ceanothus	SF-PP	Moist soils, seeps, well-drained soils.	Good	DS, NS, CS	Excellent	Rapid	Excellent	Not adapted to saturated soils but useful in planting disturbed streambanks.
Chrysothamnus nauseosus consimilis Thinleaf rubber rabbitbrush	Sage-V	Well-drained soils, sites occasionally flooded.	Good	DS, NS, CS	Excellent	Moderate	Fair	Suited to heavy saturated soils.
Cornus stolonifera stolonifera Redosier dogwood	SF-Mtn.B.	Stream edges and well-drained soils.	Good	DS, NS, CS, RC	Excellent	Rapid	Excellent	Easy to grow and establish, useful for disturbed sites, requires fresh aerated water.
Cratageus douglasii Douglas hawthorn	Asp.-Sage	Stream edges and well-drained soils.	Good	NS	Fair	Slow	Good	Slow growing, but well suited to disturbed streambanks.
Elaeagnus angustifolia Russian olive	Mtn.B.-V	Stream edges, seeps, flooded sites, and well-drained soils.	Excellent	DS, NS	Excellent	Rapid	Good	Easy to establish, can become weedy.
Elaeagnus commutata Silverberry	PJ-V	Stream edges and well-drained soils.	Excellent	NS, CS	Excellent	Rapid	Good	Easily established, grows rapidly, adapted to harsh sites.
Holodiscus discolor Rockspirea	SF-Mtn.B.	Well-drained and moist soils, occasional seeps.	Good	NC, CS	Fair	Moderate	Good	Erratic establishment, but suited to disturbed sites.

TABLE A.20 Trees and Shrubs Recommended for Planting in Disturbed Riparian Sites (*Continued*)

Species	Areas of occurrence		Establishment traits					
	Zones*	Habitat	Adaptation to disturbed sites	Methods of culture†	Seedling establishment	Growth rates	Soil stability value	Comments
Lonicera tatarica Tatarian honeysuckle	Mtn.B.-Sage	Well-drained and moist soils, occasional wet sites.	Excellent	NC, CS, DS	Excellent	Rapid	Good	Easily established, provides immediate cover, well adapted to different soil conditions.
Pachistima myrsinites Myrtle pachistima	SF-Asp.	Moist soils and seeps, requires some shade.	Fair	NS, CS	Fair	Slow	Good	Common to upland slopes, not well adapted to disturbances.
Physocarpus malvaceus Mallow ninebark	SF-Asp.	Moist and well-drained soils.	Fair	NS, CS	Fair	Moderate	Good	Requires good sites.
Populus angustifolia Narrowleaf cottonwood	Asp.-Sage	Well-drained and wet sites, edges of streams, ponds, bogs.	Good	NS, CS, RC	Good	Rapid	Good	Establishes easily, grows rapidly.
Populus fremontii fremontii Fremont cottonwood	Mtn.B.-V	Moist soils, seeps, frequently wet sites.	Good	NS, CS, RC	Good	Rapid	Good	Establishes easily, grows rapidly, furnishes good cover.
Populus tremuloides Quaking aspen	SF-Asp.	Well-drained and moist soils, occasionally occurs at edges of streams.	Fair	NS, CS, RC	Good	Rapid	Good	Considerable ecotypic differences, not well suited to highly disturbed sites, occupies wide range of moisture.
Potentilla fruticosa Bush cinquefoil	Alp.-PP	Stream edges, wet meadows.	Excellent	NS, CS	Good	Moderate	Excellent	Valuable species for riparian disturbances, establishes well and provides excellent site stability.

Prunus virginiana melanocarpa Black chokecherry	SF-PJ	Well-drained, moist soils, occasionally occurs at streams' edges.	Fair	NS, CS, RC	Good	Moderate	Good	Widely adapted, larger transplant stock establishes and grows rapidly.
Rhamnus purshiana Cascara buckthorn	SF-PP	Moist soils, frequently wet sites.	Fair	NS, CS	Fair	Moderate	Good	Limited plantings, plants perform well on disturbed sites.
Ribes aureum Golden current	Asp.-Sage	Well-drained moist sites.	Excellent	NS, CS	Excellent	Excellent	Good	Widely adapted, easily established, excellent site stability.
Rosa woodsii Woods rose	Asp.-Mtn.B.	Moist and well-drained soils, seeps and frequently streambanks.	Excellent	NS, CS, W, RC	Excellent	Moderate	Good	Widely adapted, easily established, excellent site stability, principal species for riparian disturbances.
Rubus spp.	Asp.-PP	Well-drained soils, frequently wet sites	Excellent	NS, CS, W, RC	Excellent	Moderate	Good	Well adapted to eroded sites, limited range of distribution.
Salix (see Table A.21)								
Sambucus racemosa pubens microbotrys Red elder	Asp.-PP	Moist sites, occasional seeps and streambanks.	Good	NS, CS	Fair	Moderate	Good	Adapted to restricted sites, establishes slowly on disturbed sites.
Sarcobatus vermiculatus Black greasewood	SDS-V	Sites with shallow water tables, occasionally flooded sites.	Good	NS, W	Fair	Slow	Good	Difficult to establish, well adapted to valley bottoms and salty soils.
Shepherdia argentea Silver buffaloberry	Mtn.B-V	Well-drained sites, edges of streams and ponds.	Good	NS	Good	Moderate	Good	Adapted to valley bottoms and saline soils.
Sorbus scopulina scopulina Green's mountain ash	SF-Asp.	Moist soils, occasional seeps and stream bottoms.	Fair	NS, CS	Fair	Slow	Good	Not well adapted to disturbed soils, establishes slowly.

TABLE A.20 Trees and Shrubs Recommended for Planting in Disturbed Riparian Sites (*Continued*)

	Areas of occurrence		Establishment traits					
Species	Zones*	Habitat	Adaptation to disturbed sites	Methods of culture†	Seedling establishment	Growth rates	Soil stability value	Comments
Symphoricarpos albus Common snowberry	SF-Asp.	Moist sites and well-drained soils.	Good	NS, CS, W, RC	Fair	Moderate	Excellent	Not well suited to extreme disturbed soils, once established grows well, plant large 1-0 or 2-0 stock.
Symphoricarpos occidentalis Western snowberry	SF-Mtn.B.	Moist sites, occasionally streambanks and valley bottoms.	Good	NS, CS, W, RC	Fair	Slow	Excellent	Plants not well adapted to disturbed soils, provides excellent stability and spreads well.
Symphoricarpos oreophilus Mountain snowberry	Asp.-Sage	Well-drained soils, edges of streams.	Good	NS, CS, W, RC	Fair	Slow	Excellent	Plants not well adapted to disturbed soils, provides excellent stability and spreads well.

*Alp. = alpine; SF = spruce-fir; Asp. = aspen; PP = ponderosa pine; Mtn.B. = mountain brush; PJ = pinyon-juniper; Sage = big sagebrush; SDS = salt desert shrub; V = valley bottoms.

†DS = direct seeding; RC = rooted cuttings; NS = nursery-grown seedling; CS = container-grown seeding; W = wilding.

SOURCE: Platts et al. (1987).

TABLE A.21 Willows Recommended for Planting in Riparian Sites

Species	Areas of adaptation		Origin of roots	Prevalence of roots	Number of days needed		Comments
	Zones	Habitat			Root formation	Stem formation	
Salix amygdaloids Peachleaf willow	Aspen-big sagebrush	Stream edges, pond margins, soils saturated seasonally	Callus cut	Moderate	10–20	10	Moderate rooting capabilities
Salix bebbiana Bebb willow	Spruce-fir–aspen	Edges of streams, occasionally well-drained soils	Roots throughout entire length of stem	Moderate	10	10–20	Roots freely
Salix boothii	Aspen–sagebrush	Stream edges and standing water, confined to wet soils	Roots mostly at lower one-third of stem	Abundant	10–15	10–15	Roots freely
Salix brachycarpa Barrenground willow	Subalpine–spruce-fir	Wet sites and well-drained soils	Roots throughout entire length of stem	Abundant	15–20	15–25	Roots freely
Salix drummondiana Drummond willow	Spruce-fir–upper sagebrush	Edges of streams and ponds	Roots throughout entire length of stem	Abundant	10	10	Roots freely
Salix exigua Coyote (sandbar) willow	Spruce-fir–sagebrush	Edges of streams, wet sites, sometimes well-drained soils	Roots throughout entire length of stem	Moderate	10–15	10	Easily rooted
Salix geyeriana Geyer willow	Subalpine–aspen–upper sagebrush	Edges of streams, frequent wet meadows	Roots throughout entire length of stem	Few to moderate	10	10–15	Fair rooting capabilities
Salix glauca Grayleaf willow	Subalpine–spruce-fir	Wet and dry sites, widely distributed, occupies seeps and edges of snowbanks	Roots throughout entire length of stem	Few to moderate	10	10	Requires special treatment to root

TABLE A.21 Willows Recommended for Planting in Riparian Sites (*Continued*)

Species	Areas of adaptation		Origin of roots	Prevalence of roots	Number of days needed		Comments
	Zones	Habitat			Root formation	Stem formation	
Salix lasiandra Pacific willow	Aspen-upper sagebrush	Wet soils, edges of streams and ponds	Roots throughout entire length of stem	Abundant	10	10–15	Easily rooted
Salix lasiolepis Arroyo willow	Aspen–mountain brush	Restricted to stream edges	Callus and lower one-third of stem	Few to many	10	10	Erratic rooting habits
Salix lutea Shining willow	Aspen–sagebrush	Mostly along streams, can occur on sites that remain dry for short periods	Entire stem section, most abundant at lower one-third	Moderate	10	10	Roots easily
Salix planifolia Tealeaf willow	Subalpine–aspen	Wet sites, edges of streams, wet meadows	Roots throughout entire length of stem	Few to moderate	10	10–15	Fair rooting capabilities
Salix scouleriana Scouler willow	Spruce-fir–aspen	Well-drained soils, forest understory	Callus cut	Moderate	10–15	10–15	Requires special treatment to root
Salix wolfii Wolf willow	Spruce-fir–aspen	Stream edges and ponds	Roots throughout entire length of stem	Few to moderate	10–15	10–15	Erratic rooting

SOURCE: Platts et al. (1987).

TABLE A.22 A Summary of Adaptations for Regional Plants in Coastal Marshes*

Species	At-lantic	Florida	Gulf	North Pacific	South Pacific	Great Lakes
Major species						
Smooth cordgrass *Spartina alterniflora*	1	1	1	5	5	
Saltmeadow cordgrass *Spartina patens*	1	1	1			
Pacific cordgrass *Spartina foliosa*					1	
Gulf cordgrass *Spartina spartinae*			9			
Black needlerush *Juncus roemerianus*	2	2	2			
Big cordgrass *Spartina cynosuroides*	3	3	3			
Bluejoint *Calamgrostis canadensis*						4
Common reed *Phragmites australis*	5	5	5			1
Saltgrass *Distichlis spicata*	2	2	2	2	2	
Sedge *Carex* spp.				1		
Tufted hairgrass *Deschampsia caespitosa*				1		
Seaside arrowgrass *Triglochin maritima*				4	4	
Pickleweed *Salicornia pacifica*				6	6	
Red mangrove *Rhizophora mangle*		7				
Black mangrove *Avicennia germinans*		7				
Minor species						
Sea oxeye *Borrichia frutescens*	8	8	8			
Marsh elder *Iva frutescens*	8	8	8			
Pickleweed *Salicornia pacifica*	8	8	8			
Sea blite *Suaeda* spp.	8	8	8			
Groundsel tree *Baccharis halimifolia*	8	8	8			
Dropseed *Sporobolus* spp.	8	8	8			
Panic grass *Panicum* spp.	8	8	8			
Common threesquare *Scirpus americanus*				8	8	
Jaumea *Jaumea carnosa*				8	8	

TABLE A.22 A Summary of Adaptations for Regional Plants in Coastal Marshes* (*Continued*)

Species	Atlantic	Florida	Gulf	North Pacific	South Pacific	Great Lakes
Sand spurrey *Spergularia* spp.				8	8	
Bulrush *Scirpus* spp.					8	4
Spikerush *Eleocharis* spp.					8	8
White mangrove *Laguncularia racemosa*		8				

*1 = dominant planted species; 2 = widely distributed, difficult to plant; 3 = locally abundant, difficult to plant; 4 = valuable, planting methods undeveloped; 5 = easily planted but possible pest; 6 = valuable and easily planted, usually volunteers; 7 = dominant species, better planted after initial stabilization; 8 = plantable but usually volunteers; 9 = replaces saltmeadow cordgrass on heavy-textured substrates.

SOURCE: Woodhouse (1979).

TABLE A.23 Regional Summary of Planting in Coastal Marshes—Atlantic, Gulf, Peninsular Florida, and Pacific

	Atlantic, Gulf, and Peninsular Florida			
	Smooth cordgrass (*Spartina alterniflora*)			
Propagule	Sprig[a]	Seedling	Plug	Seeds
Elevation	Low tidal range, MLW to MHW High tidal range, MTL to MHW			Upper 30–50% of tidal range
Fetch, km	< 5	< 5	< 5	< 1.0
Salinity, 0/00	< 40	< 40	< 40	< 35
Density	Marsh development: 1-m centers Stabilization: 3- to 6-dm centers			50–100 viable seeds/m^2
Date	North, early spring to early summer South, later winter to mid-summer			Spring
Fertilization	N and P first year on sandy or nutrient-deficient substrates			Somewhat more responsive than transplants

	Saltmeadow cordgrass (*Spartina patens*)		Gulf cordgrass[b] (*Spartina spartinae*)
Propagule	Sprig[c]	Seedling	Sprig[c]
Elevation	MHW to extreme high tide		MHW to extreme high tide
Density	0.5–1.0 m on centers		0.5–1.0 m on centers
Date	Spring and early summer		Spring and summer
Fertilization	N and P on sandy or nutrient-deficient substrates		None

TABLE A.23 Regional Summary of Planting in Coastal Marshes—Atlantic, Gulf, Peninsular Florida, and Pacific (*Continued*)

	Red mangrove[d] (*Rhizophora mangle*)		Black mangrove[d] (*Avicennia germinans*)	
Propagule	1–2 yr	3–6 yr	1–2 yr	3–6 yr
Elevation	MTL to MHT	MLW to MHT	MTL to extreme high tide	
Fetch, km				
Stabilized	< 5	< 10	< 8	< 12
Unstabilized	< 1	< 2	< 2	< 4
Salinity, 0/00	15–40	15–40	5–50	5–50
Density, m^2/plant	1	2–3	1	2–3
Date	Seedlings in late February through March; large plants, year round			

North Pacific

	Sedge (*Carex* spp.)	Tufted hairgrass (*Deschampsia caespitosa*)	Pickleweed (*Salicornia pacifica*)	
Propagule	Sprig[e]	Sprig[e]	Sprig	Seed
Elevation	MT to MHHW	MLHW to EHT	MLHW to EHT	
Fetch, km	< 5	< 5	< 5	< 2
Salinity, 0/00	0–35	0–35	10–50	10–50
Date	Spring	Spring	Spring	Spring
Density, m^2/plant	0.25–1.0	0.25–1.0	0.1–1.0	50–100 viable seeds/m^2

South Pacific

	Pacific cordgrass (*Spartina foliosa*)				Pickleweed (*Salicornia pacifica*)	
Propagule	Sprig[f]	Seed-ling	Plug	Seed	Sprig[g]	Seed
Elevation	MTL to MHW			MLHW to MHW	MLHW to extreme high tide	
Fetch, km	< 5	< 5	< 5	< 1	< 8	< 2
Salinity, 0/00	< 40		< 40	< 20	< 50	< 50
Date	Spring to summer			Spring	Spring and summer	Spring
Density m^2/plant	0.5–1		0.5–1	50–100 viable seeds/m^2	1 m or rooted 3 dm	50–100 viable seeds/m^2

[a]Intact single-culm plant with roots, not a fragment as is often implied by this term.
[b]Gulf coast only, on cohesive substrates.
[c]Intact multiple-culm plants with roots.
[d]Peninsular Florida only.
[e]Intact multiple-stem plant with roots.
[f]Intact single stem with roots.
[g]Rooted or unrooted cuttings.
SOURCE: Woodhouse (1979).

TABLE A.24 Regional Summary of Transplanting and Fertilization of Sand Dunes

	Planting				Fertilization	
Species	Date	Depth, cm	Stems per hill	Spacing, cm	First year, kg/ha	Maintenance
North Atlantic						
American beachgrass *Ammophila breviligulita*	Feb–Apr	20–35	1–5	45–60 or graduated	100–150 N 30–50 P_2O_5	⅓ first year to none
Bitter panicum *Panicum amarum*	Mar–May	20–35	1	In mixture	100–150 N 30–50 P_2O_5	⅓ first year to none
South Atlantic						
American beachgrass*	Nov–Mar	20–30	1–3	45–60 or graduated	100–150 N 30–50 P_2O_5	30–50 kg/ha N 1–3 yr intervals
Bitter panicum	Mar–Jun	20–35	1	45–60 or graduated	100–150 N 30–50 P_2O_5	30–50 kg/ha N 1–3 yr intervals
Sea oats *Uniola paniculata*	Feb–Apr	25–35	1	In mixture	100–150 N 30–50 P_2O_5	30–50 kg/ha N 1–3 yr intervals
Saltmeadow cordgrass *Spartina patens*	Feb–May	15–30	5–10	45–60 or graduated	100–150 N 30–50 P_2O_5	30–50 kg/ha N 1–3 yr intervals
Gulf						
Bitter panicum	Feb–Jun	20–30	1	60–90 or graduated	100 N 30 P_2O_5	According to growth
Sea oats	Jan–Feb	20–35	1	60–90 or graduated	100 N 30 P_2O_5	According to growth
North Pacific						
European beachgrass *Ammophila arenaria*	< 15°C	25–35	3–5	45 or graduated	40–60 N early April	According to growth
American beachgrass	Jan–Apr	25–35	1–3	45 or graduated	40–60 N early April	According to growth
South Pacific						
European beachgrass	< 15°C	25–35	3–5	45 or graduated	40–60 N spring	According to growth
Ice plant *Mesembryanthemum crystallinum*	Cool, wet	10–15	1	60 or broadcast	40–60 N spring	According to growth (stabilization only)
Great Lakes						
American beachgrass	Feb–May	20–35	1–3	45–60 or graduated	100–150 N 30–50 P_2O_5 and K_2O	According to growth

*Carolinas' coast only.

SOURCE: Woodhouse (1978).

TABLE A.25 Transplanting Times Recommended for Seagrasses

Species	Location	Recommended time
Eelgrass (*Zostera marina*)	Alaska and Atlantic coast north of North Carolina	March (mild winters) or May (severe winters) to late July
	Beaufort, North Carolina, south of Atlantic coast	Late September to early December
	Washington State to southern California, Pacific coast	January to May (but can be done throughout the year)
Shoalgrass(*Halodule wrightii*), widgeongrass (*Ruppia maritima*)	Gulf coast, Atlantic coast south of Cape Canaveral, Florida	Any time during year
Turtle grass (*Thalassia testudinum*), manatee grass (*Syringodium filiforme*)	Gulf coast, Atlantic coast south of Cape Canaveral, Florida	Plugs: December to April; seedlings of turtle grass: August to November as they are produced in the field

SOURCE: Phillips (1980b).

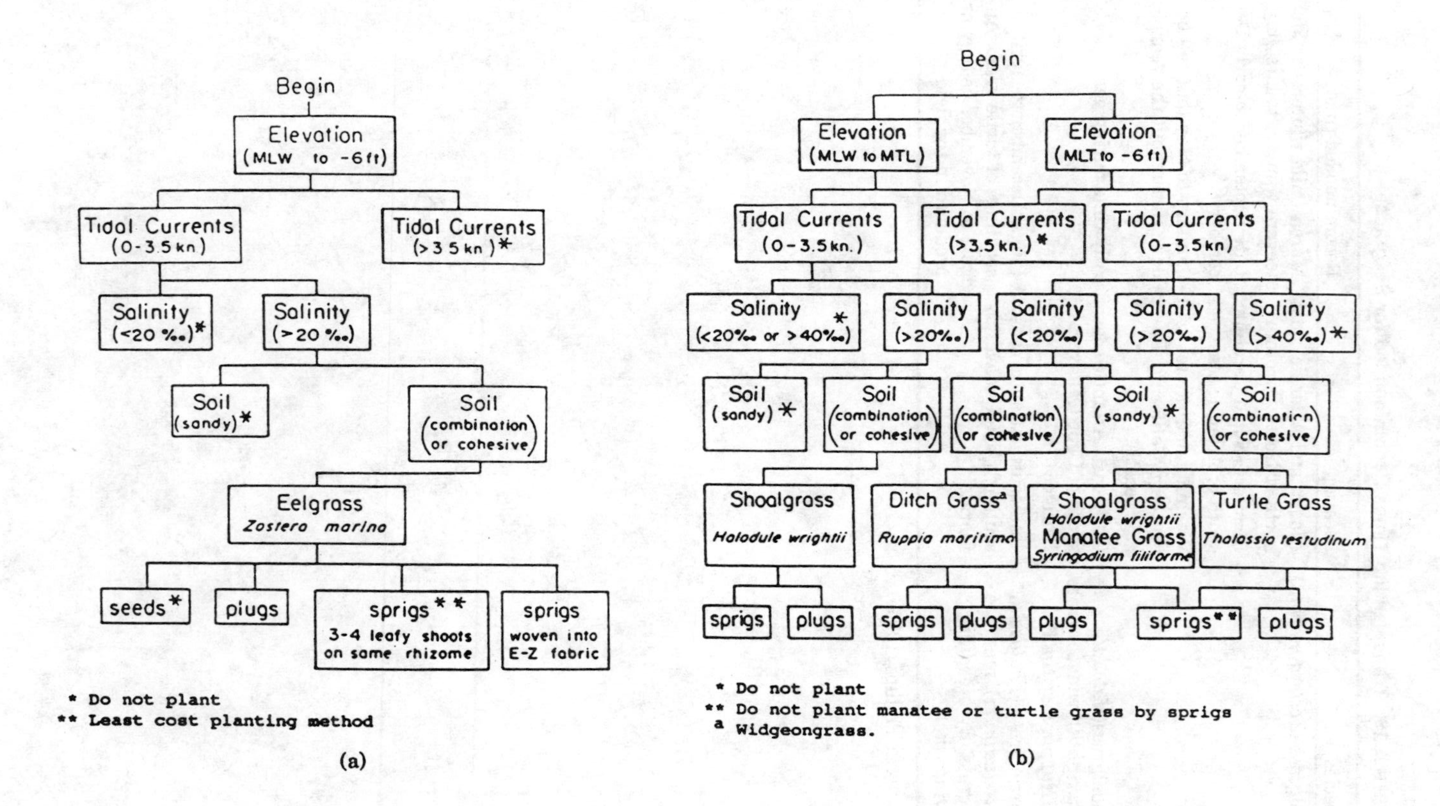
Begin
Elevation (MLW to -6 ft)
Tidal Currents (0-3.5 kn)
Tidal Currents (>3.5 kn)*
Salinity (<20 ‰)*
Salinity (>20 ‰)
Soil (sandy)*
Soil (combination or cohesive)
Eelgrass
Zostera marina
seeds*
plugs
sprigs**
3-4 leafy shoots on same rhizome
sprigs
woven into E-Z fabric
* Do not plant
** Least cost planting method
Begin
Elevation (MLW to MTL)
Elevation (MLT to -6 ft)
Tidal Currents (0-3.5 kn.)
Tidal Currents (>3.5 kn.)*
Tidal Currents (0-3.5 kn)
Salinity (<20‰ or >40‰)*
Salinity (>20‰)
Salinity (<20‰)
Salinity (>20‰)
Salinity (>40‰)*
Soil (sandy)*
Soil (combination or cohesive)
Soil (combination or cohesive)
Soil (sandy)*
Soil (combination or cohesive)
Shoalgrass
Halodule wrightii
Ditch Grass[a]
Ruppia maritima
Shoalgrass
Halodule wrightii
Manatee Grass
Syringodium filiforme
Turtle Grass
Thalassia testudinum
sprigs
plugs
sprigs
plugs
plugs
sprigs**
plugs
* Do not plant
** Do not plant manatee or turtle grass by sprigs
a Widgeongrass.

(a)

(b)

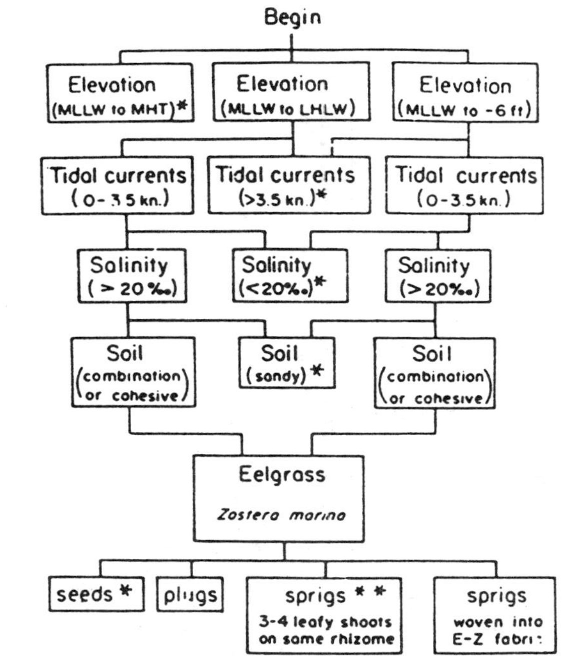

(c)

Figure A.2 Key to planting decisions for seagrasses: (*a*) Atlantic coast north of Beaufort, North Carolina; (*b*) Atlantic coast south of Beaufort, North Carolina, to Florida and along the Gulf Coast; and (*c*) Pacific coast (*Phillips 1980b*).

TABLE A.26 Nesting Habitats of Colonial Waterbirds on Artificial (Dredged) Earthen Islands

	Nesting habitats*							
Species	BS	SH	MH	DH	HSh	ShT	ShF	F
White pelican	x	x			x			
Brown pelican	x	x			x	x	x†	x†
Double-crested cormorant							x	x†
Olivaceous cormorant								x
Anhinga							x	x
Great blue heron				x	x	x†	x†	x†
Green heron				x	x†	x	x	x
Little blue heron					x	x†	x†	x†
Cattle egret					x	x	x†	x†
Reddish egret				x	x	x†	x†	x
Great egret				x	x	x	x†	x†
Snowy egret				x	x	x†	x†	x
Louisiana heron				x	x	x†	x†	x
Black-crowned night heron			x		x	x†	x†	x†
Yellow-crowned night heron					x	x	x†	x†
White-faced ibis				x	x†	x		
Glossy ibis				x	x	x	x	x
White ibis						x†	x†	x
Roseate spoonbill						x	x†	x†
Glaucous-winged gull‡		x	x†	x†	x			
Great black-backed gull		x	x					
Herring gull‡		x	x†	x	x			
Western gull‡		x	x†	x†	x			
Ring-billed gull	x	x†	x†	x	x			
Laughing gull‡			x	x†	x			
Gull-billed tern‡	x	x†	x					
Forster's tern§	x	x	x	x†				
Common tern	x	x†	x†					
Roseate tern			x	x				
Least tern	x†	x						
Royal tern	x†	x						
Sandwich tern	x†	x						
Caspian tern	x	x†	x					
Black tern§	x	x	x†	x†				
Black skimmer	x†	x†	x					

*Key to symbols: BS = bare substrate, SH = sparse herb, MH = medium herb, DH = dense herb, HSh = herb shrub, ShT = shrub thicket, ShF = shrub-forest, F = forest, x = species nests in the habitat.

†Nesting habitats used most often.

‡Objects such as clumps of vegetation, logs, drift material, cobble, etc., readily accepted in nesting habitat.

§Primarily a marsh nesting species.

SOURCE: Soots and Landin (1978).

TABLE A.27 Noncolonial Birds Nesting on Artificial (Dredged) Earthen Islands

Canada goose	Yellow-billed cuckoo
Mallard	Groove-billed ani
Black duck	Short-eared owl
Mottled duck	Common nighthawk
Gadwall	Scissor-tail flycatcher
Marsh hawk	Long-billed marsh wren
Osprey	Short-billed marsh wren
Kestrel	Fish crow
Bobwhite quail	Mockingbird
American bittern	Brown thrasher
Least bittern	Ruby-crowned kinglet
Sora	Loggerhead shrike
Black rail	Yellow warbler
Clapper rail	Chestnut-sided warbler
King rail	Prairie warbler
Common gallinule	Louisiana waterthrush
American oystercatcher	Yellowthroat
American avocet	Eastern meadowlark
Black-necked stilt	Red-winged blackbird
Piping plover	Boat-tailed grackle
Snowy plover	Great-tailed grackle
Wilson's plover	Common grackle
Kildeer	Painted bunting
Spotted sandpiper	Savannah sparrow
Willet	Grasshopper sparrow
Sooty tern	Seaside sparrow
Mourning dove	Field sparrow
Ground dove	Song sparrow

SOURCE: Soots and Landin (1978).

TABLE A.28 Common and Scientific Names of Animals

Alphabetical by common name		Alphabetical by scientific name	
Common name	Scientific name	Scientific name	Common name
Alligator	*Alligator mississippiensis*	*Actitis macularia*	Spotted sandpiper
Alligatorweed flea beetle	*Agasicles hygrophila*	*Agasicles hygrophila*	Alligatorweed flea beetle
American avocet	*Recurvirostra americana*	*Agelauis phoeniceus*	Red-winged blackbird
American bittern	*Botauru lentiginosus*	*Aix sponsa*	Wood duck
American oystercatcher	*Haematopus palliatus*	*Ajaia ajaja*	Roseate spoonbill
American widgeon (baldpate)	*Anas americana*	*Alligator mississippiensis*	Alligator
Anhinga	*Anhinga anhinga*	*Ammodramus savannarum*	Grasshopper sparrow
Ant	*Crematogaster atkinsoni*	*Ammospiza maritima*	Seaside sparrow
Arctic tern	*Sterna paradisaea*	*Anas acuta*	Pintail
Atlantic puffin	*Fratercula arctica*	*Anas americana*	American widgeon (baldpate)
Bald eagle	*Haliaeetus leucocephalus*	*Anas clypeata*	Shoveler
Barn owl	*Tyto alba*	*Anas Crecca*	Green-winged teal
Barn swallow	*Hirundo rustica*	*Anas discors*	Blue-winged teal
Barrow's goldeneye	*Bucephala islandica*	*Anas fulvigula*	Mottled duck
Bass	*Micropterus* spp.	*Anas platyrhynchos*	Mallard
Bear	*Ursus* spp.	*Anas rubripes*	Black duck
Beaver	*Castor canadensis*	*Anas strepera*	Gadwall
Black duck	*Anas rubripes*	*Anhinga anhinga*	Anhinga
Black rail	*Laterallus jamaicensis*	*Anser albifrons*	White-fronted goose
Black skimmer	*Rynchops niger*	*Ardea herodias*	Great blue heron
Black tern	*Chilidonias niger*	*Asio flammeus*	Short-eared owl
Black-backed gull	*Larus marinus*	*Aythya affinis*	Lesser scaup
Black-bellied whistling duck	*Dendrocygna autumnalis*	*Aythya americana*	Redhead
Black-crowned night heron	*Nycticorax nycticorax*	*Aythya collaris*	Ringneck duck
Black-headed gull	*Larus ridibundus*	*Aythya marila*	Greater scaup
Black-necked stilt	*Himantopus jamaicensis*	*Aythya valisineria*	Canvasback
Blue-winged teal	*Anas discors*	*Botauru lentiginosus*	American bittern
Boat-tailed grackle	*Quiscalus major*	*Branta bernicla*	Brant
Bobwhite	*Colinus virginianus*		
Botulism	*Clostridium botulinum*		

Brant	*Branta bernicla*
Brown pelican	*Pelecanus occidentalis*
Brown thrasher	*Toxostoma rufum*
Buffalo	*Ictiobus cyprinellus*
Bufflehead	*Bucephala albeola*
Bullhead	*Ictalurus* spp.
Burrowing isopod	*Sphaeroma quoyana*
California gull	*Larus californicus*
Canada goose	*Branta canadensis*
Canvasback	*Aythya valisineria*
Carp	*Cyprinus carpio*
Catfish	*Ictalurus spp.*
Cattle egret	*Bulbulcus ibis*
Chestnut-sided warbler	*Dendroica pensylvanica*
Clapper rail	*Rallus longirostris*
Cliff swallow	*Petrochelidon pyrrhonota*
Common crow	*Corvus brachyrhynchos*
Common goldeneye	*Bucephala clangula*
Common grackle	*Quiscalus quiscula*
Common loon	*Gavia immer*
Common merganser	*Mergus merganser*
Common moorhen (common gallinule)	*Gallinula chloropos*
Common nighthawk	*Chordeiles minor*
Common tern	*Sterna hirundo*
Coot	*Fulica americana*
Cormorant	*Phalacrocoracidae* spp.
Cottontail	*Sylvilagus* spp.
Coyote	*Canis latrans*
Crane	*Grus* spp.
Crayfish	*Procambarus clarkii*
Deer	*Odocoileus* spp.
Double-crested cormorant	*Phalacrocorax auritus*

Branta canadensis	Canada goose
Bucephala albeola	Bufflehead
Bucephala clangula	Common goldeneye
Bucephala islandica	Barrow's goldeneye
Bucephala spp.	Goldeneye
Bulbulcus ibis	Cattle egret
Larus californicus	California gull
Butorides striatus	Green heron
Calidris melanotos	Pectoral sandpiper
Canis latrans	Coyote
Capella gallinago	Snipe (common snipe)
Carassius auratus	Goldfish
Casmerodius albus	Great egret
Cassidix mexicanus	Great-tailed grackle
Castor canadensis	Beaver
Catophophorus semipalmatus	Willet
Charadrius alexandrinus	Snowy plover
Charadrius melodus	Piping plover
Charadrius vociferus	Killdeer
Charadrius wisonia	Wilson's plover
Chelydra serpentina	Snapping turtle
Chen caerulescens	Snow goose
Chilidonias niger	Black tern
Chordeiles minor	Common nighthawk
Circus cyaneus	Harrier (northern harrier, marshhawk)
Citellus franklini	Franklin ground squirrel
Clostridium botulinum	Botulism
Coccyzus americanus	Yellow-billed cuckoo
Colinus virginianus	Bobwhite
Columbigallina passerina	Ground dove
Corvus brachyrhynchos	Common crow

TABLE A.28 Common and Scientific Names of Animals *(Continued)*

Alphabetical by common name		Alphabetical by scientific name	
Common name	Scientific name	Scientific name	Common name
Eastern meadowlark	*Sturnella magna*	*Corvus corax*	Raven
Elder	*Somateria* spp.	*Corvus ossifragus*	Fish crow
Everglade kite	*Rostrhamus sociabilis*	*Crematogaster atkinsoni*	Ant
Field sparrow	*Spizella pusilla*	*Cristothorus platensis*	Short-billed marsh wren (sedge wren)
Fire ant	*Solenopsis invicta*		
Fish crow	*Corvus ossifragus*	*Ctenopharyngodon idella*	Grass carp
Forster's tern	*Sterna forsteri*	*Cyprinus carpio*	Carp
Fox squirrel	*Sciurus niger*	*Dendrocygna autumnalis*	Black-bellied whistling duck
Franklin ground squirrel	*Citellus franklini*	*Dendroica discolor*	Prairie warbler
Gadwall	*Anas strepera*	*Dendroica pensylvanica*	Chestnut-sided warbler
Gambusia minnow	*Gambusia affinis*	*Dendroica petechia*	Yellow warbler
Glaucous-winged gull	*Larus flaucescens*	*Dichromanassa rufescens*	Reddish egret
Glossy ibis	*Plegadis falcinellus*	*Didelphis marsupialis*	Opossum
Goldeneye	*Bucephala* spp.	*Egretta thula*	Snowy egret
Goldfish	*Carassius auratus*	*Eremophila alpestris*	Horned lark
Grackle	*Quiscalus* spp.	*Erotophaga sulcirostris*	Groove-billed ani
Grass carp	*Ctenopharyngodon idella*	*Evdocimus albus*	White ibis
Grasshopper sparrow	*Ammodramus savannarum*	*Falco sparverius*	Kestrel
Great blue heron	*Ardea herodias*	*Florida caerulea*	Little blue heron
Great egret	*Casmerodius albus*	*Fratercula arctica*	Atlantic puffin
Great-tailed grackle	*Cassidix mexicanus*	*Fulica americana*	Coot
Greater prairie chicken	*Tympanuchus cupido*	*Gallinula chloropos*	Common moorhen (common gallinule)
Greater scaup	*Aythya marila*		
Green heron	*Butorides striatus*	*Gambusia affinis*	Gambusia minnow (mosquito fish)
Green-winged teal	*Anas crecca*		
Grove-billed ani	*Erotophaga sulcirostris*	*Gasterosteus aculatus*	Threespine stickleback
		Gavia immer	Common loon
		Gelochelidon nitotica	Gull-billed tern

Ground dove	*Columbigallina passerina*
Gull	*Larus spp.*
Gull-billed tern	*Gelochelidon nitotica*
Harrier (northern harrier, marshhawk)	*Circus cyaneus*
Herring gull	*Larus argentatus*
Hooded merganser	*Lophodytes cucullatus*
Horned lark	*Eremophila alpestris*
Kestrel	*Falco sparverius*
Killdeer	*Charadrius vociferus*
King rail	*Rallus elegans*
Kingfisher	*Megaceryle alcyon*
Laughing gull	*Larus atricilla*
Leach's petrel	*Oceanodroma leucorhoa*
Least bittern	*Ixobuychus exilis*
Least tern	*Sterna albifrons*
Lesser scaup	*Aythya affinis*
Lesser yellowlegs	*Tringa flavipes*
Little blue heron	*Florida caerulea*
Loggerhead shrike	*Lanius ludovicanus*
Long-billed Marshwren	*Telmatodytes palustris*
Louisiana heron	*Hydranassa violacea*
Louisiana waterthrush	*Seiurus motacilla*
Mallard	*Anas platyrhynchos*
Manatee	*Trichechus manatus*
Mink	*Mustela vision*
Mockingbird	*Mimus polyglottos*
Mosquito fish	*Gambusia affinis*
Mottled duck	*Anas fulvigula*
Mourning dove	*Zenaida macroura*
Mullet	*Mugil* spp.

Geothypis trichas	Yellowthroat
Grus spp.	Crane
Haematopus palliatus	American oystercatcher
Haliaeetus leucocephalus	Bald eagle
Himantopus jamaicensis	Black-necked stilt
Hirundo rustica	Barn swallow
Hydranassa violacea	Louisiana heron
Ictalurus spp.	Bullhead
Ictalurus spp.	Catfish
Ictiobus cyprinellus	Buffalo
Ischadium demissium	Ribbed mussel
Ixobuychus exilis	Least bittern
Lanius ludovicanus	Loggerhead shrike
Larus argentatus	Herring gull
Larus atricilla	Laughing gull
Larus californicus	California gull
Larus delewarensis	Ring-billed gull
Larus glaucescens	Glaucous-winged gull
Larus marinus	Black-backed gull
Larus occidentalis	Western gull
Larus ridibundus	Black-headed gull
Larus spp.	Gull
Laterallus jamaicensis	Black rail
Lophodytes cucullatus	Hooded merganser
Lutra canadensis	Otter
Maleagris gallopavo	Turkey
Marmota monax	Woodchuck
Megaceryle alcyon	Kingfisher
Melanitta fussa	White-winged scoter
Melospiza melodia	Song sparrow
Mephitis mephitis	Striped skunk
Mergus merganser	Common merganser

TABLE A.28 Common and Scientific Names of Animals *(Continued)*

Alphabetical by common name		Alphabetical by scientific name	
Common name	Scientific name	Scientific name	Common name
Muskrat	*Ondatra zibethicus*	*Mergus serrator*	Red-breasted merganser
Nutria	*Myocaster coypu*	*Micropterus* spp.	Bass
Olivaceous cormorant	*Phalacrocorax olivaceus*	*Mimus polyglottos*	Mockingbird
Opossum	*Didelphis marsupialis*	*Mugil* spp.	Mullet
Osprey	*Pandion haliaetus*	*Muscivora forfic*	Scissor-tail flycatcher
Otter	*Lutra canadensis*	*Mustela vison*	Mink
Painted bunting	*Passerina ciris*	*Myocaster coypu*	Nutria
Pectoral sandpiper	*Calidris melanotos*	*Neochetina eichhorniae*	Water hyacinth beetle
Phoebe	*Sayornis phoebe*	*Nyctanassa violacea*	Yellow-crowned night heron
Pintail	*Anas acuta*	*Nycticorax nycticorax*	Black-crowned night heron
Piping plover	*Charadrius melodus*	*Oceanodroma leucorhoa*	Leach's petrel
Prairie warbler	*Dendroica discolor*	*Odocoileus* spp.	Deer
Raccoon	*Procyon lotor*	*Olor* spp.	Swan
Rail	*Rallus* spp.	*Ondatra zibethicus*	Muskrat
Rat	*Rattus* spp.	*Oxyura jamaicensis*	Ruddy duck
Raven	*Corvus corax*	*Pandion haliaetus*	Osprey
Red fox	*Vulpes vulpes*	*Passerculus sandwichensis*	Savana sparrow
Red-breasted merganser	*Mergus serrator*	*Passerina ciris*	Painted bunting
Red-winged blackbird	*Agelauis phoeniceus*	*Pelecanus erythrorhynchos*	White pelican
Reddish egret	*Dichromanassa rufescens*	*Pelecanus occidentalis*	Brown pelican
Redhead	*Aythya americana*	*Petrochelidon pyrrhonota*	Cliff swallow
Ribbed mussel	*Ischadium demissium*	*Phalacrocoracidae* spp.	Cormorant
Ring-billed gull	*Larus delewarensis*	*Phalacrocorax auritis*	Double-crested cormorant
Ringneck duck	*Aythya collaris*	*Phalacrocorax olivaceus*	Olivaceous cormorant
Ringneck pheasant	*Phasianus colchicus*	*Phasianus colchicus*	Ringneck pheasant
Robin	*Turdus migratorius*	*Physa* spp.	Snail
Roseate spoonbill	*Ajaia ajaja*	*Plegadis chihi*	White-faced ibis
Roseate tern	*Sterna dougallii*		

Royal tern	*Sterna maxima*
Ruby-crowned kinglet	*Regulus calendula*
Ruddy duck	*Oxyura jamaicensis*
Sandwich tern	*Sterna sandivicansis*
Savana sparrow	*Passerculus sandwichensis*
Scissor-tail flycatcher	*Muscivora forfic*
Seaside sparrow	*Ammospiza maritima*
Short-billed marsh wren (sedge wren)	*Cristothorus platensis*
Short-eared owl	*Asio flammeus*
Shoveler	*Anas clypeata*
Snail	*Physa* spp.
Snapping turtle	*Chelydra serpentina*
Snipe (common snipe)	*Capella gallinago*
Snow goose	*Chen caerulescens*
Snowy egret	*Egretta thula*
Snowy plover	*Charadrius alexandrinus*
Song sparrow	*Melospiza melodia*
Sooty tern	*Sterna fuscata*
Sora	*Porzana carolina*
Spotted sandpiper	*Actitis macularia*
Starling	*Sternus vulgaris*
Striped skunk	*Mephitis mephitis*
Swan	*Olor* spp.
Threespine stickleback	*Gasterosteus aculatus*
Trout	*Salmo* spp., *Salvelinus* spp.
Turkey	*Maleagris gallopavo*
Water hyacinth beetle	*Neochetina eichhorniae*
Western gull	*Larus occidentalis*
White ibis	*Evdocimus albus*
White pelican	*Pelecanus erythrorhynchos*

Plegadis falcinellus	Glossy ibis
Porzana carolina	Sora
Procambarus clarkii	Crayfish
Procyon lotor	Racoon
Quiscalus major	Boat-tailed grackle
Quiscalus quiscula	Common grackle
Quiscalus spp.	Grackle
Rallus elegans	King rail
Rallus longirostris	Clapper rail
Rallus spp.	Rail
Rattus spp.	Rat
Recurvirostra americana	American avocet
Regulus calendula	Ruby-crowned kinglet
Rostrhamus sociabilis	Everglade kite
Rynchops niger	Black skimmer
Salmo spp., *Salvelinus* spp.	Trout
Sayornis phoebe	Phoebe
Sciurus niger	Fox squirrel
Seiurus motacilla	Louisiana waterthrush
Solenopsis invicta	Fire ant
Somateria spp.	Eider
Sphaeroma quoyana	Burrowing isopod
Spizella pusilla	Field sparrow
Sterna albifrons	Least tern
Sterna dougallii	Roseate tern
Sterna forsteri	Forster's tern
Sterna fuscata	Sooty tern
Sterna hirundo	Common tern
Sterna maxima	Royal tern
Sterna paradisaea	Arctic tern
Sterna sandivicansis	Sandwich tern
Sternus vulgaris	Starling
Sturnella magna	Eastern meadowlark

TABLE A.28 Common and Scientific Names of Animals *(Continued)*

Alphabetical by common name		Alphabetical by scientific name	
Common name	Scientific name	Scientific name	Common name
		Sus scrofa	Wild boar
White-faced ibis	*Plegadis chihi*	*Sylvilagus* spp.	Cottontail
White-fronted goose	*Anser albifrons*	*Telmatodytes palustris*	Long-billed marshwren
White-winged scoter	*Melanitta fussa*	*Toxostoma rufum*	Brown thrasher
Wild boar	*Sus scrofa*	*Trichechus manatus*	Manatee
Willet	*Catophophorus semipalmatus*	*Tringa flavipes*	Lesser yellowlegs
Wilson's plover	*Charadrius wisonia*	*Turdus migratorius*	Robin
Wood duck	*Aix sponsa*	*Tympanuchus cupido*	Greater prairie chicken
Woodchuck	*Marmota monax*	*Tyto alba*	Barn owl
Yellow warbler	*Dendroica petechia*	*Ursus* spp.	Bear
Yellow-billed cuckoo	*Coccyzus americanus*	*Vulpes vulpes*	Red fox
Yellow-crowned night heron	*Nyctanassa violacea*	*Xanthocephalus xanthocephalus*	Yellow-headed blackbird
Yellow-headed blackbird	*Xanthocephalus xanthocephalus*		
Yellowthroat	*Geothypis trichas*	*Zenaida macroura*	Mourning dove

TABLE A.29 Common and Scientific Names of Plants

Alphabetical by common name		Alphabetical by scientific name	
Common name	Scientific name	Scientific name	Common name
Alder	*Alnus* spp.	*Abies* spp.	Fir
Alfalfa	*Medicago sativa*	*Abies lasiocarpa*	Subalpine fir
Alkali (saltmarsh) bulrush	*Scirpus maritimus*	*Acacia smallii*	Huisache tree
Alkali sacaton	*Sporobolus airoides*	*Acer negundo*	Boxelder
Alligatorweed	*Alternanthera philoxeroides*	*Acer rubra*	Red maple
American elm	*Ulmus americana*	*Acer saccharinum*	Silver maple
Analogne sedge	*Carex simulata*	*Acer saccharinum, Acer rubra*	Soft maple
Arrowarum (Virginia arrowarum)	*Peltandra virginica*	*Acer* spp.	Maple
Arrowhead	*Sagittaria graminea*	*Achillea millefolium lanulosa*	Western yarrow
Arrowhead	*Sagittaria heterophylla*	*Acnida cannabina*	Salt-marsh water hemp (water hemp)
Arrowhead	*Sagittaria* spp.	*Acorus calamus*	Sweet flag
Arroyo willow	*Salix lasiolepis*	*Agropyron elongatum*	Tall wheatgrass
Ash	*Fraxinus* spp.	*Agropyron intermedium*	Intermediate wheatgrass
Asiatic dayflower	*Aneilema keisak*	*Agropyron repens*	Quackgrass
Aspen	*Populus* spp.	*Agropyron smithii*	Western wheatgrass
Aster	*Aster subulatus*	*Agropyron trachycaulum*	Slender wheatgrass
Aster	*Aster* spp.	*Agrostis alba*	Redtop
Australian pine	*Casuarina equisetifolia*	*Agrostis stolonifera*	Redtop
Avens	*Geum* spp.	*Alisma plantago-aquatica*	Water plantain
Bahiagrass	*Paspalum notatum*	*Alnus rugosa*	Speckled alder
Baldcypress (cypress)	*Taxodium distichum*	*Alnus sinuata*	Sitka alder
Baltic rush	*Juncus arcticus balticus*	*Alnus tenuifolia*	Thinleaf alder
Banana (yellow) waterlily	*Nymphaea mexicana*	*Alnus* spp.	Alder
Barnyardgrass	*Echinochloa muricata*	*Alopecurus aequalis*	Foxtail
		Alopecurus pratensis	Meadow foxtail
		Alternanthera philoxeroides	Alligatorweed

TABLE A.29 Common and Scientific Names of Plants *(Continued)*

Alphabetical by common name		Alphabetical by scientific name	
Common name	Scientific name	Scientific name	Common name
Barnyardgrass (common barnyardgrass)	*Echinochloa crusgalli*	*Amarthus* spp.	Pigweed
Barrenground willow	*Salix brachycarpa*	*Ambrosia artemisiifolia*	Ragweed (common ragweed)
Basin big sagebrush	*Artemisia tridentata tridentata*	*Amelanchier alnifolia*	Saskatoon serviceberry
Bassia	*Bassia* spp.	*Ammania coccinea*	Toothcup
Bayberry	*Myrica pennsylvanica*	*Ammophila arenaria*	European beachgrass
Beach panicgrass	*Panicum anceps*	*Ammophila breviligulata*	Beachgrass (American beachgrass)
Beach pea	*Strophostyles helvola*	*Andropogon gerardi*	Big bluestem
Beachgrass (American beachgrass)	*Ammophila breviligulata*	*Andropogon scoparius*	Little bluestem
Beaked sedge	*Carex rostrata*	*Andropogon virginicus*	Broomsedge bluestem
Beakrush	*Rhynchospora corniculata*	*Andropogon* sp.	Yellow bluestem
Beakrush	*Rhynchospora* spp.	*Andropogon* spp.	Bluestem
Bebb willow	*Salix bebbiana*	*Andropogon* spp.	Broomsedge
Bedstraw	*Galium* spp.	*Aneilema keisak*	Asiatic dayflower
Beech	*Fagus grandifolia*	*Arctophila fulva*	Pendont
Beggarsticks	*Bidens* spp.	*Argrostis* spp.	Bent grass
Bellflower	*Campanula aparinoides*	*Artemisia cana viscidula*	Silver sagebrush
Bent grass	*Argrostis* spp.	*Artemisia ludoviciana*	Louisiana sagewort
Bermudagrass	*Cynodon dactylon*	*Artemesia tridentata*	Big sagebrush
Big bluestem	*Andropogon gerardi*	*Artemisia tridentata tridentata*	Basin big sagebrush
Big (giant) cordgrass (salt reed-grass)	*Spartina cynosuroides*	*Artemisia tridentata vaseyana*	Mountain big sagebrush
Big duckweed	*Spirodela polyhiza*	*Artemisia tripartita*	Tall threetip sagebrush
Big sagebrush	*Artemesia tridentata*	*Artemesia* spp.	Sagebrush
Bigelow's glasswort	*Salicornia bigelovii*	*Arundo donax*	Giant reed
Bindweed	*Polygonum cilinode*	*Asclepias incarnata*	Swamp milkweed
		Asclepias spp.	Milkweed
		Aster chilensis	Pacific aster

Birch	*Betula* spp.
Bitter panicum	*Panicum amarum*
Bitternut hickory	*Carya cordiformis*
Black alpine sedge	*Carex nigricans*
Black chokecherry	*Prunus virginiana melanocarpa*
Black greasewood	*Sarcobatus vermiculatus*
Black locust	*Robinia pseudo-acacia*
Black mangrove	*Avicennia germinans*
Black (saltmarsh) needlerush	*Juncus roemerianus*
Black sedge-root	*Carex elynoides*
Black willow	*Salix nigra*
Black-grass	*Juncus gerardi*
Blackgum (blackgum tupelo)	*Nyssa sylvatica*
Bladderwort	*Utricularia intermedia*
Bladderwort	*Utricularia purpurea*
Bladderwort	*Utricularia vulgaris*
Bladderwort	*Utricularia* spp.
Blue flag	*Iris versicolor*
Blue paloverde	*Cercidium floridum*
Bluejoint (reedgrass, bluejoint reedgrass)	*Calamagrostis canadensis*
Bluestem	*Andropogon* spp.
Bog birch	*Betula pumila*
Boneset	*Eupatorium perfoliatum*
Bottlebrush squirreltail	*Sitanion hystrix*
Boxelder	*Acer negundo*
Bramble	*Rubus* spp.
Brazilian pepper	*Schinus terebinthifolius*
Broadleaf cattail	*Typha latifolia*
Broomsedge	*Andropogon* spp.
Broomsedge bluestem	*Andropogon virginicus*
Browntop millet	*Panicum ramosum*
Buffalo grass	*Buchloe dactyloides bulbifera*

Aster exilis	Slender aster
Aster subulatus	Aster
Aster tenuifolius	Saltmarsh aster
Aster spp.	Aster
Atriplex canescens	Fourwing saltbush
Atriplex gardneri	Gardner saltbush
Atriplex lentiformis	Quailbush
Atriplex patula	Orache (fat hen)
Avicennia germinans	Black mangrove
Azolla caroliniana	Water-velvet
Baccharis halimifolia	Groundsel tree (sea myrtle)
Baccharis viminea	Mule fat
Bacopa caroliniana	Water hyssop
Bacopa spp.	Waterhyssop
Baptisia leucophaea	Wild indigo
Bassia hyssopifolia	Fivehook bassia
Bassia spp.	Bassia
Beckmannia syzigachne	Slough grass
Betula nigra	River birch
Betula occidentalis occidentalis	Water birch
Betula pumila	Bog birch
Betula spp.	Birch
Bidens spp.	Beggarticks
Borrichia frutescens	Sea oxeye
Bouteloua curtipendula	Sideoats grama
Brasenia schreberi	Watershield
Bromus carinatus	Mountain brome
Bromus erectus	Meadow brome
Bromus inermis	Smooth brome
Bromus tectorum	Cheat grass
Buchloe dactyloides bulbifera	Buffalo grass
Butomus umbellatus	Flowering rush
Cabomba caroliniana	Fanwort

TABLE A.29 Common and Scientific Names of Plants *(Continued)*

Alphabetical by common name		Alphabetical by scientific name	
Common name	Scientific name	Scientific name	Common name
Bugleweed	*Lycopus americanus*	*Calamagrostis canadensis*	Bluejoint (reedgreass, bluejoint reedgrass)
Bugleweed	*Lycopus uniflorus*	*Calamagrostis epigeois*	Chee reedgrass
Bullhead lily	*Nuphar variegatum*	*Calamagrostis inexpansa*	Northern reedgrass
Bullwhip (southern California) bulrush	*Scirpus californicus*	*Calla palustris*	Water arum
Bulrush	*Scirpus campestris*	*Callitriche heterophylla*	Waterstarwort
Bulrush	*Scirpus paludosus*	*Campanula aparinoides*	Bellflower
Bulrush	*Scirpus subterminalis*	*Campsis radicans*	Trumpetcreeper
Bulrush	*Scirpus* spp.	*Carex aquatilis*	Water sedge
Bur oak	*Quercus macrocarpa*	*Carex atherodes*	Slough sedge
Burreed	*Sparganium americanum*	*Carex aurea*	Golden sedge
Burreed	*Sparganium chlorocarpum*	*Carex disperma*	Softleaved sedge
Burreed	*Sparganium fluctuans*	*Carex douglasii*	Douglas sedge
Burreed	*Sparganium foliosa*	*Carex elynoides*	Black sedge-root
Burreed	*Sparganium gracillus*	*Carex hoodii*	Hood sedge
Burreed	*Sparganium pectinata*	*Carex lanuginosa*	Woolly sedge
Burreed	*Sparganium* spp.	*Carex lenticularis*	Kellogg sedge
Bush cinquefoil	*Potentilla fruticosa*	*Carex lyngbyei*	Lyngbye's sedge
Bushy pondweed	*Najas guadalupensis*	*Carex microptera*	Smallwing sedge
Bushy pondweed	*Najas minor*	*Carex nardina*	Hepburn sedge
Buttercup	*Ranunculus* spp.	*Carex nebrascensis*	Nebraska sedge
Butterweed groundsel	*Senecio serra*	*Carex nigricans*	Black alpine sedge
Buttonbush (common buttonbush)	*Cephalanthus occidentalis*	*Carex obnupta*	Slough sedge
		Carex obnuta	Pacific sedge
Buttonweed	*Diodia virginiana*	*Carex praegracilis*	Slim sedge
Buttonwood	*Concarpus erecta*	*Carex rostrata*	Beaked sedge

Cabbage palm	*Sabal palmetto*
Camphorweed	*Heterotheca subaxillaris*
Canada thistle	*Cirsium arvense*
Canadian waterweed	*Elodea canadensis*
Carolinian water hyssop	*Hydrotrida caroliniana*
Cascara buckthorn	*Rhamnus purshiana*
Cattail	*Typha* spp.
Chara (stonewort)	*Chara* spp.
Cheat grass	*Bromus tectorum*
Chee reedgrass	*Calamagrostis epigeois*
Cherrybark oak	*Quercus falcata pagodifolia*
Chufa (nutsedge)	*Cyperus* spp.
Cinquefoil	*Potentilla* spp.
Clasping-leaf pondweed	*Potamogeton richardsonii*
Clover	*Trifolium* spp.
Cocklebur (common cocklebur)	*Xanthium strumarium*
Cocklebur	*Xanthium* spp.
Common burhead	*Echinodorus cordifolius*
Common duckweed	*Lemna minor*
Common (soft) rush	*Juncus effusus*
Common smartweed (marsh pepper)	*Polygonum hydropiper*
Common snowberry	*Symphoricarpos albus*
Common threesquare (American bulrush)	*Scirpus americanus*
Coontail	*Ceratophyllum demersum*
Coontail	*Ceratophyllum* spp.
Cordgrass	*Spartina* spp.
Cottonwood	*Populus* spp.
Cow (basket) oak	*Quercus michauxii*

Carex rupestris	Rock sedge
Carex scirpoidea	Downy sedge
Carex simulata	Analogne sedge
Carex trichocarpa	Slough sedge
Carex vallicola	Valley sedge
Carex vulpinoidea	Fox sedge
Carex spp.	Sedge
Carya aquatica	Water hickory
Carya cordiformis	Bitternut hickory
Carya ovata	Shagbark hickory
Carya spp.	Hickory
Casuarina equisetifolia	Australian pine
Ceanothus sanguineus	Redstem ceanothus
Celtis laevigata	Sugar hackberry
Celtis occidentalis	Hackberry
Cephalanthus occidentalis	Buttonbush (common buttonbush)
Ceratophyllum demersum	Coontail
Ceratophyllum spp.	Coontail
Cercidium floridum	Blue paloverde
Chamaedaphne calyculata	Leatherleaf
Chara hornemannii	Muskgrass
Chara vulgaris	Stonewort
Chara spp.	Chara (stonewort)
Chara spp.	Muskgrass
Chenopodium album	Lambsquarter
Chenopodium rubrum	Red goosefoot
Chenopodium spp.	Goosefoot
Chrysopogon spp.	Woodlawn grass
Chrysothamnus nauseosus consimilis	Thinleaf rubber rabbitbrush

TABLE A.29 Common and Scientific Names of Plants *(Continued)*

Alphabetical by common name		Alphabetical by scientific name	
Common name	Scientific name	Scientific name	Common name
Cowlily (waterlily)	*Nuphar* spp.	*Cicuta bulbifera*	Water hemlock
Cowwheat	*Melampyrum lineare*	*Cicuta maculata*	Water hemlock
Coyote (sandbar) willow	*Salix exigua*	*Cirsium arvense*	Canada thistle
Crabgrass	*Digitaria* spp.	*Cladium jamaicensis*	Sawgrass
Creeping marshpurslane	*Ludwigia repens*	*Cladium* spp.	Sawgrass
Creeping wildrye	*Elymus triticoides*	*Cladophora* spp.	Filamentous algae
Cress	*Rorippa* spp.	*Comptonia peregrina*	Sweet fern
Crimson clover	*Trifolium incarnatum*	*Concarpus erecta*	Buttonwood
Croton	*Croton punctatus*	*Cornus florida*	Flowering dogwood
Crownvetch	*Coronilla varia*	*Cornus stolonifera*	Red-osier dogwood
Curly leaf (curly) dock	*Rumex crispus*	*Cornus* spp.	Dogwood
Curly-leaf pondweed	*Potamogeton crispus*	*Coronilla varia*	Crownvetch
Cutgrass [rice (giant) cutgrass]	*Leersia oryzoides*	*Cratageus douglasii*	Douglas hawthorn
Dallis grass	*Paspalum dilatatum*	*Croton punctatus*	Croton
Delta duck potato (arrowhead)	*Sagittaria platyphylla*	*Crypsis niliaca*	Pricklegrass
Dense waterweed	*Elodea densa*	*Cuphea* spp.	Waxweed
Desert blite	*Suaeda torreyana*	*Cynodon dactylon*	Bermudagrass
Dock	*Rumex* spp.	*Cyperus erythrorhizos*	Red-rooted sedge (redroot flatsedge)
Dog-fennel	*Eupatorium capillifolium*		
Dogwood	*Cornus* spp.	*Cyperus odoratus*	Fragrant flatsedge
Dotted smartweed	*Polygonum punctatum*	*Cyperus polystachyos*	Many-spiked flatsedge
Douglas-fir	*Pseudotsuga menziesii*	*Cyperus virens*	Umbrella sedge
Douglas hawthorn	*Cratageus douglasii*	*Cyperus* spp.	Chufa (nutsedge)
Douglas sedge	*Carex douglasii*	*Cyperus* spp.	Flat sedge
Downy sedge	*Carex scirpoidea*	*Cytisus scoparius*	Scotch broom
Dropseed	*Sporobolus* spp.	*Dactylis glomerata*	Orchard grass
Drummond rush	*Juncus drummondii*	*Danthonia spicata*	Poverty grass
Drummond willow	*Salix drummondiana*	*Decodon verticillatus*	Water willow

Duck potato	*Sagittaria cuneata*
Duck potato (arrowhead, broadleaf arrowhead)	*Sagittaria latifolia*
Duckmeal	*Wolffia* spp.
Duckweed	*Lemna* spp.
Duckweed	*Spirodela* spp.
Dwarf spikerush	*Eleocharis parvula*
Eastern cottonwood	*Populus deltoides*
Edible valerian	*Valeriana edulis*
Eelgrass	*Zostera marina*
Elm	*Ulmus* spp.
Elodea	*Elodea* spp.
European beachgrass	*Ammophila arenaria*
European glasswort	*Salicornia europea*
Evening primrose	*Oenothera humifusa*
Fall panicum	*Panicum dichotomiflorum*
False loosestrife	*Ludwigia* spp.
Fanwort	*Cabomba caroliniana*
Fescue	*Festuca* spp.
Filamentous algoe	*Cladophora* spp.
Fimbristylis	*Fimbristylis castanea*
Fir	*Abies* spp.
Fireweed	*Epilobium angustifolium*
Fireweed	*Erechtites hieracifolia*
Fivehook bassia	*Bassia hyssopifolia*
Flat sedge	*Cyperus* spp.
Flat-stemmed pondweed	*Potamogeton zosteriformes*
Floating heart	*Nymphoides* spp.
Floating leaf pondweed	*Potamogeton natans*
Floating moss	*Salvinia rotundifolia*
Flowering dogwood	*Cornus florida*
Flowering rush	*Butomus umbellatus*

Deschampsia caespitosa	Tufted hairgrass
Dianthera americana	Water willow
Digitaria sanguinalis	Hairy crabgrass
Digitaria spp.	Crabgrass
Diodia virginiana	Buttonweed
Distichlis spicata	Saltgrass (seashore saltgrass)
Distichlis stricta	Saltgrass
Dulichium arundinaceum	Three-way sedge
Echinochloa colonum	Junglerice
Echinochloa crusgalli	Barnyardgrass (common barnyardgrass)
Echinochloa crusgalli chiwapa	Japanese millet
Echinochloa crusgalli frumentacea	Japanese millet
Echinochloa muricata	Barnyardgrass
Echinochloa walteri	Walters (saltmarsh) millet
Echinochloa spp.	Millet (wild millet)
Echinodorus cordifolius	Common burhead
Eichhornia crassipes	Water hyacinth
Elaeagnus angustifolia	Russian olive
Elaeagnus commutata	Silver berry
Eleocharis acicularis	Slender spikerush
Eleocharis baldwinii	Spikerush
Eleocharis equisetoides	Spikerush
Eleocharis palustris	Spikerush
Eleocharis parvula	Dwarf spikerush
Eleocharis quadrangulata	Squarestem spikerush
Eleocharis smallii	Spikerush
Eleocharis spp.	Spikerush
Elodea canadensis	Canadian waterweed
Elodea densa	Dense waterweed
Elodea spp.	Elodea
Elymus cinereus	Great basin wildrye

TABLE A.29 Common and Scientific Names of Plants *(Continued)*

Alphabetical by common name		Alphabetical by scientific name	
Common name	Scientific name	Scientific name	Common name
Fourwing saltbush	*Atriplex canescens*	*Elymus giganteus*	Mammoth wildrye
Fowl-meadow grass	*Poa palustris*	*Elymus junceus*	Russian wildrye
Fox sedge	*Carex vulpinoidea*	*Elymus triticoides*	Creeping wildrye
Foxtail	*Alopecurus aequalis*	*Elymus virginicus*	Wild rye
Foxtail grasses	*Setaria* spp.	*Epilobium adenocaulon*	Willow-herb
Fragrant flatsedge	*Cyperus odoratus*	*Epilobium angustifolium*	Fireweed
Fragrant waterlily	*Nymphaea odorata*	*Equisetum fluviatile*	Horsetail
Frankenia	*Frankenia grandifoliz*	*Equisetum palustre*	Horsetail
Fremont cottonwood	*Populus fremontii*	*Equisetum sylvaticum*	Horsetail
Frog bit	*Limnobium spongia*	*Equisetum* spp.	Horsetail
Gardner saltbush	*Atriplex gardneri*	*Eragrostis curvula*	Weeping lovegrass
Geyer willow	*Salix geyeriana*	*Erechtites hieracifolia*	Fireweed
Giant (great) burreed	*Sparganium eurycarpum*	*Erianthus giganteus*	Plume grass
Giant foxtail	*Setaria magna*	*Erigeron canadensis*	Horseweed
Giant reed	*Arundo donax*	*Eupatorium capillifolium*	Dog-fennel
Gland cinquefoil	*Potentilla glandulosa*	*Eupatorium perfoliatum*	Boneset
Glasswort	*Salicornia* spp.	*Eupatorium serotinum*	Joe-pye-weed
Golden club	*Orontium aquaticum*	*Euphorbia polygonifolia*	Spurge
Golden current	*Ribes aureum*	*Fagus grandifolia*	Beech
Golden sedge	*Carex aurea*	*Festuca arundinacea*	Reed fescue
Goldenrod	*Solidago* spp.	*Festuca elatior*	Tall fescue
Gooding's willow	*Salix gooddingii*	*Festuca* spp.	Fescue
Goose grass	*Puccinellia phryganodes*	*Fimbristylis castanea*	Fimbristylis
Gooseberry	*Ribes* spp.	*Frankenia grandifoliz*	Frankenia
Goosefoot	*Chenopodium* spp.	*Fraxinus pennsylvanica*	Red ash
Grayleaf willow	*Salix glauca*	*Fraxinus pennsylvanica subintegerrim*	Green ash

Great basin wildrye	*Elymus cinereus*
Green ash	*Fraxinus pennsylvanica subintegerrima*
Green needlegrass	*Stipa viridula*
Green's mountain ash	*Sorbus scopulina scopulina*
Groundsel tree (sea myrtle)	*Baccharis halimifolia*
Gulf cordgrass	*Spartina spartinae*
Gum plant	*Grindelia integrifolia*
Hackberry	*Celtis occidentalis*
Hairy crabgrass	*Digitaria sanguinalis*
Halbert-leaved rose mallow	*Hibiscus militaris*
Halbert-leaved tearthumb	*Polygonum arifolium*
Hardstem bulrush (tule)	*Scirpus acutus*
Hedge nettle	*Stachys palustris*
Hepburn sedge	*Carex nardina*
Hickory	*Carya* spp.
Honey locust	*Gleditsia triacanthos*
Hood sedge	*Carex hoodii*
Horned pondweed	*Zannichellia palustris*
Horsetail	*Equisetum fluviatile*
Horsetail	*Equisetum palustre*
Horsetail	*Equisetum sylvaticum*
Horsetail	*Equisetum* spp.
Horseweed	*Erigeron canadensis*
Huisache tree	*acacia smallii*
Hydrilla	*Hydrilla verticillata*
Ice plant	*Mesembryanthemum crystallinum*
Illinois pondweed	*Potamogeton illinoensis*
Indiangrass	*Sorgastrum nutans*
Intermediate wheatgrass	*Agropyron intermedium*
Japanese millet	*Echin. crus. chiwapa*
Japanese millet	*Echinochloa crusgalli frumentacea*

Fraxinus spp.	Ash
Galium spp.	Bedstraw
Geum spp.	Avens
Gleditsia aquatica	Water locust
Gleditsia triacanthos	Honey locust
Glyceria acutiflora	Mannagrass
Glyceria borealis	Mannagrass
Glyceria fluitans	Mannagrass
Glyceria grandis	Tall (reed) mannagrass
Glyceria maxima	Mannagrass
Glyceria striata	Mannagrass
Grindelia integrifolia	Gum plant
Halodule wrightii	Shoalgrass
Halodule spp.	Seagrass
Helenium flexuosum	Sneezeweed
Heleochloa schoeroides	Swamp timothy
Heteranthera dubia	Water starwort (water star grass)
Heteranthera reniformis	Mud plantain
Heterotheca grandiflora	Telegraph weed
Heterotheca subaxillaris	Camphorweed
Hibiscus militaris	Halberd-leaved rose mallow
Hibiscus moscheutos	Marsh hibiscus
Hippuris vulgaris	Mare's tail
Holodiscus discolor	Rockspirea
Hordeum brachyantherum	Meadow barley
Hordeum jubatum	Wild barley
Hydrilla verticillata	Hydrilla
Hydrocotyle spp.	Pennywort (water pennywort)
Hydrotrida caroliniana	Carolinian water hyssop

TABLE A.29 Common and Scientific Names of Plants *(Continued)*

Alphabetical by common name		Alphabetical by scientific name	
Common name	Scientific name	Scientific name	Common name
Jaumea (saltmarsh jaumea)	*Jaumea carnosa*	*Ilex decidua*	Possumhaw holly
Jewelweed	*Impatiens biflora*	*Ilex vomitoria*	Yaupon
Joe-pye-weed	*Eupatorium serotinum*	*Impatiens biflora*	Jewelweed
Junglerice	*Echinochloa colonum*	*Ipomoea coccinea*	Morningglory
Juniper	*Juniperus* spp.	*Iris pseudacorus*	Yellow flag (yellow iris)
Kellogg sedge	*Carex lenticularis*	*Iris versicolor*	Blue flag
Kentucky bluegrass	*Poa pratensis*	*Iva frutescens*	Marsh elder
Knotweed(s)	*Polygonum* spp.	*Jaumea carnosa*	Jaumea (saltmarsh jaumea)
Ladysthumb	*Polygonum persicaria*	*Juncus arcticus balticus*	Baltic rush
Lambsquarter	*Chenopodium album*	*Juncus drummondii*	Drummond rush
Large-leaf pondweed	*Potamogeton amplifolius*	*Juncus effusus*	Common (soft) rush
leafy pondweed	*Potamogeton epihydrous*	*Juncus ensifolius*	Swordleaf rush
leafy pondweed	*Potamogeton foliosus*	*Juncus gerardi*	Black-grass
Leatherleaf	*Chamaedaphne calyculata*	*Juncus longistylis*	Longstyle rush
Lespedeza	*Lespedeza* spp.	*Juncus roemerianus*	Black (saltmarsh) needlerush
Lewis flax	*Linum lewisii*	*Juncus tenuis*	Poverty rush
Lippia (fog-fruit)	*Lippia lanceolata*	*Juncus torreyi*	Torrey rush
Littel bluestem	*Andropogon scoparius*	*Juncus* spp.	Rush
Live oak	*Quercus virginiana*	*Juniperus* spp.	Juniper
Lizard's tail	*Saururus cernuus*	*Jussiaea diffusa*	Water primrose
Lobelia (water lobelia)	*Lobelia dortmanna*	*Jussiaea leptocarpa*	Water primrose
Lodgepole pine	*Pinus contorta*	*Jussiaea* spp.	Water primrose
Longstyle rush	*Juncus longistylis*	*Justicia americana*	Water willow
Lotus	*Nelumbo lutea*	*Kochia scoparia*	Summercypress (kochia)
Louisiana sagewort	*Artemisia ludoviciana*	*Lachnanthes caroliniana*	Redroot
Low sea blite	*Suaeda maritima*	*Lachnanthes tinctoria*	Redroot
Lyngbye's sedge	*Carex lyngbyei*	*Laguncularia racemosa*	White mangrove
Maidencane	*Panicum hemitomun*	*Lantana camara*	Shrub verbena
Mallow ninebark	*Physocarpus malvaceus*	*Laportea canadensis*	Wood nettle
Mammoth wildrye	*Elymus giganteus*	*Leersia oryzoides*	Cutgrass [rice (giant) cutgrass]
Manatee grass	*Syringodium filiforme*		

Mannagrass	*Glyceria acutiflora*
Mannagrass	*Glyceria borealis*
Mannagrass	*Glyceria fluitans*
Mannagrass	*Glyceria maxima*
Mannagrass	*Glyceria striata*
Many-spiked flatsedge	*Cyperus polystachyos*
Maple	*Acer* spp.
Mare's tail	*Hippuris vulgaris*
Marsh cinquefoil	*Potentilla palustris*
Marsh elder	*Iva frutescens*
Marsh hibiscus	*Hibiscus moscheutos*
Marsh smartweed	*Polygonum coccineum*
Marsh (swamp) smartweed	*Polygonum hydropiperoides*
Marsh St. John's-wort	*Triadenum virginicum*
Marshpurslane (common marshpurslane	*Ludwigia palustris*
Marshpurslane	*Ludwigia* spp.
Meadow barley	*Hordeum brachyantherum*
Meadow brome	*Bromus erectus*
Meadow foxtail	*Alopecurus pratensis*
Meadow grass	*Poa* spp.
Meadowsweet	*Spirea alba*
Mesquite	*Prosopsis* spp.
Milkweed	*Asclepias* spp.
Millet (wild millet)	*Echinochloa* spp.
Mint	*Mentha arvensis*
Mistletoe	*Phoradendron californicum*
Monkey flower	*Mimulus ringens*
Morningglory	*Ipomoea coccinea*
Mountain big sagebrush	*Artemisia tridentata vaseyana*
Mountain brome	*Bromus carinatus*

Lemna minor	Common duckweed
Lemna spp.	Duckweed
Leptochloa fascicularis	Sprangletop
Leptochloa spp.	Sprangletop
Lespedeza cuneat	Sericea lespedeza
Lespedeza spp.	Lespedeza
Limnobium spongia	Frog bit
Limonium carolinianum	Sea lavender
Limonium vulgare	Sea lavender
Linum lewisii	Lewis flax
Lipidium virginicum	Pepper grass
Lippia lanceolata	Lippia (fog-fruit)
Liquidambar styraciflua	Sweetgum (American sweetgum)
Liriodendron tulipifera	Yellow poplar
Lobelia dortmanna	Lobelia (water lobelia)
Lolium perenne	Perennial ryegrass
Lonicera tatarica	Tatarian honeysuckle
Ludwigia palustris	Marshpurslane (common marshpurslane)
Ludwigia repens	Creeping marshpurslane
Ludwigia spp.	False loosestrife
Ludwigia spp.	Marshpurslane
Lycopus americanus	Bugleweed
Lycopus uniflorus	Bugleweed
Lycopus salicaria	Purple loosestrife
Marsilea vestita	Water clover
Medicago sativa	Alfalfa
Megalodonta beckii	Water marigold
Melampyrum lineare	Cowwheat
Melilotus alba	White sweetclover

TABLE A.29 **Common and Scientific Names of Plants** ***(Continued)***

Alphabetical by common name		Alphabetical by scientific name	
Common name	Scientific name	Scientific name	Common name
Mountain snowberry	*Symphoricarpos oreophilus*	*Melilotus officianalis*	Yellow sweetclover
Mud plantain	*Heteranthera reniformis*	*Mentha arvensis*	Mint
Mule fat	*Baccharis viminea*	*Mesembryanthemum crystallinum*	Ice plant
Muskgrass	*Chara hornemannii*		
Muskgrass	*Chara* spp.	*Mimulus ringens*	Monkey flower
Myrtle pachistima	*Pachistima myrsinites*	*Myrica californica*	Pacific wax myrtle
Naiad	*Najas marina*	*Myrica cerifera*	Wax myrtle
Naiad	*Najas* spp.	*Myrica pennsylvanica*	Bayberry
Narrow-leaved (saltmarsh) cat-tail	*Typha angustifolia*	*Myriophyllum brasiliense*	Water feather
		Myriophyllum exalbescens	Water milfoil
Narrowleaf cottonwood	*Populus angustifolia*	*Myriophyllum heterophyllum*	Water milfoil
Nebraska sedge	*Carex nebrascensis*	*Myriophyllum spicatum*	Water milfoil
Nettle	*Urtica dioica*	*Myriophyllum* spp.	Water milfoil
Nettle	*Urtica* spp.	*Najas flexilis*	Northern naiad
Nodding smartweed (curltop ladysthumb	*Plygonum lapathifolium*	*Najas guadalupensis*	Bushy pondweed
		Najas marina	Naiad
Northern naiad	*Najas flexilis*	*Najas minor*	Bushy pondweed
Northern reedgrass	*Calamagrostis inexpansa*	*Najas* spp.	Naiad
Nuttall oak	*Quescus nuttallii*	*Najas* spp.	Water nymphs
Oak	*Quercus* spp.	*Nasturtium officinale*	Watercress
Oleander	*Nerium oleander*	*Nasturtium* spp.	Watercress
Olney threesquare (Olney bulrush)	*Scirpus olneyi*	*Naumburgia thrysiflora*	Tufted loosestrife
Orache (fat hen)	*Atriplex patula*	*Nelumbo lutea*	Lotus
Orchard grass	*Dactylis glomerata*	*Nerium oleander*	Oleander
Oregon checkermallow	*Sidalcea oregana*	*Nuphar advena*	Spatterdock
Overcup oak	*Quercus lyrata*	*Nuphar lutea*	Spatterdock
Pacific aster	*Aster chilensis*		

Pacific cordgrass	*Spartina foliosa*	*Nuphar luteum*	Yellow pond lily (water lily)
Pacific sedge	*Carex obnuta*	*Nuphar macrophyllum*	White waterlily (waterlily)
Pacific wax myrtle	*Myrica californica*	*Nuphar polysepalum*	Yellow pond lily
Pacific willow	*Salix lasiandra*	*Nuphar variegatum*	Bullhead lily
Palmetto	*Serena repens*	*Nuphar* spp.	Cowlily (waterlily)
Panic grass (panicum)	*Panicum* spp.	*Nymphaea mexicana*	Banana (yellow) waterlily
Paspalum	*Paspalum* spp.	*Nymphaea odorata*	Fragrant waterlily
Peachleaf willow	*Salix amygdaloides*	*Nymphaea tuberosa*	Waterlily
Pendont	*Arctophila fulva*	*Nymphaea* spp.	Water lily (white water lily)
Pennsylvania smartweed	*Polygonum pensylvanicum*	*Nymphoides* spp.	Floating heart
Pennywort (water pennywort)	*Hydrocotyle* spp.	*Nyssa aquatica*	Tupelo (water tupelo)
Pepper grass	*Lipidium virginicum*	*Nyssa sylvatica*	Blackgum (blackgum tupelo)
Perennial ryegrass	*Lolium perenne*	*Nyssa* spp.	Tupelo/blackgum
Pickerelweed	*Pontederia cordata*	*Oenothera humifusa*	Evening primrose
Pickleweed (Pacific glasswort)	*Salicornia pacifica*	*Orontium aquaticum*	Golden club
Pigweed	*Amarthus* spp.	*Oryzopsis* spp.	Rice grass
Pin oak	*Quercus palustris*	*Pachistima myrsinites*	Myrtle pachistima
Pine	*Pinus* spp.	*Panicum amarum*	Bitter panicum
Pinyon	*Pinus edulis*	*Panicum anceps*	Beach panicgrass
Plains cottonwood	*Populus sargentii*	*Panicum dichotomiflorum*	Fall panicum
Plume grass	*Erianthus giganteus*	*Panicum hemitomun*	Maidencane
Ponderosa pine	*Pinus ponderosa*	*Panicum miliaceum*	Proso millet
Pondweed	*Potamogeton americanus*	*Panicum ramosum*	Browntop millet
Pondweed	*Potamogeton capillaceus*	*Panicum verrocosum*	Warty panicum
Pondweed	*Potamogeton compressus*	*Panicum virgatum*	Switchgrass
Pondweed	*Potamogeton heterophyllus*	*Panicum* spp.	Panic grass (panicum)
Pondweed	*Potamogeton nodosus*	*Paspalum dilatatum*	Dallis grass
Pondweed	*Potamogeton perfoliatus*	*Paspalum notatum*	Bahiagrass
Pondweed	*Potamogeton praelongus*	*Paspalum virginatum*	Seaside paspalum
Pondweed	*Potamogeton pusillus*	*Paspalum* spp.	Paspalum
Pondweed	*Potamogeton spirillus*	*Peltandra virginica*	Arrowarum (Virginia arrowarum)
Pondweed	*Potamogeton strictifolius*		
Pondweed	*Potamogeton* spp.	*Persea borbonia*	Red bay
Poplar	*Populus* spp.	*Persea* spp.	Red bay
Possumhaw holly	*Ilex decidua*		

TABLE A.29 Common and Scientific Names of Plants *(Continued)*

Alphabetical by common name		Alphabetical by scientific name	
Common name	Scientific name	Scientific name	Common name
Poverty grass	*Danthonia spicata*	*Phalaris arundinacea*	Reed canarygrass
Poverty rush	*Juncus tenuis*	*Phleum pratense*	Timothy
Prairie cordgrass	*Spartina pectinatum*	*Phoradendron californicum*	Mistletoe
Prickelgrass	*Crypsis niliaca*	*Phragmites australis*	Reed (common reed, reed grass)
Proso millet	*Panicum miliaceum*	*Physocarpus malvaceus*	Mallow ninebark
Purple Loosestrife	*Lythrum salicaria*	*Picea sitchensis*	Sitka spruce
Purple top	*Tripalsis purpurea*	*Picea* spp.	Spruce
Quackgrass	*Agropyron repens*	*Pinus australis, Pinus echinata*	Yellow pine
Quailbush	*Atriplex lentiformis*	*Pinus clausa*	Sand pine
Quaking aspen	*Populus tremuloides*	*Pinus contorta*	Lodgepole pine
Ragweed (common ragweed)	*Ambrosia artemisifolia*	*Pinus edulis*	Pinyon
Raspberry	*Rubus* spp.	*Pinus elliottii*	Slash pine
Red ash	*Fraxinus pennsylvanica*	*Pinus ponderosa*	Ponderosa pine
Red bay	*Persea borbonia*	*Pinus* spp.	Pine
Red bay	*Persea* spp.	*Pistia stratiotes*	Water lettuce
Red clover	*Trifolium pratense*	*Planera aquatica*	Saltmarsh fleabane
Red elder	*Sambucus racemosa pubens*	*Planera aquatica*	Waterelm
Red goosefoot	*Chenopodium rubrum*	*Platanus occidentalis*	Sycamore
Red mangrove	*Rhizophora mangle*	*Poa palustris*	Fowl-meadow grass
Red maple	*Acer rubra*	*Poa pratensis*	Kentucky bluegrass
Red oak	*Quercus rubrum*	*Poa secunda*	Sandberg bluegrass
Red-osier dogwood	*Cornus stolonifera*	*Poa* spp.	Meadow grass
Red-rooted sedge (redroot flatsedge)	*Cyperus erythrorhizos*	*Polygonum amphibium*	Water smartweed
		Polygonum arifolium	Halberd-leaved tearthumb
Redroot	*Lachnanthes caroliniana*	*Polygonum aviculare*	Smartweed
Redroot	*Lachnanthes tinctoria*	*Polygonum cilinode*	Bindweed
Redstem ceanothus	*Ceanothus sanguineus*	*Polygonum coccineum*	Marsh smartweed
Redtop	*Agrostis alba*	*Polygonum convolvulus*	Wild buckwheat
		Polygonum cordata	Smartweed
		Polygonum densiflorum	Southern smartweed

Redtop	*Agrostis stolonifera*
Redwood	*Sequoia sempervirens*
Reed (common reed, reed grass)	*Phragmites australis*
Reed canarygrass	*Phalaris arundinacea*
Reed fescue	*Festuca arundinacea*
Rice grass	*Oryzopsis* spp.
River birch	*Betula nigra*
River bulrush (bulrush)	*Scirpus fluviatilis*
Robbins pondweed	*Potamogeton robbinsii*
Rock sedge	*Carex rupestris*
Rockspirea	*Holodiscus discolor*
Rose	*Rosa* spp.
Rush	*Juncus* spp.
Russian olive	*Elaeagnus angustifolia*
Russian wildrye	*Elymus junceus*
Sagebrush	*Artemesia* spp.
Sago (fennelleaf) pondweed	*Potamogeton pectinatus*
Salt cedar	*Tamarix chinensis*
Salt-marsh water hemp (water hemp)	*Acnida cannabina*
Saltgrass	*Distichlis stricta*
Saltgrass (seashore saltgrass)	*Distichlis spicata*
Saltmarsh aster	*Aster tenuifolius*
Saltmarsh (robust) bulrush	*Scirpus robustus*
Saltmarsh (smooth) cordgrass	*Spartina alterniflora*
Saltmarsh fleabane	*Planera aquatica*
Saltmeadow (marshhay) cordgrass	*Spartina patens*
Saltwort	*Salsola kali*
Sand pine	*Pinus clausa*

Polygonum hydropiper	Common smartweed (marsh pepper)
Polygonum hydropiperoides	Marsh (swamp) smartweed
Polygonum lapathifolium	Nodding smartweed (curltop ladysthumb)
Polygonum muhlenbergii	Smartweed
Polygonum natans	Water smartweed
Polygonum pensylvanicum	Pennsylvania smartweed
Polygonum persicaria	Ladysthumb
Polygonum portoricense	Smartweed
Polygonum punctatum	Dotted smartweed
Polygonum sagittatum	Smartweed
Polygonum spp.	Knotweed(s)
Polygonum spp.	Smartweed
Polygonum spp.	Tearthumb
Pontederia cordata	Pickerelweed
Populus angustifolia	Narrowleaf cottonwood
Populus deltoides	Eastern cottonwood
Populus fremontii	Fremont cottonwood
Populus sargentii	Plains cottonwood
Populus tremuloides	Quaking aspen
Populus spp.	Aspen
Populus spp.	Cottonwood
Populus spp.	Poplar
Porserpinaca palustris	Smartweed
Potamogeton americanus	Pondweed
Potamogeton amplifolius	Large-leaf pondweed
Potamogeton capillaceus	Pondweed
Potamogeton compressus	Pondweed
Potamogeton crispus	Curly-leaf pondweed

TABLE A.29 Common and Scientific Names of Plants *(Continued)*

Alphabetical by common name		Alphabetical by scientific name	
Common name	Scientific name	Scientific name	Common name
Sand spurry	*Spergularia rubra*	*Potamogeton diversifolius*	Waterthread
Sandbar willow	*Salix interior*	*Potamogeton epihydrous*	Leafy pondweed
Sandberg bluegrass	*Poa secunda*	*Potamogeton foliosus*	Leafy pondweed
Saskatoon serviceberry	*Amelanchier alnifolia*	*Potamogeton gramineus*	Variable pondweed
Sawgrass	*Cladium jamaicensis*	*Potamogeton heterophyllus*	Pondweed
Sawgrass	*Cladium* spp.	*Potamogeton illinoensis*	Illinois pondweed
Scooler willow	*Salix scouleriana*	*Potamogeton natans*	Floating leaf pondweed
Scotch broom	*Cytisus scoparius*	*Potamogeton nodosus*	Pondweed
Sea blite	*Suaeda* spp.	*Potamogeton pectinatus*	Sago (fennelleaf) pondweed
Sea lavender	*Limonium carolinianum*	*Potamogeton perfoliatus*	Pondweed
Sea lavender	*Limonium vulgare*	*Potamogeton praelongus*	Pondweed
Sea oats	*Uniola paniculata*	*Potamogeton pusillus*	Pondweed
Sea oxeye	*Borrichia frutescens*	*Potamogeton richardsonii*	Clasping-leaf pondweed
Sea purslane	*Sesuvium maritimum*	*Potamogeton robbinsii*	Robbins pondweed
Seagrass	*Halodule* spp.	*Potamogeton spirillus*	Pondweed
Seaside arrowgrass	*Triglochin maritima*	*Potamogeton strictifolius*	Pondweed
Seaside paspalum	*Paspalum virginatum*	*Potamogeton zosteriformes*	Flat-stemmed pondweed
Sedge	*Carex* spp.	*Potamogeton* spp.	Pondweed
Sericea lespedeza	*Lespedeza cuneata*	*Pontentilla fruticosa*	Bush cinquefoil
Sesbania	*Sesbania exaltata*	*Potentilla glandulosa*	Gland cinquefoil
Sessile-fruited arrowhead	*Sagittaria rigida*	*Potentilla palustris*	Marsh cinquefoil
Shagbark hickory	*Carya ovata*	*Potentilla* spp.	Cinquefoil
Shining willow	*Salix lutea*	*Prosopis velutina*	Velvet mesquite
Shoalgrass	*Halodule wrightii*	*Prosopsis* spp.	Mesquite
Shrub verbena	*Lantana camara*	*Prunus virginiana melanocarpa*	Black chokecherry
Shumard oak	*Quercus shumardii*	*Pseudotsuga menziesii*	Douglas-fir
Sideoats grama	*Bouteloua curtipendula*	*Puccinellia phryganodes*	Goose grass
Silver berry	*Elaeagnus commutata*	*Quercus alba*	White oak
Silver buffaloberry	*Shepherdia argentea*	*Quercus bicolor*	Swamp white oak

Silver maple	*Acer saccharinum*
Silver sagebrush	*Artemisia cana viscidula*
Sitka alder	*Alnus sinuata*
Sitka spruce	*Picea sitchensis*
Skullcap	*Scutellaria galericulata*
Skullcap	*Scutellaria lateriflora*
Slash pine	*Pinus elliottii*
Slender aster	*Aster exilis*
Slender bulrush	*Scirpus heterochaetus*
Slender spikerush	*Eleocharis acicularis*
Slender wheatgrass	*Agropyron trachycaulum*
Slim sedge	*Carex praegracilis*
Slough grass	*Beckmannia syzigachne*
Slough sedge	*Carex atherodes*
Slough sedge	*Carex trichocarpa*
Smallwing sedge	*Carex microptera*
Smartweed	*Polygonum aviculare*
Smartweed	*Polygonum cordata*
Smartweed	*Polygonum muhlenbergii*
Smartweed	*Polygonum portoricense*
Smartweed	*Polygonum sagittatum*
Smartweed	*Polygonum* spp.
Smartweed	*Porserpinaca palustris*
Smooth brome	*Bromus inermis*
Sneezeweed	*Helenium flexuosum*
Snowberry (western snowberry)	*Symphoricarpos occidentalis*
Soft maple	*Acer saccharinum, Acer rubra*
Softleaved sedge	*Carex disperma*
Softstem bulrush (great bulrush)	*Scirpus validus*
Solomons-seal	*Smilacina racemosa*
Slough sedge	*Carex obnupta*
Southern smartweed	*Polygonum densiflorum*

Quercus falcata pagodifolia	Cherrybark oak
Quercus lyrata	Overcup oak
Quercus macrocarpa	Bur oak
Quercus michauxii	Cow (basket) oak
Quercus muehlenbergii	Swamp chestnut oak
Quercus nigra	Water oak
Quercus nuttallii	Nuttall oak
Quercus palustris	Pin oak
Quercus phellos	Willow oak
Quercus rubrum	Red oak
Quercus shumardii	Shumard oak
Quercus virginiana	Live oak
Quercus spp.	Oak
Ranunculus gmelini	Yellow watercrowfoot
Ranunculus tricophyllus	White water crowfoot
Ranunculus spp.	Buttercup
Rhamnus purshiana	Cascara buckthorn
Rhizophora mangle	Red mangrove
Rhynchospora corniculata	Beakrush
Rhynchospora spp.	Beakrush
Ribes aureum	Golden current
Ribes spp.	Gooseberry
Robinia pseudo-acacia	Black locust
Rorippa islandica	Yellow cress
Rorippa spp.	Cress
Rosa woodsii	Wood's rose
Rosa spp.	Rose
Rubus spp.	Bramble
Rubus spp.	Raspberry
Rumex crispus	Curly leaf (curly) dock
Rumex verticillatus	Swamp dock
Rumex spp.	Dock
Ruppia maritima	Widgeongrass (ditch grass)
Sabal palmetto	Cabbage palm

TABLE A.29 Common and Scientific Names of Plants *(Continued)*

Alphabetical by common name		Alphabetical by scientific name	
Common name	Scientific name	Scientific name	Common name
Southern wildrice (southern cutgrass, water millet)	*Zizaniopsis miliacea*	*Sagittaria graminea*	Arrowhead
		Sagittaria heterophylla	Arrowhead
Spatterdock	*Nuphar advena*	*Sagittaria latifolia*	Duck potato (arrowhead, broadleaf arrowhead)
Spatterdock	*Nuphar lutea*		
Speckled alder	*Alnus rugosa*	*Sagittaria platyphylla*	Delta duck potato (arowhead)
Sphagnum	*Sphagnum* spp.	*Sagittaria rigida*	Sessile-fruited arrowhead
Spikerush	*Eleocharis baldwinii*	*Sagittaria* spp.	Arrowhead
Spikerush	*Eleocharis equisetoides*	*Sagittaria cuneata*	Duck potato
Spikerush	*Eleocharis palustris*	*Salicornia bigelovii*	Bigelow's glasswort
Spikerush	*Eleocharis smallii*	*Salicornia europea*	European glasswort
Spikerush	*Eleocharis* spp.	*Salicornia pacifica*	Pickleweed (Pacific glasswort)
Sprangletop	*Leptochloa fascicularis*	*Salicornia virginica*	Woody glasswort
Sprangletop	*Leptochloa* spp.	*Salicornia* spp.	Glasswort
Spruce	*Picea* spp.	*Salix amygdaloides*	Peachleaf willow
Spurge	*Euphorbia polygonifolia*	*Salix bebbiana*	Bebb willow
Squarestem spikerush	*Eleocharis quadrangulata*	*Salix brachycarpa*	Barrenground willow
Steeplebush	*Spirea tomentosa*	*Salix drummondiana*	Drummond willow
Stonewort	*Chara vulgaris*	*Salix exigua*	Coyote (sandbar) willow
Strawberry clover	*Trifolium fragiferum*	*Salix geyeriana*	Geyer willow
Subalpine fir	*Abies lasiocarpa*	*Salix glauca*	Grayleaf willow
Sugar hackberry	*Celtis laevigata*	*Salix gooddingii*	Gooding's willow
Summercypress (kochia)	*Kochia scoparia*	*Salix interior*	Sandbar willow
Swamp chestnut oak	*Quercus muehlenbergii*	*Salix lasiandra*	Pacific willow
Swamp oak	*Rumex verticillatus*	*Salix lasiolepis*	Arroyo willow
Swamp milkweed	*Asclepias incarnata*		
Swamp timothy	*Heleochloa schoeroides*		
Swamp white oak	*Quercus bicolor*		
Sweet fern	*Comptonia peregrina*		

Sweet flag	*Acorus calamus*
Sweetgum (American sweetgum)	*Liquidambar styraciflua*
Switchgrass	*Panicum virgatum*
Swordleaf rush	*Juncus ensifolius*
Sycamore	*Platanus occidentalis*
Tall fescue	*Festuca elatior*
Tall (reed) mannagrass	*Glyceria grandis*
Tall threetip sagebrush	*Artemisia tripartita*
Tall wheatgrass	*Agropyron elongatum*
Tape-grass	*Vallisneria* spp.
Tatarian honeysuckle	*Lonicera tatarica*
Tealeaf willow	*Salix plantifolia*
Tearthumb	*Polygonum* spp.
Telegraph weed	*Heterotheca grandiflora*
Thinleaf alder	*Alnus tenuifolia*
Thinleaf rubber rabbitbrush	*Chrysothamnus nauseosus consimilis*
Three-way sedge	*Dulichium arundinaceum*
Timothy	*Phleum pratense*
Toothcup	*Ammannia coccinea*
Torrey rush	*Juncus torreyi*
Tropical (southern) cattail	*Typha domingensis*
Trumpetcreeper	*Campsis radicans*
Tufted hairgrass	*Deschampsia caespitosa*
Tufted loosestrife	*Naumburgia thrysiflora*
Tupelo (water tupelo)	*Nyssa aquatica*
Tupelo/blackgum	*Nyssa* spp.
Turtle grass	*Thalassia testudinum*
Umbrella sedge	*Cyperus virens*
Valley sedge	*Carex vallicola*
Variable pondweed	*Potamogeton gramineus*

Salix lutea	Shining willow
Salix nigra	Black willow
Salix planifolia	Tealeaf willow
Salix scouleriana	Scooler willow
Salix wolfii	Wolf willow
Salix spp.	Willow
Salsola kali	Saltwort
Salvinia rotundifolia	Floating moss
Sambucus racemosa pubens	Red elder
Sarcobatus vermiculatus	Black greasewood
Saururus cernuus	Lizard's tail
Schinus terebinthifolius	Brazilian pepper
Scirpus acutus	Hardstem bulrush (tule)
Scirpus americanus	Common threesquare (American bulrush)
Sicrpus californicus	Bullwhip (southern, California) bulrush
Scirpus campestris	Bulrush
Scirpus cyperinus	Woolgrass (woolgrass bulrush)
Scirpus fluviatilis	River bulrush (bulrush)
Scirpus heterochaetus	Slender bulrush
Scirpus maritimus	Alkali (saltmarsh) bulrush
Scirpus olneyi	Olney threesquare (Olney bulrush)
Scirpus paludosus	Bulrush
Scirpus robustus	Saltmarsh (robust) bulrush
Scirpus subterminalis	Bulrush
Scirpus validus	Softstem bulrush (great bulrush)
Scirpus spp.	Bulrush
Scolochloa festucacea	Whitetop
Scutellaria galericulata	Skullcap
Scutellaria lateriflora	Skullcap

TABLE A.29 Common and Scientific Names of Plants *(Continued)*

Alphabetical by common name		Alphabetical by scientific name	
Common name	Scientific name	Scientific name	Common name
Velvet mesquite	*Prosopis velutina*	*Senecio serra*	Butterweed groundsel
Walters (saltmarsh) millet	*Echinochloa walteri*	*Sequoia sempervirens*	Redwood
Warty panicum	*Panicum verrocosum*	*Serena repens*	Palmetto
Water arum	*Calla palustris*	*Sesbania exaltata*	Sesbania
Water chestnut	*Trapa natans*	*Sesuvium maritimum*	Sea purslane
Water clover	*Marsilea vestita*	*Setaria glauca*	Yellow foxtail
Water feather	*Myriophyllum brasiliense*	*Setaria magna*	Giant foxtail
Water hemlock	*Cicuta bulbifera*	*Setaria* spp.	Foxtail grasses
Water hemlock	*Cicuta maculata*	*Shepherdia argentea*	Silver buffaloberry
Water hickory	*Carya aquatica*	*Sidalcea oregana*	Oregon checkermallow
Water hyacinth	*Eichhornia crassipes*	*Sitanion hystrix*	Bottlebrush squirreltail
Water hyssop	*Bacopa caroliniana*	*Sium suave*	Water parsnip
Water lettuce	*Pistia stratiotes*	*Smilacina racemosa*	Solomons-seal
Water lily (white water lily)	*Nymphaea* spp.	*Solidago* spp.	Goldenrod
Water locust	*Gleditsia aquatica*	*Sorbus scopulina scopulina*	Green's mountain ash
Water marigold	*Megalodonta beckii*	*Sorgastrum nutans*	Indiangrass
Water milfoil	*Myriophyllum exalbescens*	*Sparganium americanum*	Burreed
Water milfoil	*Myriophyllum heterophyllum*	*Sparganium chlorocarpum*	Burreed
Water milfoil	*Myriophyllum spicatum*	*Sparganium eurycarpum*	Giant (great) burreed
Water milfoil	*Myriophyllum* spp.	*Sparganium fluctuans*	Burreed
Water nymphs	*Najas* spp.	*Sparganium foliosa*	Burreed
Water oak	*Quercus nigra*	*Sparganium gracillus*	Burreed
Water parsnip	*Sium suave*	*Sparganium pectinata*	Burreed
Water plantain	*Alisma plantago-aquatica*	*Sparganium* spp.	Burreed
Water primrose	*Jussiaea diffusa*	*Spartina alterniflora*	Saltmarsh (smooth) cordgrass
Water primrose	*Jussiaea leptocarpa*	*Spartina cynosuroides*	Big (giant) cordgrass (salt reed-grass)
Water primrose	*Jussiaea* spp.		
Water sedge	*Carex aquatilis*	*Spartina foliosa*	Pacific cordgrass

Common name	Scientific name
Water smartweed	*Polygonum amphibium*
Water smartweed	*Polygonum natans*
Water starwort (water star grass)	*Heteranthera dubia*
Water willow	*Decodon verticillatus*
Water willow	*Dianthera americana*
Water willow	*Justicia americana*
Water-velvet	*Azolla caroliniana*
Waterbirch	*Betula occidentalis occidentalis*
Watercress	*Nasturtium officinale*
Watercress	*Nasturtium* spp.
Waterelm	*Planera aquatica*
Waterhyssop	*Bacopa* spp.
Waterlily	*Nymphaea tuberosa*
Watershield	*Brasenia schreberi*
Waterstarwort	*Callitriche heterophylla*
Waterthread	*Potamogeton diversifolius*
Wax myrtle	*Myrica cerifera*
Waxweed	*Cuphea* spp.
Weeping lovegrass	*Eragrostis curvula*
Western wheatgrass	*Agropyron smithii*
Western yarrow	*Achillea millefolium lanulosa*
White clover	*Trifolium repens*
White mangrove	*Laguncularia racemosa*
White oak	*Quercus alba*
White sweetclover	*Melilotus alba*
White water crowfoot	*Ranunculus tricophyllus*
White waterlily (waterlily)	*Nuphar macrophyllum*
Whitetop	*Scolochloa festucacea*
Widgeongrass (ditch grass)	*Ruppia maritima*
Wild barley	*Hordeum jubatum*
Wild buckwheat	*Polygonum convolvulus*
Wild indigo	*Baptisia leucophaea*

Scientific name	Common name
Spartina patens	Saltmeadow (marshhay) cordgrass
Spartina pectinatum	Prairie cordgrass
Spartina spartinae	Gulf cordgrass
Spartina spp.	Cordgrass
Spergularia rubra	Sand spurry
Sphagnum spp.	Sphagnum
Spirea alba	Meadowsweet
Spirea tomentosa	Steeplebush
Spirodela polyrhiza	Big duckweed
Spirodela spp.	Duckweed
Sporobolus airoides	Alkali sacaton
Sporobolus spp.	Dropseed
Stachys palustris	Hedge nettle
Stipa viridula	Green needlegrass
Strophostyles helvola	Beach pea
Suaeda maritima	Low sea blite
Suaeda torreyana	Desert blite
Suaeda spp.	Sea blite
Symphoricarpos albus	Common snowberry
Symphoricarpos occidentalis	Snowberry (western snowberry)
Symphoricarpos oreophilus	Mountain snowberry
Syringodium filiforme	Manatee grass
Tamarix chinensis	Salt cedar
Taxodium distichum	Baldcypress (cypress)
Thallassia testudinum	Turtle grass
Trapa natans	Water chestnut
Triadenum virginicum	Marsh St. John's-wort
Trifolium fragiferum	Strawberry clover
Trifolium incarnatum	Crimson clover
Trifolium pratense	Red clover
Trifolium repens	White clover

TABLE A.29 Common and Scientific Names of Plants *(Continued)*

Alphabetical by common name		Alphabetical by scientific name	
Common name	Scientific name	Scientific name	Common name
Wild rye	*Elymus virginicus*	*Trifolium* spp.	Clover
Wildcelery	*Vallisneria americana*	*Triglochin maritima*	Seaside arrowgrass
Wildrice	*Zizania aquatica*	*Tripalsis purpurea*	Purple top
Willow	*Salix* spp.	*Typha angustifolia*	Narrow-leaved (saltmarsh) cattail
Willow oak	*Quercus phellos*	*Typha domingensis*	Tropical (southern) cattail
Willow-herb	*Epilobium adenocaulon*	*Typha latifolia*	Broadleaf cattail
Wolf willow	*Salix wolfii*	*Typha* spp.	Cattail
Wolffias	*Wolffia* spp.	*Ulmus americana*	American elm
Wood nettle	*Laportea canadensis*	*Ulmus* spp.	Elm
Woodawn grass	*Chrysopogon* spp.	*Uniola paniculata*	Sea oats
Wood's rose	*Rosa woodsii*	*Urtica dioica*	Nettle
Woody glasswort	*Salicornia virginica*	*Urtica* spp.	Nettle
Woolgrass (woolgrass bulrush)	*Scirpus cyperinus*	*Utricularia intermedia*	Bladderwort
Woolly sedge	*Carex lanuginosa*	*Utricularia purpurea*	Bladderwort
Yaupon	*Ilex vomitoria*	*Utricularia vulgaris*	Bladderwort
Yellow bluestem	*Andropogon* spp.	*Utricularia* spp.	Bladderwort
Yellow cress	*Rorippa islandica*	*Valeriana edulis*	Edible valerian
Yellow flat (yellow iris)	*Iris peudacorus*	*Vallisneria americana*	Wildcelery
Yellow foxtail	*Setaria glauca*	*Vallisneria* spp.	Tape-grass
Yellow pine	*Pinus australis, Pinus echinata*	*Wolffia* spp.	Duckmeal
Yellow pond lily	*Nuphar polysepalum*	*Wolffia* spp.	Wolffias
Yellow pond lily (water lily)	*Nuphar luteum*	*Xanthium strumarium*	Cocklebur (common cocklebur)
Yellow poplar	*Liriodendron tulipifera*	*Xanthium* spp.	Cocklebur
Yellow sweetclover	*Melilotus officianalis*	*Zannichellia palustris*	Horned pondweed
Yellow watercrowfoot	*Ranunculus gmelini*	*Zizania aquatica*	Wildrice
		Zizaniopsis miliacea	Southern wildrice (southern cutgrass, water millet)
		Zostera marina	Eelgrass

Alternanthera spp.
Avicennia nitida
Beckmannia spp.
Boltonia asteroides
Carex saxatalis
Centella asiatica
Cyperius strigosus
Cyrilla racemiflora
Damasomium californicum
Deschampia spp.
Eleocharis robbinsii
Eragrostis hypnoides
Eragrostis pectinacea
Eragrostis spp.
Erianthus spp.
Eriocaulon septangulare
Heliotropium spp.
Hibiscus palustris
Hydrochloa caroliniensis
Isoetes braunii
Isoetes lacustris
Juncus filiformis
Juncus macer
Lemna trisulca
Littorella uniflora
Lophotocarpus calycinus
Ludwigia peruviana
Menyanthes trifoliata
Myrica spp.
Myriophyllum alterniflorum
Myriophyllum pinnatum
Myriophyllum verticillatom
Najas gracillima
Nitella opaca

TABLE A.29 Common and Scientific Names of Plants *(Continued)*

Alphabetical by common name		Alphabetical by scientific name	
Common name	Scientific name	Scientific name	Common name
		Nuphar rubrodiscum	
		Nymphaea alba	
		Nymphoides aquaticum	
		Panicum purpurascens	
		Panicum repens	
		Paspalum boscianum	
		Paspalum distichum	
		Paspalum fruitans	
		Paspalum lividum	
		Paspalum vaginatum	
		Pluchea purpurascens	
		Potamogeton alpinus	
		Potamogeton angustifolius	
		Potamogeton filiformis	
		Potamogeton friesii	
		Potamogeton nodosus	
		Potamogeton obtusifolius	
		Potamogeton vaginatus	
		Ranunculus aquatilis	
		Ranunculus longirostris	
		Raphanus sativus	
		Riccia fluitans	
		Ricciocarpus natans	
		Sagittaria cristata	
		Sagittaria lancifolia	
		Sagittaria sagittifolia	
		Salicornia ambigua	
		Salicornia perennis	
		Salix boothii	

Schoenoplectus spp.
Scirpus expansus
Sesbania spp.
Sesuvium portulascastrum
Setaria lutescens
Setaria spp.
Sparganium minimum
Spartina bakeri
Subularia aquatica
Thalia divarcata
Torestria acuminata
Typha glauca
Wolffia columbiana

Index

ABOUT THE AUTHOR

NEIL F. PAYNE is a professor of wildlife in the College of National Resources, University of Wisconsin-Stevens Point (1975– present). He grew up in Sheboygan Falls, Wis., and received a bachelor of arts in biology from the University of Wisconsin-Madison (1961), a master of science in wildlife and forestry from Virginia Polytechnic Institute and State University (1964), and a doctor of philosophy in wildlife science from Utah State University (1975). He saw combat duty in Vietnam and was a captain with the U.S. Marine Corps (1964–67). He directed the bear and furbearer program for the Newfoundland and Labrador Wildlife Division (1967–71), and he was on the wildlife faculty at the University of Washington in Seattle (1973–75).